MW01679282

2021 NATIONAL PAINTING COST ESTIMATOR

By Dennis D. Gleason, CPE

Includes inside the back cover:

Includes Free Estimating Software Download

Inside the back cover of this book you'll find a software download certificate. To access the download, follow the instructions printed there. The download includes the National Estimator, an easy to-use estimating program with all the cost estimates in this book. The software will run on PCs using Windows XP, Vista, 7, 8, or 10 operating systems.

Quarterly price updates on the Web are free and automatic all during 2021. You'll be prompted when it's time to collect the next update. A connection to the Web is required.

Download all of Craftsman's most popular costbooks for one low price with the Craftsman Site License. http://CraftsmanSiteLicense.com

Craftsman Book Company
6058 Corte del Cedro, Carlsbad, CA 92011

Acknowledgments

The author thanks the following individuals and organizations for furnishing materials and information used in the preparation of various portions of this book.

Howard Shahan, *American Design Painting & Drywall,* Poway, CA
American Society of Professional Estimators (ASPE), Wheaton, MD
Benjamin Moore Paints, San Diego, CA
Gordon H. Brevcort, *Brevcort Consulting Associates,* Ridgewood, NJ
Luis Anguiano, *CSI Paint,* Napa, CA
Scott Williams, *CSI Paint,* San Francisco, and Napa, CA
John San Marcos, *Devoe Coatings,* San Diego Marine Hardware, San Diego, CA
Ken Hogben, *Dunn-Edwards Paints & Wallcovering,* San Francisco, CA
Randy Martin, *Dunn-Edwards Paints & Wallcovering,* La Mesa, CA
Bob Langbein, *East Bay Paint Center*, Albany, CA
Hugh Champeny, *Kelly-Moore Paint Company,* San Carlos, CA
Eli Dominguez, *Kelly-Moore Paint Company,* Pleasant Hill, CA
Dennis Cripe, *R.W. Little Co, Inc., Sandblasting,* San Diego, CA
Chris Rago, *Mark's Paint Mart*, Oakland, CA
Bruce McMullan, *McMullan & Son Painting*, San Diego, CA
Joe Garrigan, *Mr. Paints*, San Diego, CA
PPG Industries, Inc., *Pittsburgh Paints*, Torrance, CA
Carlos Jeronimo, *PPG Paints*, Santa Clara, CA
Keith Braswell, *Rent-X*, Pleasant Hill, CA
Richardson Engineering Services, Inc., Mesa, AZ
Rust-Oleum Protective Coatings, Los Angeles, CA
Squires-Belt Material Co., San Diego, CA
Steel Structures Painting Council, Pittsburgh, PA
John Meyer, *U.S. Government, Department of the Navy, Public Works*, San Diego, CA
Miguel Govea, *Vista Paint Centers,* San Diego, CA
Jerry Rittgarn, *Waco-Arise Scaffolding & Equipment*, San Diego, CA
Mark Janson, *Warehouse Paint,* Auburn, CA

Cover design by: *Jennifer Johnson*
Sheila M. Scott, Calligraphy

ISBN 978-1-57218-366-7 ISSN 1092-6852
Published September 2020 for the year 2021

List of Figures

Figure	Title	Page

How to Use This Book

Paint estimating is more of an art than a science. There's no price that's exactly right for every job and for every bidder. That's because every painting job is unique. No single material cost, no labor estimate, no pricing system fits all types of work. And just as every job varies, so do painting companies. No two painting contractors have the same productivity rates, the same labor burden, the same overhead expense and the same profit requirements.

The best paint estimates are always custom-made for a particular job. They're based on the contractor's actual productivity rate, material cost, labor cost, overhead percentage and profit expectations. No estimating book, no computerized estimating system, no estimating service can possibly account for all the variables that make every job and every painting company different. Only a skilled estimator using professional judgment and a proven estimating system can produce consistently reliable estimates on a wide variety of painting jobs.

So, Why Buy This Book?

That's easy. This is the most complete, authoritative and reliable unit cost guide ever made available to paint estimators. No matter what types of work you estimate, no matter what your costs are, this book will help produce consistently accurate painting cost estimates in dollars and cents. But it isn't a substitute for expertise. It's not a simple way to do in minutes what an experienced paint estimator might not be able to do in hours. Instead, this unit cost guide will aid you in developing a good estimate of costs for any painting operation on any project. Think of this manual as one good estimating tool. But it's not (or at least shouldn't be) the only estimating tool you'll use.

For most jobs, I expect that the figures you see here will prove to be good estimates. But anyone who understands paint estimating will understand why judgment is needed when applying figures from this manual — or any other paint estimating guide. It's your responsibility to decide which conditions on the job you're bidding are like conditions assumed in this manual, and which conditions are different. Where conditions are different, you'll need good professional judgment to arrive at a realistic estimated cost.

National Estimator '21

Inside the back cover of this book you'll find a software download certificate. To access the download, follow the instructions printed there. The download includes the National Estimator, an easy-to-use estimating program with all the cost estimates in this book. The software will run on PCs using Windows XP, Vista, 7, 8, or 10 operating systems. When the National Estimator program has been installed, click Help on the menu bar to see a list of topics that will get you up and running. Or, go online to www.craftsman-book.com and click on Support, then Tutorials, to view an interactive tutorial for National Estimator.

	Manhour productivity	Labor cost per hour	Labor burden percent	Labor burden dollars	Labor cost plus burden	Material price discount	Overhead percent	Profit
Slow (1P)	Low	$23.00	24.0%	$5.52	$28.52	20%	19.0%	16%
Medium (2P)	Average	29.50	28.9%	8.53	38.03	30%	25.0%	12%
Fast (3P)	High	36.00	35.3%	12.71	48.71	40%	31.0%	7%

Notes: These rates are for painters. Hourly rates for wallcovering are different. See page 29. Slow, Medium and Fast jobs are defined on page 13. Labor burden percentages used in this book are summarized on page 31. National Estimator uses hourly rates in the Labor cost plus burden column. National Estimator shows productivity rates (Slow, Medium and Fast) and copies the words Slow, Medium or Fast to your estimate. It also copies the crew productivity code, either 1P (Slow), 2P (Medium), or 3P (Fast) to your estimating form. National Estimator allows you to enter any percentage you select for overhead and profit.

Figure 1
The basis for painting cost estimates in this book

How to Use the Tables

The estimating tables in this book show typical costs and bid prices for every painting operation you're likely to encounter, whether paint is applied by brush, roller, mitt or spray. Selecting the right cost table and the correct application method is easy. Tables are divided into four parts:

Part I: General Painting Costs

Part II: Preparation Costs

Part III: Industrial, Institutional and Heavy Commercial Painting Costs

Part IV: Wallcovering Costs

Each section is arranged alphabetically by operation. If you have trouble finding the tables you need, use the Table of Contents at the front of the book or the Index at the back of the book.

Once you've found the right table and the appropriate application method, you have to select the correct application rate. For each of the application methods (brush, roll, mitt or spray), the tables show three application rates: "Slow," "Medium," or "Fast." That's a very important decision when using this book, because each application rate assumes different manhour productivity, material coverage, material cost per gallon, hourly labor cost, labor burden, overhead and profit. Your decision on the application rate to use (or which combination of rates to use) has to be based on your evaluation of the job, your painters and your company. That's where good common sense is needed.

Figure 1 shows crew codes, labor costs, labor burdens, material discounts, and profit for each of the three production rates for painting.

The "Slow" application rate in Figure 1 assumes lower productivity (less area covered per manhour), a lower labor cost (due to a less skilled crew), a lower labor burden (due to lower fringe benefits), a lower discount on materials (because of low volume), higher overhead (due to lower volume) and a higher profit margin (typical on small repaint or custom jobs). Figures in this "Slow" application row will apply where painters with lower skill levels are working on smaller or more difficult repaint jobs.

Look at the "Fast" row in Figure 1. These estimates will apply where a skilled crew (higher hourly rate and larger fringe benefits) is working under good supervision and good conditions (more area covered per manhour) on larger (volume discount on materials) and more competitive jobs (lower profit margin). Figures in the "Fast" application row assume high productivity and lower material coverage, (unpainted surfaces absorb more paint), like that of a residential tract job.

Each of the three application rates is described more completely later in this section.

	Pricing variables			Unit cost estimate					
	1	2	3	4	5	6	7	8	9
	Labor SF per man-hour	Material coverage SF/gallon	Material cost per gallon	Labor cost per 100 SF	Labor burden 100 SF	Material cost per 100 SF	Overhead per 100 SF	Profit per 100 SF	Total cost per 100 SF
Walls, gypsum drywall, orange peel or knock-down, roll, per 100 SF of wall area									
Flat latex, water base (material #5)									
Roll 1st coat									
Slow	400	300	36.00	5.75	1.38	12.00	3.63	3.64	26.40
Medium	538	275	31.50	5.48	1.59	11.45	4.63	2.78	25.93
Fast	675	250	27.00	5.33	1.88	10.80	5.58	1.65	25.24
Your customized figures				3.56	.85	11.45	3.97	2.38	22.21

Figure 2
Customize the tables

The Easy Case: No Adjustments

Let's suppose the "Slow" application rate fits the job you're estimating almost perfectly. Your crew's productivity is expected to be low. The labor cost will be $23.00 per hour. Labor burden (fringes, taxes and insurance) will be 24.0 percent. Discount on materials will be 20 percent. Overhead will be 19 percent and profit will be 16 percent. Then your task is easy. All of your costs match the costs in the "Slow" row. No modifications are needed. The same is true if your costs fit the "Medium" or "Fast" rows.

But that's not always going to happen. More often, the job, your crew and your company won't fit exactly into any of the three rows. What then? More evaluation is required. You'll combine costs from several application rate rows to reach an accurate bid price. I call that *customizing your costs* and it's nearly always required for an accurate estimate.

Customizing Your Costs

Every company has a different combination of worker speed and experience, taxes, benefits, spread rates, equipment needs, percentage for overhead, and profit margin. These are the cost variables in paint estimating. This book is designed so you can quickly and easily adjust estimates to reflect actual costs on the job you're estimating. It's important that you *read the rest of this section before using the cost tables in this book.* That's the only way to get from this manual all the accuracy and flexibility that's built into it.

In the remainder of this section I'll describe the assumptions I've made and the methods I used to compile the cost tables in this manual. Once you understand them, you'll be able to combine and modify costs in the estimating tables so your bids fit the job, your crew and your company as closely as possible.

When you start using the cost tables in this book, I suggest you circle numbers in the "Slow," "Medium," or "Fast" application rate rows that best fit your company and your jobs. To improve accuracy even more, write your own figures in the blank row below the "Fast" row in each table, like I've done in Figure 2.

A Practical Example

Figure 2 is part of an estimating table taken from page 228 of this book, General Painting Costs. I'm going to use it to show how to customize estimates to match

your actual costs. In Figure 2 I've circled some of the costs I plan to use in a sample estimate and calculated others.

In column 1, *Labor SF per manhour*, I've circled 675 because I feel the journeyman painter assigned to this job can paint walls at the "Fast" rate of 675 square feet per hour. That's the number I plan to use for my estimate.

In column 2, *Material coverage SF/gallon*, I've reviewed my past performance and I expect coverage will be about 275 square feet per gallon of paint. So I've circled that figure.

In column 3, *Material cost per gallon*, I've circled 31.50 for my cost per gallon for flat water base latex (including tax and an allowance for consumable supplies), based on a 30 percent discount from the retail price.

So far, so good. That completes the first three coumns, what I call the *pricing variables*. Now we can begin on the *unit cost estimate*, columns 4 through 9. Each of these columns show a price per 100 square feet of wall.

We'll start with column 4, Labor cost per 100 SF. Notice that I've entered 3.56 for this column. Here's why. Look back at Figure 1 and the "Slow" labor rate at $23.00. (See Figure 13 on page 29 for the wage rates for wallcovering.) Since I'm in a part of the country where prices, and wages, are lower than the national average, my experienced painters work for $24, closer to the "Slow" labor cost, though they produce at the "Fast" rate of 675 SF per manhour. This gives me an advantage because my labor costs are lower than those in Figure 1. To calculate the labor cost per 100 SF, divide $24.00 by 675 and multiply by 100: 24/675 = .0356 x 100 = 3.56.

In column 5, *Labor burden 100 SF*, I've entered .85. This figure is a result of my labor cost at $3.56 x 24.0 percent, my labor burden (taxes, insurance and benefits) from the "Slow" row of Figure 1. Even though the labor rate is "Fast" and the labor cost is higher than the "Slow" rate, for this example labor burden will be most like work done at the "Slow" rate because this company doesn't offer many benefits.

In column 6, *Material cost per 100 SF*, I've circled 11.45, the number in the "Medium" row. Since I've used numbers in the "Medium" row in both columns 2 and 3, I can take the figure in column 6 for material costs directly from the table, without any calculations.

In column 7, *Overhead per 100 SF*, I've calculated the overhead dollar value by adding the labor cost, labor burden and material cost then multiplying that sum by the "Medium" overhead at 25 percent: $3.56 + $.85 + $11.45 = $15.86 x .25 = $3.97.

In column 8, *Profit per 100 SF*, I've calculated the profit dollar value by adding the labor cost, labor burden, material cost and overhead then multiplying that sum by the "Medium" profit at 12 percent from Figure 1. The result is $3.56 + $.85 + $11.45 + $3.97 = $19.83 x .12 = $2.38.

Column 9, *Total cost per 100 SF*, is the bid price — it's the sum of columns 4 through 8 for each row. Because I've circled costs that fall in more than one row, I can't use any figure in column 9. Instead, I simply add the circled or calculated figures in columns 4 through 8: $3.56 + $.85 + $11.45 + $3.97 + $2.38 = $22.21. That's my bid price per 100 square feet on this job. It's the combination of costs that fit my company, my painters and the job.

Using Your Good Judgment

Of course, judgment is required when using these tables, as it is when making any estimate. For example, if your journeymen painters earn the top rate of $36.00 but work at the "Medium" production rate or slower, your labor cost per unit will be higher than the highest cost listed in column 4. An adjustment will be required.

Because figures in columns 7 and 8 are percentages of figures in columns 4, 5 and 6, you have to be careful when you blend costs from different rows. Let's look at an extreme (and unlikely) example.

Suppose you use costs from the "Slow" application row for columns 4 (5.75), 5 (1.38) and 6 (12.00) of Figure 2. The total of those three costs is $19.13. Then you decide to use overhead from the "Fast" row because your overhead is about 31 percent of cost, not 19 percent of cost as in the "Slow" row (Figure 1). "Fast" overhead is listed as $5.58 in Figure 2. The correct overhead figure is $5.93, 31 percent of the sum of "Slow" costs in columns 4, 5 and 6. Be aware of this small discrepancy and calculate figures for all the categories yourself if extreme accuracy is essential.

Converting Unit Prices

The last column in Figure 2 shows the total cost per 100 square feet of wall. Some estimating tables in this book show a total cost per 100 linear feet (such as for baseboard) or total costs per unit (such as for doors). To convert a cost per 100 square feet to a cost per square foot, move the decimal point two places to the left. Thus the cost per 100 square feet for the "Fast" rate in Figure 2 is $25.24 or about 25 cents per square foot.

General Qualifications

It's important that you understand the conditions the tables are based upon. I call these conditions the job qualifications. A qualifications statement follows each estimating table to help you understand what's included and what's excluded. Please read those *qualifications* before using costs from this manual in your estimates. The following points apply to *all* tables in this book:

Included Costs

- Minor preparation, both time and material. Normal preparation for new residential construction is included in the "Fast" row and for new commercial jobs in the "Medium" row. Minimal preparation is included for repaint jobs in the "Slow" row.
- Minimum setup and cleanup
- Equipment such as ladders, spray rigs and brushes are included in overhead for the "Fast" rate (residential tracts) or "Medium" (commercial) work. Add equipment costs at their rental rate for "Slow" (repaint) jobs.

Excluded Costs

- Equipment costs such as ladders, spray rigs, etc. for "Slow" (repaint) jobs. Add these at their rental rate whether or not you own the equipment.
- Extensive surface preparation. Add the cost of time and materials needed for more than "normal"preparation work. Also add time to remove and replace hardware and accessories, protect adjacent surfaces, and do any extensive setup, cleanup, or touchup. (See the discussion of SURRPTUCU on the next page.)
- Mobilization or demobilization
- Supervision
- Material handling, delivery, or storage
- Sample preparation
- Mixing coatings
- Excessive material waste or spillage
- Equipment rental or placement costs
- Scaffolding rental and erection costs
- Subcontract costs
- Contingency allowance
- Owner allowances
- Commissions, bonuses, overtime, premium pay for shift adjustments (evening work), travel time or per diem.
- Bonds, fees, or permits
- Additional insurance to meet owner requirements
- Work at heights above 8 feet or beyond the reach of a wand or extension pole. (See the table for High Time Difficulty Factors on page 139.)

Surface Preparation

The Preparation estimating tables that follow Part I: General Painting Costs, apply to both interior and exterior surfaces.

Surface preparation is one of the hardest parts of the job to estimate accurately. Any experienced painter can make a reasonably good estimate of the quantity of paint and time needed for application. But the amount of prep work needed will vary widely — especially for repaint jobs. Some will need very little work. Others will take more time for prep than for painting.

Preparation work for new construction jobs is relatively standard and consistent. You'll have to mask cabinets before spraying sealer on wet area walls, caulk at the baseboards, putty the nail holes in wood trim, and occasionally use a wire brush to smooth and clean a surface. The time required for this work is fairly predictable.

Labor cost for normal preparation of unpainted surfaces in new residential construction is included in the "Fast" *labor* costs and for new commercial construction in the "Medium" *labor* cost. The cost of materials for normal surface preparation on unpainted surfaces is included in the sundries allowance that's part of the "Fast" or "Medium" material cost.

But if more than normal surface prep work is needed, estimate the extra manhours and materials required and add these costs to your estimate.

Add for Repaint Preparation

The "Slow" unit costs include no surface preparation other than a quick wipedown. Preparation on a repaint job may take longer than the painting itself. That's why you have to estimate surface prep as a separate item and add that cost to your estimate.

A misjudgment in estimating preparation work can be very expensive. That's why I recommend that you bid surface preparation by the hour, using your shop rate for "time and material" jobs, or some other specified hourly rate. That protects you against cost overruns if the preparation takes longer than anticipated. But there's a danger here. Owners may be angry about the cost because they don't understand what's involved in preparation and why it takes so long. You can avoid this with a "not to exceed" bid that contains a maximum price for the prep work. Your bid should define the scope of preparation work in detail and list exactly what's included and excluded. Be sure to consider all the labor, material, and equipment costs involved.

If you have to bid repaint work, be sure to include all the miscellaneous costs. The acronym I use to identify these miscellaneous costs is SURRPTUCU: Setup (SU), Remove and Replace (RR), Protection (P), Touchup (TU) and Cleanup (CU). Add these costs to your repaint estimate if they require anything beyond minimum attention.

1) *Setup* includes unloading the vehicle, spreading the tarp and setting up the tools — everything that has to be done before prep or painting can begin.

2) Remove and replace everything that will interfere with painting, including door and cabinet hardware, the contents of cabinets, light fixtures, bathroom accessories, switch covers and outlet plates, among others.

3) *Protection* for furniture and adjacent surfaces such as floors, cabinets, plumbing or electrical fixtures, windows, and doors. Protection methods include masking, applying visqueen, laying drop cloths and applying a protective coating on windows.

4) *Touchup* time varies with the speed and quality of the painting job and how fussy the owner is. The more careful your painters are, the less touchup time needed. You can estimate touchup time accurately only if you know how well your crews perform. The Touchup table in this book is based on a percentage of total job cost.

5) *Cleanup* time is usually about the same as setup time, about 20 to 30 minutes each day for repaint jobs. Cleanup time begins when work stops for the day and ends when the crew is back in the truck and ready to go home. It includes cleaning tools, dismantling the paint shop and loading the vehicle.

Subcontractors

Painting contractors don't hire many subcontractors. But once in a while you'll need a specialist for sandblasting, waterblasting, wallcovering, scaffolding or pavement marking. Subcontract costs are not included in the estimating tables. Add the cost of any subcontract work that will be required.

Figure 3 shows some typical rates quoted by sandblasting subcontractors. Of course, prices in your area will probably be different. You could also figure sandblasting unit costs from the sandblasting estimating tables included in Part II, Preparation Costs, in this book.

Minimum charges: $611.00, scaffolding not included		Epoxy coated - add	1.34 to 1.48/SF
Additional insurance: May be required to cover adjacent personal and real property which may not be protected.		With portable equipment - add	.78 to 1.12/SF
Sandblasting water soluble paints	$1.12 to 1.28/SF	**Commercial blast** - 67% white stage	
Sandblasting oil paints	1.19 to 1.34/SF	Field welded, new, uncoated	
Sandblasting heavy mastic		ground runs	1.19 to 1.41/SF
(depends on coating thickness)	1.54 to 1.69/SF	above ground	1.48 to 2.33/SF
Sandblasting brick - light blast	1.12 to 1.28/SF	Previously painted surfaces - add	.71 to 1.28/SF
Sandblasting masonry block walls		Epoxy coated - add	1.28 to 1.48/SF
Clean up & remove grime - light	1.05 to 1.12/SF	With portable equipment - add	.91 to 1.12/SF
- heavy	1.61 to 1.76/SF	**Near white blast** - 95% white stage	
Sandblasting structural steel		Field welded, new, uncoated	
Pricing rules of thumb:		ground runs	1.41 to 1.62/SF
Pipe up to 12" O.D.	1.61 to 2.39/SF	above ground	1.62 to 2.47/SF
Structural steel up to 2 SF/LF	1.48 to 1.71/SF	Previously painted surfaces - add	.71 to 1.28/SF
Structural steel from 2 to 5 SF/LF	1.76 to 1.97/SF	Epoxy coated - add	1.28 to 1.48/SF
Structural steel over 5 SF/LF	(depends on shape)	With portable equipment - add	.91 to 1.12/SF
Tanks and vessels up to 12'0" O.D.	2.33 to 2.69/SF	**White blast** - 100% uniform white stage	
Tanks and vessels over 12'0" O.D.	2.33 to 2.69/SF	Field welded, new, uncoated	
Brush off blast - light blast (loose mill scale)		ground runs	2.12 to 2.47/SF
Field welded, new, uncoated		above ground	2.33 to 2.74/SF
ground runs	.71 to .91/SF	Previously painted surfaces - add	.71 to 1.19/SF
above ground	1.05 to 1.97/SF	Epoxy coated - add	1.28 to 1.48/SF
Previously painted surfaces - add	.71 to 1.28/SF	With portable equipment - add	.71 to 1.06/SF

Figure 3
Sandblasting pricing table

Figure 4 shows typical subcontract bids for pavement marking. Again, prices in your area may be different.

If you do much repainting, you'll probably want to buy a waterblasting rig. Even if you own the blaster, include a charge in each estimate for the equipment as though you rented it from a rental yard just for that job. Figure the unit costs for waterblasting from Part II of this book, Preparation Costs.

Consider using a waterblasting subcontractor if you don't need the service often. Figure 5 shows some typical rates for waterblasting. Make up a table like this based on quotes from subcontractors in your area. For a more detailed table, see Sandblasting in the Preparation section, page 303.

When you hire a subcontractor, make sure the quoted price includes everything that contractor has to do — all labor, material (with tax, if applicable), equipment, overhead and profit. Add your overhead and profit percentage to the subcontractor's bid price when you enter that item on the estimate.

Contingencies

Occasionally you'll add a contingency allowance on bids for repaint projects where there are unknowns that can't be forecast before work actually begins. Contingency allowances are rarely needed when estimating new construction. When necessary, the contingency amount is usually from 3 to 5 percent. It can go higher, however, if there are unusual conditions or unknowns that make it hard to produce an accurate estimate. Include a contingency allowance in your estimates only if you have reason to expect:

- An uncertain scope of work (unknown job conditions)
- An inexperienced owner or general contractor
- Incomplete drawings

Pricing rules of thumb:	
Number of parking spaces: Figure on one space per 300 SF of pavement	
Single line striping with light graphics application	$11.00 per space
Single line striping with heavy graphics application	19.20 per space
Single striping, light graphics and 3' wheel stop	27.30 per space
Single striping, heavy graphics and 3' wheel stop	35.40 per space
Equipment pricing:	
Simple "inverted spray can" approximate cost	$248.00
Professional striping machine cost range	5,100 to 5,620
Professional road/highway striper	284,000
Subcontractor pricing:	
Move on:	$167.00 to 204.00
Striping prices:	
Single line striping	$.50 to .65 per lineal foot
Bike lane striping	.65 to .76 per lineal foot
Fire lane, red curb	.65 to .76 per lineal foot
Symbol pricing:	
Templates - 8'0" template	$192.00 to 229.00 each
Arrows	43.50 to 51.00 each
Handicap symbol, one color	17.90 to 24.20 each
two color	32.00 to 38.30 each
No parking fire lane stencil	3.45 to 4.21 each
Wheel stops:	
3'0" stops	$24.20 to 30.70 each if pinned on asphalt
	32.00 to 38.30 each if glued and pinned
6'0" stops	38.30 to 46.00 each if pinned on asphalt
	46.00 to 52.50 each if glued and pinned
	(add for stops pinned to concrete)
Signs and posts:	
Sign only 12" x 18"	$53.60 to 75.40
Post mounted 12" x 18"	141.00 to 194.00
Pavement markers:	
One way pavement markers	$11.40 each
Two way pavement markers	15.30 each

Figure 4
Pavement marking pricing table

Minimum charges: $638.00, scaffolding not included	
Additional insurance: May be required to cover adjacent personal and real property	
Pricing rules of thumb:	
Up to 5,000 PSI blast	4 hour minimum $141.00/hour
5,000 to 10,000 PSI blast	8 hour minimum $204.00/hour
10,000 PSI blast	8 hour minimum $252.00/hour
Wet sandblasting	4 hour minimum $162.00/hour

Figure 5
Waterblasting pricing table

- Delays in beginning the project
- Owner involvement in supervision
- Below-standard working conditions

Don't use contingency allowances as a substitute for complete estimates. Include contingency only to cover what can't be estimated, not what you don't have time to estimate accurately.

Column Headings Defined

Take another look at Figure 2. The heading describes the surface to be coated: the type, texture, and often, condition. Sections within each surface heading are divided according to coating material, then by application method, and further into the "Slow," "Medium," and "Fast" application rates.

Column 1: Labor Productivity

This column shows units of work completed per manhour. My estimates assume that painters are experienced and motivated professionals. The labor productivity categories are shown in Figure 6.

My experience is that a painting company that can handle larger projects will have highly skilled, better qualified and more productive painters. The estimating tables also assume that repainting a surface usually takes about 35 percent more time than painting newly constructed surfaces. Much of this extra time is spent protecting adjacent areas.

Slow	Medium	Fast
Repaint jobs	New commercial projects	New residential production
Custom painting	Industrial painting	Repetitious painting
Tenant improvements	—	—
Small jobs	Medium-sized jobs	Large projects
Single units	Two to four units	Five or more units
Low production	Average production	High production
High difficulty	Average difficulty	Low difficulty
Poor conditions	Average conditions	Good conditions
High quality	Average quality	Minimum quality
Semi-skilled crew	Skilled crew	Highly skilled crew
No supervision	Some supervision	Good supervision

Figure 6
Labor productivity categories

To establish your company's production levels, ask your field superintendent to monitor the time needed to complete each task and to keep records of crew productivity. You can use the Field Production Times and Rates form on pages 419 and 420 to track your painters' productivity. Make copies of the blank form and have your field superintendent or job foreman give one to each painter on every job. Your superintendent should check the forms frequently to insure they are accurate and kept up to date. Your best guide to productivity on future jobs is productivity on jobs already completed, and this form will help you keep track of your production time. Refer back to Figure 2 on page 7. You can use the results collected on these forms to complete the customized figures row under the "Fast" operation in Figure 2 for every operation in the National Painting Cost Estimator. Examples of how to use Figure 2 are on pages 7 through 9. The more you know about your painters' performance, the more accurate your estimates will be. But don't expect your estimates and actual production to always match exactly. Painters are human beings, not robots. You can't expect them to work at the same rate at all times.

Reduced Productivity

The tables in this book assume no overtime work. Excessive overtime puts a strain on your craftsmen and reduces productivity. A few consecutive days of overtime can drag productivity down to well below average. It's good practice not to assign overtime work on more than two consecutive days.

Work efficiency is also lower when men, materials and equipment are confined in a small area or required to work in cluttered, poorly lit or dirty rooms. Painters need elbow room to work efficiently and get maximum productivity. They're also more productive in a clean environment where they can see what they're doing. It's easier — and safer — to work in a well-lighted area that's relatively clear of debris. If the work area is confined or dirty, reduce estimated productivity accordingly.

Supervision

Supervision expense is not included in the cost tables. Add the cost of supervision to your estimates.

Most supervision is done by foremen. Every crew should have a project foreman designated, usually the most experienced and reliable painter on the job. When not supervising, project foremen should be painting. Thus the project foreman is a working supervisor. Part of the foreman's time will be productive (applying coatings) and part will be nonproductive (directing the work).

If you have more than three or four jobs going at one time, you need a field superintendent. The field superintendent is the foreman's supervisor. His or her primary responsibility is to be sure that each foreman has the manpower, materials and equipment needed to get the job done. The field superintendent should monitor job progress to be sure manhour productivity and materials used are in line with estimates. Field superintendents usually are not working supervisors; all their time is nonproductive. Figure the field superintendent's salary as overhead expense, because you can't charge his salary to a specific job.

Your project foremen and field superintendent can make or break a job. The better they are, the more work will be done. You want a field superintendent who assigns the right painters to the right foreman, and a foremen who puts the right painters on the right tasks. The most experienced tradesmen should work on tasks that require more skill. Other painters should be used where less skill is needed. The project foreman is also responsible for job safety and quality control.

Your estimates will be more competitive if you can assume high productivity. That's only possible when you have good supervision, from both foremen and superintendent, and motivated crews.

Allowances for Supervision

Supervision isn't considered productive labor. A foreman isn't painting when he's scheduling, organizing a job and instructing his workers. Here are my rule-of-thumb allowances for nonproductive labor on painting jobs.

Custom homes. Allow 2.5 hours of nonproductive supervision for a home up to 1,500 square feet, 3 hours on a home between 1,500 and 2,000 square feet, 4 hours on a custom home between 2,000 and 2,500 square feet, and 5 hours on a larger home.

Model homes in a tract. One hour of nonproductive supervision for each day your crew will be on the job.

Most tract homes. One hour per house.

Higher-quality tract homes. Two hours per house.

Slow application and light coverage (Repaint jobs)	**Medium application and medium coverage (Commercial projects)**	**Fast application and heavy coverage (Residential tracts)**
Repaint jobs	Commercial projects	Residential production
Light usage	Moderate usage	Heavy usage
Low absorption	Moderate absorption	High absorption
Light application	Medium application	Heavy application
Low waste	Moderate waste	High waste
Quality paint	Standard paint	Production paint
Semi-skilled painters	Skilled crew	Highly skilled crew

Figure 7
Material coverage rates

Apartments and condos. Allow 1 hour per unit if there are 10 units or less. For 11 to 30 units, allow 0.75 hours of nonproductive time per unit. If there are more than 30 units, allow 0.5 hour per unit.

Nonproductive labor on commercial, industrial, institutional and government projects varies considerably. More complex jobs will require proportionately more nonproductive labor. Use your knowledge based on past experience to estimate supervision either as a percentage of job cost or by the square foot of floor.

Column 2: Material Coverage

The second column in the cost tables shows the estimated material coverage in units (usually square feet or linear feet) per gallon. Figure 7 shows the conditions likely to apply for each of the three material coverage rates. Every condition listed in each of these categories won't necessarily occur on every painting operation. For example, it's possible to have high waste and use low quality paint on a repaint job. But it's more likely that waste will be low and paint quality high on jobs like that.

The "Slow" (repaint) application rate assumes light coverage, "Medium" (commercial project) application rate assumes medium coverage and "Fast" (residential tract) application rate assumes heavy coverage. Light coverage is typical on "Slow" (repaint) jobs because previously painted surfaces usually absorb 10 to 15 percent less paint than an unpainted surface. All coverage rates are based on paint that's been thinned according to the manufacturer's recommendations.

Of course, coverage varies with the paint you're using and the surface you're painting. Paint manufacturers usually list the recommended coverage rate on the container label. I've listed estimated coverage rates in the tables throughout this book.

Calculating Film Thickness

Many project specifications for commercial, industrial and government jobs identify the coating (film) thickness you have to apply to each surface. The thickness is given in mils, or thousandths of an inch. One mil is 0.001 inch.

The thickness of the dry paint film depends on the percentage of solids in the paint. If you apply a gallon of paint containing 100 percent solids over 1,600 square feet, the dry film will be 1 mil thick — that is, if 100 percent of the paint adheres to the wall. But if there's 10 percent waste (because of paint that's left in the can, on brushes, or spilled), only 90 percent of the material ends up on the surface.

Slow application	Medium application	Fast application
Repaint jobs	Commercial projects	Residential tracts
Low volume	Medium volume	High volume
20% discount	30% discount	40% discount

Figure 8
Material price discounts

Here's a formula for coverage rates that makes it easy to calculate mil thickness, including the waste factor. Coverage rate equals:

$$\frac{\text{\% of solids x 1600}}{\text{mil thickness}} \times (1.00 - \text{waste factor})$$

Here's an example. Assume you're applying paint with 40 percent solids (by volume), using a roller. The waste factor is 10 percent. You need a thickness of 5 mils.

Here's the calculation for the coverage rate:

$$\frac{.40 \times 1600}{5} \times (1.00 - .10) = 115.2 \text{ per gallon}$$

You may have to apply several coats to get a thickness of 5 mils. In any case, you'll have to use one gallon of paint for each 115.2 square feet of surface.

Waste Factors

Be sure to consider waste and spillage when you figure coverage rates. Professional painters waste very little paint. They rarely kick over a five-gallon paint bucket. But there's always some waste. My material coverage formulas include a typical waste allowance for each application method, whether it's brush, roller or spray. Of course, actual waste depends on the skill of your painters no matter what application method they use.

These are the waste factors I've built into the tables:

Brush .... 3 to 5%

Roll .... 5 to 10%

Airless spray .... 20 to 25%

Conventional spray .... 25 to 35%

Changes in Paint Formulation

In the late 1970s, the California State Air Resources Board established a "model rule" for lowering the solvent in oil-based paints. They mandated replacing solvent-based paint with water-based formulas. The objective was to lower the amount of solvents escaping into the air. This change in the formulation of oil-based paints is being adopted nationwide.

Changes in paint formulation will affect coverage rates and the cost for non-flat paints. Review actual coverage rates and paint prices and make adjustments where necessary before using the estimates in this book.

Column 3: Material Pricing

The third column in the cost tables shows the cost of materials. The "Slow," "Medium," and "Fast" prices in each table are based on the discounts usually offered by suppliers for volume purchases by contractor customers. The material discounts used in this book are defined in Figure 8.

The more paint a contractor buys over a given period, the greater the discount that contractor can expect. Most paint contractors get a discount of at least 20 percent off retail. Contractors buying in heavy volume usually get discounts that approach 40 percent off retail.

Material Pricing Tables

Figures 9, 10 and 11 show the material prices I've used for each of three application rates throughout this book. In the cost estimating tables each coating is identified by a material number. To find out more about the cost of any of these coatings, refer to the material number listed in Figure 9, 10 or 11.

Material prices at 20% discount

All pricing is based on production grade material purchased in 5 gallon quantities.

		Retail price guide	Contractor price at a 20% discount	Add 15% sundries & 10% escalation	Price with sales tax at 8%	Estimating prices with tax
Interior:						
	Sealer, off white (wet area walls & ceilings)					
#1 -	Water base	33.35	26.68	33.35	36.02	36.00
#2 -	Oil base	42.80	34.24	42.80	46.22	46.20
	Undercoat (doors, casings and other paint grade wood)					
#3 -	Water base	34.20	27.36	34.20	36.94	36.90
#4 -	Oil base	44.75	35.80	44.75	48.33	48.30
	Flat latex (walls, ceilings & paint grade baseboard)					
#5 -	Water base latex paint	33.35	26.68	33.35	36.02	36.00
	Acoustic spray-on texture					
#6 -	Primer	26.55	21.24	26.55	28.67	28.70
#7 -	Finish	31.95	25.56	31.95	34.51	34.50
#8 -	Dripowder mixed (pound)	1.45	1.16	1.45	1.57	1.57
	Enamel (wet area walls & ceilings and openings)					
#9 -	Water base enamel	49.35	39.48	49.35	53.30	53.30
#10 -	Oil base enamel	56.90	45.52	56.90	61.45	61.50
	System Estimate (cabinets, bookshelves, molding, interior windows)					
#11a -	Wiping stain, oil base	57.00	45.60	57.00	61.56	61.60
#11b -	Sanding sealer, lacquer	47.15	37.72	47.15	50.92	50.90
#11c -	Lacquer, semi gloss	49.55	39.64	49.55	53.51	53.50
#11 -	Stain, seal & 2 coat lacquer SYSTEM					
	Average cost (11a + b + (2 x c))		40.65	50.81	54.87	54.90
#12 -	Shellac, clear	65.20	52.16	65.20	70.42	70.40
#13 -	Penetrating oil stain	53.60	42.88	53.60	57.89	57.90
#14 -	Penetrating stain wax (molding)	46.45	37.16	46.45	50.17	50.20
#15 -	Wax, per pound (floors)	17.25	13.80	17.25	18.63	18.60
#16 -	Glazing (mottling over enamel)	45.50	36.40	45.50	49.14	49.10
#17 -	Spray can, each (HVAC registers)	11.15	8.92	11.15	12.04	12.00
Exterior:						
	Solid body/color stain (beams, light valance, fascia, overhang, siding, plant-on trim, wood shelves)					
#18 -	Water base stain	45.30	36.24	45.30	48.92	48.90
#19 -	Oil base stain	56.40	45.12	56.40	60.91	60.90
	Semi-transparent stain (beams, siding, T & G ceiling)					
#20 -	Water base stain	46.20	36.96	46.20	49.90	49.90
#21 -	Oil base stain	53.95	43.16	53.95	58.27	58.30
#22 -	Polyurethane (exterior doors)	78.00	62.40	78.00	84.24	84.20
#23 -	Marine spar varnish, flat or gloss (exterior doors)					
	Interior or exterior	81.70	65.36	81.70	88.24	88.20

Figure 9
Material prices at 20% discount

Material prices at 20% discount (cont.)

	Retail price guide	Contractor price at a 20% discount	Add 15% sundries & 10% escalation	Price with sales tax at 8%	Estimating prices with tax
Exterior enamel (exterior doors & trim)					
#24 - Water base	50.70	40.56	50.70	54.76	54.80
#25 - Oil base	57.90	46.32	57.90	62.53	62.50
Porch & deck enamel - interior or exterior					
#26 - Water base enamel	53.30	42.64	53.30	57.56	57.60
#27 - Oil base enamel	61.65	49.32	61.65	66.58	66.60
#28 - Epoxy, 1 part, water base	67.80	54.24	67.80	73.22	73.20
#29 - Epoxy, 2 part SYSTEM	111.90	89.52	111.90	120.85	120.90
System Estimate (exterior windows)					
#30a - Wiping stain, oil base	52.90	42.32	52.90	57.13	57.10
#30b - Sanding sealer, varnish	59.05	47.24	59.05	63.77	63.80
#30c - Varnish, flat or gloss	74.95	59.96	74.95	80.95	81.00
#30 - Stain, seal & 1 coat varnish SYSTEM					
Average cost (30a + b + c))		49.84	62.30	67.28	67.30
Masonry paint (masonry, concrete, plaster)					
#31 - Water base, flat or gloss	41.50	33.20	41.50	44.82	44.80
#32 - Oil base paint	60.55	48.44	60.55	65.39	65.40
#33 - Block filler	34.35	27.48	34.35	37.10	37.10
#34 - Waterproofing, clear hydro seal	37.55	30.04	37.55	40.55	40.60
Metal primer, rust inhibitor					
#35 - Clean metal	55.30	44.24	55.30	59.72	59.70
#36 - Rusty metal	74.15	59.32	74.15	80.08	80.10
Metal finish, synthetic enamel, gloss, interior or exterior					
#37 - Off white	58.90	47.12	58.90	63.61	63.60
#38 - Colors (except orange/red)	61.40	49.12	61.40	66.31	66.30
Anti-graffiti stain eliminator					
#39 - Water base primer & sealer	45.70	36.56	45.70	49.36	49.40
#40 - Oil base primer & sealer	53.70	42.96	53.70	58.00	58.00
#41 - Polyurethane 2 part SYSTEM	173.45	138.76	173.45	187.33	187.30
Preparation:					
#42 - Caulking, per fluid ounce	0.51	0.41	0.51	0.55	0.55
Paint remover, per gallon					
#43 - Light duty	40.95	32.76	40.95	44.23	44.20
#44 - Heavy duty	51.25	41.00	51.25	55.35	55.40
#45 - Putty, per pound	7.40	5.92	7.40	7.99	8.00
#46 - Silica sand, per pound	0.61	0.49	0.61	0.66	0.66
#47 - Visqueen, 1.5 mil, 12' x 200' roll	49.65	39.72	49.65	53.62	53.60
#48 - Wood filler, per gallon	51.25	41.00	51.25	55.35	55.40

Figure 9 (continued)
Material prices at 20% discount

Material prices at 20% discount (cont.)

	Retail price guide	Contractor price at a 20% discount	Add 15% sundries & 10% escalation	Price with sales tax at 8%	Estimating prices with tax
Industrial:					
#49 - Acid wash (muriatic acid)	19.60	15.68	19.60	21.17	21.20
#50 - Aluminum base paint	101.55	81.24	101.55	109.67	109.70
Epoxy coating, 2 part SYSTEM					
#51 - Clear	148.70	118.96	148.70	160.60	160.60
#52 - White	145.35	116.28	145.35	156.98	157.00
Heat resistant enamel					
#53 - 800 to 1200 degree range	138.30	110.64	138.30	149.36	149.40
#54 - 300 to 800 degree range	126.20	100.96	126.20	136.30	136.30
#55 - Industrial bonding & penetrating oil paint	61.35	49.08	61.35	66.26	66.30
Industrial enamel, oil base, high gloss					
#56 - Light colors	60.55	48.44	60.55	65.39	65.40
#57 - Dark (OSHA) colors	72.50	58.00	72.50	78.30	78.30
#58 - Industrial waterproofing	49.70	39.76	49.70	53.68	53.70
#59 - Vinyl coating (tanks)	140.00	112.00	140.00	151.20	151.20
Wallcovering:					
Ready-mix:					
#60 - Light-weight vinyl (gal)	13.30	10.64	13.30	14.36	14.40
#61 - Heavy weight vinyl (gal)	15.10	12.08	15.10	16.31	16.30
#62 - Cellulose, clear (gal)	16.05	12.84	16.05	17.33	17.30
#63 - Vinyl to vinyl (gal)	23.30	18.64	23.30	25.16	25.20
#64 - Powdered cellulose, 2 - 4 ounces	7.80	6.24	7.80	8.42	8.40
#65 - Powdered vinyl, 2 - 4 ounces	9.35	7.48	9.35	10.10	10.10
#66 - Powdered wheat paste, 2-4 ounces	7.15	5.72	7.15	7.72	7.70

Note: Typically, powdered paste is in 2 to 4 ounce packages which will adhere 6 to 12 rolls of wallcovering.

Figure 9 (continued)
Material prices at 20% discount

Material prices at 30% discount

		Retail price guide	Contractor price at a 30% discount	Add 15% sundries & 10% escalation	Price with sales tax at 8%	Estimating prices with tax
Interior:						
	Sealer, off white (wet area walls & ceilings)					
#1 -	Water base	33.35	23.35	29.19	31.53	31.50
#2 -	Oil base	42.80	29.96	37.45	40.45	40.50
	Undercoat (doors, casings and other paint grade wood)					
#3 -	Water base	34.20	23.94	29.93	32.32	32.30
#4 -	Oil base	44.75	31.33	39.16	42.29	42.30
	Flat latex (walls, ceilings & paint grade baseboard)					
#5 -	Water base latex paint	33.35	23.35	29.19	31.53	31.50
	Acoustic spray-on texture					
#6 -	Primer	26.55	18.59	23.24	25.10	25.10
#7 -	Finish	31.95	22.37	27.96	30.20	30.20
#8 -	Dripowder mixed (pound)	1.45	1.02	1.28	1.38	1.38
	Enamel (wet area walls & ceilings and openings)					
#9 -	Water base enamel	49.35	34.55	43.19	46.65	46.70
#10 -	Oil base enamel	56.90	39.83	49.79	53.77	53.80
	System Estimate (cabinets, bookshelves, molding, interior windows)					
#11a -	Wiping stain, oil base	57.00	39.90	49.88	53.87	53.90
#11b -	Sanding sealer, lacquer	47.15	33.01	41.26	44.56	44.60
#11c -	Lacquer, semi gloss	49.55	34.69	43.36	46.83	46.80
#11 -	Stain, seal & 2 coat lacquer SYSTEM Average cost (11a + b + (2 x c))		35.57	44.46	48.02	48.00
#12 -	Shellac, clear	65.20	45.64	57.05	61.61	61.60
#13 -	Penetrating oil stain	53.60	37.52	46.90	50.65	50.70
#14 -	Penetrating stain wax (molding)	46.45	32.52	40.65	43.90	43.90
#15 -	Wax, per pound (floors)	17.25	12.08	15.10	16.31	16.30
#16 -	Glazing (mottling over enamel)	45.50	31.85	39.81	42.99	43.00
#17 -	Spray can, each (HVAC registers)	11.15	7.81	9.76	10.54	10.50
Exterior:						
	Solid body/color stain (beams, light valance, fascia, overhang, siding, plant-on trim, wood shelves)					
#18 -	Water base stain	45.30	31.71	39.64	42.81	42.80
#19 -	Oil base stain	56.40	39.48	49.35	53.30	53.30
	Semi-transparent stain (beams, siding, T & G ceiling)					
#20 -	Water base stain	46.20	32.34	40.43	43.66	43.70
#21 -	Oil base stain	53.95	37.77	47.21	50.99	51.00
#22 -	Polyurethane (exterior doors)	78.00	54.60	68.25	73.71	73.70
#23 -	Marine spar varnish, flat or gloss (exterior doors) Interior or exterior	81.70	57.19	71.49	77.21	77.20

Figure 10
Material prices at 30% discount

Material prices at 30% discount (cont.)

		Retail price guide	Contractor price at a 30% discount	Add 15% sundries & 10% escalation	Price with sales tax at 8%	Estimating prices with tax
	Exterior enamel (exterior doors & trim)					
#24 -	Water base	50.70	35.49	44.36	47.91	47.90
#25 -	Oil base	57.90	40.53	50.66	54.71	54.70
	Porch & deck enamel - interior or exterior					
#26 -	Water base enamel	53.30	37.31	46.64	50.37	50.40
#27 -	Oil base enamel	61.65	43.16	53.95	58.27	58.30
#28 -	Epoxy, 1 part, water base	67.80	47.46	59.33	64.08	64.10
#29 -	Epoxy, 2 part SYSTEM	111.90	78.33	97.91	105.74	105.70
	SYSTEM ESTIMATE (exterior windows)					
#30a -	Wiping stain, oil base	52.90	37.03	46.29	49.99	50.00
#30b -	Sanding sealer, varnish	59.05	41.34	51.68	55.81	55.80
#30c -	Varnish, flat or gloss	74.95	52.47	65.59	70.84	70.80
#30 -	Stain, seal & 1 coat varnish SYSTEM					
	Average cost (30a + b + c))		43.61	54.51	58.87	58.90
	Masonry paint (masonry, concrete, plaster)					
#31 -	Water base, flat or gloss	41.50	29.05	36.31	39.21	39.20
#32 -	Oil base paint	60.55	42.39	52.99	57.23	57.20
#33 -	Block filler	34.35	24.05	30.06	32.46	32.50
#34 -	Waterproofing, clear hydro seal	37.55	26.29	32.86	35.49	35.50
	Metal primer, rust inhibitor					
#35 -	Clean metal	55.30	38.71	48.39	52.26	52.30
#36 -	Rusty metal	74.15	51.91	64.89	70.08	70.10
	Metal finish, synthetic enamel, gloss, interior or exterior					
#37 -	Off white	58.90	41.23	51.54	55.66	55.70
#38 -	Colors (except orange/red)	61.40	42.98	53.73	58.03	58.00
	Anti-graffiti stain eliminator					
#39 -	Water base primer & sealer	45.70	31.99	39.99	43.19	43.20
#40 -	Oil base primer & sealer	53.70	37.59	46.99	50.75	50.80
#41 -	Polyurethane 2 part SYSTEM	173.45	121.42	151.78	163.92	163.90
Preparation:						
#42 -	Caulking, per fluid ounce	0.51	0.36	0.45	0.49	0.49
	Paint remover, per gallon					
#43 -	Light duty	40.95	28.67	35.84	38.71	38.70
#44 -	Heavy duty	51.25	35.88	44.85	48.44	48.40
#45 -	Putty, per pound	7.40	5.18	6.48	7.00	7.00
#46 -	Silica sand, per pound	0.61	0.43	0.54	0.58	0.58
#47 -	Visqueen, 1.5 mil, 12' x 200' roll	49.65	34.76	43.45	46.93	46.90
#48 -	Wood filler, per gallon	51.25	35.88	44.85	48.44	48.40

Figure 10 (continued)
Material prices at 30% discount

Material prices at 30% discount (cont.)

	Retail price guide	Contractor price at a 30% discount	Add 15% sundries & 10% escalation	Price with sales tax at 8%	Estimating prices with tax
Industrial:					
#49 - Acid wash (muriatic acid)	19.60	13.72	17.15	18.52	18.50
#50 - Aluminum base paint	101.55	71.09	88.86	95.97	96.00
Epoxy coating, 2 part SYSTEM					
#51 - Clear	148.70	104.09	130.11	140.52	140.50
#52 - White	145.35	101.75	127.19	137.37	137.40
Heat resistant enamel					
#53 - 800 to 1200 degree range	138.30	96.81	121.01	130.69	130.70
#54 - 300 to 800 degree range	126.20	88.34	110.43	119.26	119.30
#55 - Industrial bonding & penetrating oil paint	61.35	42.95	53.69	57.99	58.00
Industrial enamel, oil base, high gloss					
#56 - Light colors	60.55	42.39	52.99	57.23	57.20
#57 - Dark (OSHA) colors	72.50	50.75	63.44	68.52	68.50
#58 - Industrial waterproofing	49.70	34.79	43.49	46.97	47.00
#59 - Vinyl coating (tanks)	140.00	98.00	122.50	132.30	132.30
Wallcpvering:					
Ready-mix:					
#60 - Light-weight vinyl (gal)	13.30	9.31	11.64	12.57	12.60
#61 - Heavy weight vinyl (gal)	15.10	10.57	13.21	14.27	14.30
#62 - Cellulose, clear (gal)	16.05	11.24	14.05	15.17	15.20
#63 - Vinyl to vinyl (gal)	23.30	16.31	20.39	22.02	22.00
#64 - Powdered cellulose, 2 - 4 ounces	7.80	5.46	6.83	7.38	7.40
#65 - Powdered vinyl, 2 - 4 ounces	9.35	6.55	8.19	8.85	8.90
#66 - Powdered wheat paste, 2-4 ounces	7.15	5.01	6.26	6.76	6.80

Note: Typically, powdered paste is in 2 to 4 ounce packages which will adhere 6 to 12 rolls of wallcovering.

Figure 10 (continued)
Material prices at 30% discount

Material prices at 40% discount

		Retail price guide	Contractor price at a 40% discount	Add 15% sundries & 10% escalation	Price with sales tax at 8%	Estimating prices with tax
Interior:						
	Sealer, off white (wet area walls & ceilings)					
#1 -	Water base	33.35	20.01	25.01	27.01	27.00
#2 -	Oil base	42.80	25.68	32.10	34.67	34.70
	Undercoat (doors, casings and other paint grade wood)					
#3 -	Water base	34.20	20.52	25.65	27.70	27.70
#4 -	Oil base	44.75	26.85	33.56	36.24	36.20
	Flat latex (walls, ceilings & paint grade baseboard)					
#5 -	Water base latex paint	33.35	20.01	25.01	27.01	27.00
	Acoustic spray-on texture					
#6 -	Primer	26.55	15.93	19.91	21.50	21.50
#7 -	Finish	31.95	19.17	23.96	25.88	25.90
#8 -	Dripowder mixed (pound)	1.45	0.87	1.09	1.18	1.18
	Enamel (wet area walls & ceilings and openings)					
#9 -	Water base enamel	49.35	29.61	37.01	39.97	40.00
#10 -	Oil base enamel	56.90	34.14	42.68	46.09	46.10
	System Estimate (cabinets, bookshelves, molding, interior windows)					
#11a -	Wiping stain, oil base	57.00	34.20	42.75	46.17	46.20
#11b -	Sanding sealer, lacquer	47.15	28.29	35.36	38.19	38.20
#11c -	Lacquer, semi gloss	49.55	29.73	37.16	40.13	40.10
#11 -	Stain, seal & 2 coat lacquer SYSTEM					
	Average cost (11a + b + (2 x c))		30.49	38.11	41.16	41.20
#12 -	Shellac, clear	65.20	39.12	48.90	52.81	52.80
#13 -	Penetrating oil stain	53.60	32.16	40.20	43.42	43.40
#14 -	Penetrating stain wax (molding)	46.45	27.87	34.84	37.63	37.60
#15 -	Wax, per pound (floors)	17.25	10.35	12.94	13.98	14.00
#16 -	Glazing (mottling over enamel)	45.50	27.30	34.13	36.86	36.90
#17 -	Spray can, each (HVAC registers)	11.15	6.69	8.36	9.03	9.00
Exterior:						
	Solid body/color stain (beams, light valance, fascia, overhang, siding, plant-on trim, wood shelves)					
#18 -	Water base stain	45.30	27.18	33.98	36.70	36.70
#19 -	Oil base stain	56.40	33.84	42.30	45.68	45.70
	Semi-transparent stain (beams, siding, T & G ceiling)					
#20 -	Water base stain	46.20	27.72	34.65	37.42	37.40
#21 -	Oil base stain	53.95	32.37	40.46	43.70	43.70
#22 -	Polyurethane (exterior doors)	78.00	46.80	58.50	63.18	63.20
#23 -	Marine spar varnish, flat or gloss (exterior doors)					
	Interior or exterior	81.70	49.02	61.28	66.18	66.20

Figure 11
Material prices at 40% discount

Material prices at 40% discount (cont.)

		Retail price guide	Contractor price at a 40% discount	Add 15% sundries & 10% escalation	Price with sales tax at 8%	Estimating prices with tax
	Exterior enamel (exterior doors & trim)					
#24 -	Water base	50.70	30.42	38.03	41.07	41.10
#25 -	Oil base	57.90	34.74	43.43	46.90	46.90
	Porch & deck enamel - interior or exterior					
#26 -	Water base enamel	53.30	31.98	39.98	43.18	43.20
#27 -	Oil base enamel	61.65	36.99	46.24	49.94	49.90
#28 -	Epoxy, 1 part, water base	67.80	40.68	50.85	54.92	54.90
#29 -	Epoxy, 2 part SYSTEM	111.90	67.14	83.93	90.64	90.60
	System Estimate (exterior windows)					
#30a -	Wiping stain, oil base	52.90	31.74	39.68	42.85	42.90
#30b -	Sanding sealer, varnish	59.05	35.43	44.29	47.83	47.80
#30c -	Varnish, flat or gloss	74.95	44.97	56.21	60.71	60.70
#30 -	Stain, seal & 1 coat varnish SYSTEM Average cost (30a + b + c))		37.38	46.73	50.47	50.50
	Masonry paint (masonry, concrete, plaster)					
#31 -	Water base, flat or gloss	41.50	24.90	31.13	33.62	33.60
#32 -	Oil base paint	60.55	36.33	45.41	49.04	49.00
#33 -	Block filler	34.35	20.61	25.76	27.82	27.80
#34 -	Waterproofing, clear hydro seal	37.55	22.53	28.16	30.41	30.40
	Metal primer, rust inhibitor					
#35 -	Clean metal	55.30	33.18	41.48	44.80	44.80
#36 -	Rusty metal	74.15	44.49	55.61	60.06	60.10
	Metal finish, synthetic enamel, gloss, interior or exterior					
#37 -	Off white	58.90	35.34	44.18	47.71	47.70
#38 -	Colors (except orange/red)	61.40	36.84	46.05	49.73	49.70
	Anti-graffiti stain eliminator					
#39 -	Water base primer & sealer	45.70	27.42	34.28	37.02	37.00
#40 -	Oil base primer & sealer	53.70	32.22	40.28	43.50	43.50
#41 -	Polyurethane 2 part SYSTEM	173.45	104.07	130.09	140.50	140.50
Preparation:						
#42 -	Caulking, per fluid ounce	0.51	0.31	0.39	0.42	0.42
	Paint remover, per gallon					
#43 -	Light duty	40.95	24.57	30.71	33.17	33.20
#44 -	Heavy duty	51.25	30.75	38.44	41.52	41.50
#45 -	Putty, per pound	7.40	4.44	5.55	5.99	6.00
#46 -	Silica sand, per pound	0.61	0.37	0.46	0.50	0.50
#47 -	Visqueen, 1.5 mil, 12' x 200' roll	49.65	29.79	37.24	40.22	40.20
#48 -	Wood filler, per gallon	51.25	30.75	38.44	41.52	41.50

Figure 11 (continued)
Material prices at 40% discount

Material prices at 40% discount (cont.)

	Retail price guide	Contractor price at a 40% discount	Add 15% sundries & 10% escalation	Price with sales tax at 8%	Estimating prices with tax
Industrial:					
#49 - Acid wash (muriatic acid)	19.60	11.76	14.70	15.88	15.90
#50 - Aluminum base paint	101.55	60.93	76.16	82.25	82.30
Epoxy coating, 2 part SYSTEM					
#51 - Clear	148.70	89.22	111.53	120.45	120.50
#52 - White	145.35	87.21	109.01	117.73	117.70
Heat resistant enamel					
#53 - 800 to 1200 degree range	138.30	82.98	103.73	112.03	112.00
#54 - 300 to 800 degree range	126.20	75.72	94.65	102.22	102.20
#55 - Industrial bonding & penetrating oil paint	61.35	36.81	46.01	49.69	49.70
Industrial enamel, oil base, high gloss					
#56 - Light colors	60.55	36.33	45.41	49.04	49.00
#57 - Dark (OSHA) colors	72.50	43.50	54.38	58.73	58.70
#58 - Industrial waterproofing	49.70	29.82	37.28	40.26	40.30
#59 - Vinyl coating (tanks)	140.00	84.00	105.00	113.40	113.40
Wallcovering:					
Ready-mix:					
#60 - Light-weight vinyl (gal)	13.30	7.98	9.98	10.78	10.80
#61 - Heavy weight vinyl (gal)	15.10	9.06	11.33	12.24	12.20
#62 - Cellulose, clear (gal)	16.05	9.63	12.04	13.00	13.00
#63 - Vinyl to vinyl (gal)	23.30	13.98	17.48	18.88	18.90
#64 - Powdered cellulose, 2 - 4 ounces	7.80	4.68	5.85	6.32	6.30
#65 - Powdered vinyl, 2 - 4 ounces	9.35	5.61	7.01	7.57	7.60
#66 - Powdered wheat paste, 2-4 ounces	7.15	4.29	5.36	5.79	5.80

Note: Typically, powdered paste is in 2 to 4 ounce packages which will adhere 6 to 12 rolls of wallcovering.

Figure 11 (continued)
Material prices at 40% discount

Figure 9 shows prices at a 20 percent discount off retail. It applies to "Slow" work and assumes light coverage on a previously painted surface. These costs would be typical for a lower-volume company handling mostly repaint or custom work.

Figure 10 reflects a 30 percent discount. It applies to "Medium" work and assumes medium coverage, as in commercial work.

Figure 11 is the 40 percent discount table. It applies to "Fast" work and assumes heavier coverage typically required on unpainted surfaces in new construction. This discount is usually available only to large, high-volume painting companies that purchase materials in large quantities.

Here's an explanation of the columns in Figures 9, 10 and 11:

Retail price guide: This is an average based on a survey of up to a dozen paint manufacturers or distributors, for standard grade, construction-quality paint, purchased in five gallon quantities.

Material pricing and discount percentages will vary from supplier to supplier and from area to area. Always keep your supplier's current price list handy. It should show your current cost for all the coatings and supplies you use. Also post a list of all suppliers, their phone numbers, and the salesperson's name beside your phone.

Prices change frequently. Paint quality, your supplier's discount programs, their marketing strategy and competition from other paint manufacturers will influence the price you pay. Never guess about paint prices — especially about less commonly used coatings. Don't assume that a product you haven't used before costs about the same as similar products. It might not. A heavy-duty urethane finish, for example, will cost about twice as much as a heavy-duty vinyl coating. If you don't know that, your profit for the job can disappear very quickly.

Prices at discount: The retail price, less the appropriate discount.

Allowance for sundries: It's not practical to figure the cost of every sheet of sandpaper and every rag you'll use on a job. And there's no way to accurately predict how many jobs you'll get out of each brush or roller pole, roller handle, ladder, or drop cloth. But don't let that keep you from including an allowance for these important costs in your estimates. If you leave them out, it's the same as estimating the cost of those items as zero. That's a 100 percent miss. Too many of those, and you're out of the painting business. It's better to estimate any amount than to omit some costs entirely.

Figure 12 is a sundries inventory checklist. Use it to keep track of the actual cost of expendable tools and equipment.

I've added 15 percent to the paint cost to cover expendable tools and supplies. This is enough for sundries on most jobs. There is one exception, however. On repaint jobs where there's extensive prep work, the cost of sundries may be more than 15 percent of the paint cost. When preparation work is extensive, figure the actual cost of supplies. Then add to the estimate that portion of the sundries cost that exceeds 15 percent of the paint cost. You might have to double the normal sundries allowance. When it comes to prep work, make sure your estimate covers all your supplies.

Price with sales tax at 8 percent: This column increases the material cost, including sundries, by 8 percent to cover sales tax. If sales tax in your area is more or less than 8 percent, you can adjust the material cost, or use the price that's closest to your actual cost.

In most cases contractors have to pay sales tax. If you don't pay the tax yourself, you may have to collect it from the building owner or general contractor and remit it to the state taxing authority. In either case, include sales tax in your estimate.

Estimating prices with tax: The figures in the last column of Figures 9 through 11 are rounded to the nearest dime unless the total is under a dollar. Those prices are rounded to the nearest penny.

This system for pricing materials isn't exact. But it's quick, easy and flexible. Compare your current material costs with costs in Figures 9, 10 and 11. If your costs are more than a few percent higher or lower than my costs, make a note on the blank line below "Fast" in the estimating tables.

Sundry Inventory Checklist

Suppliers: D-Dumphy Paints
F-Fisher Paints
S-Superior Paints
P-Pioneer Paints

Supplier	Product number	Product	Inventory quantity	Unit	Cost	7/21	7/27	8/2	8/10
D	# —	Bender paint pads	3	Each	$ 5.25				
D	#792	Brush - 3" nylon Peacock	2	Each	$ 27.80		1		
D	#783	Brush - 4" nylon Scooter	2	Each	$ 41.20			1	
D	#115	Brush - 5" nylon Pacer	2	Each	$ 69.90			1	
D	#784	Brush - 3" bristle	2	Each	$ 25.80			1	
D	#2170	Caulking bags	2	Each	$ 5.57				
D	Latex	Caulking-DAP Acrylic latex	12	Each	$ 2.94		12		
D	#2172	Caulking gun (Newborn)	2	Each	$ 10.50		1		
P	# —	Hydraulic fluid	2	Qt	$ 11.92				
P	# —	Lemon oil	2	Pint	$ 5.96		1		
F	# —	Masking paper 18" wide	3	Roll	$ 30.50				
F	Anchor	Masking tape 11/2"	24	Roll	$ 4.46		12		12
P	#2176	Lacquer - 5 gallons	2	5's	$ 135.00			1	
P	#2173	Sanding sealer - 5 gallons	2	5's	$ 129.00		1		
P	#9850	Resin sealer - 5 gallons	2	5's	$ 116.00				
P	#131	PVA sealer (clear) - 5 gallons	2	5's	$ 122.00		1		
F	#8500	Particle masks 100/box	1	Box	$ 20.00			1	
P	# —	Putty (Crawfords)	3	Qt	$ 13.50		2		
F	#R-10	Respirators	1	Each	$ 55.70				1
F	#R-49	Respirator cartridges 20/box	2	Box	$ 64.70				
F	#R-51	Respirator filters 20/box	2	Box	$ 46.20			1	
P	# —	Rags - 10 pound sack	2	Sack	$ 33.20				
F	#AR 691	Roller covers 9" x 3/4"	6	Each	$ 6.14		2		
F	#AR 692	Roller covers 9" x 3/8"	6	Each	$ 6.27	3			2
F	#AR 671	Roller covers 7" x 3/4"	3	Each	$ 5.09			1	
F	#AR 672	Roller covers 7" x 3/8"	3	Each	$ 5.57		1		

Figure 12
Sundry inventory checklist

Supplier	Product number	Product	Inventory quantity	Unit	Cost	7/21	7/27	8/2	8/10
F	#AR 611	Roller covers mini	3	Each	$ 4.28			1	
F	#95	Roller frames 9"	6	Each	$ 7.89	1	2		
F	#75	Roller frames 7"	5	Each	$ 7.62	3		3	
F	#TSR	Roller frames mini	2	Each	$ 4.41				
D	#40	Roller poles 4' wood tip	3	Each	$ 3.89		1		
D	#10	Roller poles 6' wood tip	10	Each	$ 6.02			2	
P	# 1	Roller pole tips metal	2	Each	$ 4.82			2	
P	# —	Sandpaper (120C production)	2	Slve	$ 71.80				1
P	# —	Sandpaper (220A trimite)	2	Slve	$ 55.80				
P	# —	Sandpaper (220A garnet)	1	Slve	$ 50.80		1		
D	# —	Spackle (Synkloid)	3	Qt	$ 7.83	1		1	
D	#42/61	Spray bombs (black^B/white^W)	12	Each	$ 4.56	^{B}12			^{W}12
F	# —	Spray gun tips #3 or #4	10	Each	$ 11.30			3	
F	#2762	Spray gun couplers	10	Each	$ 3.08			5	
F	#S-71	Spray socks 48/box	1	Box	$ 24.60				
D	#5271	Stip fill	1	Gal	$ 13.20			1	
D	#5927	Strainer bags	2	Each	$ 2.14	1			
D	#JT-21	Staples - 5/16"	2	Box	$ 3.48				
P	50 Gal	Thinner, lacquer	1	Drum	$ 618.00				
P	50 Gal	Thinner, paint	1	Drum	$ 308.00				1
P	# —	Thinner, shellac (alcohol)	1	Gal	$ 14.44				
D	# —	Visqueen 1.5 mil 12' x 200'	3	Roll	$ 36.10				
D	#5775	Work pots (2 gal. plastic)	3	Each	$ 4.15		1		2
	#				$				
	#				$				
	#				$				
	#				$				
		Order date:				7/21	7/27	8/2	8/10
		Ordered by: (initials)				jj	jj	jj	jj
		Purchase order no.				0352	0356	0361	0371

Figure 12 (continued)
Sundry inventory checklist

	Residential Wallcovering				Commercial Wallcovering				Flexible Wood Wallcovering			
Production Rate	**Computer Program Crew Code**	**Labor Cost per Hour**	**Labor Burden per Hour**	**Labor Cost + Burden**	**Computer Program Crew Code**	**Labor Cost per Hour**	**Labor Burden per Hour**	**Labor Cost + Burden**	**Computer Program Crew Code**	**Labor Cost per Hour**	**Labor Burden per Hour**	**Labor Cost + Burden**
Slow	1W	$22.50	$5.40	$27.90	4W	$21.50	$5.16	$26.66	7W	$22.00	$5.28	$27.28
Medium	2W	29.00	8.38	37.38	5W	27.50	7.95	35.45	8W	28.25	8.16	36.41
Fast	3W	35.50	12.53	45.03	6W	33.50	11.83	45.33	9W	34.50	12.18	46.68

Figure 13
Hourly wage rates for wallcovering application

Price Escalation

Escalation is the change in prices between the time you bid a job and the time you pay for labor and materials. Painting contractors seldom include escalation clauses in their bids because they don't expect lengthy delays. That's why escalation isn't included as a separate item in the estimating forms, Figures 18 and 19.

Any minor price escalation will be covered by the 15 percent added to material prices for sundries. But don't rely on that small cushion to absorb major inflationary cost increases. Plan ahead if prices are rising. In that case, add 10% of your material costs as an escalation factor and include this figure as a separate line item in the estimate.

Many formal construction contracts include an escalator clause that allows the contractor to recover for cost increases during the time of construction — especially if there was an unreasonable delay through no fault of the subcontractor. This clause may give you the right to collect for increases in both labor and material costs.

If work is delayed after you've been awarded the contract, you may be able to recover for cost increases under the escalator clause. This is more likely on public projects than on private jobs. Also, if there's a significant delay due to weather, you may have a good argument for adjusting the contract amount.

You can protect yourself against escalation if you include an expiration date on your bids. If the contract award is delayed beyond your expiration date, you can review your costs and make necessary adjustments. But be careful here. Increase the bid too much and you'll probably lose the contract. So raise your bid only if necessary, and then only by the amount of the actual cost increases. Don't try to make a killing on the job just because the bid prices have expired.

Column 4: Labor Cost

Column 4 in Figure 2 on page 7 shows the labor cost per unit. This figure is based on the productivity rate in column 1 and the wage rate in Figure 1. The wage rate for "Slow" (repaint) work is assumed to be $23.00 per hour. The wage rate for "Medium" (commercial) work is $29.50 per hour. The wage rate for "Fast" (residential tract) work is $36.00 per hour. Wage rates for wallcovering are different (Figure 13).

Wage Rates Vary

Wages vary from city to city. Recently I saw a survey of hourly union rates for painters in U.S. cities. The lowest rate shown was $17.91 an hour for painters in Raleigh, North Carolina. The highest rate was $46.15 for painters in Nome, Alaska. You might ask, "Why don't all the painters in Raleigh move to Nome?"

I don't know the answer, except to suggest that painters aren't starving in Raleigh. Nor are they getting rich in Nome. Working conditions and the cost of living are very different in those two cities. However, on private jobs using non-union tradesmen, wage rates usually don't vary as much from city to city. The wage you pay depends on the demand for painting and how many painters are available for work.

Wages also change over time. For example, wage rates increased between 2009 and 2019. The national average union wage (including fringes) for painters in large cities went from $34.62 in 2009 to $37.82 per hour in 2019. In 2019, the average union wage for commercial work increased to as high as $52.69 per hour. Always base your estimates on the actual wages you'll pay your ***most experienced*** painters.

Wages for Higher Skilled Specialists

Wages also vary with a workers' skill, dependability and with job difficulty. Generally higher paid painters are more productive than lower paid painters. Here's a chart to determine how much more per hour to estimate for supervision and for painting and surface preparation specialists. These figures are in addition to the basic journeyman rate.

Foremen .... $2.00 to 6.00

Field superintendents .... $9.00 to 12.00

Swing stage brush painters,
spray painters, or paperhangers $1.00

Iron, steel and bridge painters
(ground work) $2.00

Sandblasters, iron, steel, or
bridge painters (swing stage) $4.00

Steeplejacks $5.00

Most government and defense painting contracts require compliance with the Davis Bacon Act, which specifies that contractors pay at least the prevailing wage for each trade in the area where the job is located.

Calculate Your Labor Rate

Use the wage rate in Figure 1 ($23.00, $29.50 or $36.00 for "Slow," "Medium," or "Fast") that's appropriate for your company. Or, use a rate somewhere in between the rates listed. If you use your own wage rate, divide the hourly wage by the labor productivity (such as square feet per manhour in column 1). That's your labor cost per unit. Multiply by 100 if the units used are 100 linear feet or 100 square feet. ($22 ÷ 400 x 100 = $5.50.)

Column 5: Labor Burden

For each dollar of wages your company pays, at least another 28 cents has to be paid in payroll tax and for insurance. That's part of your labor burden. The rest is fringe benefits such as vacation pay, health benefits and pension plans.

Federal taxes are the same for all employers. State taxes vary from state to state. Fringe benefits vary the most. Generally, larger companies with more skilled painters offer considerably more fringe benefits than smaller companies.

In the estimating tables, the labor burden percentage varies with the application rate. For "Slow" (repaint) work, it's assumed to be 24.0 percent of $23.00 or $5.52 per hour. For "Medium" (commercial) work, the estimating tables use 28.90 percent of $29.50 or $8.53 per hour. For "Fast" (residential tract) work, the labor burden is 35.3 percent of $36.00 or $12.71 per hour.

Figure 14 shows how the labor burden percentages were compiled for each application rate.

FICA — Social Security tax: This is the portion paid by employers and is set by federal law. A similar amount is withheld from each employee's wage and deposited with a Federal Reserve bank by the employer.

FUTA — Federal Unemployment Insurance tax: Paid entirely by the employer and set by federal law. No portion is deducted from employee wages.

SUI — State Unemployment Insurance: Varies from state to state.

WCI — Workers' Compensation Insurance: Provides benefits for employees in case of injury on the job. Workers' comp is required by state law. Rates vary by state, job description and the loss experience of the employer.

Liab. Ins. — Liability Insurance: Covers injury or damage done to the public by employees. Comprehensive contractor's liability insurance includes current operations, completed operations, bodily injury, property damage, protective and contractual coverages with a $1,000,000 policy limit.

	Fixed burden					Fringe benefits					
	FICA	**FUTA**	**SUI**	**WCI**	**Liab. Ins.**	**Vac**	**Med**	**Life**	**Pension**	**Training**	**Total**
Slow	7.65%	0.6%	3.0%	5.5%	6.25%	0	1.0%	0	0	0	24.00%
Medium	7.65%	0.6%	4.5%	6.5%	6.65%	.5%	2.0%	.25%	.25%	0	28.90%
Fast	7.65%	0.6%	6.0%	8.5%	7.05%	1.5%	3.0%	.25%	0.5%	.25%	35.30%

Figure 14
Labor burden percentages

Fringe benefits: *Vac* is vacation *pay*. Med is medical insurance. *Life* is life insurance contribution. Pension is a pension plan contribution. *Training* is an apprentice training fund.

Vacation, life, pension and training payments depend on the agreement between employers and employees. These are voluntary contributions if not required by a collective bargaining agreement. Smaller companies are less likely to provide these benefits. The cost of fringe benefits in a painting company can range from zero to more than 10 percent of wages.

Column 6: Material Cost per Unit

This column is the result of dividing column 3 (material cost) by column 2 (material coverage) for each application rate. For example, in Figure 2 in the "Medium" row, a material cost of $31.50 is divided by material coverage of 275, then multiplied by 100 to arrive at $11.45 per 100 square feet. That's the figure listed for "Medium" in column 6.

Column 7: Overhead

The overhead rate for "Slow" (repaint) jobs is assumed to be 19 percent. For "Medium" (commercial projects), overhead is 25 percent. For "Fast" (residential tracts), overhead is 31 percent. The overhead cost per unit in each row is calculated by adding the labor cost per unit, labor burden per unit, and material cost per unit and then multiplying by the appropriate overhead percentage.

There are two types of overhead, direct overhead and indirect overhead. Only indirect overhead is included in the "Overhead" column of the estimating cost tables. Enter your direct overhead costs on a separate line on your take-off sheet.

Direct overhead is job site overhead, expenses you charge to a specific job. Examples include performance bonds, special insurance premiums, or rental of a job site storage trailer. These expenses are not included in the estimating tables and have to be added to your estimates. On many jobs, there may be little or no direct overhead.

Indirect overhead is office overhead, expenses that aren't related to any particular job and that tend to continue whether the volume of work increases or decreases. Examples are non-trade salaries, office rent, vehicles, sales and financial expenses, insurance, taxes and licenses.

The percentage of income spent on overhead is assumed to be lower for high volume companies and higher for low volume companies. A large company working many projects at the same time can spread overhead costs over many projects — charging a smaller percentage of overhead to each job. The more jobs, the lower the overhead per job — assuming overhead doesn't increase faster than business volume.

On the other hand, a small business may have to absorb all overhead on a single job. Even painting contractors who work out of their homes have overhead expenses.

Here's one overhead expense every paint contractor has and that you might overlook: the cost of estimating jobs. That's part of the salary cost of the employee who does the estimating.

Figure Overhead Carefully

Estimating indirect (office) overhead isn't as easy as estimating labor and material. There aren't as many clear-cut answers. That's why indirect overhead is often underestimated. Don't make that mistake in your estimates. Underestimating overhead is the same as giving up part of your profit. After all, indirect overhead expenses are real costs, just like paint, labor and taxes.

In large painting companies, management accumulates indirect overhead costs and translates them into a percentage the estimator should add to the costs of each job. In smaller companies, the estimator should keep a record of indirect overhead expenses. With a good record of overhead expense, you can calculate your overhead percentage for future periods very accurately. Then it's easy to add a percentage for indirect overhead costs into your estimate.

Computing Your Overhead Percentage

Here's how to decide which overhead rate to use in the cost estimating tables:

1) List all your overhead expenses for at least the last six months; a year would be better. You need overhead cost information that goes back far enough to eliminate the effect of seasonal changes in business volume

   If your company is new, estimate your annual overhead by projecting overhead costs for the first full year. For example, if you've been in business for five months and overhead has been $5,500 so far, you can expect annual overhead to be about $13,200 ($5,500 divided by 5 and multiplied by 12).

2) Here's how to calculate your indirect overhead percentage:

$$\frac{\text{Annual indirect overhead}}{\text{Annual job expenses}} = \text{Overhead \%}$$

Calculate your indirect overhead by adding together your real (or anticipated) annual expenses for the following:

Salaries. Include what you pay for all employees except trade workers, plus payroll-related expenses for all employees.

Office and shop expense. Rent or mortgage, utilities, furniture and equipment, maintenance, office supplies and postage, storage sheds, warehouses, fences or yard maintenance.

Vehicles. Lease or purchase payments, maintenance, repairs and fuel.

Sales promotion. Advertising, entertainment and sales-related travel.

Taxes. Property tax and income tax, and sales tax (if not included in your material prices).

Licenses. Contractor's and business licenses.

Insurance. General liability, property and vehicle policies.

Interest expense. Loan interest and bank charges. Also consider loss of interest on payments retained by the general contractor until the job is finished.

Miscellaneous expenses. Depreciation and amortization on building and vehicles, bad debts, legal and accounting fees, and educational expenses.

Direct overhead is easier to figure. It's all job expenses except tradesman labor, payroll taxes and insurance, materials, equipment, subcontracts, and contingency expenses. Permits, bonds, fees and special insurance policies for property owners are also examples of direct overhead. Add the direct overhead expense on the appropriate lines in your estimate. Direct overhead is not included in the estimating tables in this manual.

Field Equipment May Be Part of Overhead

As you may have noticed, there's no equipment cost column in the estimating tables. Instead, field equipment expense is included in the overhead percentage for "Fast" and "Medium" work but not "Slow" work.

Equipment Rental Rates

Use the following rates only as a guide. They may not be accurate for your area.
Verify equipment rental rates at your local yard.

	Rental Day	Rental Week	Rental Month
Acoustical sprayer	63.80	191.00	477.00
Air compressors			
Electric or gasoline, wheel mounted			
5 CFM, 1.5 HP, electric	38.40	117.00	295.00
8 CFM, 1.5 HP, electric	46.00	135.00	338.00
10 CFM, 5.5 HP, gasoline	52.50	156.00	391.00
15 CFM, shop type, electric	58.70	177.00	443.00
50 CFM, shop type, electric	77.90	233.00	581.00
100 CFM, gasoline	106.00	316.00	792.00
125 CFM, gasoline	119.00	359.00	897.00
150 CFM, gasoline	134.00	401.00	1,000.00
175 CFM, gasoline	147.00	443.00	1,110.00
190 CFM, gasoline	161.00	485.00	1,220.00
Diesel, wheel mounted			
to 159 CFM	119.00	359.00	1,070.00
160 to 249 CFM	147.00	442.00	1,320.00
250 to 449 CFM	218.00	654.00	1,950.00
450 to 749 CFM	324.00	971.00	2,920.00
750 to 1199 CFM	443.00	1,320.00	3,980.00
1200 CFM & over	647.00	1,940.00	7,230.00
Air hose - with coupling, 50' lengths			
1/4" I.D.	8.42	25.60	63.80
3/8" I.D.	9.95	29.40	74.00
1/2" I.D.	11.20	32.10	84.20
5/8" I.D.	12.80	38.30	95.70
3/4" I.D.	14.10	42.10	106.00
1" I.D.	15.40	46.00	116.00
1-1/2" I.D.	22.50	67.70	170.00
Boomlifts			
3' x 4' to 3' x 8' basket			
20' two wheel drive	197.00	591.00	1,770.00
30' two wheel drive	238.00	718.00	2,150.00
40' four wheel drive	274.00	823.00	2,470.00
50' - 1000 lb.	453.00	1,350.00	4,050.00
Telescoping and articulating booms, self propelled, gas or diesel powered, 2-wheel drive			
21' to 30' high	281.00	843.00	2,520.00
31' to 40' high	352.00	1,050.00	3,160.00
41' to 50' high	458.00	1,380.00	4,130.00
51' to 60' high	561.00	1,690.00	5,070.00
Burner, paint	16.90	51.00	126.00

	Rental Day	Rental Week	Rental Month
Dehumidifier - 5000 Btu, 89 lb, 8.7 amp	77.90	233.00	581.00
Ladders			
Aluminum extension			
16' to 36'	42.10	126.00	316.00
40' to 60'	63.80	190.00	477.00
Step - fiberglass or wood			
6'	11.20	33.80	84.20
8'	14.10	42.10	106.00
10'	16.90	50.70	126.00
12'	19.70	59.10	148.00
14'	22.50	67.70	169.00
16'	28.10	84.20	211.00
20'	36.60	109.00	274.00
Ladder jacks - No guardrail.	11.20	28.10	70.20
Masking paper dispenser	28.10	70.20	176.00
Painter's pic (walkboards); No guardrail. (Also known as airplane planks, toothpicks and banana boards)			
16' long	11.20	33.80	84.20
20' long	22.50	67.70	169.00
24' long	28.10	84.20	211.00
28' long	33.80	101.00	252.00
32' long	39.60	119.00	295.00
Planks - plain end microlam scaffold plank			
9" wide	14.10	42.10	106.00
10" wide	16.90	50.70	126.00
12" wide	19.70	59.10	148.00
Pressure washers (See Water pressure washers)			
Sandblast compressor and hopper			
To 250 PSI	84.20	252.00	634.00
Over 250 to 300 PSI	120.00	359.00	897.00
Over 600 to 1000 PSI	154.00	464.00	1,160.00
Sandblast machines			
150 lb pot with hood, 175 CFM compressor	324.00	971.00	2,440.00
300 lb pot with hood, 325 CFM compressor	579.00	1,720.00	4,340.00
600 lb pot with hood, 600 CFM compressor	1,050.00	3,150.00	7,860.00

Figure 15

Typical equipment purchase and rental prices

	Rental Day	Rental Week	Rental Month
Sandblast hoses - 50' lengths, coupled			
3/8" I.D.	14.10	42.10	107.00
3/4" I.D.	19.70	59.10	147.00
1" I.D.	25.20	75.40	190.00
1-1/4" I.D.	28.10	84.20	212.00
1-1/2" I.D.	31.00	93.10	233.00
Sandblast accessories			
Nozzles, all types	25.20	76.70	190.00
Hood, air-fed	39.40	119.00	296.00
Valves, remote control (deadman, all sizes)	42.10	126.00	316.00
Sanders			
Belt - 3"	19.70	59.10	148.00
Belt - 4" x 24"	23.90	71.40	180.00
Disc - 7"	31.00	93.10	233.00
Finish sander, 6"	16.90	50.70	126.00
Floor edger, 7" disk, 29#, 15 amp.	28.10	84.20	211.00
Floor sander, 8" drum, 118#, 14 amp.	63.40	189.00	477.00
Palm sander, 4" x 4"	14.10	42.10	106.00
Palm sander, 4-1/2" x 9-1/4"	16.90	50.70	126.00
Scaffolding, rolling stage, caster mounted, 30" wide by 7' or 10' long			
4' to 6' reach	56.10	112.00	225.00
7' to 11' reach	70.20	141.00	281.00
12' to 16' reach	98.30	197.00	394.00
17' to 21' reach	134.00	267.00	534.00
22' to 26' reach	148.00	295.00	591.00
27' to 30' reach	161.00	323.00	647.00
Casters - each	14.10	28.10	42.10
Scissor lifts			
Electric powered, rolling with 2' x 3' platform, 650 lb capacity			
30' high	106.00	316.00	949.00
40' high	183.00	550.00	1,640.00
50' high	211.00	634.00	1,900.00
Rolling, self-propelled, hydraulic, electric powered			
to 20' high	154.00	626.00	1,400.00
21' to 30' high	190.00	569.00	1,710.00
31' to 40' high	238.00	718.00	2,150.00
Rolling, self-propelled, hydraulic, diesel powered			
to 20' high	176.00	529.00	1,590.00
21' to 30' high	218.00	654.00	1,950.00
31' to 40' high	281.00	845.00	2,520.00
Spray rigs			
Airless pumps, complete with gun and 50' of line			
Titan 447, 7/8 HP, electric	98.30	296.00	887.00
Titan 660, 1 HP, electric	112.00	338.00	1,010.00
Gasoline, .75 gpm	120.00	359.00	1,110.00
Emulsion pumps			
65 gal, 5 HP engine	98.30	296.00	887.00
200 gal, 5 HP engine	112.00	334.00	1,010.00
Emulsion airless, 1.25 gpm, gasoline	120.00	359.00	1,110.00
Conventional pumps, gas, portable			
High pressure, low vol. (HVLP)	63.80	190.00	570.00
8 CFM complete	84.20	252.00	759.00
17 CFM complete	92.00	274.00	823.00
85 CFM complete	106.00	316.00	949.00
150 CFM complete	154.00	464.00	1,400.00
Spray rig accessories: 6' wand	9.95	29.60	74.00
Striper, paint (parking lot striping)			
Aerosol	28.10	84.20	211.00
Pressure regulated	41.00	119.00	296.00
Swing stage, rental			
Any length drop, motor operated, excluding safety gear and installation or dismantling. Note: Must be set up by a professional to ensure safety.			
Swing stage	141.00	421.00	1,260.00
Basket	71.40	212.00	633.00
Bosun's chair	71.40	212.00	633.00
Swing stage safety gear, purchase only			
Safety harness (126.00)			
4' lanyard with locking snap at each end (92.00)			
DBI rope grab for 5/8" safety line (98.30)			
Komet rope grab for 3/4" safety line (141.00)			
Texturing equipment			
Texturing gun - w/ hopper, no compressor	7.14	21.20	63.80
Texturing mud paddle mixer	9.95	29.50	89.50
Texturing outfit - 1 HP w/ gun, 50' hose, 75 PSI	15.50	46.40	139.00
Wallpaper hanging kit	23.90	71.40	215.00
Wallpaper steamer			
Electric, small, 10 amp	28.10	84.20	252.00
Electric, 15 amp	42.10	126.00	380.00
Pressurized, electric	53.40	161.00	482.00
Water pressure washer (pressure washer, water blaster, power washer)			
1000 PSI, electric, 15 amp	63.80	190.00	569.00
2000 PSI, gas	106.00	316.00	949.00
2500 PSI, gas	112.00	338.00	1,010.00
3500 PSI, gas	124.00	372.00	1,110.00

Figure 15 (continued)
Typical equipment purchase and rental prices

New Construction and Commercial Work: The overhead percentage for "Fast" (residential tract) work and "Medium" (commercial) projects *includes* equipment costs such as ladders, spray equipment, and masking paper holders. Those items are used on many jobs, not just one specific job. The overhead allowance covers equipment purchase payments, along with maintenance, repairs and fuel. If you have to rent equipment for a specific new construction project, add that rental expense as a separate cost item in your estimate.

Repaint Jobs: Overhead rates for "Slow" (repaint) work do *not* include equipment costs. When you estimate a repaint job, any small or short-term job, or a job that uses only a small quantity of materials, *add* the cost of equipment at the rental rate — even if the equipment is owned by your company.

Rental yards quote daily, weekly and monthly equipment rental rates. Figure 15 shows typical rental costs for painting equipment. Your actual equipment costs may be different. Here's a suggestion that can save you more than a few minutes on the telephone collecting rental rates. Make up a blank form like Figure 15 and give it to your favorite rental equipment suppliers. Ask each supplier to fill in current rental costs. Use the completed forms until you notice that rates have changed. Then ask for a new set of rental rates.

Commissions and Bonuses

Any commissions or bonuses you have to pay on a job aren't included in the estimating tables. You must add these expenses to your bid.

Painting contractors rarely have a sales staff, so there won't be sales commissions to pay on most jobs. There's one exception, however. Most room addition and remodeling contractors have salespeople. And many of their remodeling projects exclude painting. In fact, their contract may specify that the owner is responsible for the painting. These jobs may be a good source of leads for a painting contractor. Develop a relationship with the remodeling contractor's sales staff (with the remodeling contractor's approval, of course). If you have to pay a sales commission for the referral, this is direct overhead and has to be added to the estimate.

Some painting contractors pay their estimators a bonus of 1 to 3 percent per job in addition to their salary. If you offer an incentive like this, add the cost to your estimate, again as a direct overhead item.

An Example of Overhead

Here's an example of how overhead is added into an estimate. A painting company completed 20 new housing projects in the last year. Average revenue per project was $50,000. Gross receipts were $1,000,000 and the company made a 5 percent profit.

Gross income	$1,000,000
Less the profit earned (5%)	- 50,000
Gross expenses	950,000
Less total direct job cost	- 825,000
Indirect overhead expense	125,000

$$\frac{125{,}000 \text{ (overhead cost)}}{825{,}000 \text{ (direct job cost)}} = 0.1515 \text{ or } 15.15\%$$

When you've calculated indirect overhead as a percentage of direct job cost, add that percentage to your estimates. If you leave indirect overhead out of your estimates, you've left out some very significant costs.

Column 8: Profit

The estimating tables assume that profit on "Slow" (repaint) jobs is 16 percent, profit on "Medium" (commercial) projects is 12 percent and profit on "Fast" (residential tract) jobs is 7 percent. Calculate the profit per unit by first adding together the costs in columns 4 (labor cost per unit), column 5 (labor burden per unit), column 6 (material costs per unit), and column 7 (overhead per unit). Then multiply the total by the appropriate profit percentage to find the profit per unit.

It's my experience that larger companies with larger projects can survive with a smaller profit percentage. Stiff competition for high volume tract work forces bidders to trim their profit margin. Many smaller companies doing custom work earn a higher profit margin because they produce better quality work, have fewer jobs, and face less competition.

Risk factor	Normal profit (assume 10%)		Difficulty factor		Proposed profit range
High risk	10%	x	1.5 to 3.5	=	15% to 35%
Average risk	10%	x	1.3 to 1.4	=	13% to 14%
Moderate risk	10%	x	1.0 to 1.2	=	10% to 12%
Low risk	10%	x	0.5 to 0.9	=	5% to 9%

Figure 16
Risk factors and profit margin

Profit and Risk

Profit is usually proportionate to risk. The more risk, the greater the potential profit has to be to attract bidders. Smaller companies handling custom or repaint work have more risk of a major cost overrun because there are many more variables in that type of work. It's usually safe to estimate a smaller profit on new work because new work tends to be more predictable. The risk of loss smaller.

How do you define risk? Here's my definition: Risk is the *headache factor*, the number and size of potential problems you could face in completing the project. Repaint jobs have more unknowns, so they're a greater risk. And dealing with an indecisive or picky homeowner can be the greatest headache of all. You may need to use a profit margin even higher than the 15 to 35 range indicated for high-risk work in Figure 16.

Tailoring Your Profit Margin

Of course, your profit margin has to be based on the job, your company and the competition. But don't cut your profit to the bone just to get more work. Instead, review your bid to see if there are reasons why the standard costs wouldn't apply.

I use the term *standard base* bid to refer to my usual charge for all the estimated costs, including my standard profit. Before submitting any bid, spend a minute or two deciding whether your standard base bid will apply.

Risk Factors

Your assessment of the difficulty of the job may favor assigning a risk factor that could be used to modify your profit percentage. The higher the risk, the higher potential profit should be. My suggestions are in Figure 16.

As you might expect, opinions on difficulty factors can vary greatly. There's a lot of knowledge involved. You need experience and good judgment to apply these factors effectively.

Bidding Variables

Of course, your profit may be affected by an error in evaluating the job risk factor. You can greatly reduce the risk by accurately evaluating the bidding variables in Figure 17. Make adjustments to your standard base bid for example, if you expect your crews to be more or less efficient on this project, or if you expect competition to be intense. If there are logical reasons to modify your standard base bid, make those changes.

But remember, if you adjust your standard base bid, you're not changing your profit margin. You're only allowing for cost variables in the job. Adjust your standard base costs for unusual labor productivity, material or equipment cost changes, or because of unusual overhead conditions. Review the following bidding variables when deciding how to adjust your standard base bid.

Reputations and Attitudes
- Owner
- Architect
- General Contractor
- Lender
- Inspector

The Site
- Location (distance from shop and suppliers)
- Accessibility
- Working conditions
- Security requirements
- Safety considerations

The Project
- Building type
- Project size
- Your financial limits
- Start date
- Weather conditions
- Manpower availability and capability

Competition
- Number bidding
- Their strength, size and competence

Desire for the work

Figure 17
Bidding variables

The Bottom Line

The profit margin you include in estimates depends on the way you do business, the kind of work you do, and your competition. Only you can decide what percentage is right for your bids. Don't take another paint estimator's advice on the "correct" profit margin. There's no single correct answer. Use your own judgment. But here are some typical profit margins for the kinds of work most painting contractors do.

Repaints:	Custom	20 to 35%
	Average	15 to 20%
Commercial or industrial		10 to 15%
New residential:	1-4 units	10 to 12%
	5 or more	5 to 7%
Government work		5 to 7%

Column 9: Total Cost

The costs in Column 9 of Figure 2, and all the estimating tables in this book, are the totals per unit for each application rate in columns 4, 5, 6, 7, and 8. That includes labor, labor burden, material cost, overhead and profit.

Sample Estimate

Figure 18 is a sample repaint estimate, using the slow production rate, for a small house with many amenities. The final bid total is the bid price. Figure 19 is a blank estimating form for your use.

Date	1/7/21	**Due date**	1/15/21
Customer	Dan Gleason	**Job name**	Gleason Repaint
Address	3333 A Street	**Job location**	3333 A Street
City/State/Zip	Yourtown, USA 77777	**Estimate #**	14-012
Phone	(619) 555-1212	**Total square feet**	1,020 SF (5 rooms)
Estimated by	CHS	**Checked by**	Jack

Interior Costs

	Operation	Material	Application Method	Dimensions	Quantity SF/LF/Each		Unit Cost		Total Cost	Formula Page
1	Ceilings - T & G	Semi-Trans-WB	R + B	17.5x15.3x1.3	348 SF	x	.43	= $	150.00	86
2	Beams to 13'H	Solid Body-WB	R + B	17.5 x 7	122.5 LF	x	2.47	= $	303.00	45
3	Ceilings - GYP. Drywal	Orange Peel-Flat	R	127 + 127	254 SF	x	.29	= $	74.00	65
4	Ceilings - GYP. Drywall	Sealer-WB	R	75 + 15 + 40	130 SF	x	.28	= $	36.00	65
5	Ceilings - GYP. Drywall	Enamel-WB	R	75 + 15 + 40	130 SF	x	.35	= $	46.00	65
6	Walls - GYP. Drywall	Orange Peel-Flat	R	675+392+392	1,459 SF	x	.26	= $	379.00	228
7	Walls - Above 8' (clip)	Orange Peel-Flat	R	70+85=155x1.3	201.5 SF	x	.26	= $	52.00	228
8	Walls - GYP. Drywall	Sealer-WB	R	280+128+208	616 SF	x	.30	= $	185.00	228
9	Walls - GYP. Drywall	Enamel-WB	R	280+128+208	616 SF	x	.39	= $	240.00	228
10	Doors-Flush	Undercoat-WB	R + B	Opening Count	10 Ea	=	161.26	= $	161.00	108
11	Doors-Flush	Enamel-WB	R + B	Opening Count	10 Ea	=	183.11	= $	183.00	108
12	Baseboard - Prime	Flat w/walls	R + B	64 + 49 + 49	162 LF	x	.11	= $	18.00	43
13	Baseboard - Finish	Enamel-WB	B	11+16+35	62 LF	x	.50	= $	31.00	43
14	Railing - W.I.-Preprimed	Enamel/Off-white	B	42" High	15 LF	x	2.72	= $	41.00	180
15	Valance-Light-2" x 8"	Solid Body Stain	B	2 x 8	10 LF	x	2.06	= $	21.00	224
16	Registers	Spray Can	Spray	1,020 SF Home	1,020 SF	x	.07	= $	71.00	182
17						x		= $		
18						x		= $		

Total Interior Costs (includes overhead and profit) = $ 1,991.00

Exterior Costs

	Operation	Material	Application Method	Dimensions	Quantity SF/LF/Each		Unit Cost		Total Cost	Formula Page
1	Roof Jacks - 1 Story	Finish-enamel	B	1 Story	1 House	x	29.37	= $	29.00	183
2	S.M. Diverter-3" W	Finish-enamel	B	14	14 LF	x	.27	= $	4.00	198
3	S.M. Vents & Flashing	Finish-enamel	B	1 Story	1 House	x	62.67	= $	63.00	199
4	Fascia - 2 x 8	Solid-water	Roll	66 + 59	125 LF	x	.89	= $	111.00	120
5	Overhang - 24"	Solid-water	R + B	(132+76)x1.5	312 SF	x	.79	= $	246.00	160
6	Siding - R.S. Wood	Solid-water	Roll	(1/2x24x4.5)x2	108 SF	x	.55	= $	59.00	210
7	Plaster / Stucco	Masonry - WB	Roll	255+255+204+204	918 SF	x	.47	= $	431.00	169
8	Door - Panel (Entry)	Enam 2 coats-WB	R + B	Entry	1 Ea	x	74.03	= $	74.00	101
9	Door - Flush	Enam 2 coats-WB	R + B	Exterior	1 Ea	x	30.50	= $	31.00	98
10	Plant-On Trim - 2 x 4	Solid-water	R + B	66 + 62 + 52	180 LF	x	.71	= $	128.00	162
11	Pass Through-Preprimed	Finish-enamel	B	10	10 LF	x	2.04	= $	20.00	162
12	Pot Shelf	Solid-water	R + B	27	27 LF	x	2.29	= $	62.00	172
13						x		= $		
14						x		= $		
15						x		= $		
16						x		= $		
17						x		= $		
18						x		= $		

Total Exterior Costs (includes overhead and profit) = $ 1,258.00

Figure 18
Sample painting estimate

Preparation Costs

	Operation	Dimensions	Quantity SF/LF/Each		Unit cost Per SF		Total cost	Formula Page
1	Sand/Putty Wood Ceil (Siding x 1.3)	17.5 x 15.3 x 1.3	348 SF	x	.22	= $	77.00	300
2	Sand and Putty Int. Wall	675 + 392 + 392	1,459 SF	x	.21	= $	306.00	300
3	Lt. Sand Doors/Frames (Enamel)	14 Ea x 21 SF x 2 Sides	588 SF	x	.26	= $	153.00	301
4	Wash Int. Walls/Ceil-Enamel	280 + 128 + 208	616 SF	x	.21	= $	129.00	313
5	Waterblast Exterior Stucco	125 + 210 + 108 + 918	1,361 SF	x	.06	= $	82.00	315
6	Sand and Putty Ext. Trim	125 + 210 + 108	443 SF	x	.39	= $	173.00	300
7	Caulk Ext. Windows-1/8" gap	20 + 15 + 10 + 20 + 12	77 SF	x	.71	= $	55.00	298
8				x		= $		
9				x		= $		
10				x		= $		
	Total Preparation Costs (includes overhead and profit)					= $	975.00	

SURRPTUCU Costs

Operation	Description	Labor hours	Labor cost (at $28.52)	Approximate material cost	Totals	Formula Page
Set**U**p	2 Days @1/day	2.0	57.04	—	57.00	6
Remove/**R**eplace	Hardware & Plates	1.25	35.65	—	36.00	6
Protection	Furniture & Floors	2.0	57.04	25.00	82.00	6

Touch**U**p is applied as a percentage of the total costs. See *Extensions*

Operation	Description	Labor hours	Labor cost	Approximate material cost	Totals	Formula Page
Clean**U**p	2 Days @1/day	2.0	57.04	—	57.00	6

Equipment Costs

Equipment description	Rental days	Daily cost	Total cost	Formula Page
Pressure Washer	1	106.00	$ 106.00	34
Ladders, 6', 2 Ea	1	22.40	$ 22.00	33
Palm Sander 4" x 4"	1	14.10	$ 14.00	34
			$	
			$	
			$	
		Total Equipment Costs	$ 142.00	

Subcontractor Costs

Trade	Bid Amount
Pavement marking	$ 0
Sandblasting	$ 0
Scaffolding	$ 0
Wallcovering	$ 0
Waterblasting	$ 0
Other ______	$ 0
Other ______	$ 0
Other ______	$ 0
Total Subcontractor Costs	$ —

Extensions

Supervision (2 Hr.)	$ 57.00
Setup	$ 57.00
Remove/replace	$ 36.00
Protection	$ 82.00
Cleanup	$ 57.00
Equipment	$ 142.00
Subcontracts	$ 0
Commissions	$ 0
Other costs	$ 0
Subtotal	$ 431.00
Overhead (19 %)	$ 82.00
Profit (16 %)	$ 69.00
Subtotal	$ 151.00
Preparation	$ 975.00
Interior total	$ 1,991.00
Exterior total	$ 1,258.00
Subtotal	$ 4,224.00
Touchup (10 %)	$ 422.00
Contingency (0 %)	$ 0
Total base bid	$ 5,228.00
Adjustment (-2 %)	$ <-105.00>
Final bid total	$ 5,123.00
Price per SF (1020)	$ 5.02
Price per room (5)	$ 1,025.00

Figure 18 (continued)
Sample painting estimate

Date ________	Due date ________
Customer ________	Job name ________
Address ________	Job location ________
City/State/Zip ________	Estimate # ________
Phone ________	Total square feet ________
Estimated by ________	Checked by ________

Interior Costs

	Operation	Material	Application Method	Dimensions	Quantity SF/LF/Each		Unit Cost		Total Cost
1						x		= $	
2						x		= $	
3						x		= $	
4						x		= $	
5						x		= $	
6						x		= $	
7						x		= $	
8						x		= $	
9						x		= $	
10						x		= $	
11						x		= $	
12						x		= $	
13						x		= $	
14						x		= $	
15						x		= $	
16						x		= $	
17						x		= $	
18						x		= $	

Total Interior Costs (includes overhead and profit) = $ ________

Exterior Costs

	Operation	Material	Application Method	Dimensions	Quantity SF/LF/Each		Unit Cost		Total Cost
1						x		= $	
2						x		= $	
3						x		= $	
4						x		= $	
5						x		= $	
6						x		= $	
7						x		= $	
8						x		= $	
9						x		= $	
10						x		= $	
11						x		= $	
12						x		= $	
13						x		= $	
14						x		= $	
15						x		= $	
16						x		= $	
17						x		= $	
18						x		= $	

Total Exterior Costs (includes overhead and profit) = $ ________

Figure 19
Blank painting estimate

Preparation Costs

	Operation	Dimensions	Quantity SF/LF/Each		Unit cost		Total cost
1				x		= $	
2				x		= $	
3				x		= $	
4				x		= $	
5				x		= $	
6				x		= $	
7				x		= $	
8				x		= $	
9				x		= $	
10				x		= $	

Total Preparation Costs (includes overhead and profit) = $ ______

SURRPTUCU Costs

Operation	Description	Labor hours	Labor cost (at______)	Material cost	Totals
Set**U**p					
Remove/**R**eplace					
Protection					

Touch**U**p is applied as a percentage of the total costs. See *Extensions*

Operation	Description	Labor hours	Labor cost	Material cost	Totals
Clean**U**p					

Equipment Costs

Equipment description	Rental days	Daily cost	Total cost
			$
			$
			$
			$
			$
			$
		Total Equipment Costs	$

Subcontractor Costs

Trade	Bid Amount
Pavement marking	$
Sandblasting	$
Scaffolding	$
Wallcovering	$
Waterblasting	$
Other ______	$
Other ______	$
Other ______	$
Total Subcontractor Costs	$

Extensions

Supervision (______)	$
Setup	$
Remove/replace	$
Protection	$
Cleanup	$
Equipment	$
Subcontracts	$
Commissions	$
Other costs	$
Subtotals	$
Overhead (______%)	$
Profit (_____%)	$
Subtotal	$
Preparation	$
Interior total	$
Exterior total	$
Subtotal	$
Touchup (_____%)	$
Contingency (____%)	$
Total base bid	$
Adjustment (_____%)	$
Final bid total	$
Price per SF (_____)	$
Price per room (____)	$

Figure 19 (continued)
Blank painting estimate

Part I

General Painting Costs

Contents

	Labor LF per manhour	Material coverage LF/gallon	Material cost per gallon	Labor cost per 100 LF	Labor burden 100 LF	Material cost per 100 LF	Overhead per 100 LF	Profit per 100 LF	Total price per 100 LF
Baseboard, per linear foot									
Roll 1 coat with walls, brush touchup, paint grade base									
Flat latex, water base, (material #5)									
Slow	900	800	36.00	2.56	.61	4.50	1.46	1.46	10.59
Medium	1200	750	31.50	2.46	.70	4.20	1.84	1.11	10.31
Fast	1500	700	27.00	2.40	.86	3.86	2.20	.65	9.97
Enamel, water base (material #9)									
Slow	600	750	53.30	3.83	.93	7.11	2.25	2.26	16.38
Medium	800	725	46.70	3.69	1.06	6.44	2.80	1.68	15.67
Fast	1000	700	40.00	3.60	1.27	5.71	3.28	.97	14.83
Enamel, oil base (material #10)									
Slow	600	750	61.50	3.83	.93	8.20	2.46	2.47	17.89
Medium	800	725	53.80	3.69	1.06	7.42	3.05	1.83	17.05
Fast	1000	700	46.10	3.60	1.27	6.59	3.55	1.05	16.06
Brush 1 coat, cut-in, paint grade base									
Enamel, water base (material #9)									
Slow	100	700	53.30	23.00	5.52	7.61	6.86	6.88	49.87
Medium	120	675	46.70	24.58	7.10	6.92	9.65	5.79	54.04
Fast	140	650	40.00	25.71	9.07	6.15	12.69	3.75	57.37
Enamel, oil base (material #10)									
Slow	100	700	61.50	23.00	5.52	8.79	7.09	7.10	51.50
Medium	120	675	53.80	24.58	7.10	7.97	9.91	5.95	55.51
Fast	140	650	46.10	25.71	9.07	7.09	12.98	3.84	58.69
Spray 1 coat, stain in boneyard, stain grade base									
Wiping stain (material #11a)									
Slow	--	--	--	--	--	--	--	--	--
Medium	1500	1750	53.90	1.97	.58	3.08	1.41	.84	7.88
Fast	2000	1500	46.20	1.80	.64	3.08	1.71	.51	7.74

Use these figures for 1-1/2 inch to 3 inch baseboard stock, painted or stained on one side. Measurements are based on linear feet of baseboard. Paint grade base is painted after it is installed but stain grade base is usually stained in a boneyard. Typically, finger joint stock is paint grade and butt joint stock is stain grade. These figures include minimal preparation time and material. Add for extensive preparation. "Slow" work is based on an hourly wage of $23.00, "Medium" work on an hourly wage of $29.50, and "Fast" work on an hourly wage of $36.00. Other qualifications that apply to this table are on page 9.

	Labor SF per manhour	Material coverage SF/gallon	Material cost per gallon	Labor cost per 100 SF	Labor burden 100 SF	Material cost per 100 SF	Overhead per 100 SF	Profit per 100 SF	Total price per 100 SF
Baseboard, per square foot of floor area									
Roll 1 coat with walls, brush touchup, paint grade base									
Flat latex, water base, (material #5)									
Slow	2500	1500	36.00	.92	.22	2.40	.67	.67	4.88
Medium	2750	1250	31.50	1.07	.30	2.52	.98	.59	5.46
Fast	3000	1000	27.00	1.20	.41	2.70	1.34	.40	6.05
Enamel, water base (material #9)									
Slow	2000	1000	53.30	1.15	.28	5.33	1.28	1.29	9.33
Medium	2200	900	46.70	1.34	.37	5.19	1.73	1.04	9.67
Fast	2400	800	40.00	1.50	.55	5.00	2.18	.64	9.87
Enamel, oil base (material #10)									
Slow	2000	1000	61.50	1.15	.28	6.15	1.44	1.44	10.46
Medium	2200	900	53.80	1.34	.37	5.98	1.93	1.16	10.78
Fast	2400	800	46.10	1.50	.55	5.76	2.41	.71	10.93
Brush 1 coat, cut-in, paint grade base									
Enamel, water base (material #9)									
Slow	500	1500	53.30	4.60	1.10	3.55	1.76	1.76	12.77
Medium	550	1350	46.70	5.36	1.56	3.46	2.59	1.56	14.53
Fast	600	1200	40.00	6.00	2.13	3.33	3.55	1.05	16.06
Enamel, oil base (material #10)									
Slow	500	1500	61.50	4.60	1.10	4.10	1.86	1.87	13.53
Medium	550	1350	53.80	5.36	1.56	3.99	2.73	1.64	15.28
Fast	600	1200	46.10	6.00	2.13	3.84	3.71	1.10	16.78
Spray 1 coat, stain in boneyard, stain grade base									
Wiping stain (material #11a)									
Slow	--	--	--	--	--	--	--	--	--
Medium	4000	1350	53.90	.74	.21	3.99	1.24	.74	6.92
Fast	5000	1200	46.20	.72	.25	3.85	1.49	.44	6.75

Baseboard measurements are based on square feet of floor area. Use these figures for 1-1/2 inch to 3 inch stock, painted or stained on one side. Stain grade base is to be stained in a boneyard. Typically, finger joint stock is paint grade and butt joint stock is stain grade. These figures include minimal preparation time and material. Add for extensive preparation. "Slow" work is based on an hourly wage of $23.00, "Medium" work on an hourly wage of $29.50, and "Fast" work on an hourly wage of $36.00. Other qualifications that apply to this table are on page 9.

	Labor LF per manhour	Material coverage LF/gallon	Material cost per gallon	Labor cost per 100 LF	Labor burden 100 LF	Material cost per 100 LF	Overhead per 100 LF	Profit per 100 LF	Total price per 100 LF
Beams, per linear foot, heights to 13 feet									
Solid body stain, water base (material #18)									
Roll & brush each coat									
Slow	35	50	48.90	65.71	15.77	97.80	34.06	34.13	247.47
Medium	40	45	42.80	73.75	21.33	95.11	47.54	28.53	266.26
Fast	45	40	36.70	80.00	28.23	91.75	62.00	18.34	280.32
Solid body stain, oil base (material #19)									
Roll & brush each coat									
Slow	35	50	60.90	65.71	15.77	121.80	38.62	38.70	280.60
Medium	40	45	53.30	73.75	21.33	118.44	53.38	32.03	298.93
Fast	45	40	45.70	80.00	28.23	114.25	68.97	20.40	311.85
Semi-transparent stain, water base (material #20)									
Roll & brush each coat									
Slow	40	55	49.90	57.50	13.80	90.73	30.79	30.85	223.67
Medium	45	50	43.70	65.56	18.94	87.40	42.98	25.79	240.67
Fast	50	45	37.40	72.00	25.42	83.11	55.96	16.55	253.04
Semi-transparent stain, oil base (material #21)									
Roll & brush each coat									
Slow	40	55	58.30	57.50	13.80	106.00	33.69	33.76	244.75
Medium	45	50	51.00	65.56	18.94	102.00	46.63	27.98	261.11
Fast	50	45	43.70	72.00	25.42	97.11	60.30	17.84	272.67

Beam measurements are based on linear feet of installed 4" x 6" to 8" x 14" beams. High time difficulty factors are already figured into the formulas. "Slow" work is based on an hourly wage of $23.00, "Medium" work on an hourly wage of $29.50, and "Fast" work on an hourly wage of $36.00. Other qualifications that apply to this table are on page 9.

	Labor LF per manhour	Material coverage LF/gallon	Material cost per gallon	Labor cost per 100 LF	Labor burden 100 LF	Material cost per 100 LF	Overhead per 100 LF	Profit per 100 LF	Total price per 100 LF
Beams, per linear foot, heights from 13 to 17 feet									
Solid body stain, water base (material #18)									
Roll & brush each coat									
Slow	24	50	48.90	95.83	23.01	97.80	41.16	41.25	299.05
Medium	27	45	42.80	109.26	31.60	95.11	58.99	35.39	330.35
Fast	30	40	36.70	120.00	42.35	91.75	78.77	23.30	356.17
Solid body stain, oil base (material #19)									
Roll & brush each coat									
Slow	24	50	60.90	95.83	23.01	121.80	45.72	45.82	332.18
Medium	27	45	53.30	109.26	31.60	118.44	64.82	38.89	363.01
Fast	30	40	45.70	120.00	42.35	114.25	85.75	25.37	387.72
Semi-transparent stain, water base (material #20)									
Roll & brush each coat									
Slow	28	55	49.90	82.14	19.70	90.73	36.59	36.67	265.83
Medium	31	50	43.70	95.16	27.52	87.40	52.52	31.51	294.11
Fast	34	45	37.40	105.88	37.38	83.11	70.17	20.76	317.30
Semi-transparent stain, oil base (material #21)									
Roll & brush each coat									
Slow	28	55	58.30	82.14	19.70	106.00	39.49	39.57	286.90
Medium	31	50	51.00	95.16	27.52	102.00	56.17	33.70	314.55
Fast	34	45	43.70	105.88	37.38	97.11	74.51	22.04	336.92

Beam measurements are based on linear feet of installed 4" x 6" to 8" x 14" beams. High time difficulty factors are already figured into the formulas. "Slow" work is based on an hourly wage of $23.00, "Medium" work on an hourly wage of $29.50, and "Fast" work on an hourly wage of $36.00. Other qualifications that apply to this table are on page 9.

	Labor LF per manhour	Material coverage LF/gallon	Material cost per gallon	Labor cost per 100 LF	Labor burden 100 LF	Material cost per 100 LF	Overhead per 100 LF	Profit per 100 LF	Total price per 100 LF
Beams, per linear foot, heights from 18 to 19 feet									
Solid body stain, water base (material #18)									
Roll & brush each coat									
Slow	16	50	48.90	143.75	34.50	97.80	52.45	52.56	381.06
Medium	18	45	42.80	163.89	47.40	95.11	76.59	45.95	428.94
Fast	20	40	36.70	180.00	63.55	91.75	103.94	30.75	469.99
Solid body stain, oil base (material #19)									
Roll & brush each coat									
Slow	16	50	60.90	143.75	34.50	121.80	57.01	57.13	414.19
Medium	18	45	53.30	163.89	47.40	118.44	82.42	49.45	461.60
Fast	20	40	45.70	180.00	63.55	114.25	110.91	32.81	501.52

	Labor LF per manhour	Material coverage LF/gallon	Material cost per gallon	Labor cost per 100 LF	Labor burden 100 LF	Material cost per 100 LF	Overhead per 100 LF	Profit per 100 LF	Total price per 100 LF
Semi-transparent stain, water base (material #20)									
Roll & brush each coat									
Slow	19	55	49.90	121.05	29.05	90.73	45.76	45.85	332.44
Medium	21	50	43.70	140.48	40.62	87.40	67.12	40.27	375.89
Fast	23	45	37.40	156.52	55.27	83.11	91.41	27.04	413.35
Semi-transparent stain, oil base (material #21)									
Roll & brush each coat									
Slow	19	55	58.30	121.05	29.05	106.00	48.66	48.76	353.52
Medium	21	50	51.00	140.48	40.62	102.00	70.77	42.46	396.33
Fast	23	45	43.70	156.52	55.27	97.11	95.75	28.32	432.97

Beam measurements are based on linear feet of installed 4" x 6" to 8" x 14" beams. High time difficulty factors are already figured into the formulas. "Slow" work is based on an hourly wage of $23.00, "Medium" work on an hourly wage of $29.50, and "Fast" work on an hourly wage of $36.00. Other qualifications that apply to this table are on page 9.

	Labor LF per manhour	Material coverage LF/gallon	Material cost per gallon	Labor cost per 100 LF	Labor burden 100 LF	Material cost per 100 LF	Overhead per 100 LF	Profit per 100 LF	Total price per 100 LF
Beams, per linear foot, heights from 20 to 21 feet									
Solid body stain, water base (material #18)									
Roll & brush each coat									
Slow	12	50	48.90	191.67	45.99	97.80	63.74	63.87	463.07
Medium	14	45	42.80	210.71	60.94	95.11	91.68	55.01	513.45
Fast	16	40	36.70	225.00	79.44	91.75	122.82	36.33	555.34
Solid body stain, oil base (material #19)									
Roll & brush each coat									
Slow	12	50	60.90	191.67	45.99	121.80	68.30	68.44	496.20
Medium	14	45	53.30	210.71	60.94	118.44	97.51	58.51	546.11
Fast	16	40	45.70	225.00	79.44	114.25	129.79	38.39	586.87
Semi-transparent stain, water base (material #20)									
Roll & brush each coat									
Slow	14	55	49.90	164.29	39.43	90.73	55.95	56.06	406.46
Medium	16	50	43.70	184.38	53.31	87.40	81.27	48.76	455.12
Fast	18	45	37.40	200.00	70.63	83.11	109.65	32.44	495.83
Semi-transparent stain, oil base (material #21)									
Roll & brush each coat									
Slow	14	55	58.30	164.29	39.43	106.00	58.85	58.97	427.54
Medium	16	50	51.00	184.38	53.31	102.00	84.92	50.95	475.56
Fast	18	45	43.70	200.00	70.63	97.11	113.99	33.72	515.45

Beam measurements are based on linear feet of installed 4" x 6" to 8" x 14" beams. High time difficulty factors are already figured into the formulas. "Slow" work is based on an hourly wage of $23.00, "Medium" work on an hourly wage of $29.50, and "Fast" work on an hourly wage of $36.00. Other qualifications that apply to this table are on page 9.

	Labor SF per manhour	Material coverage SF/gallon	Material cost per gallon	Labor cost per 100 SF	Labor burden 100 SF	Material cost per 100 SF	Overhead per 100 SF	Profit per 100 SF	Total price per 100 SF
Bookcases and shelves, paint grade, brush application									
Undercoat, water base (material #3)									
Roll & brush 1 coat									
Slow	25	300	36.90	92.00	22.08	12.30	24.01	24.06	174.45
Medium	30	280	32.30	98.33	28.42	11.54	34.57	20.74	193.60
Fast	35	260	27.70	102.86	36.30	10.65	46.44	13.74	209.99
Undercoat, oil base (material #4)									
Roll & brush 1 coat									
Slow	25	340	48.30	92.00	22.08	14.21	24.38	24.43	177.10
Medium	30	318	42.30	98.33	28.42	13.30	35.01	21.01	196.07
Fast	35	295	36.20	102.86	36.30	12.27	46.95	13.89	212.27
Split coat (1/2 undercoat + 1/2 enamel), water base (material #3 + #9)									
Roll & brush each coat									
Slow	40	350	45.10	57.50	13.80	12.89	16.00	16.03	116.22
Medium	45	328	39.50	65.56	18.94	12.04	24.14	14.48	135.16
Fast	50	305	33.85	72.00	25.42	11.10	33.64	9.95	152.11
Split coat (1/2 undercoat + 1/2 enamel), oil base (material #4 + #10)									
Roll & brush each coat									
Slow	40	350	54.90	57.50	13.80	15.69	16.53	16.56	120.08
Medium	45	328	48.05	65.56	18.94	14.65	24.79	14.87	138.81
Fast	50	305	41.15	72.00	25.42	13.49	34.38	10.17	155.46
Enamel, water base (material #9)									
Roll & brush 1st finish coat									
Slow	35	340	53.30	65.71	15.77	15.68	18.46	18.50	134.12
Medium	40	318	46.70	73.75	21.33	14.69	27.44	16.46	153.67
Fast	45	295	40.00	80.00	28.23	13.56	37.76	11.17	170.72
Roll & brush 2nd or additional finish coats									
Slow	40	350	53.30	57.50	13.80	15.23	16.44	16.48	119.45
Medium	45	328	46.70	65.56	18.94	14.24	24.69	14.81	138.24
Fast	50	305	40.00	72.00	25.42	13.11	34.26	10.14	154.93
Enamel, oil base (material #10)									
Roll & brush 1st finish coat									
Slow	35	340	61.50	65.71	15.77	18.09	18.92	18.96	137.45
Medium	40	318	53.80	73.75	21.33	16.92	28.00	16.80	156.80
Fast	45	295	46.10	80.00	28.23	15.63	38.40	11.36	173.62
Roll & brush 2nd or additional finish coats									
Slow	40	350	61.50	57.50	13.80	17.57	16.89	16.92	122.68
Medium	45	318	53.80	65.56	18.94	16.92	25.36	15.21	141.99
Fast	50	305	46.10	72.00	25.42	15.11	34.88	10.32	157.73

Bookcase and shelf estimates are based on overall dimensions (length times width) to 8 feet high and include painting all exposed surfaces (including stiles, interior shelves and backs). For heights above 8 feet, use the High Time Difficulty Factors on page 139. "Slow" work is based on an hourly wage of $23.00, "Medium" work on an hourly wage of $29.50, and "Fast" work on an hourly wage of $36.00. Other qualifications that apply to this table are on page 9.

	Labor SF per manhour	Material coverage SF/gallon	Material cost per gallon	Labor cost per 100 SF	Labor burden 100 SF	Material cost per 100 SF	Overhead per 100 SF	Profit per 100 SF	Total price per 100 SF
Bookcases and shelves, paint grade, spray application									
Undercoat, water base (material #3)									
Spray 1 coat									
Slow	150	145	36.90	15.33	3.69	25.45	8.45	8.47	61.39
Medium	165	133	32.30	17.88	5.17	24.29	11.84	7.10	66.28
Fast	175	120	27.70	20.57	7.24	23.08	15.78	4.67	71.34
Undercoat, oil base (material #4)									
Spray 1 coat									
Slow	150	145	48.30	15.33	3.69	33.31	9.94	9.96	72.23
Medium	165	133	42.30	17.88	5.17	31.80	13.71	8.23	76.79
Fast	175	120	36.20	20.57	7.24	30.17	17.98	5.32	81.28
Split coat (1/2 undercoat + 1/2 enamel), water base (material #3 + #9)									
Spray each coat									
Slow	245	195	45.10	9.39	2.25	23.13	6.61	6.62	48.00
Medium	270	183	39.50	10.93	3.14	21.58	8.92	5.35	49.92
Fast	295	170	33.85	12.20	4.31	19.91	11.29	3.34	51.05
Split coat (1/2 undercoat + 1/2 enamel), oil base (material #4 + #10)									
Spray each coat									
Slow	245	195	54.90	9.39	2.25	28.15	7.56	7.58	54.93
Medium	270	183	48.05	10.93	3.14	26.26	10.09	6.05	56.47
Fast	295	170	41.15	12.20	4.31	24.21	12.62	3.73	57.07
Enamel, water base (material #9)									
Spray 1st finish coat									
Slow	225	170	53.30	10.22	2.44	31.35	8.36	8.38	60.75
Medium	250	158	46.70	11.80	3.41	29.56	11.19	6.72	62.68
Fast	275	145	40.00	13.09	4.64	27.59	14.04	4.15	63.51
Spray 2nd or additional finish coats									
Slow	245	195	53.30	9.39	2.25	27.33	7.40	7.42	53.79
Medium	270	183	46.70	10.93	3.14	25.52	9.90	5.94	55.43
Fast	295	170	40.00	12.20	4.31	23.53	12.41	3.67	56.12
Enamel, oil base (material #10)									
Spray 1st finish coat									
Slow	225	170	61.50	10.22	2.44	36.18	9.28	9.30	67.42
Medium	250	158	53.80	11.80	3.41	34.05	12.32	7.39	68.97
Fast	275	145	46.10	13.09	4.64	31.79	15.35	4.54	69.41
Spray 2nd or additional finish coats									
Slow	245	195	61.50	9.39	2.25	31.54	8.20	8.22	59.60
Medium	270	183	53.80	10.93	3.14	29.40	10.87	6.52	60.86
Fast	295	170	46.10	12.20	4.31	27.12	13.53	4.00	61.16

Bookcase and shelf estimates are based on overall dimensions (length times width) to 8 feet high and include painting all exposed surfaces (including stiles, interior shelves and backs). For heights above 8 feet, use the High Time Difficulty Factors on page 139. "Slow" work is based on an hourly wage of $23.00, "Medium" work on an hourly wage of $29.50, and "Fast" work on an hourly wage of $36.00. Other qualifications that apply to this table are on page 9.

	Labor SF per manhour	Material coverage SF/gallon	Material cost per gallon	Labor cost per 100 SF	Labor burden 100 SF	Material cost per 100 SF	Overhead per 100 SF	Profit per 100 SF	Total price per 100 SF
Bookcases and shelves, stain grade									
Stain, seal & lacquer (7 step process)									
STEP 1: Sand & putty;									
Slow	100	--	--	23.00	5.52	--	5.42	5.43	39.37
Medium	125	--	--	23.60	6.82	--	7.61	4.56	42.59
Fast	150	--	--	24.00	8.49	--	10.07	2.98	45.54
STEP 2 & 3: Stain (material #11a) & wipe									
Brush 1 coat & wipe									
Slow	75	500	61.60	30.67	7.35	12.32	9.57	9.59	69.50
Medium	85	475	53.90	34.71	10.01	11.35	14.02	8.41	78.50
Fast	95	450	46.20	37.89	13.40	10.27	19.08	5.64	86.28
Spray 1 coat & wipe									
Slow	300	175	61.60	7.67	1.83	35.20	8.49	8.51	61.70
Medium	400	138	53.90	7.38	2.13	39.06	12.14	7.29	68.00
Fast	500	100	46.20	7.20	2.54	46.20	17.34	5.13	78.41
STEP 4: Sanding sealer (material #11b)									
Brush 1 coat									
Slow	130	550	50.90	17.69	4.24	9.25	5.93	5.94	43.05
Medium	140	525	44.60	21.07	6.08	8.50	8.92	5.35	49.92
Fast	150	500	38.20	24.00	8.49	7.64	12.43	3.68	56.24
Spray 1 coat									
Slow	375	175	50.90	6.13	1.48	29.09	6.97	6.99	50.66
Medium	475	138	44.60	6.21	1.81	32.32	10.08	6.05	56.47
Fast	575	100	38.20	6.26	2.22	38.20	14.47	4.28	65.43
STEP 5: Sand lightly									
Slow	175	--	--	13.14	3.14	--	3.10	3.10	22.48
Medium	225	--	--	13.11	3.78	--	4.23	2.54	23.66
Fast	275	--	--	13.09	4.64	--	5.49	1.62	24.84
STEP 6 & 7: Lacquer (material #11c), 2 coats									
Brush 1st coat									
Slow	140	400	53.50	16.43	3.93	13.38	6.41	6.43	46.58
Medium	185	375	46.80	15.95	4.62	12.48	8.26	4.96	46.27
Fast	245	350	40.10	14.69	5.18	11.46	9.72	2.87	43.92
Brush 2nd coat									
Slow	155	425	53.50	14.84	3.56	12.59	5.89	5.90	42.78
Medium	208	413	46.80	14.18	4.11	11.33	7.40	4.44	41.46
Fast	260	400	40.10	13.85	4.90	10.03	8.92	2.64	40.34
Spray 1st coat									
Slow	340	175	53.50	6.76	1.62	30.57	7.40	7.42	53.77
Medium	458	138	46.80	6.44	1.85	33.91	10.55	6.33	59.08
Fast	575	100	40.10	6.26	2.22	40.10	15.06	4.45	68.09

	Labor SF per manhour	Material coverage SF/gallon	Material cost per gallon	Labor cost per 100 SF	Labor burden 100 SF	Material cost per 100 SF	Overhead per 100 SF	Profit per 100 SF	Total price per 100 SF
Spray 2nd coat									
Slow	430	200	53.50	5.35	1.30	26.75	6.34	6.36	46.10
Medium	530	163	46.80	5.57	1.62	28.71	8.97	5.38	50.25
Fast	630	125	40.10	5.71	2.03	32.08	7.56	7.58	54.96
Complete 7 step stain, seal & lacquer process (material #11)									
Brush all coats									
Slow	30	160	54.90	76.67	18.39	34.31	24.58	24.63	178.58
Medium	35	150	48.00	84.29	24.36	32.00	26.72	26.78	194.15
Fast	40	140	41.20	90.00	31.78	29.43	37.80	22.68	211.69
Spray all coats									
Slow	65	60	54.90	35.38	8.48	91.50	33.84	20.31	189.51
Medium	83	48	48.00	35.54	10.29	100.00	27.70	27.76	201.29
Fast	100	35	41.20	36.00	12.71	117.71	41.61	24.96	232.99
Shellac, clear (material #12)									
Brush each coat									
Slow	205	570	70.40	11.22	2.70	12.35	4.99	5.00	36.26
Medium	230	545	61.60	12.83	3.71	11.30	6.96	4.18	38.98
Fast	255	520	52.80	14.12	4.97	10.15	9.07	2.68	40.99
Varnish, flat or gloss (material #30c)									
Brush each coat									
Slow	175	450	81.00	13.14	3.14	18.00	6.52	6.53	47.33
Medium	200	438	70.80	14.75	4.27	16.16	8.79	5.28	49.25
Fast	225	425	60.70	16.00	5.63	14.28	11.14	3.29	50.34
Penetrating stain wax (material #14) & polish									
Brush 1st coat									
Slow	150	595	50.20	15.33	3.69	8.44	5.22	5.23	37.91
Medium	175	558	43.90	16.86	4.86	7.87	7.40	4.44	41.43
Fast	200	520	37.60	18.00	6.36	7.23	9.79	2.90	44.28
Brush 2nd or additional coats									
Slow	175	600	50.20	13.14	3.14	8.37	6.17	3.70	34.52
Medium	200	575	43.90	14.75	4.27	7.63	5.06	5.07	36.78
Fast	225	550	37.60	16.00	5.63	6.84	7.12	4.27	39.86

Bookcase and shelf estimates are based on overall dimensions (length times width) to 8 feet high and include painting all exposed surfaces (including stiles, interior shelves and backs). For heights above 8 feet, use the High Time Difficulty Factors on page 139. "Slow" work is based on an hourly wage of $23.00, "Medium" work on an hourly wage of $29.50, and "Fast" work on an hourly wage of $36.00. Other qualifications that apply to this table are on page 9.

	Labor SF per manhour	Material coverage SF/gallon	Material cost per gallon	Labor cost per 100 SF	Labor burden 100 SF	Material cost per 100 SF	Overhead per 100 SF	Profit per 100 SF	Total price per 100 SF
Cabinet backs, paint grade, brush									
Flat latex, water base (material #5)									
Brush each coat									
Slow	100	300	36.00	23.00	5.52	12.00	7.70	7.72	55.94
Medium	150	275	31.50	19.67	5.70	11.45	9.20	5.52	51.54
Fast	200	250	27.00	18.00	6.36	10.80	10.90	3.22	49.28
Enamel, water base (material #9)									
Brush each coat									
Slow	80	275	53.30	28.75	6.90	19.38	10.46	10.48	75.97
Medium	130	250	46.70	22.69	6.56	18.68	11.98	7.19	67.10
Fast	175	225	40.00	20.57	7.24	17.78	14.14	4.18	63.91
Enamel, oil base (material #10)									
Brush each coat									
Slow	80	275	61.50	28.75	6.90	22.36	11.02	11.04	80.07
Medium	130	250	53.80	22.69	6.56	21.52	12.69	7.62	71.08
Fast	175	225	46.10	20.57	7.24	20.49	14.98	4.43	67.71

Cabinet back estimates are based on overall dimensions (length times width) to 8 feet high and include painting the inside back wall of paint grade or stain grade cabinets. ADD for preparation time. For heights above 8 feet, use the High Time Difficulty Factors on page 139. Measurements are based on total area of cabinet faces. "Slow" work is based on an hourly wage of $23.00, "Medium" work on an hourly wage of $29.50, and "Fast" work on an hourly wage of $36.00. Other qualifications that apply to this table are on page 9.

	Labor SF per manhour	Material coverage SF/gallon	Material cost per gallon	Labor cost per 100 SF	Labor burden 100 SF	Material cost per 100 SF	Overhead per 100 SF	Profit per 100 SF	Total price per 100 SF
Cabinet faces, stain grade									
Complete 7 step stain, seal & 2 coat lacquer system (material #11)									
Brush all coats									
Slow	20	190	54.90	115.00	27.60	28.89	32.58	32.65	236.72
Medium	35	178	48.00	84.29	24.36	26.97	33.91	20.34	189.87
Fast	50	165	41.20	72.00	25.42	24.97	37.94	11.22	171.55
Spray all coats									
Slow	85	67	54.90	27.06	6.48	81.94	21.94	21.99	159.41
Medium	110	51	48.00	26.82	7.75	94.12	32.17	19.30	180.16
Fast	135	35	41.20	26.67	9.42	117.71	47.67	14.10	215.57

Cabinet face estimates are based on overall dimensions (length times width) to 8 feet high. Use these figures to estimate finishing the faces of stain grade kitchen, bar, linen, pullman or vanity cabinets. ADD for preparation time. For heights above 8 feet, use the High Time Difficulty Factors on page 139. Measurements are based on total area of cabinet faces. "Slow" work is based on an hourly wage of $23.00, "Medium" work on an hourly wage of $29.50, and "Fast" work on an hourly wage of $36.00. Other qualifications that apply to this table are on page 9.

	Labor SF per manhour	Material coverage SF/gallon	Material cost per gallon	Labor cost per 100 SF	Labor burden 100 SF	Material cost per 100 SF	Overhead per 100 SF	Profit per 100 SF	Total price per 100 SF
Cabinets, paint grade, roll and brush									
Undercoat, water base (material #3)									
Roll & brush, 1 coat									
Slow	75	260	36.90	30.67	7.35	14.19	9.92	9.94	72.07
Medium	93	250	32.30	31.72	9.16	12.92	13.45	8.07	75.32
Fast	110	240	27.70	32.73	11.55	11.54	17.30	5.12	78.24
Undercoat, oil base (material #4)									
Roll & brush, 1 coat									
Slow	75	275	48.30	30.67	7.35	17.56	10.56	10.58	76.72
Medium	93	268	42.30	31.72	9.16	15.78	14.17	8.50	79.33
Fast	110	250	36.20	32.73	11.55	14.48	18.22	5.39	82.37
Split coat (1/2 undercoat + 1/2 enamel), water base (material #3 + #9)									
Roll & brush each coat									
Slow	95	310	45.10	24.21	5.82	14.55	8.47	8.49	61.54
Medium	113	298	39.50	26.11	7.55	13.26	11.73	7.04	65.69
Fast	130	285	33.85	27.69	9.77	11.88	15.30	4.52	69.16
Split coat (1/2 undercoat + 1/2 enamel), oil base (material #4 + #10)									
Roll & brush each coat									
Slow	95	310	54.90	24.21	5.82	17.71	9.07	9.09	65.90
Medium	113	298	48.05	26.11	7.55	16.12	12.45	7.47	69.70
Fast	130	285	41.15	27.69	9.77	14.44	16.09	4.76	72.75
Enamel, water base (material #9)									
Roll & brush 1st finish coat									
Slow	85	300	53.30	27.06	6.48	17.77	9.75	9.77	70.83
Medium	103	288	46.70	28.64	8.29	16.22	13.29	7.97	74.41
Fast	120	275	40.00	30.00	10.58	14.55	17.09	5.06	77.28
Roll & brush 2nd or additional finish coats									
Slow	95	310	53.30	24.21	5.82	17.19	8.97	8.99	65.18
Medium	113	298	46.70	26.11	7.55	15.67	12.33	7.40	69.06
Fast	130	285	40.00	27.69	9.77	14.04	15.97	4.72	72.19
Enamel, oil base (material #10)									
Roll & brush 1st finish coat									
Slow	85	300	61.50	27.06	6.48	20.50	10.27	10.29	74.60
Medium	103	288	53.80	28.64	8.29	18.68	13.90	8.34	77.85
Fast	120	275	46.10	30.00	10.58	16.76	17.78	5.26	80.38
Roll & brush 2nd or additional finish coats									
Slow	95	310	61.50	24.21	5.82	19.84	9.47	9.49	68.83
Medium	113	298	53.80	26.11	7.55	18.05	12.93	7.76	72.40
Fast	130	285	46.10	27.69	9.77	16.18	16.63	4.92	75.19

Cabinet estimates are based on overall dimensions (length times width) to 8 feet high and include painting the cabinet face, back of doors, stiles and rails. See Cabinet backs for painting the inside back wall of the cabinets. Use these figures to estimate paint grade kitchen cabinets. Use the Opening Count Method to estimate paint grade pullmans, vanities, bars or linen cabinets. For heights above 8 feet, apply the High Time Difficulty Factors to labor costs and the labor burden cost categories and add these figures to the total cost. Measurements are based on total area of cabinet faces. "Slow" work is based on an hourly wage of $23.00, "Medium" work on an hourly wage of $29.50, and "Fast" work on an hourly wage of $36.00. Other qualifications that apply to this table are on page 9.

	Labor SF per manhour	Material coverage SF/gallon	Material cost per gallon	Labor cost per 100 SF	Labor burden 100 SF	Material cost per 100 SF	Overhead per 100 SF	Profit per 100 SF	Total price per 100 SF
Cabinets, paint grade, spray application									
Undercoat, water base (material #3)									
Spray 1 coat									
Slow	125	125	36.90	18.40	4.42	29.52	9.94	9.96	72.24
Medium	140	113	32.30	21.07	6.08	28.58	13.94	8.36	78.03
Fast	155	100	27.70	23.23	8.19	27.70	18.33	5.42	82.87
Undercoat, oil base (material #4)									
Spray 1 coat									
Slow	125	135	48.30	18.40	4.42	35.78	11.13	11.16	80.89
Medium	140	123	42.30	21.07	6.08	34.39	15.39	9.23	86.16
Fast	155	110	36.20	23.23	8.19	32.91	19.95	5.90	90.18
Split coat (1/2 undercoat + 1/2 enamel), water base (material #3 + #9)									
Spray each coat									
Slow	200	175	45.10	11.50	2.76	25.77	7.61	7.62	55.26
Medium	225	163	39.50	13.11	3.78	24.23	10.28	6.17	57.57
Fast	250	150	33.85	14.40	5.08	22.57	13.04	3.86	58.95
Split coat (1/2 undercoat + 1/2 enamel), oil base (material #4 + #10)									
Spray each coat									
Slow	200	175	54.90	11.50	2.76	31.37	8.67	8.69	62.99
Medium	225	163	48.05	13.11	3.78	29.48	11.60	6.96	64.93
Fast	250	150	41.15	14.40	5.08	27.43	14.54	4.30	65.75
Enamel, water base (material #9)									
Spray 1st finish coat									
Slow	185	150	53.30	12.43	3.00	35.53	9.68	9.70	70.34
Medium	210	138	46.70	14.05	4.05	33.84	12.99	7.79	72.72
Fast	235	125	40.00	15.32	5.43	32.00	16.35	4.84	73.94
Spray 2nd or additional finish coats									
Slow	200	175	53.30	11.50	2.76	30.46	8.50	8.52	61.74
Medium	225	163	46.70	13.11	3.78	28.65	11.39	6.83	63.76
Fast	250	150	40.00	14.40	5.08	26.67	14.31	4.23	64.69
Enamel, oil base (material #10)									
Spray 1st finish coat									
Slow	185	160	61.50	12.43	3.00	38.44	10.23	10.25	74.35
Medium	210	148	53.80	14.05	4.05	36.35	13.62	8.17	76.24
Fast	235	135	46.10	15.32	5.43	34.15	17.01	5.03	76.94
Spray 2nd or additional finish coats									
Slow	200	185	61.50	11.50	2.76	33.24	9.03	9.04	65.57
Medium	225	173	53.80	13.11	3.78	31.10	12.00	7.20	67.19
Fast	250	160	46.10	14.40	5.08	28.81	14.97	4.43	67.69

Cabinet estimates are based on overall dimensions (length times width) to 8 feet high and include painting the cabinet face, back of doors, stiles and rails. See Cabinet backs for painting the inside back wall of the cabinets. Use these figures to estimate paint grade kitchen cabinets. Use the Opening Count Method to estimate paint grade pullmans, vanities, bars or linen cabinets. For heights above 8 feet, apply the High Time Difficulty Factors to labor costs and the labor burden cost categories and add these figures to the total cost. Measurements are based on total area of cabinet faces. "Slow" work is based on an hourly wage of $23.00, "Medium" work on an hourly wage of $29.50, and "Fast" work on an hourly wage of $36.00. Other qualifications that apply to this table are on page 9.

	Labor SF per manhour	Material coverage SF/gallon	Material cost per gallon	Labor cost per 100 SF	Labor burden 100 SF	Material cost per 100 SF	Overhead per 100 SF	Profit per 100 SF	Total price per 100 SF
Cabinets, stain grade									
Stain, seal & 2 coats lacquer system (7 step process)									
STEP 1: Sand & putty;									
Slow	125	--	--	18.40	4.42	--	4.34	4.35	31.51
Medium	150	--	--	19.67	5.70	--	6.34	3.80	35.51
Fast	175	--	--	20.57	7.24	--	8.63	2.55	38.99
STEP 2 & 3: Stain (material #11a) & wipe									
Brush 1 coat & wipe									
Slow	65	450	61.60	35.38	8.48	13.69	10.94	10.96	79.45
Medium	75	400	53.90	39.33	11.36	13.48	16.05	9.63	89.85
Fast	85	350	46.20	42.35	14.93	13.20	21.86	6.47	98.81
Spray 1 coat & wipe									
Slow	250	175	61.60	9.20	2.21	35.20	8.86	8.88	64.35
Medium	350	138	53.90	8.43	2.45	39.06	12.48	7.49	69.91
Fast	450	100	46.20	8.00	2.81	46.20	17.68	5.23	79.92
STEP 4: Sanding sealer (material #11b)									
Brush 1 coat									
Slow	110	450	50.90	20.91	5.01	11.31	7.08	7.09	51.40
Medium	120	425	44.60	24.58	7.10	10.49	10.54	6.33	59.04
Fast	130	400	38.20	27.69	9.77	9.55	14.57	4.31	65.89
Spray 1 coat									
Slow	330	175	50.90	6.97	1.67	29.09	7.17	7.18	52.08
Medium	430	138	44.60	6.86	2.00	32.32	10.29	6.17	57.64
Fast	530	100	38.20	6.79	2.42	38.20	14.69	4.35	66.45
STEP 5: Sand lightly									
Slow	200	--	--	11.50	2.76	--	2.71	2.72	19.69
Medium	250	--	--	11.80	3.41	--	3.80	2.28	21.29
Fast	300	--	--	12.00	4.22	--	5.03	1.49	22.74
STEP 6 & 7: Lacquer (material #11c), 2 coats									
Brush 1st coat									
Slow	120	375	53.50	19.17	4.59	14.27	7.23	7.24	52.50
Medium	165	350	46.80	17.88	5.17	13.37	9.11	5.46	50.99
Fast	215	325	40.10	16.74	5.91	12.34	10.85	3.21	49.05
Brush 2nd coat									
Slow	130	400	53.50	17.69	4.24	13.38	6.71	6.72	48.74
Medium	173	388	46.80	17.05	4.93	12.06	8.51	5.11	47.66
Fast	225	375	40.10	16.00	5.63	10.69	10.03	2.97	45.32
Spray 1st coat									
Slow	275	150	53.50	8.36	2.02	35.67	8.75	8.77	63.57
Medium	388	100	46.80	7.60	2.21	46.80	14.15	8.49	79.25
Fast	500	75	40.10	7.20	2.54	53.47	19.60	5.80	88.61

	Labor SF per manhour	Material coverage SF/gallon	Material cost per gallon	Labor cost per 100 SF	Labor burden 100 SF	Material cost per 100 SF	Overhead per 100 SF	Profit per 100 SF	Total price per 100 SF
Spray 2nd coat									
Slow	350	200	53.50	6.57	1.59	26.75	6.63	6.64	48.18
Medium	475	163	46.80	6.21	1.81	28.71	9.18	5.51	51.42
Fast	600	125	40.10	6.00	2.13	32.08	12.46	3.69	56.36
Complete 7 step stain, seal & 2 coat lacquer system (material #11)									
Brush all coats									
Slow	20	125	54.90	115.00	27.60	43.92	35.44	35.51	257.47
Medium	25	113	48.00	118.00	34.12	42.48	48.65	29.19	272.44
Fast	30	100	41.20	120.00	42.35	41.20	63.10	18.67	285.32
Spray all coats									
Slow	40	40	54.90	57.50	13.80	137.25	39.62	39.71	287.88
Medium	50	30	48.00	59.00	17.06	160.00	59.01	35.41	330.48
Fast	60	21	41.20	60.00	21.20	196.19	85.98	25.43	388.80
Shellac, clear (material #12)									
Brush each coat									
Slow	175	525	70.40	13.14	3.14	13.41	7.43	4.46	41.58
Medium	200	513	61.60	14.75	4.27	12.01	9.62	2.84	43.49
Fast	225	500	52.80	16.00	5.63	10.56	9.99	2.95	45.13
Varnish, flat or gloss (material #30c)									
Brush each coat									
Slow	155	475	81.00	14.84	3.56	17.05	10.99	3.25	49.69
Medium	180	463	70.80	16.39	4.75	15.29	6.92	6.93	50.28
Fast	205	450	60.70	17.56	6.21	13.49	9.31	5.59	52.16
Penetrating stain wax (material #14) & polish									
Brush 1st coat									
Slow	125	575	50.20	18.40	4.42	8.73	9.78	2.89	44.22
Medium	150	538	43.90	19.67	5.70	8.16	6.37	6.38	46.28
Fast	175	500	37.60	20.57	7.24	7.52	8.84	5.30	49.47
Brush 2nd or additional coats									
Slow	150	600	50.20	15.33	3.69	8.37	8.49	2.51	38.39
Medium	175	575	43.90	16.86	4.86	7.63	9.10	2.69	41.14
Fast	200	550	37.60	18.00	6.36	6.84	5.93	5.94	43.07

Cabinet estimates are based on overall dimensions (length times width) to 8 feet high. Use these figures to estimate stain grade kitchen, bar, linen, pullman or vanity cabinets. For the stain, seal and lacquer process, the figures include finishing both sides of cabinet doors, stiles and rails with a fog coat of stain on shelves and the wall behind the cabinet (cabinet back). See Cabinet backs for painting the inside back wall of the cabinets. For heights above 8 feet, use the High Time Difficulty Factors on page 139. Measurements are based on total area of cabinet faces. "Slow" work is based on an hourly wage of $23.00, "Medium" work on an hourly wage of $29.50, and "Fast" work on an hourly wage of $36.00. Other qualifications that apply to this table are on page 9.

	Labor SF per manhour	Material coverage SF/gallon	Material cost per gallon	Labor cost per 100 SF	Labor burden 100 SF	Material cost per 100 SF	Overhead per 100 SF	Profit per 100 SF	Total price per 100 SF
Ceiling panels, suspended, fiber panels in T-bar frames, brush application									
Flat latex, water base (material #5)									
Brush 1st coat									
Slow	80	260	36.00	28.75	6.90	13.85	9.41	9.43	68.34
Medium	110	230	31.50	26.82	7.75	13.70	12.07	7.24	67.58
Fast	140	200	27.00	25.71	9.07	13.50	14.97	4.43	67.68
Brush 2nd or additional coats									
Slow	130	300	36.00	17.69	4.24	12.00	6.45	6.46	46.84
Medium	150	275	31.50	19.67	5.70	11.45	9.20	5.52	51.54
Fast	170	250	27.00	21.18	7.46	10.80	12.23	3.62	55.29
Enamel, water base (material #9)									
Brush 1st coat									
Slow	65	260	53.30	35.38	8.48	20.50	12.23	12.26	88.85
Medium	100	230	46.70	29.50	8.53	20.30	14.58	8.75	81.66
Fast	125	200	40.00	28.80	10.17	20.00	18.28	5.41	82.66
Brush 2nd or additional coats									
Slow	115	300	53.30	20.00	4.81	17.77	8.09	8.11	58.78
Medium	135	275	46.70	21.85	6.33	16.98	11.29	6.77	63.22
Fast	155	250	40.00	23.23	8.19	16.00	14.70	4.35	66.47
Enamel, oil base (material #10)									
Brush 1st coat									
Slow	65	250	61.50	35.38	8.48	24.60	13.01	13.04	94.51
Medium	95	213	53.80	31.05	9.00	25.26	16.32	9.79	91.42
Fast	125	175	46.10	28.80	10.17	26.34	20.25	5.99	91.55
Brush 2nd or additional coats									
Slow	115	275	61.50	20.00	4.81	22.36	8.96	8.98	65.11
Medium	135	260	53.80	21.85	6.33	20.69	12.21	7.33	68.41
Fast	155	240	46.10	23.23	8.19	19.21	15.70	4.64	70.97

Ceiling panel estimates are based on overall dimensions (length times width) to 8 feet high. Do not make deductions for openings in the ceiling area that are under 100 square feet. For heights above 8 feet, use the High Time Difficulty Factors on page 139. "Slow" work is based on an hourly wage of $23.00, "Medium" work on an hourly wage of $29.50, and "Fast" work on an hourly wage of $36.00. Other qualifications that apply to this table are on page 9.

	Labor SF per manhour	Material coverage SF/gallon	Material cost per gallon	Labor cost per 100 SF	Labor burden 100 SF	Material cost per 100 SF	Overhead per 100 SF	Profit per 100 SF	Total price per 100 SF
Ceiling panels, suspended, fiber panels in T-bar frame, roll application									
Flat latex, water base (material #5)									
Roll 1st coat									
Slow	150	270	36.00	15.33	3.69	13.33	6.14	6.16	44.65
Medium	215	235	31.50	13.72	3.96	13.40	7.77	4.66	43.51
Fast	280	200	27.00	12.86	4.53	13.50	9.58	2.83	43.30
Roll 2nd or additional coats									
Slow	225	280	36.00	10.22	2.44	12.86	4.85	4.86	35.23
Medium	288	260	31.50	10.24	2.96	12.12	6.33	3.80	35.45
Fast	350	240	27.00	10.29	3.64	11.25	7.80	2.31	35.29
Enamel, water base (material #9)									
Roll 1st coat									
Slow	135	250	53.30	17.04	4.09	21.32	8.07	8.08	58.60
Medium	200	220	46.70	14.75	4.27	21.23	10.06	6.04	56.35
Fast	265	190	40.00	13.58	4.78	21.05	12.22	3.61	55.24
Roll 2nd or additional finish coats									
Slow	210	280	53.30	10.95	2.63	19.04	6.20	6.21	45.03
Medium	273	260	46.70	10.81	3.11	17.96	7.97	4.78	44.63
Fast	335	240	40.00	10.75	3.81	16.67	9.68	2.86	43.77
Enamel, oil base (material #10)									
Roll 1st coat									
Slow	135	240	61.50	17.04	4.09	25.63	8.88	8.90	64.54
Medium	200	230	53.80	14.75	4.27	23.39	10.60	6.36	59.37
Fast	265	210	46.10	13.58	4.78	21.95	12.50	3.70	56.51
Roll 2nd or additional finish coats									
Slow	210	275	61.50	10.95	2.63	22.36	6.83	6.84	49.61
Medium	273	250	53.80	10.81	3.11	21.52	8.86	5.32	49.62
Fast	335	230	46.10	10.75	3.81	20.04	10.72	3.17	48.49

Ceiling panel estimates are based on overall dimensions (length times width) to 8 feet high. Do not make deductions for openings in the ceiling area that are under 100 square feet. For heights above 8 feet, use the High Time Difficulty Factors on page 139. "Slow" work is based on an hourly wage of $23.00, "Medium" work on an hourly wage of $29.50, and "Fast" work on an hourly wage of $36.00. Other qualifications that apply to this table are on page 9.

	Labor SF per manhour	Material coverage SF/gallon	Material cost per gallon	Labor cost per 100 SF	Labor burden 100 SF	Material cost per 100 SF	Overhead per 100 SF	Profit per 100 SF	Total price per 100 SF
Ceiling panels, suspended, fiber panels in T-bar frame, spray application									
Flat latex, water base (material #5)									
Spray 1st coat									
Slow	300	250	36.00	7.67	1.83	14.40	4.54	4.55	32.99
Medium	345	238	31.50	8.55	2.48	13.24	6.07	3.64	33.98
Fast	390	225	27.00	9.23	3.24	12.00	7.59	2.25	34.31
Spray 2nd or additional coats									
Slow	500	270	36.00	4.60	1.10	13.33	3.62	3.62	26.27
Medium	545	260	31.50	5.41	1.55	12.12	4.77	2.86	26.71
Fast	590	250	27.00	6.10	2.13	10.80	5.91	1.75	26.69
Enamel, water base (material #9)									
Spray 1st coat									
Slow	275	250	53.30	8.36	2.02	21.32	6.02	6.03	43.75
Medium	325	238	46.70	9.08	2.63	19.62	7.83	4.70	43.86
Fast	375	225	40.00	9.60	3.41	17.78	9.54	2.82	43.15
Spray 2nd or additional coats									
Slow	450	275	53.30	5.11	1.22	19.38	4.89	4.90	35.50
Medium	500	263	46.70	5.90	1.71	17.76	6.34	3.81	35.52
Fast	550	250	40.00	6.55	2.32	16.00	7.71	2.28	34.86
Enamel, oil base (material #10)									
Spray 1st coat									
Slow	275	240	61.50	8.36	2.02	25.63	6.84	6.85	49.70
Medium	325	220	53.80	9.08	2.63	24.45	9.04	5.42	50.62
Fast	375	200	46.10	9.60	3.41	23.05	11.17	3.30	50.53
Spray 2nd or additional coats									
Slow	450	250	61.50	5.11	1.22	24.60	5.88	5.89	42.70
Medium	500	238	53.80	5.90	1.71	22.61	7.56	4.53	42.31
Fast	550	225	46.10	6.55	2.32	20.49	9.10	2.69	41.15

Ceiling panel estimates are based on overall dimensions (length times width) to 8 feet high. Do not make deductions for openings in the ceiling area that are under 100 square feet. For heights above 8 feet, use the High Time Difficulty Factors on page 139. "Slow" work is based on an hourly wage of $23.00, "Medium" work on an hourly wage of $29.50, and "Fast" work on an hourly wage of $36.00. Other qualifications that apply to this table are on page 9.

	Labor SF per manhour	Material coverage SF/gallon	Material cost per gallon	Labor cost per 100 SF	Labor burden 100 SF	Material cost per 100 SF	Overhead per 100 SF	Profit per 100 SF	Total price per 100 SF
Ceiling pans, metal, exterior enamel finish									
Enamel, water base (material #24)									
Brush each coat									
Slow	80	450	54.80	28.75	6.90	12.18	9.09	9.11	66.03
Medium	100	388	47.90	29.50	8.53	12.35	12.60	7.56	70.54
Fast	125	325	41.10	28.80	10.17	12.65	16.00	4.73	72.35
Enamel, oil base (material #25)									
Brush each coat									
Slow	80	400	62.50	28.75	6.90	15.63	9.74	9.76	70.78
Medium	103	338	54.70	28.64	8.29	16.18	13.28	7.97	74.36
Fast	125	275	46.90	28.80	10.17	17.05	17.37	5.14	78.53
Enamel, water base (material #24)									
Roll each coat									
Slow	175	425	54.80	13.14	3.14	12.89	5.54	5.56	40.27
Medium	200	368	47.90	14.75	4.27	13.02	8.01	4.80	44.85
Fast	225	300	41.10	16.00	5.63	13.70	10.96	3.24	49.53
Enamel, oil base (material #25)									
Roll each coat									
Slow	175	375	62.50	13.14	3.14	16.67	6.26	6.28	45.49
Medium	200	313	54.70	14.75	4.27	17.48	9.12	5.47	51.09
Fast	225	250	46.90	16.00	5.63	18.76	12.53	3.71	56.63
Enamel, water base (material #24)									
Spray each coat									
Slow	550	380	54.80	4.18	1.01	14.42	3.72	3.73	27.06
Medium	600	370	47.90	4.92	1.43	12.95	4.82	2.89	27.01
Fast	650	260	41.10	5.54	1.96	15.81	7.23	2.14	32.68
Enamel, oil base (material #25)									
Spray each coat									
Slow	550	330	62.50	4.18	1.01	18.94	4.58	4.59	33.30
Medium	600	270	54.70	4.92	1.43	20.26	6.65	3.99	37.25
Fast	650	210	46.90	5.54	1.96	22.33	9.25	2.74	41.82

Ceiling panel estimates are based on overall dimensions (length times width) to 8 feet high. Do not make deductions for openings in the ceiling area that are under 100 square feet. For heights above 8 feet, use the High Time Difficulty Factors on page 139. "Slow" work is based on an hourly wage of $23.00, "Medium" work on an hourly wage of $29.50, and "Fast" work on an hourly wage of $36.00. Other qualifications that apply to this table are on page 9.

	Labor SF per manhour	Material coverage SF/gallon	Material cost per gallon	Labor cost per 100 SF	Labor burden 100 SF	Material cost per 100 SF	Overhead per 100 SF	Profit per 100 SF	Total price per 100 SF
Ceilings, acoustic spray-on texture on gypsum drywall									
Acoustic spray-on texture, primer (material #6)									
Spray prime coat									
Slow	250	100	28.70	9.20	2.21	28.70	7.62	7.64	55.37
Medium	300	90	25.10	9.83	2.83	27.89	10.14	6.08	56.77
Fast	350	80	21.50	10.29	3.64	26.88	12.65	3.74	57.20
Acoustic spray-on texture, finish (material #7)									
Spray 1st finish coat									
Slow	400	180	34.50	5.75	1.38	19.17	5.00	5.01	36.31
Medium	450	170	30.20	6.56	1.88	17.76	6.56	3.93	36.69
Fast	500	160	25.90	7.20	2.54	16.19	8.04	2.38	36.35
Spray 2nd or additional finish coats									
Slow	500	200	34.50	4.60	1.10	17.25	4.36	4.37	31.68
Medium	550	188	30.20	5.36	1.56	16.06	5.74	3.45	32.17
Fast	600	175	25.90	6.00	2.13	14.80	7.11	2.10	32.14

Ceiling texture estimates are based on overall dimensions (length times width) to 8 feet high. Do not make deductions for openings in the ceiling area that are under 100 square feet. For heights above 8 feet, use the High Time Difficulty Factors on page 139. "Slow" work is based on an hourly wage of $23.00, "Medium" work on an hourly wage of $29.50, and "Fast" work on an hourly wage of $36.00. Other qualifications that apply to this table are on page 9.

	Labor SF per manhour	Material coverage SF/pound	Material cost per pound	Labor cost per 100 SF	Labor burden 100 SF	Material cost per 100 SF	Overhead per 100 SF	Profit per 100 SF	Total price per 100 SF
Ceilings, stipple finish texture paint, light, on drywall									
Stipple finish texture paint, Drypowder mix, light coverage (material #8)									
Spray each coat									
Slow	225	10.0	1.57	10.22	2.44	15.70	5.39	5.40	39.15
Medium	250	7.5	1.38	11.80	3.41	18.40	8.40	5.04	47.05
Fast	275	5.0	1.18	13.09	4.64	23.60	12.81	3.79	57.93

Estimates for stipple finish texture paint are based on overall dimensions (length times width) to 8 feet high. Do not make deductions for openings in the ceiling area that are under 100 square feet. For heights above 8 feet, use the High Time Difficulty Factors on page 139. "Slow" work is based on an hourly wage of $23.00, "Medium" work on an hourly wage of $29.50, and "Fast" work on an hourly wage of $36.00. Other qualifications that apply to this table are on page 9.

	Labor SF per manhour	Material coverage SF/gallon	Material cost per gallon	Labor cost per 100 SF	Labor burden 100 SF	Material cost per 100 SF	Overhead per 100 SF	Profit per 100 SF	Total price per 100 SF
Ceilings, gypsum drywall, anti-graffiti stain eliminator									
Water base primer and pigmented sealer (material #39)									
Roll & brush each coat									
Slow	350	450	49.40	6.57	1.59	10.98	3.63	3.64	26.41
Medium	375	425	43.20	7.87	2.28	10.16	5.08	3.05	28.44
Fast	400	400	37.00	9.00	3.18	9.25	6.64	1.96	30.03
Oil base primer and pigmented sealer (material #40)									
Roll & brush each coat									
Slow	350	400	58.00	6.57	1.59	14.50	4.30	4.31	31.27
Medium	375	388	50.80	7.87	2.28	13.09	5.81	3.48	32.53
Fast	400	375	43.50	9.00	3.18	11.60	7.37	2.18	33.33
Polyurethane 2 part system (material #41)									
Roll & brush each coat									
Slow	300	400	187.30	7.67	1.83	46.83	10.70	10.73	77.76
Medium	325	375	163.90	9.08	2.63	43.71	13.85	8.31	77.58
Fast	350	350	140.50	10.29	3.64	40.14	16.76	4.96	75.79

Ceiling estimates are based on overall dimensions (length times width) to 8 feet high. Do not make deductions for openings in the ceiling area that are under 100 square feet. For heights above 8 feet, use the High Time Difficulty Factors on page 139. "Slow" work is based on an hourly wage of $23.00, "Medium" work on an hourly wage of $29.50, and "Fast" work on an hourly wage of $36.00. Other qualifications that apply to this table are on page 9.

	Labor SF per manhour	Material coverage SF/gallon	Material cost per gallon	Labor cost per 100 SF	Labor burden 100 SF	Material cost per 100 SF	Overhead per 100 SF	Profit per 100 SF	Total price per 100 SF

Ceilings, gypsum drywall, orange peel or knock-down texture, brush

	Labor SF per manhour	Material coverage SF/gallon	Material cost per gallon	Labor cost per 100 SF	Labor burden 100 SF	Material cost per 100 SF	Overhead per 100 SF	Profit per 100 SF	Total price per 100 SF
Flat latex, water base (material #5)									
Brush 1st coat									
Slow	150	300	36.00	15.33	3.69	12.00	5.89	5.90	42.81
Medium	175	288	31.50	16.86	4.86	10.94	8.17	4.90	45.73
Fast	200	275	27.00	18.00	6.36	9.82	10.59	3.13	47.90
Brush 2nd coat									
Slow	175	350	36.00	13.14	3.14	10.29	5.05	5.06	36.68
Medium	200	338	31.50	14.75	4.27	9.32	7.08	4.25	39.67
Fast	225	325	27.00	16.00	5.63	8.31	9.29	2.75	41.98
Brush 3rd or additional coats									
Slow	200	400	36.00	11.50	2.76	9.00	4.42	4.43	32.11
Medium	225	375	31.50	13.11	3.78	8.40	6.33	3.80	35.42
Fast	250	350	27.00	14.40	5.08	7.71	8.43	2.49	38.11
Sealer, water base (material #1)									
Brush prime coat									
Slow	175	300	36.00	13.14	3.14	12.00	5.38	5.39	39.05
Medium	200	288	31.50	14.75	4.27	10.94	7.49	4.49	41.94
Fast	225	275	27.00	16.00	5.63	9.82	9.76	2.89	44.10
Sealer, oil base (material #2)									
Brush prime coat									
Slow	175	250	46.20	13.14	3.14	18.48	6.61	6.62	47.99
Medium	200	238	40.50	14.75	4.27	17.02	9.01	5.40	50.45
Fast	225	225	34.70	16.00	5.63	15.42	11.49	3.40	51.94
Enamel, water base (material #9)									
Brush 1st finish coat									
Slow	150	350	53.30	15.33	3.69	15.23	6.51	6.52	47.28
Medium	175	338	46.70	16.86	4.86	13.82	8.89	5.33	49.76
Fast	200	325	40.00	18.00	6.36	12.31	11.36	3.36	51.39
Brush 2nd or additional finish coats									
Slow	175	400	53.30	13.14	3.14	13.33	5.63	5.64	40.88
Medium	200	375	46.70	14.75	4.27	12.45	7.87	4.72	44.06
Fast	225	350	40.00	16.00	5.63	11.43	10.25	3.03	46.34

	Labor SF per manhour	Material coverage SF/gallon	Material cost per gallon	Labor cost per 100 SF	Labor burden 100 SF	Material cost per 100 SF	Overhead per 100 SF	Profit per 100 SF	Total price per 100 SF
Enamel, oil base (material #10)									
Brush 1st finish coat									
Slow	150	325	61.50	15.33	3.69	18.92	7.21	7.22	52.37
Medium	175	313	53.80	16.86	4.86	17.19	9.73	5.84	54.48
Fast	200	300	46.10	18.00	6.36	15.37	12.31	3.64	55.68
Brush 2nd or additional finish coats									
Slow	150	400	61.50	15.33	3.69	15.38	6.53	6.55	47.48
Medium	175	375	53.80	16.86	4.86	14.35	9.02	5.41	50.50
Fast	200	350	46.10	18.00	6.36	13.17	11.63	3.44	52.60
Epoxy coating, white (material #52)									
Brush 1st coat									
Slow	125	350	157.00	18.40	4.42	44.86	12.86	12.89	93.43
Medium	150	325	137.40	19.67	5.70	42.28	16.91	10.14	94.70
Fast	175	300	117.70	20.57	7.24	39.23	20.79	6.15	93.98
Brush 2nd or additional coats									
Slow	175	375	157.00	13.14	3.14	41.87	11.05	11.07	80.27
Medium	200	350	137.40	14.75	4.27	39.26	14.57	8.74	81.59
Fast	225	325	117.70	16.00	5.63	36.22	17.94	5.31	81.10

Ceiling estimates are based on overall dimensions (length times width) to 8 feet high. Do not make deductions for openings in the ceiling area that are under 100 square feet. For heights above 8 feet, use the High Time Difficulty Factors on page 139. ADD for cutting-in at ceilings and protecting adjacent surfaces if ceilings alone are being painted, not walls. Otherwise, figure any cutting-in time with walls. See the notes under the wall formulas for clarification. "Slow" work is based on an hourly wage of $23.00, "Medium" work on an hourly wage of $29.50, and "Fast" work on an hourly wage of $36.00. Other qualifications that apply to this table are on page 9.

	Labor SF per manhour	Material coverage SF/gallon	Material cost per gallon	Labor cost per 100 SF	Labor burden 100 SF	Material cost per 100 SF	Overhead per 100 SF	Profit per 100 SF	Total price per 100 SF
Ceilings, gypsum drywall, orange peel or knock-down texture, roll									
Flat latex, water base (material #5)									
Roll 1st coat									
Slow	325	300	36.00	7.08	1.70	12.00	3.95	3.96	28.69
Medium	350	275	31.50	8.43	2.45	11.45	5.58	3.35	31.26
Fast	375	250	27.00	9.60	3.41	10.80	7.37	2.18	33.36
Roll 2nd coat									
Slow	350	325	36.00	6.57	1.59	11.08	3.65	3.66	26.55
Medium	375	313	31.50	7.87	2.28	10.06	5.05	3.03	28.29
Fast	400	300	27.00	9.00	3.18	9.00	6.57	1.94	29.69
Roll 3rd or additional coats									
Slow	400	350	36.00	5.75	1.38	10.29	3.31	3.32	24.05
Medium	425	338	31.50	6.94	2.00	9.32	4.57	2.74	25.57
Fast	450	325	27.00	8.00	2.81	8.31	5.93	1.75	26.80
Sealer, water base (material #1)									
Roll prime coat									
Slow	350	300	36.00	6.57	1.59	12.00	3.83	3.84	27.83
Medium	375	275	31.50	7.87	2.28	11.45	5.40	3.24	30.24
Fast	400	250	27.00	9.00	3.18	10.80	7.12	2.11	32.21
Sealer, oil base (material #2)									
Roll prime coat									
Slow	350	275	46.20	6.57	1.59	16.80	4.74	4.75	34.45
Medium	375	250	40.50	7.87	2.28	16.20	6.59	3.95	36.89
Fast	400	225	34.70	9.00	3.18	15.42	8.56	2.53	38.69
Enamel, water base (material #9)									
Roll 1st finish coat									
Slow	325	325	53.30	7.08	1.70	16.40	4.78	4.79	34.75
Medium	350	313	46.70	8.43	2.45	14.92	6.45	3.87	36.12
Fast	375	300	40.00	9.60	3.41	13.33	8.16	2.41	36.91
Roll 2nd or additional finish coats									
Slow	375	350	53.30	6.13	1.48	15.23	4.34	4.35	31.53
Medium	400	338	46.70	7.38	2.13	13.82	5.83	3.50	32.66
Fast	425	325	40.00	8.47	2.98	12.31	7.37	2.18	33.31

	Labor SF per manhour	Material coverage SF/gallon	Material cost per gallon	Labor cost per 100 SF	Labor burden 100 SF	Material cost per 100 SF	Overhead per 100 SF	Profit per 100 SF	Total price per 100 SF
Enamel, oil base (material #10)									
Roll 1st finish coat									
Slow	325	300	61.50	7.08	1.70	20.50	5.56	5.57	40.41
Medium	350	275	53.80	8.43	2.45	19.56	7.61	4.56	42.61
Fast	375	250	46.10	9.60	3.41	18.44	9.74	2.88	44.07
Roll 2nd or additional finish coats									
Slow	375	300	61.50	6.13	1.48	20.50	5.34	5.35	38.80
Medium	400	288	53.80	7.38	2.13	18.68	7.05	4.23	39.47
Fast	425	275	46.10	8.47	2.98	16.76	8.75	2.59	39.55
Epoxy coating, white (material #52)									
Roll 1st coat									
Slow	300	300	157.00	7.67	1.83	52.33	11.75	11.77	85.35
Medium	325	288	137.40	9.08	2.63	47.71	14.85	8.91	83.18
Fast	350	275	117.70	10.29	3.64	42.80	17.58	5.20	79.51
Roll 2nd or additional coats									
Slow	350	300	157.00	6.57	1.59	52.33	11.49	11.52	83.50
Medium	375	288	137.40	7.87	2.28	47.71	14.46	8.68	81.00
Fast	400	275	117.70	9.00	3.18	42.80	17.04	5.04	77.06

Ceiling estimates are based on overall dimensions (length times width) to 8 feet high. Do not make deductions for openings in the ceiling area that are under 100 square feet. For heights above 8 feet, use the High Time Difficulty Factors on page 139. ADD for cutting-in at ceilings and protecting adjacent surfaces if ceilings alone are being painted, not walls. Otherwise, figure any cutting-in time with walls. See the notes under the wall formulas for clarification. "Slow" work is based on an hourly wage of $23.00, "Medium" work on an hourly wage of $29.50, and "Fast" work on an hourly wage of $36.00. Other qualifications that apply to this table are on page 9.

	Labor SF per manhour	Material coverage SF/gallon	Material cost per gallon	Labor cost per 100 SF	Labor burden 100 SF	Material cost per 100 SF	Overhead per 100 SF	Profit per 100 SF	Total price per 100 SF
Ceilings, gypsum drywall, orange peel or knock-down texture, spray									
Flat latex, water base (material #5)									
Spray 1st coat									
Slow	650	225	36.00	3.54	.85	16.00	3.87	3.88	28.14
Medium	750	200	31.50	3.93	1.13	15.75	5.21	3.12	29.14
Fast	850	175	27.00	4.24	1.51	15.43	6.56	1.94	29.68
Spray 2nd coat									
Slow	775	250	36.00	2.97	.71	14.40	3.44	3.44	24.96
Medium	875	225	31.50	3.37	.97	14.00	4.59	2.75	25.68
Fast	975	200	27.00	3.69	1.33	13.50	5.73	1.70	25.95
Spray 3rd or additional coats									
Slow	825	275	36.00	2.79	.66	13.09	3.14	3.15	22.83
Medium	925	250	31.50	3.19	.92	12.60	4.18	2.51	23.40
Fast	1025	225	27.00	3.51	1.26	12.00	5.19	1.54	23.50
Sealer, water base (material #1)									
Spray prime coat									
Slow	700	225	36.00	3.29	.79	16.00	3.82	3.82	27.72
Medium	800	200	31.50	3.69	1.06	15.75	5.13	3.08	28.71
Fast	900	175	27.00	4.00	1.41	15.43	6.46	1.91	29.21
Sealer, oil base (material #2)									
Spray prime coat									
Slow	700	200	46.20	3.29	.79	23.10	5.16	5.17	37.51
Medium	800	188	40.50	3.69	1.06	21.54	6.58	3.95	36.82
Fast	900	175	34.70	4.00	1.41	19.83	7.82	2.31	35.37
Enamel, water base (material #9)									
Spray 1st finish coat									
Slow	725	250	53.30	3.17	.77	21.32	4.80	4.81	34.87
Medium	825	225	46.70	3.58	1.02	20.76	6.34	3.81	35.51
Fast	925	200	40.00	3.89	1.37	20.00	7.83	2.32	35.41
Spray 2nd or additional finish coat									
Slow	775	275	53.30	2.97	.71	19.38	4.38	4.39	31.83
Medium	875	250	46.70	3.37	.97	18.68	5.76	3.45	32.23
Fast	975	225	40.00	3.69	1.33	17.78	7.06	2.09	31.95

	Labor SF per manhour	Material coverage SF/gallon	Material cost per gallon	Labor cost per 100 SF	Labor burden 100 SF	Material cost per 100 SF	Overhead per 100 SF	Profit per 100 SF	Total price per 100 SF
Enamel, oil base (material #10)									
Spray 1st finish coat									
Slow	725	225	61.50	3.17	.77	27.33	5.94	5.95	43.16
Medium	825	213	53.80	3.58	1.02	25.26	7.47	4.48	41.81
Fast	925	200	46.10	3.89	1.37	23.05	8.78	2.60	39.69
Spray 2nd or additional finish coat									
Slow	775	250	61.50	2.97	.71	24.60	5.37	5.38	39.03
Medium	875	238	53.80	3.37	.97	22.61	6.74	4.04	37.73
Fast	975	225	46.10	3.69	1.33	20.49	7.90	2.34	35.75

Ceiling estimates are based on overall dimensions (length times width) to 8 feet high. Do not make deductions for openings in the ceiling area that are under 100 square feet. For heights above 8 feet, use the High Time Difficulty Factors on page 139. ADD for cutting-in at ceilings and protecting adjacent surfaces if ceilings alone are being painted, not walls. Otherwise, figure any cutting-in and protection time with the walls. See the notes under the wall formulas for clarification. "Slow" work is based on an hourly wage of $23.00, "Medium" work on an hourly wage of $29.50, and "Fast" work on an hourly wage of $36.00. Other qualifications that apply to this table are on page 9.

	Labor SF per manhour	Material coverage SF/gallon	Material cost per gallon	Labor cost per 100 SF	Labor burden 100 SF	Material cost per 100 SF	Overhead per 100 SF	Profit per 100 SF	Total price per 100 SF
Ceilings, gypsum drywall, sand finish or skip trowel texture, brush									
Flat latex, water base (material #5)									
Brush 1st coat									
Slow	175	325	36.00	13.14	3.14	11.08	5.20	5.21	37.77
Medium	200	313	31.50	14.75	4.27	10.06	7.27	4.36	40.71
Fast	225	300	27.00	16.00	5.63	9.00	9.50	2.81	42.94
Brush 2nd coat									
Slow	200	400	36.00	11.50	2.76	9.00	4.42	4.43	32.11
Medium	238	375	31.50	12.39	3.58	8.40	6.09	3.66	34.12
Fast	275	350	27.00	13.09	4.64	7.71	7.88	2.33	35.65
Brush 3rd or additional coats									
Slow	225	425	36.00	10.22	2.44	8.47	4.02	4.03	29.18
Medium	263	400	31.50	11.22	3.23	7.88	5.59	3.35	31.27
Fast	300	375	27.00	12.00	4.22	7.20	7.27	2.15	32.84
Sealer, water base (material #1)									
Brush prime coat									
Slow	200	325	36.00	11.50	2.76	11.08	4.81	4.82	34.97
Medium	225	313	31.50	13.11	3.78	10.06	6.74	4.04	37.73
Fast	250	300	27.00	14.40	5.08	9.00	8.83	2.61	39.92
Sealer, oil base (material #2)									
Brush prime coat									
Slow	200	325	46.20	11.50	2.76	14.22	5.41	5.42	39.31
Medium	225	313	40.50	13.11	3.78	12.94	7.46	4.48	41.77
Fast	250	300	34.70	14.40	5.08	11.57	9.63	2.85	43.53
Enamel, water base (material #9)									
Brush 1st finish coat									
Slow	200	400	53.30	11.50	2.76	13.33	5.24	5.25	38.08
Medium	225	375	46.70	13.11	3.78	12.45	7.34	4.40	41.08
Fast	250	350	40.00	14.40	5.08	11.43	9.58	2.83	43.32
Brush 2nd or additional finish coats									
Slow	225	425	53.30	10.22	2.44	12.54	4.79	4.80	34.79
Medium	263	400	46.70	11.22	3.23	11.68	6.54	3.92	36.59
Fast	300	375	40.00	12.00	4.22	10.67	8.34	2.47	37.70

	Labor SF per manhour	Material coverage SF/gallon	Material cost per gallon	Labor cost per 100 SF	Labor burden 100 SF	Material cost per 100 SF	Overhead per 100 SF	Profit per 100 SF	Total price per 100 SF
Enamel, oil base (material #10)									
Brush 1st finish coat									
Slow	200	375	61.50	11.50	2.76	16.40	5.83	5.84	42.33
Medium	225	350	53.80	13.11	3.78	15.37	8.07	4.84	45.17
Fast	250	325	46.10	14.40	5.08	14.18	10.43	3.09	47.18
Brush 2nd or additional finish coats									
Slow	225	400	61.50	10.22	2.44	15.38	5.33	5.34	38.71
Medium	263	375	53.80	11.22	3.23	14.35	7.20	4.32	40.32
Fast	300	350	46.10	12.00	4.22	13.17	9.12	2.70	41.21
Epoxy coating, white (material #52)									
Brush 1st coat									
Slow	150	375	157.00	15.33	3.69	41.87	11.57	11.59	84.05
Medium	175	350	137.40	16.86	4.86	39.26	15.25	9.15	85.38
Fast	225	325	117.70	16.00	5.63	36.22	17.94	5.31	81.10
Brush 2nd or additional coats									
Slow	175	400	157.00	13.14	3.14	39.25	10.55	10.57	76.65
Medium	200	375	137.40	14.75	4.27	36.64	13.91	8.35	77.92
Fast	225	350	117.70	16.00	5.63	33.63	17.14	5.07	77.47

Ceiling estimates are based on overall dimensions (length times width) to 8 feet high. Do not make deductions for openings in the ceiling area that are under 100 square feet. For heights above 8 feet, use the High Time Difficulty Factors on page 139. ADD for cutting-in at ceilings and protecting adjacent surfaces if ceilings alone are being painted, not walls. Otherwise, figure any cutting-in time with walls. See the notes under the wall formulas for clarification. "Slow" work is based on an hourly wage of $23.00, "Medium" work on an hourly wage of $29.50, and "Fast" work on an hourly wage of $36.00. Other qualifications that apply to this table are on page 9.

	Labor SF per manhour	Material coverage SF/gallon	Material cost per gallon	Labor cost per 100 SF	Labor burden 100 SF	Material cost per 100 SF	Overhead per 100 SF	Profit per 100 SF	Total price per 100 SF
Ceilings, gypsum drywall, sand finish or skip trowel texture, roll									
Flat latex, water base (material #5)									
Roll 1st coat									
Slow	300	325	36.00	7.67	1.83	11.08	3.91	3.92	28.41
Medium	350	300	31.50	8.43	2.45	10.50	5.34	3.21	29.93
Fast	400	275	27.00	9.00	3.18	9.82	6.82	2.02	30.84
Roll 2nd coat									
Slow	350	350	36.00	6.57	1.59	10.29	3.50	3.51	25.46
Medium	388	338	31.50	7.60	2.21	9.32	4.78	2.87	26.78
Fast	425	325	27.00	8.47	2.98	8.31	6.13	1.81	27.70
Roll 3rd or additional coats									
Slow	425	350	36.00	5.41	1.29	10.29	3.23	3.24	23.46
Medium	450	338	31.50	6.56	1.88	9.32	4.45	2.67	24.88
Fast	475	325	27.00	7.58	2.70	8.31	5.76	1.70	26.05
Sealer, water base (material #1)									
Roll prime coat									
Slow	325	325	36.00	7.08	1.70	11.08	3.77	3.78	27.41
Medium	375	300	31.50	7.87	2.28	10.50	5.16	3.10	28.91
Fast	425	275	27.00	8.47	2.98	9.82	6.60	1.95	29.82
Sealer, oil base (material #2)									
Roll prime coat									
Slow	325	300	46.20	7.08	1.70	15.40	4.59	4.60	33.37
Medium	375	275	40.50	7.87	2.28	14.73	6.22	3.73	34.83
Fast	425	250	34.70	8.47	2.98	13.88	7.86	2.32	35.51
Enamel, water base (material #9)									
Roll 1st finish coat									
Slow	325	350	53.30	7.08	1.70	15.23	4.56	4.57	33.14
Medium	363	338	46.70	8.13	2.33	13.82	6.08	3.65	34.01
Fast	400	325	40.00	9.00	3.18	12.31	7.59	2.25	34.33
Roll 2nd or additional finish coats									
Slow	400	350	53.30	5.75	1.38	15.23	4.25	4.26	30.87
Medium	425	338	46.70	6.94	2.00	13.82	5.69	3.42	31.87
Fast	450	325	40.00	8.00	2.81	12.31	7.17	2.12	32.41
Enamel, oil base (material #10)									
Roll 1st finish coat									
Slow	325	325	61.50	7.08	1.70	18.92	5.26	5.27	38.23
Medium	363	313	53.80	8.13	2.33	17.19	6.92	4.15	38.72
Fast	400	300	46.10	9.00	3.18	15.37	8.54	2.53	38.62

	Labor SF per manhour	Material coverage SF/gallon	Material cost per gallon	Labor cost per 100 SF	Labor burden 100 SF	Material cost per 100 SF	Overhead per 100 SF	Profit per 100 SF	Total price per 100 SF
Roll 2nd or additional finish coats									
Slow	400	350	61.50	5.75	1.38	17.57	4.69	4.70	34.09
Medium	425	338	53.80	6.94	2.00	15.92	6.22	3.73	34.81
Fast	450	325	46.10	8.00	2.81	14.18	7.75	2.29	35.03
Epoxy coating, white (material #52)									
Roll 1st coat									
Slow	300	350	157.00	7.67	1.83	44.86	10.33	10.35	75.04
Medium	350	325	137.40	8.43	2.45	42.28	13.29	7.97	74.42
Fast	375	300	117.70	9.60	3.41	39.23	16.19	4.79	73.22
Roll 2nd or additional coats									
Slow	375	375	157.00	6.13	1.48	41.87	9.40	9.42	68.30
Medium	400	350	137.40	7.38	2.13	39.26	12.19	7.32	68.28
Fast	425	325	117.70	8.47	2.98	36.22	14.78	4.37	66.82

Ceiling estimates are based on overall dimensions (length times width) to 8 feet high. Do not make deductions for openings in the ceiling area that are under 100 square feet. For heights above 8 feet, use the High Time Difficulty Factors on page 139. ADD for cutting-in at ceilings and protecting adjacent surfaces if ceilings alone are being painted, not walls. Otherwise, figure any cutting-in time with walls. See the notes under the wall formulas for clarification. "Slow" work is based on an hourly wage of $23.00, "Medium" work on an hourly wage of $29.50, and "Fast" work on an hourly wage of $36.00. Other qualifications that apply to this table are on page 9.

	Labor SF per manhour	Material coverage SF/gallon	Material cost per gallon	Labor cost per 100 SF	Labor burden 100 SF	Material cost per 100 SF	Overhead per 100 SF	Profit per 100 SF	Total price per 100 SF
Ceilings, gypsum drywall, sand finish or skip trowel texture, spray									
Flat latex, water base (material #5)									
Spray 1st coat									
Slow	700	275	36.00	3.29	.79	13.09	3.26	3.27	23.70
Medium	800	250	31.50	3.69	1.06	12.60	4.34	2.60	24.29
Fast	900	225	27.00	4.00	1.41	12.00	5.40	1.60	24.41
Spray 2nd coat									
Slow	800	325	36.00	2.88	.69	11.08	2.78	2.79	20.22
Medium	900	300	31.50	3.28	.94	10.50	3.68	2.21	20.61
Fast	1000	275	27.00	3.60	1.27	9.82	4.55	1.35	20.59
Spray 3rd or additional coats									
Slow	850	325	36.00	2.71	.66	11.08	2.74	2.75	19.94
Medium	950	313	31.50	3.11	.88	10.06	3.52	2.11	19.68
Fast	1050	300	27.00	3.43	1.20	9.00	4.23	1.25	19.11
Sealer, water base (material #1)									
Spray prime coat									
Slow	750	275	36.00	3.07	.72	13.09	3.21	3.22	23.31
Medium	850	250	31.50	3.47	1.02	12.60	4.27	2.56	23.92
Fast	950	225	27.00	3.79	1.32	12.00	5.31	1.57	23.99
Sealer, oil base (material #2)									
Spray prime coat									
Slow	750	225	46.20	3.07	.72	20.53	4.62	4.63	33.57
Medium	850	213	40.50	3.47	1.02	19.01	5.87	3.52	32.89
Fast	950	200	34.70	3.79	1.32	17.35	6.97	2.06	31.49
Enamel, water base (material #9)									
Spray 1st finish coat									
Slow	750	325	53.30	3.07	.72	16.40	3.84	3.85	27.88
Medium	850	300	46.70	3.47	1.02	15.57	5.01	3.01	28.08
Fast	950	275	40.00	3.79	1.32	14.55	6.10	1.80	27.56
Spray 2nd or additional finish coat									
Slow	800	325	53.30	2.88	.69	16.40	3.79	3.80	27.56
Medium	900	313	46.70	3.28	.94	14.92	4.79	2.87	26.80
Fast	1000	300	40.00	3.60	1.27	13.33	5.64	1.67	25.51

	Labor SF per manhour	Material coverage SF/gallon	Material cost per gallon	Labor cost per 100 SF	Labor burden 100 SF	Material cost per 100 SF	Overhead per 100 SF	Profit per 100 SF	Total price per 100 SF
Enamel, oil base (material #10)									
Spray 1st finish coat									
Slow	750	300	61.50	3.07	.72	20.50	4.62	4.63	33.54
Medium	850	288	53.80	3.47	1.02	18.68	5.79	3.47	32.43
Fast	950	275	46.10	3.79	1.32	16.76	6.79	2.01	30.67
Spray 2nd or additional finish coat									
Slow	800	325	61.50	2.88	.69	18.92	4.27	4.28	31.04
Medium	900	313	53.80	3.28	.94	17.19	5.36	3.21	29.98
Fast	1000	300	46.10	3.60	1.27	15.37	6.27	1.86	28.37

Ceiling estimates are based on overall dimensions (length times width) to 8 feet high. Do not make deductions for openings in the ceiling area that are under 100 square feet. For heights above 8 feet, use the High Time Difficulty Factors on page 139. ADD for cutting-in at ceilings and protecting adjacent surfaces if ceilings alone are being painted, not walls. Otherwise, figure any cutting-in and protection time with the walls. See the notes under the wall formulas for clarification. "Slow" work is based on an hourly wage of $23.00, "Medium" work on an hourly wage of $29.50, and "Fast" work on an hourly wage of $36.00. Other qualifications that apply to this table are on page 9.

	Labor SF per manhour	Material coverage SF/gallon	Material cost per gallon	Labor cost per 100 SF	Labor burden 100 SF	Material cost per 100 SF	Overhead per 100 SF	Profit per 100 SF	Total price per 100 SF
Ceilings, gypsum drywall, smooth finish, brush									
Flat latex, water base (material #5)									
Brush 1st coat									
Slow	175	325	36.00	13.14	3.14	11.08	5.20	5.21	37.77
Medium	200	313	31.50	14.75	4.27	10.06	7.27	4.36	40.71
Fast	225	300	27.00	16.00	5.63	9.00	9.50	2.81	42.94
Brush 2nd coat									
Slow	225	400	36.00	10.22	2.44	9.00	4.12	4.13	29.91
Medium	250	375	31.50	11.80	3.41	8.40	5.90	3.54	33.05
Fast	275	350	27.00	13.09	4.64	7.71	7.88	2.33	35.65
Brush 3rd or additional coats									
Slow	250	425	36.00	9.20	2.21	8.47	3.78	3.79	27.45
Medium	275	400	31.50	10.73	3.11	7.88	5.43	3.26	30.41
Fast	300	375	27.00	12.00	4.22	7.20	7.27	2.15	32.84
Sealer, water base (material #1)									
Brush prime coat									
Slow	200	325	36.00	11.50	2.76	11.08	4.81	4.82	34.97
Medium	225	313	31.50	13.11	3.78	10.06	6.74	4.04	37.73
Fast	250	300	27.00	14.40	5.08	9.00	8.83	2.61	39.92
Sealer, oil base (material #2)									
Brush prime coat									
Slow	200	350	46.20	11.50	2.76	13.20	5.22	5.23	37.91
Medium	225	338	40.50	13.11	3.78	11.98	7.22	4.33	40.42
Fast	250	325	34.70	14.40	5.08	10.68	9.35	2.77	42.28
Enamel, water base (material #9)									
Brush 1st finish coat									
Slow	200	400	53.30	11.50	2.76	13.33	5.24	5.25	38.08
Medium	225	375	46.70	13.11	3.78	12.45	7.34	4.40	41.08
Fast	250	350	40.00	14.40	5.08	11.43	9.58	2.83	43.32
Brush 2nd and additional finish coats									
Slow	225	425	53.30	10.22	2.44	12.54	4.79	4.80	34.79
Medium	250	400	46.70	11.80	3.41	11.68	6.72	4.03	37.64
Fast	275	375	40.00	13.09	4.64	10.67	8.80	2.60	39.80
Enamel, oil base (material #10)									
Brush 1st finish coat									
Slow	200	400	61.50	11.50	2.76	15.38	5.63	5.64	40.91
Medium	225	388	53.80	13.11	3.78	13.87	7.69	4.62	43.07
Fast	250	375	46.10	14.40	5.08	12.29	9.85	2.91	44.53

	Labor SF per manhour	Material coverage SF/gallon	Material cost per gallon	Labor cost per 100 SF	Labor burden 100 SF	Material cost per 100 SF	Overhead per 100 SF	Profit per 100 SF	Total price per 100 SF
Brush 2nd or additional finish coats									
Slow	225	425	61.50	10.22	2.44	14.47	5.16	5.17	37.46
Medium	250	413	53.80	11.80	3.41	13.03	7.06	4.24	39.54
Fast	275	400	46.10	13.09	4.64	11.53	9.06	2.68	41.00
Epoxy coating, white (material #52)									
Brush 1st coat									
Slow	175	425	157.00	13.14	3.14	36.94	10.11	10.13	73.46
Medium	200	400	137.40	14.75	4.27	34.35	13.34	8.00	74.71
Fast	225	375	117.70	16.00	5.63	31.39	16.44	4.86	74.32
Brush 2nd or additional coats									
Slow	200	450	157.00	11.50	2.76	34.89	9.34	9.36	67.85
Medium	225	425	137.40	13.11	3.78	32.33	12.31	7.38	68.91
Fast	250	400	117.70	14.40	5.08	29.43	15.16	4.48	68.55

Ceiling estimates are based on overall dimensions (length times width) to 8 feet high. Do not make deductions for openings in the ceiling area that are under 100 square feet. For heights above 8 feet, use the High Time Difficulty Factors on page 139. ADD for cutting-in at ceilings and protecting adjacent surfaces if ceilings alone are being painted, not walls. Otherwise, figure any cutting-in time with walls. See the notes under the wall formulas for clarification. "Slow" work is based on an hourly wage of $23.00, "Medium" work on an hourly wage of $29.50, and "Fast" work on an hourly wage of $36.00. Other qualifications that apply to this table are on page 9.

	Labor SF per manhour	Material coverage SF/gallon	Material cost per gallon	Labor cost per 100 SF	Labor burden 100 SF	Material cost per 100 SF	Overhead per 100 SF	Profit per 100 SF	Total price per 100 SF
Ceilings, gypsum drywall, smooth finish, roll									
Flat latex, water base (material #5)									
Roll 1st coat									
Slow	325	350	36.00	7.08	1.70	10.29	3.62	3.63	26.32
Medium	375	325	31.50	7.87	2.28	9.69	4.96	2.97	27.77
Fast	425	300	27.00	8.47	2.98	9.00	6.34	1.88	28.67
Roll 2nd coat									
Slow	375	375	36.00	6.13	1.48	9.60	3.27	3.28	23.76
Medium	413	363	31.50	7.14	2.06	8.68	4.47	2.68	25.03
Fast	450	350	27.00	8.00	2.81	7.71	5.74	1.70	25.96
Roll 3rd or additional coats									
Slow	425	400	36.00	5.41	1.29	9.00	2.98	2.99	21.67
Medium	450	388	31.50	6.56	1.88	8.12	4.15	2.49	23.20
Fast	475	375	27.00	7.58	2.70	7.20	5.41	1.60	24.49
Sealer, water base (material #1)									
Roll prime coat									
Slow	350	350	36.00	6.57	1.59	10.29	3.50	3.51	25.46
Medium	400	325	31.50	7.38	2.13	9.69	4.80	2.88	26.88
Fast	450	300	27.00	8.00	2.81	9.00	6.14	1.82	27.77
Sealer, oil base (material #2)									
Roll prime coat									
Slow	350	300	46.20	6.57	1.59	15.40	4.47	4.48	32.51
Medium	400	288	40.50	7.38	2.13	14.06	5.89	3.54	33.00
Fast	450	275	34.70	8.00	2.81	12.62	7.27	2.15	32.85
Enamel, water base, (material #9)									
Roll 1st finish coat									
Slow	350	375	53.30	6.57	1.59	14.21	4.25	4.26	30.88
Medium	400	363	46.70	7.38	2.13	12.87	5.60	3.36	31.34
Fast	450	350	40.00	8.00	2.81	11.43	6.90	2.04	31.18
Roll 2nd or additional finish coats									
Slow	425	400	53.30	5.41	1.29	13.33	3.81	3.82	27.66
Medium	450	388	46.70	6.56	1.88	12.04	5.13	3.08	28.69
Fast	475	375	40.00	7.58	2.70	10.67	6.49	1.92	29.36
Enamel, oil base (material #10)									
Roll 1st finish coat									
Slow	350	350	61.50	6.57	1.59	17.57	4.89	4.90	35.52
Medium	400	338	53.80	7.38	2.13	15.92	6.36	3.81	35.60
Fast	450	325	46.10	8.00	2.81	14.18	7.75	2.29	35.03

	Labor SF per manhour	Material coverage SF/gallon	Material cost per gallon	Labor cost per 100 SF	Labor burden 100 SF	Material cost per 100 SF	Overhead per 100 SF	Profit per 100 SF	Total price per 100 SF
Roll 2nd or additional finish coats									
Slow	425	375	61.50	5.41	1.29	16.40	4.39	4.40	31.89
Medium	450	363	53.80	6.56	1.88	14.82	5.82	3.49	32.57
Fast	475	350	46.10	7.58	2.70	13.17	7.26	2.15	32.86
Epoxy coating, white (material #52)									
Roll 1st finish coat									
Slow	325	400	157.00	7.08	1.70	39.25	9.13	9.15	66.31
Medium	363	375	137.40	8.13	2.33	36.64	11.78	7.07	65.95
Fast	400	350	117.70	9.00	3.18	33.63	14.20	4.20	64.21
Roll 2nd or additional finish coats									
Slow	400	425	157.00	5.75	1.38	36.94	8.37	8.39	60.83
Medium	425	400	137.40	6.94	2.00	34.35	10.83	6.50	60.62
Fast	450	375	117.70	8.00	2.81	31.39	13.09	3.87	59.16
Stipple finish									
Slow	200	--	--	11.50	2.76	--	2.71	2.72	19.69
Medium	225	--	--	13.11	3.78	--	4.23	2.54	23.66
Fast	250	--	--	14.40	5.08	--	6.04	1.79	27.31

Ceiling estimates are based on overall dimensions (length times width) to 8 feet high. Do not make deductions for openings in the ceiling area that are under 100 square feet. For heights above 8 feet, use the High Time Difficulty Factors on page 139. ADD for cutting-in at ceilings and protecting adjacent surfaces if ceilings alone are being painted, not walls. Otherwise, figure any cutting-in time with walls. See the notes under the wall formulas for clarification. "Slow" work is based on an hourly wage of $23.00, "Medium" work on an hourly wage of $29.50, and "Fast" work on an hourly wage of $36.00. Other qualifications that apply to this table are on page 9.

	Labor SF per manhour	Material coverage SF/gallon	Material cost per gallon	Labor cost per 100 SF	Labor burden 100 SF	Material cost per 100 SF	Overhead per 100 SF	Profit per 100 SF	Total price per 100 SF
Ceilings, gypsum drywall, smooth finish, spray									
Flat latex, water base (material #5)									
Spray 1st coat									
Slow	750	300	36.00	3.07	.72	12.00	3.00	3.01	21.80
Medium	850	275	31.50	3.47	1.02	11.45	3.98	2.39	22.31
Fast	950	250	27.00	3.79	1.32	10.80	4.94	1.46	22.31
Spray 2nd coat									
Slow	850	350	36.00	2.71	.66	10.29	2.59	2.60	18.85
Medium	950	325	31.50	3.11	.88	9.69	3.43	2.06	19.17
Fast	1050	300	27.00	3.43	1.20	9.00	4.23	1.25	19.11
Spray 3rd or additional coats									
Slow	900	350	36.00	2.56	.61	10.29	2.56	2.56	18.58
Medium	1000	338	31.50	2.95	.85	9.32	3.28	1.97	18.37
Fast	1100	325	27.00	3.27	1.16	8.31	3.95	1.17	17.86
Sealer, water base (material #1)									
Spray prime coat									
Slow	800	300	36.00	2.88	.69	12.00	2.96	2.96	21.49
Medium	900	275	31.50	3.28	.94	11.45	3.92	2.35	21.94
Fast	1000	250	27.00	3.60	1.27	10.80	4.86	1.44	21.97
Sealer, oil base (material #2)									
Spray prime coat									
Slow	800	250	46.20	2.88	.69	18.48	4.19	4.20	30.44
Medium	900	238	40.50	3.28	.94	17.02	5.31	3.19	29.74
Fast	1000	225	34.70	3.60	1.27	15.42	6.29	1.86	28.44
Enamel, water base (material #9)									
Spray 1st finish coat									
Slow	800	350	53.30	2.88	.69	15.23	3.57	3.58	25.95
Medium	900	325	46.70	3.28	.94	14.37	4.65	2.79	26.03
Fast	1000	300	40.00	3.60	1.27	13.33	5.64	1.67	25.51
Spray 2nd or additional finish coats									
Slow	850	350	53.30	2.71	.66	15.23	3.53	3.54	25.67
Medium	950	338	46.70	3.11	.88	13.82	4.46	2.67	24.94
Fast	1050	325	40.00	3.43	1.20	12.31	5.25	1.55	23.74
Enamel, oil base (material #10)									
Spray 1st finish coat									
Slow	800	300	61.50	2.88	.69	20.50	4.57	4.58	33.22
Medium	900	280	53.80	3.28	.94	19.21	5.86	3.52	32.81
Fast	1000	260	46.10	3.60	1.27	17.73	7.01	2.07	31.68
Spray 2nd or additional finish coats									
Slow	850	325	61.50	2.71	.66	18.92	4.23	4.24	30.76
Medium	950	313	53.80	3.11	.88	17.19	5.30	3.18	29.66
Fast	1050	300	46.10	3.43	1.20	15.37	6.20	1.83	28.03

Ceiling estimates are based on overall dimensions (length times width) to 8 feet high. Do not make deductions for openings in the ceiling area that are under 100 square feet. For heights above 8 feet, use the High Time Difficulty Factors on page 139. ADD for cutting-in at ceilings and protecting adjacent surfaces if ceilings alone are being painted, not walls. Otherwise, figure any cutting-in and protection time with the walls. "Slow" work is based on an hourly wage of $23.00, "Medium" work on an hourly wage of $29.50, and "Fast" work on an hourly wage of $36.00. Other qualifications that apply to this table are on page 9.

	Labor SF per manhour	Material coverage SF/gallon	Material cost per gallon	Labor cost per 100 SF	Labor burden 100 SF	Material cost per 100 SF	Overhead per 100 SF	Profit per 100 SF	Total price per 100 SF
Ceilings, tongue & groove, paint grade, brush									
Flat latex, water base (material #5)									
Brush 1st coat									
Slow	55	300	36.00	41.82	10.03	12.00	12.13	12.16	88.14
Medium	65	288	31.50	45.38	13.11	10.94	17.36	10.41	97.20
Fast	75	275	27.00	48.00	16.93	9.82	23.18	6.86	104.79
Brush 2nd coat									
Slow	65	350	36.00	35.38	8.48	10.29	10.29	10.31	74.75
Medium	75	338	31.50	39.33	11.36	9.32	15.01	9.00	84.02
Fast	85	325	27.00	42.35	14.93	8.31	20.34	6.02	91.95
Brush 3rd or additional coats									
Slow	80	375	36.00	28.75	6.90	9.60	8.60	8.62	62.47
Medium	90	363	31.50	32.78	9.47	8.68	12.73	7.64	71.30
Fast	100	350	27.00	36.00	12.71	7.71	17.49	5.17	79.08
Sealer, water base (material #1)									
Brush prime coat									
Slow	60	300	36.00	38.33	9.21	12.00	11.31	11.33	82.18
Medium	70	288	31.50	42.14	12.20	10.94	16.32	9.79	91.39
Fast	80	275	27.00	45.00	15.89	9.82	21.92	6.48	99.11
Sealer, oil base (material #2)									
Brush prime coat									
Slow	60	350	46.20	38.33	9.21	13.20	11.54	11.56	83.84
Medium	70	338	40.50	42.14	12.20	11.98	16.58	9.95	92.85
Fast	80	325	34.70	45.00	15.89	10.68	22.19	6.56	100.32
Enamel, water base (material #9)									
Brush 1st finish coat									
Slow	65	300	53.30	35.38	8.48	17.77	11.71	11.74	85.08
Medium	75	288	46.70	39.33	11.36	16.22	16.73	10.04	93.68
Fast	85	275	40.00	42.35	14.93	14.55	22.27	6.59	100.69
Brush 2nd or additional finish coats									
Slow	80	375	53.30	28.75	6.90	14.21	9.47	9.49	68.82
Medium	90	363	46.70	32.78	9.47	12.87	13.78	8.27	77.17
Fast	100	350	40.00	36.00	12.71	11.43	18.64	5.51	84.29

	Labor SF per manhour	Material coverage SF/gallon	Material cost per gallon	Labor cost per 100 SF	Labor burden 100 SF	Material cost per 100 SF	Overhead per 100 SF	Profit per 100 SF	Total price per 100 SF
Enamel, oil base (material #10)									
Brush 1st finish coat									
Slow	55	375	61.50	41.82	10.03	16.40	12.97	13.00	94.22
Medium	65	363	53.80	45.38	13.11	14.82	18.33	11.00	102.64
Fast	75	350	46.10	48.00	16.93	13.17	24.21	7.16	109.47
Brush 2nd or additional finish coats									
Slow	70	425	61.50	32.86	7.90	14.47	10.49	10.51	76.23
Medium	80	413	53.80	36.88	10.66	13.03	15.14	9.09	84.80
Fast	90	400	46.10	40.00	14.12	11.53	20.35	6.02	92.02

Ceiling estimates are based on overall dimensions (length times width) to 8 feet high. Do not make deductions for openings in the ceiling area that are under 100 square feet. Figure painting of wood ceilings separate from wood beams. For heights above 8 feet, use the High Time Difficulty Factors on page 139. ADD for protecting adjacent surfaces if ceilings alone are painted, not the walls. Otherwise, figure any cutting-in and protection time with the walls. "Slow" work is based on an hourly wage of $23.00, "Medium" work on an hourly wage of $29.50, and "Fast" work on an hourly wage of $36.00. Other qualifications that apply to this table are on page 9.

	Labor SF per manhour	Material coverage SF/gallon	Material cost per gallon	Labor cost per 100 SF	Labor burden 100 SF	Material cost per 100 SF	Overhead per 100 SF	Profit per 100 SF	Total price per 100 SF
Ceilings, tongue & groove, paint grade, roll and brush									
Flat latex, water base (material #5)									
Roll 1st coat									
Slow	110	275	36.00	20.91	5.01	13.09	7.41	7.43	53.85
Medium	130	263	31.50	22.69	6.56	11.98	10.31	6.18	57.72
Fast	150	250	27.00	24.00	8.49	10.80	13.41	3.97	60.67
Roll 2nd coat									
Slow	140	325	36.00	16.43	3.93	11.08	5.98	5.99	43.41
Medium	155	313	31.50	19.03	5.50	10.06	8.65	5.19	48.43
Fast	170	300	27.00	21.18	7.46	9.00	11.67	3.45	52.76
Roll 3rd or additional coats									
Slow	190	350	36.00	12.11	2.89	10.29	4.81	4.82	34.92
Medium	200	338	31.50	14.75	4.27	9.32	7.08	4.25	39.67
Fast	210	325	27.00	17.14	6.05	8.31	9.77	2.89	44.16
Sealer, water base (material #1)									
Roll prime coat									
Slow	100	275	36.00	23.00	5.52	13.09	7.91	7.92	57.44
Medium	120	263	31.50	24.58	7.10	11.98	10.92	6.55	61.13
Fast	150	250	27.00	24.00	8.49	10.80	13.41	3.97	60.67
Sealer, oil base (material #2)									
Roll prime coat									
Slow	100	350	46.20	23.00	5.52	13.20	7.93	7.94	57.59
Medium	120	325	40.50	24.58	7.10	12.46	11.04	6.62	61.80
Fast	150	300	34.70	24.00	8.49	11.57	13.65	4.04	61.75
Enamel, water base (material #9)									
Roll 1st finish coat									
Slow	130	325	53.30	17.69	4.24	16.40	7.28	7.30	52.91
Medium	145	313	46.70	20.34	5.90	14.92	10.29	6.17	57.62
Fast	160	300	40.00	22.50	7.94	13.33	13.57	4.01	61.35
Roll 2nd or additional finish coats									
Slow	180	350	53.30	12.78	3.08	15.23	5.91	5.92	42.92
Medium	190	338	46.70	15.53	4.47	13.82	8.46	5.08	47.36
Fast	200	325	40.00	18.00	6.36	12.31	11.36	3.36	51.39

	Labor SF per manhour	Material coverage SF/gallon	Material cost per gallon	Labor cost per 100 SF	Labor burden 100 SF	Material cost per 100 SF	Overhead per 100 SF	Profit per 100 SF	Total price per 100 SF
Enamel, oil base (material #10)									
Roll 1st finish coat									
Slow	130	375	61.50	17.69	4.24	16.40	7.28	7.30	52.91
Medium	145	363	53.80	20.34	5.90	14.82	10.26	6.16	57.48
Fast	160	350	46.10	22.50	7.94	13.17	13.52	4.00	61.13
Roll 2nd or additional finish coats									
Slow	180	400	61.50	12.78	3.08	15.38	5.93	5.95	43.12
Medium	190	388	53.80	15.53	4.47	13.87	8.47	5.08	47.42
Fast	200	375	46.10	18.00	6.36	12.29	11.36	3.36	51.37

Ceiling estimates are based on overall dimensions (length times width) to 8 feet high. Do not make deductions for openings in the ceiling area that are under 100 square feet. Figure painting of wood ceilings separate from wood beams. For heights above 8 feet, use the High Time Difficulty Factors on page 139. ADD for protecting adjacent surfaces if ceilings alone are painted, not the walls. Otherwise, figure any cutting-in and protection time with the walls. "Slow" work is based on an hourly wage of $23.00, "Medium" work on an hourly wage of $29.50, and "Fast" work on an hourly wage of $36.00. Other qualifications that apply to this table are on page 9.

	Labor SF per manhour	Material coverage SF/gallon	Material cost per gallon	Labor cost per 100 SF	Labor burden 100 SF	Material cost per 100 SF	Overhead per 100 SF	Profit per 100 SF	Total price per 100 SF
Ceilings, tongue & groove, paint grade, spray									
Flat latex, water base (material #5)									
Spray 1st coat									
Slow	300	180	36.00	7.67	1.83	20.00	5.61	5.62	40.73
Medium	360	155	31.50	8.19	2.38	20.32	7.72	4.63	43.24
Fast	420	125	27.00	8.57	3.02	21.60	10.29	3.04	46.52
Spray 2nd coat									
Slow	420	250	36.00	5.48	1.31	14.40	4.03	4.04	29.26
Medium	470	225	31.50	6.28	1.82	14.00	5.52	3.31	30.93
Fast	520	200	27.00	6.92	2.43	13.50	7.09	2.10	32.04
Spray 3rd or additional coats									
Slow	520	325	36.00	4.42	1.06	11.08	3.15	3.15	22.86
Medium	570	300	31.50	5.18	1.48	10.50	4.30	2.58	24.04
Fast	620	275	27.00	5.81	2.03	9.82	5.48	1.62	24.76
Sealer, water base (material #1)									
Spray prime coat									
Slow	320	180	36.00	7.19	1.74	20.00	5.49	5.51	39.93
Medium	380	155	31.50	7.76	2.24	20.32	7.58	4.55	42.45
Fast	440	125	27.00	8.18	2.88	21.60	10.13	3.00	45.79
Sealer, oil base (material #2)									
Spray prime coat									
Slow	320	200	46.20	7.19	1.74	23.10	6.08	6.10	44.21
Medium	380	190	40.50	7.76	2.24	21.32	7.83	4.70	43.85
Fast	440	180	34.70	8.18	2.88	19.28	9.41	2.78	42.53
Enamel, water base (material #9)									
Spray 1st finish coat									
Slow	400	250	53.30	5.75	1.38	21.32	5.41	5.42	39.28
Medium	450	225	46.70	6.56	1.88	20.76	7.31	4.38	40.89
Fast	500	200	40.00	7.20	2.54	20.00	9.22	2.73	41.69
Spray 2nd or additional finish coat									
Slow	500	325	53.30	4.60	1.10	16.40	4.20	4.21	30.51
Medium	550	300	46.70	5.36	1.56	15.57	5.62	3.37	31.48
Fast	600	275	40.00	6.00	2.13	14.55	7.03	2.08	31.79

	Labor SF per manhour	Material coverage SF/gallon	Material cost per gallon	Labor cost per 100 SF	Labor burden 100 SF	Material cost per 100 SF	Overhead per 100 SF	Profit per 100 SF	Total price per 100 SF
Enamel, oil base (material #10)									
Spray 1st finish coat									
Slow	400	270	61.50	5.75	1.38	22.78	5.68	5.69	41.28
Medium	450	250	53.80	6.56	1.88	21.52	7.50	4.50	41.96
Fast	500	230	46.10	7.20	2.54	20.04	9.23	2.73	41.74
Spray 2nd or additional finish coat									
Slow	500	325	61.50	4.60	1.10	18.92	4.68	4.69	33.99
Medium	550	313	53.80	5.36	1.56	17.19	6.03	3.62	33.76
Fast	600	300	46.10	6.00	2.13	15.37	7.28	2.15	32.93

Ceiling estimates are based on overall dimensions (length times width) to 8 feet high. Do not make deductions for openings in the ceiling area that are under 100 square feet. Figure painting of wood ceilings separate from wood beams. For heights above 8 feet, use the High Time Difficulty Factors on page 139. ADD for masking-off or cutting-in at wall-to-ceiling intersections and for protecting adjacent surfaces as necessary. "Slow" work is based on an hourly wage of $23.00, "Medium" work on an hourly wage of $29.50, and "Fast" work on an hourly wage of $36.00. Other qualifications that apply to this table are on page 9.

	Labor SF per manhour	Material coverage SF/gallon	Material cost per gallon	Labor cost per 100 SF	Labor burden 100 SF	Material cost per 100 SF	Overhead per 100 SF	Profit per 100 SF	Total price per 100 SF
Ceilings, tongue & groove, stain grade, roll and brush									
Semi-transparent stain, water base (material #20)									
Roll & brush each coat									
Slow	200	300	49.90	11.50	2.76	16.63	5.87	5.88	42.64
Medium	240	275	43.70	12.29	3.57	15.89	7.93	4.76	44.44
Fast	280	250	37.40	12.86	4.53	14.96	10.03	2.97	45.35
Semi-transparent stain, oil base (material #21)									
Roll & brush each coat									
Slow	200	280	58.30	11.50	2.76	20.82	6.67	6.68	48.43
Medium	240	260	51.00	12.29	3.57	19.62	8.87	5.32	49.67
Fast	280	240	43.70	12.86	4.53	18.21	11.04	3.27	49.91
Ceilings, tongue & groove, stain grade, spray application									
Semi-transparent stain, water base (material #20)									
Spray each coat									
Slow	300	220	49.90	7.67	1.83	22.68	6.12	6.13	44.43
Medium	350	200	43.70	8.43	2.45	21.85	8.18	4.91	45.82
Fast	400	180	37.40	9.00	3.18	20.78	10.22	3.02	46.20
Semi-transparent stain, oil base (material #21)									
Spray each coat									
Slow	300	200	58.30	7.67	1.83	29.15	7.35	7.36	53.36
Medium	350	188	51.00	8.43	2.45	27.13	9.50	5.70	53.21
Fast	400	175	43.70	9.00	3.18	24.97	11.52	3.41	52.08
Stain, seal and 2 coat lacquer system (7 step process)									
STEP 1: Sand & putty									
Slow	100	--	--	23.00	5.52	--	5.42	5.43	39.37
Medium	125	--	--	23.60	6.82	--	7.61	4.56	42.59
Fast	150	--	--	24.00	8.49	--	10.07	2.98	45.54
STEP 2 & 3: Wiping stain, oil base (material #11a) & wipe									
Roll & brush, 1 coat & wipe									
Slow	75	300	61.60	30.67	7.35	20.53	11.13	11.15	80.83
Medium	100	275	53.90	29.50	8.53	19.60	14.41	8.64	80.68
Fast	125	250	46.20	28.80	10.17	18.48	17.81	5.27	80.53
Spray, 1 coat & wipe									
Slow	275	150	61.60	8.36	2.02	41.07	9.77	9.79	71.01
Medium	200	125	53.90	14.75	4.27	43.12	15.53	9.32	86.99
Fast	325	100	46.20	11.08	3.92	46.20	18.97	5.61	85.78

	Labor SF per manhour	Material coverage SF/gallon	Material cost per gallon	Labor cost per 100 SF	Labor burden 100 SF	Material cost per 100 SF	Overhead per 100 SF	Profit per 100 SF	Total price per 100 SF
STEP 4: Sanding sealer (material #11b)									
Brush, 1 coat									
Slow	125	325	50.90	18.40	4.42	15.66	7.31	7.33	53.12
Medium	150	300	44.60	19.67	5.70	14.87	10.06	6.03	56.33
Fast	175	275	38.20	20.57	7.24	13.89	12.93	3.83	58.46
Spray, 1 coat									
Slow	350	150	50.90	6.57	1.59	33.93	8.00	8.01	58.10
Medium	400	125	44.60	7.38	2.13	35.68	11.30	6.78	63.27
Fast	450	100	38.20	8.00	2.81	38.20	15.20	4.50	68.71
STEP 5: Sand lightly									
Slow	175	--	--	13.14	3.14	--	3.10	3.10	22.48
Medium	225	--	--	13.11	3.78	--	4.23	2.54	23.66
Fast	275	--	--	13.09	4.64	--	5.49	1.62	24.84
STEP 6 & 7: Lacquer, 2 coats (material #11c)									
Brush, 1st coat									
Slow	150	350	53.50	15.33	3.69	15.29	6.52	6.53	47.36
Medium	200	338	46.80	14.75	4.27	13.85	8.22	4.93	46.02
Fast	275	325	40.10	13.09	4.64	12.34	9.32	2.76	42.15
Brush, 2nd coat									
Slow	200	400	53.50	11.50	2.76	13.38	5.25	5.26	38.15
Medium	250	375	46.80	11.80	3.41	12.48	6.92	4.15	38.76
Fast	325	350	40.10	11.08	3.92	11.46	8.20	2.43	37.09
Spray, 1st coat									
Slow	425	275	53.50	5.41	1.29	19.45	4.97	4.98	36.10
Medium	525	250	46.80	5.62	1.61	18.72	6.49	3.89	36.33
Fast	625	225	40.10	5.76	2.03	17.82	7.94	2.35	35.90
Spray, 2nd coat									
Slow	475	300	53.50	4.84	1.18	17.83	4.53	4.54	32.92
Medium	588	275	46.80	5.02	1.45	17.02	5.87	3.52	32.88
Fast	650	250	40.10	5.54	1.96	16.04	7.30	2.16	33.00

	Labor SF per manhour	Material coverage SF/gallon	Material cost per gallon	Labor cost per 100 SF	Labor burden 100 SF	Material cost per 100 SF	Overhead per 100 SF	Profit per 100 SF	Total price per 100 SF
Complete 7 step stain, seal & 2 coat lacquer system (material #11)									
Brush all coats									
Slow	30	150	54.90	76.67	18.39	36.60	25.02	25.07	181.75
Medium	35	138	48.00	84.29	24.36	34.78	35.86	21.51	200.80
Fast	40	125	41.20	90.00	31.78	32.96	47.97	14.19	216.90
Spray all coats									
Slow	60	50	54.90	38.33	9.21	109.80	29.89	29.96	217.19
Medium	70	40	48.00	42.14	12.20	120.00	43.58	26.15	244.07
Fast	80	30	41.20	45.00	15.89	137.33	61.45	18.18	277.85

Ceiling estimates are based on overall dimensions (length times width) to 8 feet high. Do not make deductions for openings in the ceiling area that are under 100 square feet. Figure painting of wood ceilings separate from wood beams. For heights above 8 feet, use the High Time Difficulty Factors on page 139. ADD for masking-off or cutting-in at wall-to-ceiling intersections and for protecting adjacent surfaces as necessary. "Slow" work is based on an hourly wage of $23.00, "Medium" work on an hourly wage of $29.50, and "Fast" work on an hourly wage of $36.00. Other qualifications that apply to this table are on page 9.

	Labor SF per manhour	Material coverage SF/gallon	Material cost per gallon	Labor cost per 100 SF	Labor burden 100 SF	Material cost per 100 SF	Overhead per 100 SF	Profit per 100 SF	Total price per 100 SF
Closet pole, stain grade									
Penetrating oil stain, (material #13)									
Brush & wipe, 1 coat									
Slow	40	225	57.90	57.50	13.80	25.73	18.44	18.48	133.95
Medium	50	213	50.70	59.00	17.06	23.80	24.96	14.98	139.80
Fast	60	200	43.40	60.00	21.20	21.70	31.89	9.43	144.22

To stain poles in new construction, apply stain before installation. On repaints, remove the pole before staining. When estimating by the Opening Count Method, count one opening for each 10 linear feet of pole. "Slow" work is based on an hourly wage of $23.00, "Medium" work on an hourly wage of $29.50, and "Fast" work on an hourly wage of $36.00. Other qualifications that apply to this table are on page 9.

	Labor LF per manhour	Material coverage LF/gallon	Material cost per gallon	Labor cost per 100 LF	Labor burden 100 LF	Material cost per 100 LF	Overhead per 100 LF	Profit per 100 LF	Total price per 100 LF
Closet shelf & pole, paint grade									
Undercoat, water base (material #3)									
Brush 1 coat									
Slow	17	80	36.90	135.29	32.46	46.13	40.64	40.72	295.24
Medium	22	70	32.30	134.09	38.76	46.14	54.75	32.85	306.59
Fast	33	60	27.70	109.09	38.50	46.17	60.07	17.77	271.60
Undercoat, oil base (material #4)									
Brush 1 coat									
Slow	17	90	48.30	135.29	32.46	53.67	42.07	42.16	305.65
Medium	22	80	42.30	134.09	38.76	52.88	56.43	33.86	316.02
Fast	33	65	36.20	109.09	38.50	55.69	63.02	18.64	284.94
Split coat (1/2 undercoat + 1/2 enamel), water base (material #3 & #9)									
Brush 1 coat									
Slow	16	80	45.10	143.75	34.50	56.38	44.58	44.67	323.88
Medium	21	70	39.50	140.48	40.62	56.43	59.38	35.63	332.54
Fast	32	60	33.85	112.50	39.72	56.42	64.68	19.13	292.45
Split coat (1/2 undercoat + 1/2 enamel), oil base (material #4 & #10)									
Brush 1 coat									
Slow	16	90	54.90	143.75	34.50	61.00	45.46	45.55	330.26
Medium	21	80	48.05	140.48	40.62	60.06	60.29	36.17	337.62
Fast	32	65	41.15	112.50	39.72	63.31	66.81	19.76	302.10
Enamel, water base (material #9)									
Brush each coat									
Slow	15	80	53.30	153.33	36.81	66.63	48.78	48.89	354.44
Medium	20	70	46.70	147.50	42.65	66.71	64.21	38.53	359.60
Fast	30	60	40.00	120.00	42.35	66.67	71.00	21.00	321.02
Enamel, oil base (material #10)									
Brush each coat									
Slow	15	90	61.50	153.33	36.81	68.33	49.11	49.21	356.79
Medium	20	80	53.80	147.50	42.65	67.25	64.35	38.61	360.36
Fast	30	65	46.10	120.00	42.35	70.92	72.32	21.39	326.98

Use these costs for painting the wardrobe closet shelves and poles with an undercoat and enamel system. If painting wardrobe closet shelves and poles with flat latex paint along with walls, use the Opening Count Method described under Doors, interior openings. Measurements are based on linear feet (LF) of shelves and poles. "Slow" work is based on an hourly wage of $23.00, "Medium" work on an hourly wage of $29.50, and "Fast" work on an hourly wage of $36.00. Other qualifications that apply to this table are on page 9.

	Labor LF per manhour	Material coverage LF/gallon	Material cost per gallon	Labor cost per 100 LF	Labor burden 100 LF	Material cost per 100 LF	Overhead per 100 LF	Profit per 100 LF	Total price per 100 LF
Closet shelves, paint grade									
Undercoat, water base (material #3)									
Brush 1 coat									
Slow	30	100	36.90	76.67	18.39	36.90	25.07	25.13	182.16
Medium	37	90	32.30	79.73	23.07	35.89	34.67	20.80	194.16
Fast	44	80	27.70	81.82	28.90	34.63	45.05	13.33	203.73
Undercoat, oil base (material #4)									
Brush 1 coat									
Slow	30	110	48.30	76.67	18.39	43.91	26.41	26.46	191.84
Medium	37	100	42.30	79.73	23.07	42.30	36.27	21.76	203.13
Fast	44	90	36.20	81.82	28.90	40.22	46.79	13.84	211.57
Split coat (1/2 undercoat + 1/2 enamel), water base (material #3 & #9)									
Brush 1 coat									
Slow	27	100	45.10	85.19	20.45	45.10	28.64	28.70	208.08
Medium	35	90	39.50	84.29	24.36	43.89	38.14	22.88	213.56
Fast	42	80	33.85	85.71	30.27	42.31	49.07	14.51	221.87
Split coat (1/2 undercoat + 1/2 enamel), oil base (material #4 & #10)									
Brush 1 coat									
Slow	27	110	54.90	85.19	20.45	49.91	29.55	29.62	214.72
Medium	35	100	48.05	84.29	24.36	48.05	39.18	23.51	219.39
Fast	42	90	41.15	85.71	30.27	45.72	50.12	14.83	226.65
Enamel, water base (material #9)									
Brush each coat									
Slow	25	100	53.30	92.00	22.08	53.30	31.80	31.87	231.05
Medium	33	90	46.70	89.39	25.84	51.89	41.78	25.07	233.97
Fast	40	80	40.00	90.00	31.78	50.00	53.25	15.75	240.78
Enamel, oil base (material #10)									
Brush each coat									
Slow	25	110	61.50	92.00	22.08	55.91	32.30	32.37	234.66
Medium	33	100	53.80	89.39	25.84	53.80	42.26	25.35	236.64
Fast	40	90	46.10	90.00	31.78	51.22	53.63	15.86	242.49

Use these costs for painting the wardrobe closet shelves with an undercoat and enamel system. If painting wardrobe closet shelves with flat latex paint along with walls, use the Opening Count Method described under Doors, interior openings. Measurements are based on linear feet (LF) of shelves. "Slow" work is based on an hourly wage of $23.00, "Medium" work on an hourly wage of $29.50, and "Fast" work on an hourly wage of $36.00. Other qualifications that apply to this table are on page 9.

	Labor LF per manhour	Material coverage LF/gallon	Material cost per gallon	Labor cost per 100 LF	Labor burden 100 LF	Material cost per 100 LF	Overhead per 100 LF	Profit per 100 LF	Total price per 100 LF
Closets, molding at perimeter, paint grade									
Undercoat, water base (material #3)									
Brush 1 coat									
Slow	80	225	36.90	28.75	6.90	16.40	9.89	9.91	71.85
Medium	95	213	32.30	31.05	9.00	15.16	13.80	8.28	77.29
Fast	110	200	27.70	32.73	11.55	13.85	18.02	5.33	81.48
Undercoat, oil base (material #4)									
Brush 1 coat									
Slow	80	250	48.30	28.75	6.90	19.32	10.44	10.47	75.88
Medium	95	238	42.30	31.05	9.00	17.77	14.45	8.67	80.94
Fast	110	220	36.20	32.73	11.55	16.45	18.83	5.57	85.13
Split coat (1/2 undercoat + 1/2 enamel), water base (material #3 & #9)									
Brush 1 coat									
Slow	75	225	45.10	30.67	7.35	20.04	11.03	11.06	80.15
Medium	90	213	39.50	32.78	9.47	18.54	15.20	9.12	85.11
Fast	105	200	33.85	34.29	12.08	16.93	19.63	5.81	88.74
Split coat (1/2 undercoat + 1/2 enamel), oil base (material #4 & #10)									
Brush 1 coat									
Slow	75	250	54.90	30.67	7.35	21.96	11.40	11.42	82.80
Medium	90	238	48.05	32.78	9.47	20.19	15.61	9.37	87.42
Fast	105	220	41.15	34.29	12.08	18.70	20.18	5.97	91.22
Enamel, water base (material #9)									
Brush each coat									
Slow	70	225	53.30	32.86	7.90	23.69	12.24	12.27	88.96
Medium	85	213	46.70	34.71	10.01	21.92	16.67	10.00	93.31
Fast	100	200	40.00	36.00	12.71	20.00	21.30	6.30	96.31
Enamel, oil base (material #10)									
Brush each coat									
Slow	70	250	61.50	32.86	7.90	24.60	12.42	12.44	90.22
Medium	85	238	53.80	34.71	10.01	22.61	16.84	10.10	94.27
Fast	100	220	46.10	36.00	12.71	20.95	21.59	6.39	97.64

Use these costs for molding around wardrobe closets. Measurements are based on linear feet (LF) of molding. "Slow" work is based on an hourly wage of $23.00, "Medium" work on an hourly wage of $29.50, and "Fast" work on an hourly wage of $36.00. Other qualifications that apply to this table are on page 9.

	Labor LF per manhour	Material coverage LF/gallon	Material cost per gallon	Labor cost per 100 LF	Labor burden 100 LF	Material cost per 100 LF	Overhead per 100 LF	Profit per 100 LF	Total price per 100 LF
Corbels, wood trim, stain grade, average size 4" x 8"									
Solid body stain, water base (material #18)									
Brush each coat									
Slow	15	50	48.90	153.33	36.81	97.80	54.71	54.82	397.47
Medium	20	48	42.80	147.50	42.65	89.17	69.83	41.90	391.05
Fast	25	45	36.70	144.00	50.84	81.56	85.68	25.34	387.42
Solid body stain, oil base (material #19)									
Brush each coat									
Slow	15	55	60.90	153.33	36.81	110.73	57.16	57.28	415.31
Medium	20	53	53.30	147.50	42.65	100.57	72.68	43.61	407.01
Fast	25	50	45.70	144.00	50.84	91.40	88.73	26.25	401.22
Semi-transparent stain, water base (material #20)									
Brush each coat									
Slow	18	55	49.90	127.78	30.68	90.73	47.34	47.44	343.97
Medium	22	53	43.70	134.09	38.76	82.45	63.82	38.29	357.41
Fast	28	50	37.40	128.57	45.37	74.80	77.12	22.81	348.67
Semi-transparent stain, oil base (material #21)									
Brush each coat									
Slow	18	60	58.30	127.78	30.68	97.17	48.57	48.67	352.87
Medium	22	58	51.00	134.09	38.76	87.93	65.19	39.12	365.09
Fast	28	55	43.70	128.57	45.37	79.45	78.56	23.24	355.19

Use these costs for painting corbels averaging 4" x 8" in size. Measurements are based on linear feet (LF) of corbels that are painted or stained with a different material or color than the surface they extend from. "Slow" work is based on an hourly wage of $23.00, "Medium" work on an hourly wage of $29.50, and "Fast" work on an hourly wage of $36.00. Other qualifications that apply to this table are on page 9.

	Labor LF per manhour	Material coverage LF/gallon	Material cost per gallon	Labor cost per 100 LF	Labor burden 100 LF	Material cost per 100 LF	Overhead per 100 LF	Profit per 100 LF	Total price per 100 LF
Cutting-in, horizontal, interior or exterior									
Cutting-in, horizontal, heights less than 6'8"									
Brush application									
Slow	23	--	--	100.00	24.00	--	23.56	23.61	171.17
Medium	35	--	--	85.51	24.74	--	27.56	16.53	154.34
Fast	56	--	--	63.77	22.50	--	26.75	7.91	120.93
Cutting-in, horizontal, heights from 6'8" to 9'0" (1.3 High Time Difficulty Factor included)									
Brush application									
Slow	18	--	--	130.00	31.20	--	30.63	30.69	222.52
Medium	27	--	--	111.16	32.14	--	35.82	21.49	200.61
Fast	43	--	--	82.90	29.28	--	34.77	10.29	157.24

	Labor LF per manhour	Material coverage LF/gallon	Material cost per gallon	Labor cost per 100 LF	Labor burden 100 LF	Material cost per 100 LF	Overhead per 100 LF	Profit per 100 LF	Total price per 100 LF
Cutting-in, horizontal, heights from 9'0" to 13'0" (1.6 High Time Difficulty Factor included)									
Brush application									
Slow	14	--	--	160.00	38.41	--	37.70	37.78	273.89
Medium	22	--	--	136.81	39.57	--	44.09	26.45	246.92
Fast	35	--	--	102.03	36.01	--	42.80	12.66	193.50
Cutting-in, horizontal, heights from 13'0" to 17'0" (1.9 High Time Difficulty Factor included)									
Brush application									
Slow	12	--	--	190.00	45.60	--	44.76	44.86	325.22
Medium	18	--	--	162.46	46.97	--	52.35	31.41	293.19
Fast	30	--	--	121.16	42.80	--	50.82	15.03	229.81
Cutting-in, horizontal, heights from 17'0" to 19'0" (2.2 High Time Difficulty Factor included)									
Brush application									
Slow	10	--	--	220.00	52.79	--	51.83	51.94	376.56
Medium	16	--	--	188.12	54.40	--	60.62	36.37	339.51
Fast	26	--	--	140.29	49.53	--	58.84	17.41	266.07
Cutting-in, horizontal, heights from 19'0" to 21'0" (2.5 High Time Difficulty Factor included)									
Brush application									
Slow	9	--	--	250.00	60.01	--	58.90	59.02	427.93
Medium	14	--	--	213.77	61.80	--	68.89	41.33	385.79
Fast	23	--	--	159.42	56.27	--	66.87	19.78	302.34

Cutting-in, vertical, interior or exterior

	Labor LF per manhour	Material coverage LF/gallon	Material cost per gallon	Labor cost per 100 LF	Labor burden 100 LF	Material cost per 100 LF	Overhead per 100 LF	Profit per 100 LF	Total price per 100 LF
Cutting-in, vertical, heights less than 6'8"									
Brush application									
Slow	29	--	--	79.31	19.03	--	18.68	18.72	135.74
Medium	41	--	--	72.66	21.01	--	23.42	14.05	131.14
Fast	58	--	--	62.07	21.91	--	26.03	7.70	117.71
Cutting-in, vertical, heights from 6'8" to 9'0" (1.3 High Time Difficulty Factor included)									
Brush application									
Slow	22	--	--	103.10	24.76	--	24.29	24.34	176.49
Medium	31	--	--	94.46	27.31	--	30.44	18.26	170.47
Fast	45	--	--	80.69	28.47	--	33.84	10.01	153.01
Cutting-in, vertical, heights from 9'0" to 13'0" (1.6 High Time Difficulty Factor included)									
Brush application									
Slow	18	--	--	126.90	30.44	--	29.90	29.96	217.20
Medium	25	--	--	116.26	33.62	--	37.47	22.48	209.83
Fast	36	--	--	99.31	35.08	--	41.65	12.32	188.36
Cutting-in, vertical, heights from 13'0" to 17'0" (1.9 High Time Difficulty Factor included)									
Brush application									
Slow	15	--	--	150.69	36.17	--	35.50	35.58	257.94
Medium	21	--	--	138.05	39.93	--	44.49	26.69	249.16
Fast	31	--	--	117.93	41.64	--	49.46	14.63	223.66

	Labor LF per manhour	Material coverage LF/gallon	Material cost per gallon	Labor cost per 100 LF	Labor burden 100 LF	Material cost per 100 LF	Overhead per 100 LF	Profit per 100 LF	Total price per 100 LF
Cutting-in, vertical, heights from 17'0" to 19'0" (2.2 High Time Difficulty Factor included)									
Brush application									
Slow	13	--	--	174.48	41.87	--	41.11	41.20	298.66
Medium	18	--	--	159.85	46.23	--	51.51	30.91	288.50
Fast	26	--	--	136.55	48.21	--	57.27	16.94	258.97
Cutting-in, vertical, heights from 19'0" to 21'0" (2.5 High Time Difficulty Factor included)									
Brush application									
Slow	12	--	--	198.28	47.59	--	46.72	46.81	339.40
Medium	16	--	--	181.65	52.54	--	58.54	35.12	327.85
Fast	23	--	--	155.17	54.77	--	65.08	19.25	294.27

Use these figures when cutting-in by hand with a brush at vertical walls, horizontal ceilings, at baseboards, around door frames, etc. when different colors or different sheens (i.e. flat vs. semi-gloss) are used on the adjacent surfaces. "Slow" work is based on an hourly wage of $23.00, "Medium" work on an hourly wage of $29.50, and "Fast" work on an hourly wage of $36.00.

"Slow" applies to residential repaints with heavy texture. "Slow" also applies when cutting in freehand, i.e. using a brush to build-up a bead of paint and running the cut line 3/16" to 1/8" from the adjacent surface, typically horizontal cutting-in at ceilings. "Medium" applies to residential or commercial repaints with light to medium texture. "Medium" speed also applies when using masking tape to form the cut line, then using light brush coats so excess paint doesn't seep under the tape. This application is common around doors, windows at woodwork, and can be used vertically at wall to wall intersections. "Fast" applies to new construction with smooth wall or orange peel texture. "Fast" speed also applies when using a paint guide, either metal or plastic. The paint guide is typically used for vertical wall-to-wall applications, but can be used for horizontal use at baseboards, over doors, etc. As when using masking tape at medium speed, a series of light brush coats works best so paint doesn't seep under the guide tool.

Tip: At baseboards, it's sometimes best to use a stiff putty knife to pull the baseboard away from the wall, and use a Bender paint pad to paint the wall behind the baseboard. Then, while the base is away from the wall, paint the top edge of the base and reinstall the baseboard.

Notes:

1 - Material consumption for cutting-in is minimal or zero (0) since the material cost is actually calculated in the wall painting or ceiling painting line item.

2 - High Time Difficulty Factors are built into these figures to allow for up and down time and moving ladders or scaffolding.

3 - Horizontal cutting-in is typically more difficult and consumes more time than vertical cutting-in, as the figures indicate.

ADD for scribing and back-painting (scraping the texture to achieve a clean line and back-painting to cover the scraped intersection) at a wall-to-ceiling or wall-to-wall corner intersection. See Preparation section of this book.

ADD for time to switch paint if there is more than one color in any room, i.e. ceiling color (1), wall color (2), accent wall color (3). In this example, add time to switch paint twice.

Deck overhang, wood

Multiply the horizontal surface area by 1.5 to allow for painting floor joists and use the overhang table for areas greater than 2.5 feet wide to determine pricing.

Deck surfaces, steps, stair treads & porches, wood

Measure the surface area and apply the prices for smooth siding.

	Labor LF per manhour	Material coverage LF/gallon	Material cost per gallon	Labor cost per 100 LF	Labor burden 100 LF	Material cost per 100 LF	Overhead per 100 LF	Profit per 100 LF	Total price per 100 LF
Door frames and trim only, per 100 linear feet									
Undercoat, water base (material #3)									
Brush 1 coat									
Slow	220	500	36.90	10.45	2.53	7.38	3.86	3.87	28.09
Medium	270	465	32.30	10.93	3.14	6.95	5.26	3.16	29.44
Fast	320	425	27.70	11.25	4.00	6.52	6.74	1.99	30.50
Undercoat, oil base (material #4)									
Brush 1 coat									
Slow	220	550	48.30	10.45	2.53	8.78	4.13	4.14	30.03
Medium	270	510	42.30	10.93	3.14	8.29	5.60	3.36	31.32
Fast	320	465	36.20	11.25	4.00	7.78	7.13	2.11	32.27
Split coat (1/2 undercoat + 1/2 enamel), water base (material #3 & #9)									
Brush 1 coat									
Slow	205	560	45.10	11.22	2.70	8.05	4.17	4.18	30.32
Medium	240	515	39.50	12.29	3.57	7.67	5.88	3.53	32.94
Fast	275	475	33.85	13.09	4.64	7.13	7.70	2.28	34.84
Split coat (1/2 undercoat + 1/2 enamel), oil base (material #4 & #10)									
Brush 1 coat									
Slow	205	610	54.90	11.22	2.70	9.00	4.35	4.36	31.63
Medium	240	565	48.05	12.29	3.57	8.50	6.09	3.65	34.10
Fast	275	525	41.15	13.09	4.64	7.84	7.92	2.34	35.83
Enamel, water base (material #9)									
Brush each coat									
Slow	185	560	53.30	12.43	3.00	9.52	4.74	4.75	34.44
Medium	220	515	46.70	13.41	3.89	9.07	6.59	3.95	36.91
Fast	255	475	40.00	14.12	4.97	8.42	8.53	2.52	38.56
Enamel, oil base (material #10)									
Brush each coat									
Slow	185	610	61.50	12.43	3.00	10.08	4.84	4.85	35.20
Medium	220	565	53.80	13.41	3.89	9.52	6.70	4.02	37.54
Fast	255	525	46.10	14.12	4.97	8.78	8.64	2.56	39.07

Use these figures for painting door frames and wood trim on all sides when doors are not to be painted. ADD for additional preparation time, as needed, including setup, remove & replace (hardware), protection (mask walls and visqueen floors), putty, caulking, sanding and cleanup. When doors are painted along with the frames and trim, use the Opening Count Method under Doors, interior openings, and/or the exterior door costs under Doors, exterior. Measurements for door frames and trim are based on linear feet (LF) of the frame. Prices are for about 17 linear feet of frame at each opening. "Slow" work is based on an hourly wage of $23.00, "Medium" work on an hourly wage of $29.50, and "Fast" work on an hourly wage of $36.00. Other qualifications that apply to this table are on page 9.

	Openings per manhour	Openings per gallon	Material cost per gallon	Labor cost per opening	Labor burden opening	Material cost per opening	Overhead per opening	Profit per opening	Total price per opening
Door frames and trim only, per opening									
Undercoat, water base (material #3)									
Brush 1 coat									
Slow	13	30	36.90	1.77	.43	1.23	.65	.65	4.73
Medium	16	28	32.30	1.84	.56	1.15	.88	.53	4.96
Fast	18	25	27.70	2.00	.73	1.11	1.18	.35	5.37
Undercoat, oil base (material #4)									
Brush 1 coat									
Slow	13	33	48.30	1.77	.43	1.46	.69	.69	5.04
Medium	16	31	42.30	1.84	.56	1.36	.93	.56	5.25
Fast	18	28	36.20	2.00	.73	1.29	1.24	.37	5.63
Split coat (1/2 undercoat + 1/2 enamel), water base (material #3 & #9)									
Brush 1 coat									
Slow	12	33	45.10	1.92	.45	1.37	.71	.71	5.16
Medium	14	31	39.50	2.11	.59	1.27	1.00	.60	5.57
Fast	16	28	33.85	2.25	.82	1.21	1.32	.39	5.99
Split coat (1/2 undercoat + 1/2 enamel), oil base (material #4 & #10)									
Brush 1 coat									
Slow	12	36	54.90	1.92	.45	1.53	.74	.74	5.38
Medium	14	34	48.05	2.11	.59	1.41	1.03	.62	5.76
Fast	16	31	41.15	2.25	.82	1.33	1.35	.40	6.15
Enamel, water base (material #9)									
Brush each coat									
Slow	11	33	53.30	2.09	.51	1.62	.80	.80	5.82
Medium	13	31	46.70	2.27	.66	1.51	1.11	.67	6.22
Fast	15	28	40.00	2.40	.86	1.43	1.45	.43	6.57
Enamel, oil base (material #10)									
Brush each coat									
Slow	11	36	61.50	2.09	.51	1.71	.82	.82	5.95
Medium	13	34	53.80	2.27	.66	1.58	1.13	.68	6.32
Fast	15	31	46.10	2.40	.86	1.49	1.47	.43	6.65

Use these figures for painting door frames and wood trim on all sides when doors are not to be painted. ADD for additional preparation time, as needed, including setup, remove & replace (hardware), protection (mask walls and visqueen floors); putty, caulking, sanding and cleanup. When doors are painted along with the frames and trim, use the Opening Count Method under Doors, interior openings, and/or the exterior door costs under Doors, exterior. These costs are based on a count of the openings requiring paint. Prices are for about 17 linear feet of frame at each opening. "Slow" work is based on an hourly wage of $23.00, "Medium" work on an hourly wage of $29.50, and "Fast" work on an hourly wage of $36.00. Other qualifications that apply to this table are on page 9.

Doors, exterior

The tables that follow include costs for both time and material needed to apply two coats of a high quality finish to all six sides of each exterior door, finish the jamb and trim, and lay-off each door smoothly. These costs are in addition to those shown under Doors, interior openings, for both the *Opening Count Method* and the *Per Door Method,* which include one coat of undercoat and one coat of enamel for each exterior door along with the interior doors. New exterior paint grade doors actually receive two coats of exterior enamel. New exterior stain grade doors actually receive a coat of stain, sealer, and then a coat of either marine spar varnish or polyurethane finish. The following two examples give total cost to finish a flush or a panel (entry) exterior door with polyurethane, and includes the cost for the two coats from the interior take-off.

Example #1 Flush Doors

OPENING COUNT METHOD

Included in the interior take-off		
Opening Count Method, interior undercoat cost - 1 coat, slow	$ 19.82	(page 103)
Opening Count Method, interior enamel cost - 1 coat, slow	$ 21.63	(page 103)
Included in the exterior take-off		
Exterior, flush, two coat system, polyurethane cost - 2 coats, slow	$ 50.80	(page 98)
Total to finish the exterior door	$ 92.25	

OR

PER DOOR METHOD

Included in the interior take-off		
Per Door Method, interior, flush, undercoat cost - 1 coat, slow	$ 19.67	(page 109)
Per Door Method, interior enamel cost - 1 coat, slow	$ 18.25	(page 109)
Included in the exterior take-off		
Exterior, flush, two coat system, polyurethane cost - 2 coats, slow	$ 50.80	(page 98)
Total to finish the exterior door	$ 88.72	

Under this system, much of the cost to paint the exterior door is included in the interior take-off. When counting interior doors, be sure you include all the exterior doors whether you use either the *Opening Count Method* or the *Per Door Method* to estimate doors.

Example #2 Panel (Entry) Doors

OPENING COUNT METHOD

Included in the interior take-off		
Opening Count Method, interior undercoat cost - 1 coat, slow	$ 19.82	(page 103)
Opening Count Method, interior enamel cost - 1 coat, slow	$ 21.63	(page 103)
Included in the exterior take-off		
Exterior, panel (entry) two coat system, polyurethane cost - 2 coats, slow	$ 94.58	(page 101)
Total to finish the exterior door	$ 136.03	

OR

PER DOOR METHOD

Included in the interior take-off		
Per Door Method, interior, flush, undercoat cost - 1 coat, slow	$ 19.67	(page 109)
Per Door Method, interior enamel cost - 1 coat, slow	$ 18.25	(page 109)
Included in the exterior take-off		
Exterior, panel (entry) two coat system, polyurethane cost - 2 coats, slow	$ 94.58	(page 101)
Total to finish the exterior door	$ 132.50	

	Manhours per door	Doors per gallon	Material cost per gallon	Labor cost per door	Labor burden door	Material cost per door	Overhead per door	Profit per door	Total price per door
Doors, exterior, flush, two coat system									
Exterior enamel, 2 coat system, water base (material #24)									
Roll & brush 2 coats									
Slow	0.5	7	54.80	11.50	2.76	7.83	4.20	4.21	30.50
Medium	0.4	6	47.90	11.80	3.41	7.98	5.80	3.48	32.47
Fast	0.3	5	41.10	10.80	3.81	8.22	7.08	2.09	32.00
Exterior enamel, 2 coat system, oil base (material #25)									
Roll & brush 2 coats									
Slow	0.5	7	62.50	11.50	2.76	8.93	4.41	4.42	32.02
Medium	0.4	6	54.70	11.80	3.41	9.12	6.08	3.65	34.06
Fast	0.3	5	46.90	10.80	3.81	9.38	7.44	2.20	33.63
Polyurethane (material #22)									
Brush 2 coats									
Slow	0.7	5.0	84.20	16.10	3.86	16.84	6.99	7.01	50.80
Medium	0.6	4.5	73.70	17.70	5.12	16.38	9.80	5.88	54.88
Fast	0.5	4.0	63.20	18.00	6.36	15.80	12.45	3.68	56.29
Marine spar varnish, flat or gloss (material #23)									
Brush 2 coats									
Slow	0.6	6	88.20	13.80	3.31	14.70	6.04	6.06	43.91
Medium	0.5	5	77.20	14.75	4.27	15.44	8.61	5.17	48.24
Fast	0.4	4	66.20	14.40	5.08	16.55	11.17	3.30	50.50
ADD - Preparation for spar varnish									
Steel wool buff	0.2	--	--	5.90	1.71	--	1.90	1.14	10.65
Wax application	0.2	--	--	5.90	1.71	--	1.90	1.14	10.65

Use these figures for painting two coats on flush exterior doors, other than entry doors. These costs are to be included with the exterior door take-off and are in addition to the costs that are included in the interior take-off as explained by the example in the previous section. ADD for additional preparation time, as needed, including setup, remove & replace (hardware), protection (mask walls and visqueen floors); putty, caulking, sanding and cleanup. Add minimum preparation time for varnishing as indicated above. "Slow" work is based on an hourly wage of $23.00, "Medium" work on an hourly wage of $29.50, and "Fast" work on an hourly wage of $36.00. Other qualifications that apply to this table are on page 9.

	Manhours per door	Doors per gallon	Material cost per gallon	Labor cost per door	Labor burden door	Material cost per door	Overhead per door	Profit per door	Total price per door
Doors, exterior, French, two coat system									
Exterior enamel, 2 coat system, water base (material #24)									
Roll & brush 2 coats									
Slow	1.0	12	54.80	23.00	5.52	4.57	6.29	6.30	45.68
Medium	0.8	10	47.90	23.60	6.82	4.79	8.80	5.28	49.29
Fast	0.6	8	41.10	21.60	7.63	5.14	10.65	3.15	48.17
Exterior enamel, 2 coat system, oil base (material #25)									
Roll & brush 2 coats									
Slow	1.0	12	62.50	23.00	5.52	5.21	6.41	6.42	46.56
Medium	0.8	10	54.70	23.60	6.82	5.47	8.97	5.38	50.24
Fast	0.6	8	46.90	21.60	7.63	5.86	10.87	3.22	49.18
Polyurethane (material #22)									
Brush 2 coats									
Slow	1.5	8.0	84.20	34.50	8.28	10.53	10.13	10.15	73.59
Medium	1.3	7.5	73.70	38.35	11.09	9.83	14.82	8.89	82.98
Fast	1.0	7.0	63.20	36.00	12.71	9.03	17.90	5.29	80.93
Marine spar varnish, flat or gloss (material #23)									
Brush 2 coats									
Slow	1.2	12	88.20	27.60	6.62	7.35	7.90	7.92	57.39
Medium	1.0	10	77.20	29.50	8.53	7.72	11.44	6.86	64.05
Fast	0.8	8	66.20	28.80	10.17	8.28	14.65	4.33	66.23
ADD - Preparation for spar varnish									
Steel wool buff	0.3	--	--	8.85	2.56	--	2.85	1.71	15.97
Wax application	0.3	--	--	8.85	2.56	--	2.85	1.71	15.97

Use these figures for painting two coats on exterior French doors that have 10 to 15 lites. These costs are to be included with the exterior door take-off and are in addition to the costs that are included in the interior take-off as explained by the example in the previous section. ADD for additional preparation time, as needed, including setup, remove & replace (hardware), protection (mask walls and visqueen floors); putty, caulking, sanding and cleanup. Add minimum preparation time for varnishing as indicated above. "Slow" work is based on an hourly wage of $23.00, "Medium" work on an hourly wage of $29.50, and "Fast" work on an hourly wage of $36.00. Other qualifications that apply to this table are on page 9.

	Manhours per door	Doors per gallon	Material cost per gallon	Labor cost per door	Labor burden door	Material cost per door	Overhead per door	Profit per door	Total price per door
Doors, exterior, louvered, two coat system									
Exterior enamel, 2 coat system, water base (material #24)									
Roll & brush 2 coats									
Slow	1.4	7	54.80	32.20	7.73	7.83	9.07	9.09	65.92
Medium	1.1	6	47.90	32.45	9.38	7.98	12.45	7.47	69.73
Fast	0.7	5	41.10	25.20	8.90	8.22	13.12	3.88	59.32
Exterior enamel, 2 coat system, oil base (material #25)									
Roll & brush 2 coats									
Slow	1.4	7	62.50	32.20	7.73	8.93	9.28	9.30	67.44
Medium	1.1	6	54.70	32.45	9.38	9.12	12.74	7.64	71.33
Fast	0.7	5	46.90	25.20	8.90	9.38	13.48	3.99	60.95
Polyurethane (material #22)									
Brush 2 coats									
Slow	1.7	5.0	84.20	39.10	9.38	16.84	12.41	12.44	90.17
Medium	1.5	4.5	73.70	44.25	12.80	16.38	18.36	11.01	102.80
Fast	1.2	4.0	63.20	43.20	15.25	15.80	23.02	6.81	104.08
Marine spar varnish, flat or gloss (material #23)									
Brush 2 coats									
Slow	1.6	7	88.20	36.80	8.83	12.60	11.06	11.09	80.38
Medium	1.3	6	77.20	38.35	11.09	12.87	15.58	9.35	87.24
Fast	0.9	5	66.20	32.40	11.44	13.24	17.69	5.23	80.00
ADD - Preparation for spar varnish									
Steel wool buff	0.4	--	--	11.80	3.41	--	3.80	2.28	21.29
Wax application	0.4	--	--	11.80	3.41	--	3.80	2.28	21.29

Use these figures for painting two coats on exterior louvered doors. These costs are to be included with the exterior door take-off and are in addition to the costs that are included in the interior take-off as explained by the example in the previous section. ADD for additional preparation time, as needed, including setup, remove & replace (hardware), protection (mask walls and visqueen floors); putty, caulking, sanding and cleanup. Add minimum preparation time for varnishing as indicated above. "Slow" work is based on an hourly wage of $23.00, "Medium" work on an hourly wage of $29.50, and "Fast" work on an hourly wage of $36.00. Other qualifications that apply to this table are on page 9.

	Manhours per door	Doors per gallon	Material cost per gallon	Labor cost per door	Labor burden door	Material cost per door	Overhead per door	Profit per door	Total price per door

Doors, exterior, panel (entry), two coat system

	Manhours per door	Doors per gallon	Material cost per gallon	Labor cost per door	Labor burden door	Material cost per door	Overhead per door	Profit per door	Total price per door
Exterior enamel, 2 coat system, water base (material #24)									
Roll & brush 2 coats									
Slow	1.4	4	54.80	32.20	7.73	13.70	10.19	10.21	74.03
Medium	1.1	3	47.90	32.45	9.38	15.97	14.45	8.67	80.92
Fast	0.8	2	41.10	28.80	10.17	20.55	18.45	5.46	83.43
Exterior enamel, 2 coat system, oil base (material #25)									
Roll & brush 2 coats									
Slow	1.4	4	62.50	32.20	7.73	15.63	10.56	10.58	76.70
Medium	1.1	3	54.70	32.45	9.38	18.23	15.02	9.01	84.09
Fast	0.8	2	46.90	28.80	10.17	23.45	19.35	5.72	87.49
Polyurethane (material #22)									
Brush 2 coats									
Slow	1.7	4	84.20	39.10	9.38	21.05	13.21	13.24	95.98
Medium	1.5	3	73.70	44.25	12.80	24.57	20.40	12.24	114.26
Fast	1.2	2	63.20	43.20	15.25	31.60	27.92	8.26	126.23
Marine spar varnish, flat or gloss (material #23)									
Brush 2 coats									
Slow	1.4	4	88.20	32.20	7.73	22.05	11.78	11.80	85.56
Medium	1.1	3	77.20	32.45	9.38	25.73	16.89	10.13	94.58
Fast	0.8	2	66.20	28.80	10.17	33.10	22.34	6.61	101.02
ADD - Preparation for spar varnish									
Steel wool buff	0.3	--	--	8.85	2.56	--	2.85	1.71	15.97
Wax application	0.3	--	--	8.85	2.56	--	2.85	1.71	15.97

Use these figures for painting two coats on typical exterior paneled doors. These costs are included with the exterior door take-off and are in addition to the costs that are included in the interior take-off as explained by the example in the previous section. ADD for additional preparation time, as needed, including setup, remove & replace (hardware), protection (mask walls and visqueen floors); putty, caulking, sanding and cleanup. Add minimum preparation time for varnishing as indicated above. "Slow" work is based on an hourly wage of $23.00, "Medium" work on an hourly wage of $29.50, and "Fast" work on an hourly wage of $36.00. Other qualifications that apply to this table are on page 9.

Doors, *Opening Count Method*

Many painting companies estimate paint grade doors (including jambs and frames), wood windows, pullmans, linens, bookcases, wine racks and other interior surfaces that take an undercoat and enamel finish by the "opening." Based on Figure 20 below, each opening is considered to take the same time regardless of whether it's a door, window, pullman, etc. These figures are based on the number of openings finished per 8-hour day and the material required per opening. The Opening Count Method of estimating involves counting the quantity of all openings (including all exterior doors) based on the opening allowance table at Figure 20 below. After you determine the number of openings, use the following table in accumulated multiples of 10 for applying undercoat and enamel. The undercoat process is based on 11 to 13 openings per gallon and enamel is based on 10 to 12 openings per gallon. As an example, using the medium rate for water based material on 12 openings with 1 coat of undercoat and 1 coat of enamel, add the 10 opening figures for each coat to the 2 opening figures for each coat as follows:

Interior Take-off

	Undercoat	Enamel	
10 openings	165.76	187.92	(page 108)
2 openings	34.32	38.40	(page 103)
12 openings	200.08	226.32	

Item		Opening Count
Closets	Molding at closet perimeter	Count 1 opening per 25'0" length
	Poles, stain	Count 1 opening per 10'0" length
	Shelf & pole (undercoat or enamel)	Count 1 opening per 6'0" length
	Shelves (undercoat or enamel)	Count 1 opening per 10'0" length
Doors	Bifold doors & frames	Count 1 opening per door
	Dutch doors & frames	Count 2 openings per door
	Entry doors & frames	Count 1 opening per door
	Forced air unit doors & frames	Count 1 opening per door
	French doors & frames	Count 1.5 openings per door
	Linen doors with face frame	Count 1 opening per 2'0" width
	Louvered bifold doors & frames	Count 1 opening per door panel
	false	Count 1 opening per door panel
	real	Count 1 opening per door or per 1'6" width
	Passage doors & frames	
	flush	Count 1 opening per door
	paneled	Count 1.25 openings per door
	Wardrobe doors	Count 1 opening per door
	Split coat operation, doors & frames	Count 1 opening per door
	Tipoff operation (doors only)	Count .5 opening per door
Pullman cabinets		Count 1 opening per lavatory or per 4'0" width
Windows, wood		Count 1 opening per 6 SF of window

Figure 20
Interior opening count allowance table

	Manhours per opening	Gallons per opening	Material cost per gallon	Labor cost per opening	Labor burden opening	Material cost per opening	Overhead per opening	Profit per opening	Total price per opening
Doors, interior openings, *Opening Count Method*									
1 opening total									
Undercoat, water base (material #3)									
Roll & brush 1 coat									
Slow	0.4	0.080	36.90	9.20	2.21	2.95	2.73	2.73	19.82
Medium	0.3	0.085	32.30	8.85	2.56	2.75	3.54	2.12	19.82
Fast	0.2	0.090	27.70	7.20	2.54	2.49	3.79	1.12	17.14
Undercoat, oil base (material #4)									
Roll & brush 1 coat									
Slow	0.4	0.080	48.30	9.20	2.21	3.86	2.90	2.91	21.08
Medium	0.3	0.085	42.30	8.85	2.56	3.60	3.75	2.25	21.01
Fast	0.2	0.090	36.20	7.20	2.54	3.26	4.03	1.19	18.22
Enamel, water base (material #9)									
Roll & brush 1 coat									
Slow	0.4	0.08	53.30	9.20	2.21	4.26	2.98	2.98	21.63
Medium	0.3	0.09	46.70	8.85	2.56	4.20	3.90	2.34	21.85
Fast	0.2	0.10	40.00	7.20	2.54	4.00	4.26	1.26	19.26
Enamel, oil base (material #10)									
Roll & brush 1 coat									
Slow	0.4	0.08	61.50	9.20	2.21	4.92	3.10	3.11	22.54
Medium	0.3	0.09	53.80	8.85	2.56	4.84	4.06	2.44	22.75
Fast	0.2	0.10	46.10	7.20	2.54	4.61	4.45	1.32	20.12
2 openings total									
Undercoat, water base (material #3)									
Roll & brush 1 coat									
Slow	0.7	0.16	36.90	16.10	3.86	5.90	4.91	4.92	35.69
Medium	0.5	0.17	32.30	14.75	4.27	5.49	6.13	3.68	34.32
Fast	0.3	0.18	27.70	10.80	3.81	4.99	6.08	1.80	27.48
Undercoat, oil base (material #4)									
Roll & brush 1 coat									
Slow	0.7	0.16	48.30	16.10	3.86	7.73	5.26	5.27	38.22
Medium	0.5	0.17	42.30	14.75	4.27	7.19	6.55	3.93	36.69
Fast	0.3	0.18	36.20	10.80	3.81	6.52	6.55	1.94	29.62
Enamel, water base (material #9)									
Roll & brush 1 coat									
Slow	0.7	0.17	53.30	16.10	3.86	9.06	5.51	5.52	40.05
Medium	0.5	0.18	46.70	14.75	4.27	8.41	6.86	4.11	38.40
Fast	0.3	0.20	40.00	10.80	3.81	8.00	7.01	2.07	31.69

	Manhours per opening	Gallons per opening	Material cost per gallon	Labor cost per opening	Labor burden opening	Material cost per opening	Overhead per opening	Profit per opening	Total price per opening
Enamel, oil base (material #10)									
Roll & brush 1 coat									
Slow	0.7	0.17	61.50	16.10	3.86	10.46	5.78	5.79	41.99
Medium	0.5	0.18	53.80	14.75	4.27	9.68	7.17	4.30	40.17
Fast	0.3	0.20	46.10	10.80	3.81	9.22	7.39	2.19	33.41
3 openings total									
Undercoat, water base (material #3)									
Roll & brush 1 coat									
Slow	1.00	0.23	36.90	23.00	5.52	8.49	7.03	7.05	51.09
Medium	0.75	0.25	32.30	22.13	6.39	8.08	9.15	5.49	51.24
Fast	0.50	0.27	27.70	18.00	6.36	7.48	9.87	2.92	44.63
Undercoat, oil base (material #4)									
Roll & brush 1 coat									
Slow	1.00	0.23	48.30	23.00	5.52	11.11	7.53	7.55	54.71
Medium	0.75	0.25	42.30	22.13	6.39	10.58	9.78	5.87	54.75
Fast	0.50	0.27	36.20	18.00	6.36	9.77	10.58	3.13	47.84
Enamel, water base (material #9)									
Roll & brush 1 coat									
Slow	1.00	0.25	53.30	23.00	5.52	13.33	7.95	7.97	57.77
Medium	0.75	0.27	46.70	22.13	6.39	12.61	10.29	6.17	57.59
Fast	0.50	0.30	40.00	18.00	6.36	12.00	11.27	3.33	50.96
Enamel, oil base (material #10)									
Roll & brush 1 coat									
Slow	1.00	0.25	61.50	23.00	5.52	15.38	8.34	8.36	60.60
Medium	0.75	0.27	53.80	22.13	6.39	14.53	10.77	6.46	60.28
Fast	0.50	0.30	46.10	18.00	6.36	13.83	11.84	3.50	53.53
4 openings total									
Undercoat, water base (material #3)									
Roll & brush 1 coat									
Slow	1.3	0.31	36.90	29.90	7.18	11.44	9.22	9.24	66.98
Medium	1.0	0.33	32.30	29.50	8.53	10.66	12.17	7.30	68.16
Fast	0.7	0.36	27.70	25.20	8.90	9.97	13.66	4.04	61.77
Undercoat, oil base (material #4)									
Roll & brush 1 coat									
Slow	1.3	0.31	48.30	29.90	7.18	14.97	9.89	9.91	71.85
Medium	1.0	0.33	42.30	29.50	8.53	13.96	13.00	7.80	72.79
Fast	0.7	0.36	36.20	25.20	8.90	13.03	14.61	4.32	66.06

	Manhours per opening	Gallons per opening	Material cost per gallon	Labor cost per opening	Labor burden opening	Material cost per opening	Overhead per opening	Profit per opening	Total price per opening
Enamel, water base (material #9)									
Roll & brush 1 coat									
Slow	1.3	0.33	53.30	29.90	7.18	17.59	10.39	10.41	75.47
Medium	1.0	0.36	46.70	29.50	8.53	16.81	13.71	8.23	76.78
Fast	0.7	0.40	40.00	25.20	8.90	16.00	15.53	4.59	70.22
Enamel, oil base (material #10)									
Roll & brush 1 coat									
Slow	1.3	0.33	61.50	29.90	7.18	20.30	10.90	10.92	79.20
Medium	1.0	0.36	53.80	29.50	8.53	19.37	14.35	8.61	80.36
Fast	0.7	0.40	46.10	25.20	8.90	18.44	16.29	4.82	73.65
5 openings total									
Undercoat, water base (material #3)									
Roll & brush 1 coat									
Slow	1.6	0.38	36.90	36.80	8.83	14.02	11.33	11.36	82.34
Medium	1.3	0.41	32.30	38.35	11.09	13.24	15.67	9.40	87.75
Fast	0.9	0.45	27.70	32.40	11.44	12.47	17.46	5.16	78.93
Undercoat, oil base (material #4)									
Roll & brush 1 coat									
Slow	1.6	0.38	48.30	36.80	8.83	18.35	12.16	12.18	88.32
Medium	1.3	0.41	42.30	38.35	11.09	17.34	16.69	10.02	93.49
Fast	0.9	0.45	36.20	32.40	11.44	16.29	18.64	5.51	84.28
Enamel, water base (material #9)									
Roll & brush 1 coat									
Slow	1.6	0.42	53.30	36.80	8.83	22.39	12.92	12.95	93.89
Medium	1.3	0.46	46.70	38.35	11.09	21.48	17.73	10.64	99.29
Fast	0.9	0.50	40.00	32.40	11.44	20.00	19.79	5.85	89.48
Enamel, oil base (material #10)									
Roll & brush 1 coat									
Slow	1.6	0.42	61.50	36.80	8.83	25.83	13.58	13.61	98.65
Medium	1.3	0.46	53.80	38.35	11.09	24.75	18.55	11.13	103.87
Fast	0.9	0.50	46.10	32.40	11.44	23.05	20.74	6.13	93.76
6 openings total									
Undercoat, water base (material #3)									
Roll & brush 1 coat									
Slow	1.90	0.46	36.90	43.70	10.49	16.97	13.52	13.55	98.23
Medium	1.45	0.50	32.30	42.78	12.36	16.15	17.82	10.69	99.80
Fast	1.00	0.54	27.70	36.00	12.71	14.96	19.74	5.84	89.25

	Manhours per opening	Gallons per opening	Material cost per gallon	Labor cost per opening	Labor burden opening	Material cost per opening	Overhead per opening	Profit per opening	Total price per opening
Undercoat, oil base (material #4)									
Roll & brush 1 coat									
Slow	1.90	0.46	48.30	43.70	10.49	22.22	14.52	14.55	105.48
Medium	1.45	0.50	42.30	42.78	12.36	21.15	19.07	11.44	106.80
Fast	1.00	0.54	36.20	36.00	12.71	19.55	21.16	6.26	95.68
Enamel, water base (material #9)									
Roll & brush 1 coat									
Slow	1.90	0.50	53.30	43.70	10.49	26.65	15.36	15.39	111.59
Medium	1.45	0.55	46.70	42.78	12.36	25.69	20.21	12.12	113.16
Fast	1.00	0.60	40.00	36.00	12.71	24.00	22.54	6.67	101.92
Enamel, oil base (material #10)									
Roll & brush 1 coat									
Slow	1.90	0.50	61.50	43.70	10.49	30.75	16.14	16.17	117.25
Medium	1.45	0.55	53.80	42.78	12.36	29.59	21.18	12.71	118.62
Fast	1.00	0.60	46.10	36.00	12.71	27.66	23.67	7.00	107.04
7 openings total									
Undercoat, water base (material #3)									
Roll & brush 1 coat									
Slow	2.2	0.54	36.90	50.60	12.14	19.93	15.71	15.74	114.12
Medium	1.7	0.59	32.30	50.15	14.50	19.06	20.93	12.56	117.20
Fast	1.2	0.64	27.70	43.20	15.25	17.73	23.62	6.99	106.79
Undercoat, oil base (material #4)									
Roll & brush 1 coat									
Slow	2.2	0.54	48.30	50.60	12.14	26.08	16.88	16.91	122.61
Medium	1.7	0.59	42.30	50.15	14.50	24.96	22.40	13.44	125.45
Fast	1.2	0.64	36.20	43.20	15.25	23.17	25.30	7.48	114.40
Enamel, water base (material #9)									
Roll & brush 1 coat									
Slow	2.2	0.58	53.30	50.60	12.14	30.91	17.79	17.83	129.27
Medium	1.7	0.64	46.70	50.15	14.50	29.89	23.63	14.18	132.35
Fast	1.2	0.70	40.00	43.20	15.25	28.00	26.80	7.93	121.18
Enamel, oil base (material #10)									
Roll & brush 1 coat									
Slow	2.2	0.58	61.50	50.60	12.14	35.67	18.70	18.74	135.85
Medium	1.7	0.64	53.80	50.15	14.50	34.43	24.77	14.86	138.71
Fast	1.2	0.70	46.10	43.20	15.25	32.27	28.12	8.32	127.16

	Manhours per opening	Gallons per opening	Material cost per gallon	Labor cost per opening	Labor burden opening	Material cost per opening	Overhead per opening	Profit per opening	Total price per opening
8 openings total									
Undercoat, water base (material #3)									
Roll & brush 1 coat									
Slow	2.50	0.62	36.90	57.50	13.80	22.88	17.89	17.93	130.00
Medium	1.95	0.67	32.30	57.53	16.63	21.64	23.95	14.37	134.12
Fast	1.40	0.73	27.70	50.40	17.79	20.22	27.41	8.11	123.93
Undercoat, oil base (material #4)									
Roll & brush 1 coat									
Slow	2.50	0.62	48.30	57.50	13.80	29.95	19.24	19.28	139.77
Medium	1.95	0.67	42.30	57.53	16.63	28.34	25.63	15.38	143.51
Fast	1.40	0.73	36.20	50.40	17.79	26.43	29.33	8.68	132.63
Enamel, water base (material #9)									
Roll & brush 1 coat									
Slow	2.50	0.67	53.30	57.50	13.80	35.71	20.33	20.37	147.71
Medium	1.95	0.74	46.70	57.53	16.63	34.56	27.18	16.31	152.21
Fast	1.40	0.80	40.00	50.40	17.79	32.00	31.06	9.19	140.44
Enamel, oil base (material #10)									
Roll & brush 1 coat									
Slow	2.50	0.67	61.50	57.50	13.80	41.21	21.38	21.42	155.31
Medium	1.95	0.74	53.80	57.53	16.63	39.81	28.49	17.10	159.56
Fast	1.40	0.80	46.10	50.40	17.79	36.88	32.57	9.63	147.27
9 openings total									
Undercoat, water base (material #3)									
Roll & brush 1 coat									
Slow	2.80	0.69	36.90	64.40	15.46	25.46	20.01	20.05	145.38
Medium	2.15	0.75	32.30	63.43	18.33	24.23	26.50	15.90	148.39
Fast	1.50	0.81	27.70	54.00	19.07	22.44	29.61	8.76	133.88
Undercoat, oil base (material #4)									
Roll & brush 1 coat									
Slow	2.80	0.69	48.30	64.40	15.46	33.33	21.51	21.55	156.25
Medium	2.15	0.75	42.30	63.43	18.33	31.73	28.37	17.02	158.88
Fast	1.50	0.81	36.20	54.00	19.07	29.32	31.74	9.39	143.52
Enamel, water base (material #9)									
Roll & brush 1 coat									
Slow	2.80	0.75	53.30	64.40	15.46	39.98	22.77	22.82	165.43
Medium	2.15	0.82	46.70	63.43	18.33	38.29	30.01	18.01	168.07
Fast	1.50	0.90	40.00	54.00	19.07	36.00	33.81	10.00	152.88

	Manhours per opening	Gallons per opening	Material cost per gallon	Labor cost per opening	Labor burden opening	Material cost per opening	Overhead per opening	Profit per opening	Total price per opening
Enamel, oil base (material #10)									
Roll & brush 1 coat									
Slow	2.80	0.75	61.50	64.40	15.46	46.13	23.94	23.99	173.92
Medium	2.15	0.82	53.80	63.43	18.33	44.12	31.47	18.88	176.23
Fast	1.50	0.90	46.10	54.00	19.07	41.49	35.51	10.50	160.57
10 openings total									
Undercoat, water base (material #3)									
Roll & brush 1 coat									
Slow	3.1	0.77	36.90	71.30	17.11	28.41	22.20	22.24	161.26
Medium	2.4	0.84	32.30	70.80	20.47	27.13	29.60	17.76	165.76
Fast	1.7	0.90	27.70	61.20	21.61	24.93	33.40	9.88	151.02
Undercoat, oil base (material #4)									
Roll & brush 1 coat									
Slow	3.1	0.77	48.30	71.30	17.11	37.19	23.86	23.91	173.37
Medium	2.4	0.84	42.30	70.80	20.47	35.53	31.70	19.02	177.52
Fast	1.7	0.90	36.20	61.20	21.61	32.58	35.77	10.58	161.74
Enamel, water base (material #9)									
Slow	3.1	0.83	53.30	71.30	17.11	44.24	25.20	25.26	183.11
Medium	2.4	0.92	46.70	70.80	20.47	42.96	33.56	20.13	187.92
Fast	1.7	1.00	40.00	61.20	21.61	40.00	38.07	11.26	172.14
Enamel, oil base (material #10)									
Roll & brush 1 coat									
Slow	3.1	0.83	61.50	71.30	17.11	51.05	26.50	26.55	192.51
Medium	2.4	0.92	53.80	70.80	20.47	49.50	35.19	21.11	197.07
Fast	1.7	1.00	46.10	61.20	21.61	46.10	39.96	11.82	180.69

Use these figures for painting interior doors, pullmans, linens and other surfaces described in Figure 20 on page 102. ADD for additional preparation time, as needed, including setup, remove & replace (hardware), protection (mask walls and visqueen floors); putty, caulking, sanding and cleanup. "Slow" work is based on an hourly wage of $23.00, "Medium" work on an hourly wage of $29.50, and "Fast" work on an hourly wage of $36.00. Other qualifications that apply to this table are on page 9.

	Manhours per door	Doors per gallon	Material cost per gallon	Labor cost per door	Labor burden door	Material cost per door	Overhead per door	Profit per door	Total price per door
Doors, interior, flush, paint grade, roll & brush, based on per door method									
Undercoat, water base (material #3)									
Roll & brush 1 coat									
Slow	0.40	13.0	36.90	9.20	2.21	2.84	2.71	2.71	19.67
Medium	0.30	11.5	32.30	8.85	2.56	2.81	3.56	2.13	19.91
Fast	0.20	10.0	27.70	7.20	2.54	2.77	3.88	1.15	17.54
Undercoat, oil base (material #4)									
Roll & brush 1 coat									
Slow	0.40	13.0	48.30	9.20	2.21	3.72	2.87	2.88	20.88
Medium	0.30	11.5	42.30	8.85	2.56	3.68	3.77	2.26	21.12
Fast	0.20	10.0	36.20	7.20	2.54	3.62	4.14	1.23	18.73
Enamel, water base (material #9)									
Roll & brush 1st finish coat									
Slow	0.33	14.0	53.30	7.59	1.82	3.81	2.51	2.52	18.25
Medium	0.25	12.5	46.70	7.38	2.13	3.74	3.31	1.99	18.55
Fast	0.17	11.0	40.00	6.12	2.16	3.64	3.70	1.09	16.71
Roll & brush additional finish coats									
Slow	0.25	15.0	53.30	5.75	1.38	3.55	2.03	2.03	14.74
Medium	0.20	13.5	46.70	5.90	1.71	3.46	2.77	1.66	15.50
Fast	0.15	12.0	40.00	5.40	1.91	3.33	3.30	.98	14.92
Enamel, oil base (material #10)									
Roll & brush 1st finish coat									
Slow	0.33	14.0	61.50	7.59	1.82	4.39	2.62	2.63	19.05
Medium	0.25	12.5	53.80	7.38	2.13	4.30	3.45	2.07	19.33
Fast	0.17	11.0	46.10	6.12	2.16	4.19	3.87	1.14	17.48
Roll & brush additional finish coats									
Slow	0.25	15.0	61.50	5.75	1.38	4.10	2.13	2.14	15.50
Medium	0.20	13.5	53.80	5.90	1.71	3.99	2.90	1.74	16.24
Fast	0.15	12.0	46.10	5.40	1.91	3.84	3.46	1.02	15.63

This table is not to be confused with the Opening Count Method for estimating doors. Use these figures for painting both interior and exterior doors. These costs are in addition to the costs for applying the finish coats to exterior doors as explained on page 97. These figures include coating all six (6) sides of each door along with the frame and the jamb on both sides. These figures include minimal preparation time. ADD for additional preparation time, as needed, including setup, remove & replace (hardware), protection (mask walls and visqueen floors); putty, caulking, sanding and cleanup. "Slow" work is based on an hourly wage of $23.00, "Medium" work on an hourly wage of $29.50, and "Fast" work on an hourly wage of $36.00. Other qualifications that apply to this table are on page 9.

Doors, interior, flush, paint grade, spray application, based on per door method

	Manhours per door	Doors per gallon	Material cost per gallon	Labor cost per door	Labor burden door	Material cost per door	Overhead per door	Profit per door	Total price per door
Undercoat, water base (material #3)									
Spray 1 coat									
Slow	0.10	17	36.90	2.30	.55	2.17	.95	.96	6.93
Medium	0.09	16	32.30	2.66	.76	2.02	1.36	.82	7.62
Fast	0.08	15	27.70	2.88	1.02	1.85	1.78	.53	8.06
Undercoat, oil base (material #4)									
Spray 1 coat									
Slow	0.10	17	48.30	2.30	.55	2.84	1.08	1.08	7.85
Medium	0.09	16	42.30	2.66	.76	2.64	1.52	.91	8.49
Fast	0.08	15	36.20	2.88	1.02	2.41	1.96	.58	8.85
Enamel, water base (material #9)									
Spray 1st finish coat									
Slow	0.09	18	53.30	2.07	.50	2.96	1.05	1.05	7.63
Medium	0.08	17	46.70	2.36	.68	2.75	1.45	.87	8.11
Fast	0.07	16	40.00	2.52	.89	2.50	1.83	.54	8.28
Spray 2nd or additional finish coats									
Slow	0.08	19	53.30	1.84	.44	2.81	.97	.97	7.03
Medium	0.07	18	46.70	2.07	.59	2.59	1.32	.79	7.36
Fast	0.06	17	40.00	2.16	.76	2.35	1.63	.48	7.38
Enamel, oil base (material #10)									
Spray 1st finish coat									
Slow	0.09	18	61.50	2.07	.50	3.42	1.14	1.14	8.27
Medium	0.08	17	53.80	2.36	.68	3.16	1.55	.93	8.68
Fast	0.07	16	46.10	2.52	.89	2.88	1.95	.58	8.82
Spray 2nd or additional finish coats									
Slow	0.08	19	61.50	1.84	.44	3.24	1.05	1.05	7.62
Medium	0.07	18	53.80	2.07	.59	2.99	1.42	.85	7.92
Fast	0.06	17	46.10	2.16	.76	2.71	1.75	.52	7.90

This table is not to be confused with the Opening Count Method for estimating doors. Use these figures for painting both interior and exterior doors. These costs are in addition to the costs for applying the finish coats to exterior doors as explained on page 97. These figures include coating all six (6) sides of each door along with the frame and the jamb on both sides. These figures include minimal preparation time. ADD for additional preparation time, as needed, including setup, remove & replace (hardware), protection (mask walls and visqueen floors); putty, caulking, sanding and cleanup. "Slow" work is based on an hourly wage of $23.00, "Medium" work on an hourly wage of $29.50, and "Fast" work on an hourly wage of $36.00. Other qualifications that apply to this table are on page 9.

	Manhours per door	Doors per gallon	Material cost per gallon	Labor cost per door	Labor burden door	Material cost per door	Overhead per door	Profit per door	Total price per door

Doors, interior, flush, stain grade, spray application, based on per door method

	Manhours per door	Doors per gallon	Material cost per gallon	Labor cost per door	Labor burden door	Material cost per door	Overhead per door	Profit per door	Total price per door
Complete 7 step stain, seal & 2 coat lacquer system (material #11)									
Spray all coats									
Slow	0.90	6	54.90	20.70	4.97	9.15	6.62	6.63	48.07
Medium	0.80	5	48.00	23.60	6.82	9.60	10.01	6.00	56.03
Fast	0.70	4	41.20	25.20	8.90	10.30	13.76	4.07	62.23

This table is not to be confused with the Opening Count Method for estimating doors. Use these figures for painting both interior and exterior doors. These costs are in addition to the costs for applying the finish coats to exterior doors as explained on page 97. These figures include coating all six (6) sides of each door along with the frame and the jamb on both sides. These figures include minimal preparation time. ADD for preparation time, as needed, including setup, remove & replace (hardware), protection (mask walls and visqueen floors); putty, caulking, sanding and cleanup. "Slow" work is based on an hourly wage of $23.00, "Medium" work on an hourly wage of $29.50, and "Fast" work on an hourly wage of $36.00. Other qualifications that apply to this table are on page 9.

Doors, interior, French, paint grade, roll & brush, based on per door method

	Manhours per door	Doors per gallon	Material cost per gallon	Labor cost per door	Labor burden door	Material cost per door	Overhead per door	Profit per door	Total price per door
Undercoat, water base (material #3)									
Roll & brush 1 coat									
Slow	0.45	14	36.90	10.35	2.48	2.64	2.94	2.95	21.36
Medium	0.38	13	32.30	11.21	3.24	2.48	4.23	2.54	23.70
Fast	0.30	12	27.70	10.80	3.81	2.31	5.25	1.55	23.72
Undercoat, oil base (material #4)									
Roll & brush 1 coat									
Slow	0.45	14	48.30	10.35	2.48	3.45	3.09	3.10	22.47
Medium	0.38	13	42.30	11.21	3.24	3.25	4.43	2.66	24.79
Fast	0.30	12	36.20	10.80	3.81	3.02	5.47	1.62	24.72
Enamel, water base (material #9)									
Roll & brush 1st finish coat									
Slow	0.43	15	53.30	9.89	2.37	3.55	3.00	3.01	21.82
Medium	0.35	14	46.70	10.33	2.98	3.34	4.17	2.50	23.32
Fast	0.28	13	40.00	10.08	3.56	3.08	5.18	1.53	23.43
Roll & brush 2nd or additional finish coats									
Slow	0.40	16	53.30	9.20	2.21	3.33	2.80	2.81	20.35
Medium	0.33	15	46.70	9.74	2.81	3.11	3.92	2.35	21.93
Fast	0.25	14	40.00	9.00	3.18	2.86	4.66	1.38	21.08

	Manhours per door	Doors per gallon	Material cost per gallon	Labor cost per door	Labor burden door	Material cost per door	Overhead per door	Profit per door	Total price per door
Enamel, oil base (material #10)									
Roll & brush 1st finish coat									
Slow	0.43	15	61.50	9.89	2.37	4.10	3.11	3.12	22.59
Medium	0.35	14	53.80	10.33	2.98	3.84	4.29	2.57	24.01
Fast	0.28	13	46.10	10.08	3.56	3.55	5.33	1.58	24.10
Roll & brush 2nd or additional finish coats									
Slow	0.40	16	61.50	9.20	2.21	3.84	2.90	2.90	21.05
Medium	0.33	15	53.80	9.74	2.81	3.59	4.04	2.42	22.60
Fast	0.25	14	46.10	9.00	3.18	3.29	4.80	1.42	21.69

This table is not to be confused with the Opening Count Method for estimating doors. Use these figures for painting both interior and exterior doors. These costs are in addition to the costs for applying the finish coats to exterior doors as explained on page 97. These figures include coating all six (6) sides of each door along with the frame and the jamb on both sides. These figures include minimal preparation time. ADD for additional preparation time, as needed, including setup, remove & replace (hardware), protection (mask walls and visqueen floors); putty, caulking, sanding and cleanup. "Slow" work is based on an hourly wage of $23.00, "Medium" work on an hourly wage of $29.50, and "Fast" work on an hourly wage of $36.00. Other qualifications that apply to this table are on page 9.

Doors, interior, French, stain grade, spray application, based on per door method

	Manhours per door	Doors per gallon	Material cost per gallon	Labor cost per door	Labor burden door	Material cost per door	Overhead per door	Profit per door	Total price per door
Complete 7 step stain, seal & 2 coat lacquer system (material #11)									
Spray all coats									
Slow	1.50	12	54.90	34.50	8.28	4.58	9.00	9.02	65.38
Medium	1.25	11	48.00	36.88	10.66	4.36	12.98	7.79	72.67
Fast	1.00	10	41.20	36.00	12.71	4.12	16.38	4.84	74.05

This table is not to be confused with the Opening Count Method for estimating doors. Use these figures for painting both interior and exterior doors. These costs are in addition to the costs for applying the finish coats to exterior doors as explained on page 97. These figures include coating all six (6) sides of each door along with the frame and the jamb on both sides. These figures include minimal preparation time. ADD for additional preparation time, as needed, including setup, remove & replace (hardware), protection (mask walls and visqueen floors); putty, caulking, sanding and cleanup. "Slow" work is based on an hourly wage of $23.00, "Medium" work on an hourly wage of $29.50, and "Fast" work on an hourly wage of $36.00. Other qualifications that apply to this table are on page 9.

	Manhours per door	Doors per gallon	Material cost per gallon	Labor cost per door	Labor burden door	Material cost per door	Overhead per door	Profit per door	Total price per door

Doors, interior, louvered, paint grade, roll & brush, based on per door method

	Manhours per door	Doors per gallon	Material cost per gallon	Labor cost per door	Labor burden door	Material cost per door	Overhead per door	Profit per door	Total price per door
Undercoat, water base (material #3)									
Roll & brush 1 coat									
Slow	0.67	8	36.90	15.41	3.70	4.61	4.51	4.52	32.75
Medium	0.54	7	32.30	15.93	4.61	4.61	6.29	3.77	35.21
Fast	0.40	6	27.70	14.40	5.08	4.62	7.47	2.21	33.78
Undercoat, oil base (material #4)									
Roll & brush 1 coat									
Slow	0.67	8	48.30	15.41	3.70	6.04	4.78	4.79	34.72
Medium	0.54	7	42.30	15.93	4.61	6.04	6.64	3.99	37.21
Fast	0.40	6	36.20	14.40	5.08	6.03	7.91	2.34	35.76
Enamel, water base (material #9)									
Roll & brush 1st finish coat									
Slow	0.50	9	53.30	11.50	2.76	5.92	3.83	3.84	27.85
Medium	0.42	8	46.70	12.39	3.58	5.84	5.45	3.27	30.53
Fast	0.33	7	40.00	11.88	4.19	5.71	6.75	2.00	30.53
Roll & brush 2nd or additional finish coats									
Slow	0.40	10	53.30	9.20	2.21	5.33	3.18	3.19	23.11
Medium	0.30	9	46.70	8.85	2.56	5.19	4.15	2.49	23.24
Fast	0.20	8	40.00	7.20	2.54	5.00	4.57	1.35	20.66
Enamel, oil base (material #10)									
Roll & brush 1st finish coat									
Slow	0.50	9	61.50	11.50	2.76	6.83	4.01	4.02	29.12
Medium	0.42	8	53.80	12.39	3.58	6.73	5.68	3.41	31.79
Fast	0.33	7	46.10	11.88	4.19	6.59	7.02	2.08	31.76
Roll & brush 2nd or additional finish coats									
Slow	0.40	10	61.50	9.20	2.21	6.15	3.34	3.34	24.24
Medium	0.30	9	53.80	8.85	2.56	5.98	4.35	2.61	24.35
Fast	0.20	8	46.10	7.20	2.54	5.76	4.81	1.42	21.73

This table is not to be confused with the Opening Count Method for estimating doors. Use these figures for painting both interior and exterior doors. These costs are in addition to the costs for applying the finish coats to exterior doors as explained on page 97. These figures include coating all six (6) sides of each door along with the frame and the jamb on both sides. These figures include minimal preparation time. ADD for additional preparation time, as needed, including setup, remove & replace (hardware), protection (mask walls and visqueen floors); putty, caulking, sanding and cleanup. "Slow" work is based on an hourly wage of $23.00, "Medium" work on an hourly wage of $29.50, and "Fast" work on an hourly wage of $36.00. Other qualifications that apply to this table are on page 9.

	Manhours per door	Doors per gallon	Material cost per gallon	Labor cost per door	Labor burden door	Material cost per door	Overhead per door	Profit per door	Total price per door

Doors, interior, louvered, paint grade, spray application, based on per door method

	Manhours per door	Doors per gallon	Material cost per gallon	Labor cost per door	Labor burden door	Material cost per door	Overhead per door	Profit per door	Total price per door
Undercoat, water base (material #3)									
Spray 1 coat									
Slow	0.17	12	36.90	3.91	.94	3.08	1.51	1.51	10.95
Medium	0.14	11	32.30	4.13	1.19	2.94	2.07	1.24	11.57
Fast	0.12	10	27.70	4.32	1.53	2.77	2.67	.79	12.08
Undercoat, oil base (material #4)									
Spray 1 coat									
Slow	0.17	12	48.30	3.91	.94	4.03	1.69	1.69	12.26
Medium	0.14	11	42.30	4.13	1.19	3.85	2.29	1.38	12.84
Fast	0.12	10	36.20	4.32	1.53	3.62	2.93	.87	13.27
Enamel, water base (material #9)									
Spray 1st finish coat									
Slow	0.13	13	53.30	2.99	.72	4.10	1.48	1.49	10.78
Medium	0.11	12	46.70	3.25	.93	3.89	2.02	1.21	11.30
Fast	0.09	11	40.00	3.24	1.14	3.64	2.49	.74	11.25
Spray 2nd or additional finish coats									
Slow	0.10	14	53.30	2.30	.55	3.81	1.27	1.27	9.20
Medium	0.09	13	46.70	2.66	.76	3.59	1.76	1.05	9.82
Fast	0.08	12	40.00	2.88	1.02	3.33	2.24	.66	10.13
Enamel, oil base (material #10)									
Spray 1st finish coat									
Slow	0.13	13	61.50	2.99	.72	4.73	1.60	1.61	11.65
Medium	0.11	12	53.80	3.25	.93	4.48	2.17	1.30	12.13
Fast	0.09	11	46.10	3.24	1.14	4.19	2.66	.79	12.02
Spray 2nd or additional finish coats									
Slow	0.10	14	61.50	2.30	.55	4.39	1.38	1.38	10.00
Medium	0.09	13	53.80	2.66	.76	4.14	1.89	1.14	10.59
Fast	0.08	12	46.10	2.88	1.02	3.84	2.40	.71	10.85

This table is not to be confused with the Opening Count Method for estimating doors. Use these figures for painting both interior and exterior doors. These costs are in addition to the costs for applying the finish coats to exterior doors as explained on page 97. These figures include coating all six (6) sides of each door along with the frame and the jamb on both sides. These figures include minimal preparation time. ADD for additional preparation time, as needed, including setup, remove & replace (hardware), protection (mask walls and visqueen floors); putty, caulking, sanding and cleanup. "Slow" work is based on an hourly wage of $23.00, "Medium" work on an hourly wage of $29.50, and "Fast" work on an hourly wage of $36.00. Other qualifications that apply to this table are on page 9.

	Manhours per door	Doors per gallon	Material cost per gallon	Labor cost per door	Labor burden door	Material cost per door	Overhead per door	Profit per door	Total price per door

Doors, interior, louvered, stain grade, spray application, based on per door method

Complete 7 step stain, seal & 2 coat lacquer system (material #11)

Spray all coats									
Slow	1.70	5	54.90	39.10	9.38	10.98	11.30	11.32	82.08
Medium	1.45	4	48.00	42.78	12.36	12.00	16.79	10.07	94.00
Fast	1.20	3	41.20	43.20	15.25	13.73	22.38	6.62	101.18

This table is not to be confused with the Opening Count Method for estimating doors. Use these figures for painting both interior and exterior doors. These costs are in addition to the costs for applying the finish coats to exterior doors as explained on page 97. These figures include coating all six (6) sides of each door along with the frame and the jamb on both sides. These figures include minimal preparation time. ADD for additional preparation time, as needed, including setup, remove & replace (hardware), protection (mask walls and visqueen floors); putty, caulking, sanding and cleanup. "Slow" work is based on an hourly wage of $23.00, "Medium" work on an hourly wage of $29.50, and "Fast" work on an hourly wage of $36.00. Other qualifications that apply to this table are on page 9.

	Manhours per door	Doors per gallon	Material cost per gallon	Labor cost per door	Labor burden door	Material cost per door	Overhead per door	Profit per door	Total price per door
Doors, interior, panel, paint grade, roll & brush, based on per door method									
Undercoat, water base (material #3)									
Roll & brush 1 coat									
Slow	0.50	8	36.90	11.50	2.76	4.61	3.59	3.59	26.05
Medium	0.33	7	32.30	9.74	2.81	4.61	4.29	2.57	24.02
Fast	0.25	6	27.70	9.00	3.18	4.62	5.21	1.54	23.55
Undercoat, oil base (material #4)									
Roll & brush 1 coat									
Slow	0.50	8	48.30	11.50	2.76	6.04	3.86	3.87	28.03
Medium	0.33	7	42.30	9.74	2.81	6.04	4.65	2.79	26.03
Fast	0.25	6	36.20	9.00	3.18	6.03	5.65	1.67	25.53
Enamel, water base (material #9)									
Roll & brush 1st finish coat									
Slow	0.40	11	53.30	9.20	2.21	4.85	3.09	3.10	22.45
Medium	0.30	10	46.70	8.85	2.56	4.67	4.02	2.41	22.51
Fast	0.20	9	40.00	7.20	2.54	4.44	4.40	1.30	19.88
Roll & brush 2nd or additional finish coats									
Slow	0.33	12	53.30	7.59	1.82	4.44	2.63	2.64	19.12
Medium	0.25	11	46.70	7.38	2.13	4.25	3.44	2.06	19.26
Fast	0.17	10	40.00	6.12	2.16	4.00	3.81	1.13	17.22
Enamel, oil base (material #10)									
Roll & brush 1st finish coat									
Slow	0.40	11	61.50	9.20	2.21	5.59	3.23	3.24	23.47
Medium	0.30	10	53.80	8.85	2.56	5.38	4.20	2.52	23.51
Fast	0.20	9	46.10	7.20	2.54	5.12	4.61	1.36	20.83
Roll & brush 2nd or additional finish coats									
Slow	0.33	12	61.50	7.59	1.82	5.13	2.76	2.77	20.07
Medium	0.25	11	53.80	7.38	2.13	4.89	3.60	2.16	20.16
Fast	0.17	10	46.10	6.12	2.16	4.61	4.00	1.18	18.07

This table is not to be confused with the Opening Count Method for estimating doors. Use these figures for painting both interior and exterior doors. These costs are in addition to the costs for applying the finish coats to exterior doors as explained on page 97. These figures include coating all six (6) sides of each door along with the frame and the jamb on both sides. These figures include minimal preparation time. ADD for additional preparation time, as needed, including setup, remove & replace (hardware), protection (mask walls and visqueen floors); putty, caulking, sanding and cleanup. "Slow" work is based on an hourly wage of $23.00, "Medium" work on an hourly wage of $29.50, and "Fast" work on an hourly wage of $36.00. Other qualifications that apply to this table are on page 9.

	Manhours per door	Doors per gallon	Material cost per gallon	Labor cost per door	Labor burden door	Material cost per door	Overhead per door	Profit per door	Total price per door
Doors, interior, panel, paint grade, spray application, based on per door method									
Undercoat, water base (material #3)									
Spray 1 coat									
Slow	0.13	15	36.90	2.99	.72	2.46	1.17	1.17	8.51
Medium	0.11	14	32.30	3.25	.93	2.31	1.63	.98	9.10
Fast	0.10	13	27.70	3.60	1.27	2.13	2.17	.64	9.81
Undercoat, oil base (material #4)									
Spray 1 coat									
Slow	0.13	15	48.30	2.99	.72	3.22	1.32	1.32	9.57
Medium	0.11	14	42.30	3.25	.93	3.02	1.80	1.08	10.08
Fast	0.10	13	36.20	3.60	1.27	2.78	2.37	.70	10.72
Enamel, water base (material #9)									
Spray 1st finish coat									
Slow	0.09	16	53.30	2.07	.50	3.33	1.12	1.12	8.14
Medium	0.08	15	46.70	2.36	.68	3.11	1.54	.92	8.61
Fast	0.08	14	40.00	2.88	1.02	2.86	2.10	.62	9.48
Spray 2nd or additional finish coats									
Slow	0.08	17	53.30	1.84	.44	3.14	1.03	1.03	7.48
Medium	0.08	16	46.70	2.36	.68	2.92	1.49	.89	8.34
Fast	0.07	15	40.00	2.52	.89	2.67	1.88	.56	8.52
Enamel, oil base (material #10)									
Spray 1st finish coat									
Slow	0.09	16	61.50	2.07	.50	3.84	1.22	1.22	8.85
Medium	0.08	15	53.80	2.36	.68	3.59	1.66	.99	9.28
Fast	0.08	14	46.10	2.88	1.02	3.29	2.23	.66	10.08
Spray 2nd or additional finish coats									
Slow	0.08	17	61.50	1.84	.44	3.62	1.12	1.12	8.14
Medium	0.08	16	53.80	2.36	.68	3.36	1.60	.96	8.96
Fast	0.07	15	46.10	2.52	.89	3.07	2.01	.59	9.08

This table is not to be confused with the Opening Count Method for estimating doors. Use these figures for painting both interior and exterior doors. These costs are in addition to the costs for applying the finish coats to exterior doors as explained on page 97. These figures include coating all six (6) sides of each door along with the frame and the jamb on both sides. These figures include minimal preparation time. ADD for additional preparation time, as needed, including setup, remove & replace (hardware), protection (mask walls and visqueen floors); putty, caulking, sanding and cleanup. "Slow" work is based on an hourly wage of $23.00, "Medium" work on an hourly wage of $29.50, and "Fast" work on an hourly wage of $36.00. Other qualifications that apply to this table are on page 9.

	Manhours per door	Doors per gallon	Material cost per gallon	Labor cost per door	Labor burden door	Material cost per door	Overhead per door	Profit per door	Total price per door

Doors, interior, panel, stain grade, spray application, based on per door method

	Manhours per door	Doors per gallon	Material cost per gallon	Labor cost per door	Labor burden door	Material cost per door	Overhead per door	Profit per door	Total price per door
Complete 7 step stain, seal & 2 coat lacquer system (material #11)									
Spray all coats									
Slow	1.25	6	54.90	28.75	6.90	9.15	8.51	8.53	61.84
Medium	1.08	5	48.00	31.86	9.21	9.60	12.67	7.60	70.94
Fast	0.90	4	41.20	32.40	11.44	10.30	16.78	4.96	75.88

This table is not to be confused with the Opening Count Method for estimating doors. Use these figures for painting both interior and exterior doors. These costs are in addition to the costs for applying the finish coats to exterior doors as explained on page 97. These figures include coating all six (6) sides of each door along with the frame and the jamb on both sides. These figures include minimal preparation time. ADD for additional preparation time, as needed, including setup, remove & replace (hardware), protection (mask walls and visqueen floors); putty, caulking, sanding and cleanup. "Slow" work is based on an hourly wage of $23.00, "Medium" work on an hourly wage of $29.50, and "Fast" work on an hourly wage of $36.00. Other qualifications that apply to this table are on page 9.

Fascia, 2" x 4", brush one coat stain "to cover"

	Labor LF per manhour	Material coverage LF/gallon	Material cost per gallon	Labor cost per 100 LF	Labor burden 100 LF	Material cost per 100 LF	Overhead per 100 LF	Profit per 100 LF	Total price per 100 LF
Solid body stain, water or oil base (material #18 or #19)									
Brush each coat									
Slow	80	170	54.90	28.75	6.90	32.29	12.91	12.94	93.79
Medium	105	160	48.05	28.10	8.10	30.03	16.56	9.94	92.73
Fast	130	150	41.20	27.69	9.77	27.47	20.13	5.95	91.01
Semi-transparent stain, water or oil base (material #20 or #21)									
Brush each coat									
Slow	95	195	54.10	24.21	5.82	27.74	10.97	11.00	79.74
Medium	120	185	47.35	24.58	7.10	25.59	14.32	8.59	80.18
Fast	145	175	40.55	24.83	8.78	23.17	17.60	5.21	79.59

Use these figures for brushing stain on 2" x 4" fascia board on two sides - the face and the lower edge. The back side is calculated with the overhang (eaves) operation. These figures include one full coat and any touchup required to meet the "to cover" specification. They also include minimum preparation time. ADD for additional preparation time, as needed, including setup, remove & replace (hardware), protection (masking and visqueen); putty, caulking, sanding and cleanup. Measurements are based on continuous linear feet. "Slow" work is based on an hourly wage of $23.00, "Medium" work on an hourly wage of $29.50, and "Fast" work on an hourly wage of $36.00. Other qualifications that apply to this table are on page 9.

	Labor LF per manhour	Material coverage LF/gallon	Material cost per gallon	Labor cost per 100 LF	Labor burden 100 LF	Material cost per 100 LF	Overhead per 100 LF	Profit per 100 LF	Total price per 100 LF
Fascia, 2" x 4", roll one coat stain "to cover"									
Solid body stain, water or oil base (material #18 or #19)									
Roll each coat									
Slow	180	140	54.90	12.78	3.08	39.21	10.46	10.48	76.01
Medium	205	130	48.05	14.39	4.17	36.96	13.88	8.33	77.73
Fast	230	120	41.20	15.65	5.54	34.33	17.21	5.09	77.82
Semi-transparent stain, water or oil base (material #20 or #21)									
Roll each coat									
Slow	200	160	54.10	11.50	2.76	33.81	9.13	9.15	66.35
Medium	225	150	47.35	13.11	3.78	31.57	12.12	7.27	67.85
Fast	250	140	40.55	14.40	5.08	28.96	15.02	4.44	67.90

Use these figures for rolling stain on 2" x 4" fascia board on two sides - the face and the lower edge. The back side is calculated with the overhang (eaves) operation. These figures include one full coat and any touchup required to meet the "to cover" specification. They also include minimum preparation time. ADD time for extensive preparation. Measurements are based on continuous linear feet. "Slow" work is based on an hourly wage of $23.00, "Medium" work on an hourly wage of $29.50, and "Fast" work on an hourly wage of $36.00. Other qualifications that apply to this table are on page 9.

	Labor LF per manhour	Material coverage LF/gallon	Material cost per gallon	Labor cost per 100 LF	Labor burden 100 LF	Material cost per 100 LF	Overhead per 100 LF	Profit per 100 LF	Total price per 100 LF
Fascia, 2" x 4", spray one coat stain "to cover"									
Solid body stain, water or oil base (material #18 or #19)									
Spray each coat									
Slow	275	110	54.90	8.36	2.02	49.91	11.45	11.48	83.22
Medium	325	100	48.05	9.08	2.63	48.05	14.94	8.96	83.66
Fast	375	90	41.20	9.60	3.41	45.78	18.22	5.39	82.40
Semi-transparent stain, water or oil base (material #20 or #21)									
Spray each coat									
Slow	300	125	54.10	7.67	1.83	43.28	10.03	10.05	72.86
Medium	350	115	47.35	8.43	2.45	41.17	13.01	7.81	72.87
Fast	400	105	40.55	9.00	3.18	38.62	15.75	4.66	71.21

Use these figures for spraying stain on 2" x 4" fascia board on two sides - the face and the lower edge. The back side is calculated with the overhang (eaves) operation. These figures include one full coat and any touchup required to meet the "to cover" specification. They also include minimum preparation time. ADD time for extensive preparation. Measurements are based on continuous linear feet. "Slow" work is based on an hourly wage of $23.00, "Medium" work on an hourly wage of $29.50, and "Fast" work on an hourly wage of $36.00. Other qualifications that apply to this table are on page 9.

	Labor LF per manhour	Material coverage LF/gallon	Material cost per gallon	Labor cost per 100 LF	Labor burden 100 LF	Material cost per 100 LF	Overhead per 100 LF	Profit per 100 LF	Total price per 100 LF

Fascia, 2" x 6" to 2" x 10", brush one coat stain "to cover"

Solid body stain, water or oil base (material #18 or #19)									
Brush each coat									
Slow	70	140	54.90	32.86	7.90	39.21	15.19	15.22	110.38
Medium	90	130	48.05	32.78	9.47	36.96	19.80	11.88	110.89
Fast	110	120	41.20	32.73	11.55	34.33	24.37	7.21	110.19
Semi-transparent stain, water or oil base (material #20 or #21)									
Brush each coat									
Slow	85	165	54.10	27.06	6.48	32.79	12.60	12.63	91.56
Medium	105	155	47.35	28.10	8.10	30.55	16.69	10.02	93.46
Fast	125	145	40.55	28.80	10.17	27.97	20.75	6.14	93.83

Use these figures for brushing stain on 2" x 6" to 2" x 10" fascia board on two sides - the face and the lower edge. The back side is calculated with the overhang (eaves) operation. These figures include one full coat and any touchup required to meet the "to cover" specification. They also include minimum preparation time. ADD time for extensive preparation. Measurements are based on continuous linear feet. "Slow" work is based on an hourly wage of $23.00, "Medium" work on an hourly wage of $29.50, and "Fast" work on an hourly wage of $36.00. Other qualifications that apply to this table are on page 9.

	Labor LF per manhour	Material coverage LF/gallon	Material cost per gallon	Labor cost per 100 LF	Labor burden 100 LF	Material cost per 100 LF	Overhead per 100 LF	Profit per 100 LF	Total price per 100 LF

Fascia, 2" x 6" to 2" x 10", roll one coat stain "to cover"

Solid body stain, water or oil base (material #18 or #19)									
Roll each coat									
Slow	150	120	54.90	15.33	3.69	45.75	12.30	12.33	89.40
Medium	175	110	48.05	16.86	4.86	43.68	16.35	9.81	91.56
Fast	200	100	41.20	18.00	6.36	41.20	20.32	6.01	91.89
Semi-transparent stain, water or oil base (material #20 or #21)									
Roll each coat									
Slow	170	140	54.10	13.53	3.24	38.64	10.53	10.55	76.49
Medium	195	130	47.35	15.13	4.38	36.42	13.98	8.39	78.30
Fast	220	120	40.55	16.36	5.80	33.79	17.34	5.13	78.42

Use these figures for rolling stain on 2" x 6" to 2" x 10" fascia board on two sides - the face and the lower edge. The back side is calculated with the overhang (eaves) operation. These figures include one full coat and any touchup required to meet the "to cover" specification. They also include minimum preparation time. ADD time for extensive preparation. Measurements are based on continuous linear feet. "Slow" work is based on an hourly wage of $23.00, "Medium" work on an hourly wage of $29.50, and "Fast" work on an hourly wage of $36.00. Other qualifications that apply to this table are on page 9.

	Labor LF per manhour	Material coverage LF/gallon	Material cost per gallon	Labor cost per 100 LF	Labor burden 100 LF	Material cost per 100 LF	Overhead per 100 LF	Profit per 100 LF	Total price per 100 LF

Fascia, 2" x 6" to 2" x 10", spray one coat stain "to cover"

	Labor LF per manhour	Material coverage LF/gallon	Material cost per gallon	Labor cost per 100 LF	Labor burden 100 LF	Material cost per 100 LF	Overhead per 100 LF	Profit per 100 LF	Total price per 100 LF
Solid body stain, water or oil base (material #18 or #19)									
Spray each coat									
Slow	225	90	54.90	10.22	2.44	61.00	14.00	14.03	101.69
Medium	300	80	48.05	9.83	2.83	60.06	18.18	10.91	101.81
Fast	350	70	41.20	10.29	3.64	58.86	22.56	6.67	102.02
Semi-transparent stain, water or oil base (material #20 or #21)									
Spray each coat									
Slow	250	105	54.10	9.20	2.21	51.52	11.96	11.98	86.87
Medium	313	95	47.35	9.42	2.71	49.84	15.50	9.30	86.77
Fast	375	85	40.55	9.60	3.41	47.71	18.82	5.57	85.11

Use these figures for spraying stain on 2" x 6" to 2" x 10" fascia board on two sides - the face and the lower edge. The back side is calculated with the overhang (eaves) operation. These figures include one full coat and any touchup required to meet the "to cover" specification. They also include minimum preparation. ADD time for extensive preparation. Measurements are based on continuous linear feet. "Slow" work is based on an hourly wage of $23.00, "Medium" work on an hourly wage of $29.50, and "Fast" work on an hourly wage of $36.00. Other qualifications that apply to this table are on page 9.

Fascia, 2" x 12", brush one coat stain "to cover"

	Labor LF per manhour	Material coverage LF/gallon	Material cost per gallon	Labor cost per 100 LF	Labor burden 100 LF	Material cost per 100 LF	Overhead per 100 LF	Profit per 100 LF	Total price per 100 LF
Solid body stain, water or oil base (material #18 or #19)									
Brush each coat									
Slow	60	100	54.90	38.33	9.21	54.90	19.46	19.50	141.40
Medium	80	90	48.05	36.88	10.66	53.39	25.23	15.14	141.30
Fast	100	80	41.20	36.00	12.71	51.50	31.07	9.19	140.47
Semi-transparent stain, water or oil base (material #20 or #21)									
Brush each coat									
Slow	75	125	54.10	30.67	7.35	43.28	15.45	15.48	112.23
Medium	95	115	47.35	31.05	9.00	41.17	20.30	12.18	113.70
Fast	105	105	40.55	34.29	12.08	38.62	26.35	7.80	119.14

Use these figures for brushing stain on 2" x 12" fascia board on two sides - the face and the lower edge. The back side is calculated with the overhang (eaves) operation. These figures include one full coat and any touchup required to meet the "to cover" specification. They also include minimum preparation. ADD time for extensive preparation. Measurements are based on continuous linear feet. "Slow" work is based on an hourly wage of $23.00, "Medium" work on an hourly wage of $29.50, and "Fast" work on an hourly wage of $36.00. Other qualifications that apply to this table are on page 9.

	Labor LF per manhour	Material coverage LF/gallon	Material cost per gallon	Labor cost per 100 LF	Labor burden 100 LF	Material cost per 100 LF	Overhead per 100 LF	Profit per 100 LF	Total price per 100 LF

Fascia, 2" x 12", roll one coat stain "to cover"

Solid body stain, water or oil base (material #18 or #19)

Roll each coat

Slow	110	90	54.90	20.91	5.01	61.00	16.52	16.55	119.99
Medium	130	75	48.05	22.69	6.56	64.07	23.33	14.00	130.65
Fast	150	60	41.20	24.00	8.49	68.67	31.35	9.27	141.78

Semi-transparent stain, water or oil base (material #20 or #21)

Roll each coat

Slow	130	110	54.10	17.69	4.24	49.18	13.51	13.54	98.16
Medium	150	95	47.35	19.67	5.70	49.84	18.80	11.28	105.29
Fast	170	80	40.55	21.18	7.46	50.69	24.60	7.28	111.21

Use these figures for rolling stain on 2" x 12" fascia board on two sides - the face and the lower edge. The back side is calculated with the overhang (eaves) operation. These figures include one full coat and any touchup required to meet the "to cover" specification. They also include minimum preparation. ADD time for extensive preparation. Measurements are based on continuous linear feet. "Slow" work is based on an hourly wage of $23.00, "Medium" work on an hourly wage of $29.50, and "Fast" work on an hourly wage of $36.00. Other qualifications that apply to this table are on page 9.

	Labor LF per manhour	Material coverage LF/gallon	Material cost per gallon	Labor cost per 100 LF	Labor burden 100 LF	Material cost per 100 LF	Overhead per 100 LF	Profit per 100 LF	Total price per 100 LF

Fascia, 2" x 12", spray one coat stain "to cover"

Solid body stain, water or oil base (material #18 or #19)

Spray each coat

Slow	200	60	54.90	11.50	2.76	91.50	20.09	20.14	145.99
Medium	263	50	48.05	11.22	3.23	96.10	27.64	16.58	154.77
Fast	325	40	41.20	11.08	3.92	103.00	36.58	10.82	165.40

Semi-transparent stain, water or oil base (material #20 or #21)

Spray each coat

Slow	225	75	54.10	10.22	2.44	72.13	16.11	16.15	117.05
Medium	288	65	47.35	10.24	2.96	72.85	21.51	12.91	120.47
Fast	350	55	40.55	10.29	3.64	73.73	27.17	8.04	122.87

Use these figures for spraying stain on 2" x 12" fascia board on two sides - the face and the lower edge. The back side is calculated with the overhang (eaves) operation. These figures include one full coat and any touchup required to meet the "to cover" specification. They also include minimum preparation. ADD time for extensive preparation. Measurements are based on continuous linear feet. "Slow" work is based on an hourly wage of $23.00, "Medium" work on an hourly wage of $29.50, and "Fast" work on an hourly wage of $36.00. Other qualifications that apply to this table are on page 9.

	Labor SF per manhour	Material coverage SF/gallon	Material cost per gallon	Labor cost per 100 SF	Labor burden 100 SF	Material cost per 100 SF	Overhead per 100 SF	Profit per 100 SF	Total price per 100 SF

Fence, chain link or wire mesh

Solid body stain, water or oil base (material #18 or #19)

	Labor SF per manhour	Material coverage SF/gallon	Material cost per gallon	Labor cost per 100 SF	Labor burden 100 SF	Material cost per 100 SF	Overhead per 100 SF	Profit per 100 SF	Total price per 100 SF
Brush 1st coat									
Slow	90	600	54.90	25.56	6.13	9.15	7.76	7.78	56.38
Medium	110	550	48.05	26.82	7.75	8.74	10.83	6.50	60.64
Fast	125	500	41.20	28.80	10.17	8.24	14.64	4.33	66.18
Brush 2nd coat									
Slow	130	650	54.90	17.69	4.24	8.45	5.77	5.79	41.94
Medium	145	600	48.05	20.34	5.90	8.01	8.56	5.13	47.94
Fast	160	550	41.20	22.50	7.94	7.49	11.76	3.48	53.17
Roll 1st coat									
Slow	260	575	54.90	8.85	2.13	9.55	3.90	3.91	28.34
Medium	275	525	48.05	10.73	3.11	9.15	5.75	3.45	32.19
Fast	290	475	41.20	12.41	4.39	8.67	7.89	2.33	35.69
Roll 2nd coat									
Slow	280	625	54.90	8.21	1.97	8.78	3.60	3.61	26.17
Medium	300	575	48.05	9.83	2.83	8.36	5.26	3.15	29.43
Fast	320	525	41.20	11.25	4.00	7.85	7.15	2.12	32.37

Use these figures for chain link or wire mesh fencing. The figures are based on painting both sides to meet the "to cover" specification. These figures include minimum preparation time. ADD time for extensive preparation. To calculate the area, base measurements on the square feet (length times width) of one side of the fence then multiply by a difficulty factor of 3 and use the figures in the above table. For example, if the fence is 100' long x 3' high, the area is 300 SF. Multiply 300 SF x 3 to arrive at 900 SF which is the total to be used with this table. "Slow" work is based on an hourly wage of $23.00, "Medium" work on an hourly wage of $29.50, and "Fast" work on an hourly wage of $36.00. Other qualifications that apply to this table are on page 9.

Fence, wood

For *solid plank fence,* measure the surface area of one side and multiply by 2 to find the area for both sides. Then use the cost table for Siding, exterior. For good neighbor fence (planks on alternate sides of the rail), measure the surface area of one side and multiply by 2 to find the area for both sides. Then multiply by a difficulty factor of 1.5 and use the pricing table for Siding, exterior.

	Labor SF per manhour	Material coverage SF/gallon	Material cost per gallon	Labor cost per 100 SF	Labor burden 100 SF	Material cost per 100 SF	Overhead per 100 SF	Profit per 100 SF	Total price per 100 SF
Fence, picket, brush application									
Solid body or semi-transparent stain, water base (material #18 or #20)									
Brush 1st coat									
Slow	75	400	49.40	30.67	7.35	12.35	9.57	9.59	69.53
Medium	113	388	43.25	26.11	7.55	11.15	11.20	6.72	62.73
Fast	150	375	37.05	24.00	8.49	9.88	13.13	3.88	59.38
Brush 2nd or additional coats									
Slow	120	450	49.40	19.17	4.59	10.98	6.60	6.62	47.96
Medium	145	438	43.25	20.34	5.90	9.87	9.02	5.41	50.54
Fast	170	425	37.05	21.18	7.46	8.72	11.59	3.43	52.38
Solid body or semi-transparent stain, oil base (material #19 or #21)									
Brush 1st coat									
Slow	75	450	59.60	30.67	7.35	13.24	9.74	9.76	70.76
Medium	113	438	52.15	26.11	7.55	11.91	11.39	6.84	63.80
Fast	150	425	44.70	24.00	8.49	10.52	13.33	3.94	60.28
Brush 2nd or additional coats									
Slow	120	500	59.60	19.17	4.59	11.92	6.78	6.80	49.26
Medium	145	488	52.15	20.34	5.90	10.69	9.23	5.54	51.70
Fast	170	475	44.70	21.18	7.46	9.41	11.80	3.49	53.34

For picket fence, measure the overall area of one side and multiply by 4 for painting both sides. Then apply these cost figures. "Slow" work is based on an hourly wage of $23.00, "Medium" work on an hourly wage of $29.50, and "Fast" work on an hourly wage of $36.00. Other qualifications that apply to this table are on page 9.

	Labor SF per manhour	Material coverage SF/gallon	Material cost per gallon	Labor cost per 100 SF	Labor burden 100 SF	Material cost per 100 SF	Overhead per 100 SF	Profit per 100 SF	Total price per 100 SF

Fence, picket, roll application

Solid body or semi-transparent stain, water base (material #18 or #20)

Roll 1st coat									
Slow	120	360	49.40	19.17	4.59	13.72	7.12	7.14	51.74
Medium	145	343	43.25	20.34	5.90	12.61	9.71	5.82	54.38
Fast	170	325	37.05	21.18	7.46	11.40	12.42	3.67	56.13
Roll 2nd or additional coats									
Slow	200	400	49.40	11.50	2.76	12.35	5.06	5.07	36.74
Medium	225	388	43.25	13.11	3.78	11.15	7.01	4.21	39.26
Fast	250	375	37.05	14.40	5.08	9.88	9.10	2.69	41.15

Solid body or semi-transparent stain, oil base (material #19 or #21)

Roll 1st coat									
Slow	120	400	59.60	19.17	4.59	14.90	7.35	7.36	53.37
Medium	145	388	52.15	20.34	5.90	13.44	9.92	5.95	55.55
Fast	170	375	44.70	21.18	7.46	11.92	12.58	3.72	56.86
Roll 2nd or additional coats									
Slow	200	450	59.60	11.50	2.76	13.24	5.23	5.24	37.97
Medium	225	438	52.15	13.11	3.78	11.91	7.20	4.32	40.32
Fast	250	425	44.70	14.40	5.08	10.52	9.30	2.75	42.05

For picket fence, measure the overall area of one side and multiply by 4 for painting both sides. Then apply these cost figures. "Slow" work is based on an hourly wage of $23.00, "Medium" work on an hourly wage of $29.50, and "Fast" work on an hourly wage of $36.00. Other qualifications that apply to this table are on page 9.

	Labor SF per manhour	Material coverage SF/gallon	Material cost per gallon	Labor cost per 100 SF	Labor burden 100 SF	Material cost per 100 SF	Overhead per 100 SF	Profit per 100 SF	Total price per 100 SF
Fence, picket, spray application									
Solid body or semi-transparent stain, water base (material #18 or #20)									
Spray 1st coat									
Slow	400	300	49.40	5.75	1.38	16.47	4.48	4.49	32.57
Medium	500	275	43.25	5.90	1.71	15.73	5.84	3.50	32.68
Fast	600	250	37.05	6.00	2.13	14.82	7.11	2.10	32.16
Spray 2nd or additional coats									
Slow	500	350	49.40	4.60	1.10	14.11	3.76	3.77	27.34
Medium	600	325	43.25	4.92	1.43	13.31	4.91	2.95	27.52
Fast	700	300	37.05	5.14	1.83	12.35	5.98	1.77	27.07
Solid body or semi-transparent stain, oil base (material #19 or #21)									
Spray 1st coat									
Slow	400	350	59.60	5.75	1.38	17.03	4.59	4.60	33.35
Medium	500	325	52.15	5.90	1.71	16.05	5.92	3.55	33.13
Fast	600	300	44.70	6.00	2.13	14.90	7.14	2.11	32.28
Spray 2nd or additional coats									
Slow	500	425	59.60	4.60	1.10	14.02	3.75	3.76	27.23
Medium	600	400	52.15	4.92	1.43	13.04	4.85	2.91	27.15
Fast	700	375	44.70	5.14	1.83	11.92	5.85	1.73	26.47

For picket fence, measure the overall area of one side and multiply by 4 for painting both sides. Then apply these cost figures. "Slow" work is based on an hourly wage of $23.00, "Medium" work on an hourly wage of $29.50, and "Fast" work on an hourly wage of $36.00. Other qualifications that apply to this table are on page 9.

	Labor SF per manhour	Material coverage SF/gallon	Material cost per gallon	Labor cost per 100 SF	Labor burden 100 SF	Material cost per 100 SF	Overhead per 100 SF	Profit per 100 SF	Total price per 100 SF
Fireplace masonry, interior, smooth surface masonry									
Masonry paint, water base (material #31)									
Brush each coat									
Slow	70	140	44.80	32.86	7.90	32.00	13.82	13.85	100.43
Medium	75	130	39.20	39.33	11.36	30.15	20.21	12.13	113.18
Fast	80	120	33.60	45.00	15.89	28.00	27.56	8.15	124.60
Masonry paint, oil base (material #32)									
Brush each coat									
Slow	70	165	65.40	32.86	7.90	39.64	15.27	15.31	110.98
Medium	75	155	57.20	39.33	11.36	36.90	21.90	13.14	122.63
Fast	80	145	49.00	45.00	15.89	33.79	29.35	8.68	132.71
Masonry paint, water base (material #31)									
Roll each coat									
Slow	140	120	44.80	16.43	3.93	37.33	10.96	10.99	79.64
Medium	150	110	39.20	19.67	5.70	35.64	15.25	9.15	85.41
Fast	160	100	33.60	22.50	7.94	33.60	19.85	5.87	89.76
Masonry paint, oil base (material #32)									
Roll each coat									
Slow	140	140	65.40	16.43	3.93	46.71	12.75	12.77	92.59
Medium	150	130	57.20	19.67	5.70	44.00	17.34	10.40	97.11
Fast	160	120	49.00	22.50	7.94	40.83	22.09	6.54	99.90
Masonry paint, water base (material #31)									
Spray each coat									
Slow	400	105	44.80	5.75	1.38	42.67	9.46	9.48	68.74
Medium	450	100	39.20	6.56	1.88	39.20	11.92	7.15	66.71
Fast	500	90	33.60	7.20	2.54	37.33	14.59	4.32	65.98
Masonry paint, oil base (material #32)									
Spray each coat									
Slow	400	125	65.40	5.75	1.38	52.32	11.30	11.32	82.07
Medium	450	120	57.20	6.56	1.88	47.67	14.03	8.42	78.56
Fast	500	110	49.00	7.20	2.54	44.55	16.83	4.98	76.10

Measurements are based on square feet of the surface area (length times width) to be painted. For fireplace exteriors, use the Masonry cost table which applies. "Slow" work is based on an hourly wage of $23.00, "Medium" work on an hourly wage of $29.50, and "Fast" work on an hourly wage of $36.00. Other qualifications that apply to this table are on page 9.

	Labor LF per manhour	Material coverage LF/gallon	Material cost per gallon	Labor cost per 100 LF	Labor burden 100 LF	Material cost per 100 LF	Overhead per 100 LF	Profit per 100 LF	Total price per 100 LF

Fireplace trim, wood, roll & brush each coat

Mantel, rough sawn 4' x 12"

Solid body or semi-transparent stain, water or oil base (Material #18 or #19 or #20 or #21)

Roll & brush each coat

Slow	15	50	54.50	153.33	36.81	109.00	56.83	56.95	412.92
Medium	18	45	47.70	163.89	47.40	106.00	79.31	47.59	444.19
Fast	20	40	40.88	180.00	63.55	102.20	107.18	31.70	484.63

Plant-on trim, interior

Solid body or semi-transparent stain, water or oil base (Material #18 or #19 or #20 or #21)

Roll & brush each coat

Slow	75	135	54.50	30.67	7.35	40.37	14.90	14.93	108.22
Medium	80	130	47.70	36.88	10.66	36.69	21.06	12.63	117.92
Fast	85	125	40.88	42.35	14.93	32.70	27.90	8.25	126.13

Siding, interior, tongue & groove

Solid body or semi-transparent stain, water or oil base (Material #18 or #19 or #20 or #21)

Roll & brush each coat

Slow	50	100	54.50	46.00	11.04	54.50	21.19	21.24	153.97
Medium	75	95	47.70	39.33	11.36	50.21	25.23	15.14	141.27
Fast	100	90	40.88	36.00	12.71	45.42	29.18	8.63	131.94

"Slow" work is based on an hourly wage of $23.00, "Medium" work on an hourly wage of $29.50, and "Fast" work on an hourly wage of $36.00. Other qualifications that apply to this table are on page 9.

	Manhours per box	Material coverage gallons/box	Material cost per gallon	Labor cost per box	Labor burden box	Material cost per box	Overhead per box	Profit per box	Total price per box

Firewood boxes, wood, brush each coat

Boxes of rough sawn wood, 3'0" x 3'0" x 3'0" deep

Solid body or semi-transparent stain, water or oil base (Material #18 or #19 or #20 or #21)

Roll & brush each coat

Slow	0.40	0.20	54.50	9.20	2.21	10.90	4.24	4.25	30.80
Medium	0.35	0.23	47.70	10.33	2.98	10.97	6.07	3.64	33.99
Fast	0.30	0.25	40.88	10.80	3.81	10.22	7.70	2.28	34.81

"Slow" work is based on an hourly wage of $23.00, "Medium" work on an hourly wage of $29.50, and "Fast" work on an hourly wage of $36.00. Other qualifications that apply to this table are on page 9.

	Labor SF per manhour	Material coverage SF/gallon	Material cost per gallon	Labor cost per 100 SF	Labor burden 100 SF	Material cost per 100 SF	Overhead per 100 SF	Profit per 100 SF	Total price per 100 SF
Floors, concrete, brush, interior or exterior									
Masonry (concrete) paint, water base (material #31)									
Brush 1st coat									
Slow	90	250	44.80	25.56	6.13	17.92	9.43	9.45	68.49
Medium	145	238	39.20	20.34	5.90	16.47	10.67	6.40	59.78
Fast	200	225	33.60	18.00	6.36	14.93	12.18	3.60	55.07
Brush 2nd coat									
Slow	125	375	44.80	18.40	4.42	11.95	6.61	6.62	48.00
Medium	200	325	39.20	14.75	4.27	12.06	7.77	4.66	43.51
Fast	275	275	33.60	13.09	4.64	12.22	9.28	2.74	41.97
Brush 3rd or additional coats									
Slow	150	335	44.80	15.33	3.69	13.37	6.15	6.16	44.70
Medium	225	310	39.20	13.11	3.78	12.65	7.39	4.43	41.36
Fast	300	285	33.60	12.00	4.22	11.79	8.69	2.57	39.27
Masonry (concrete) paint, oil base (material #32)									
Brush 1st coat									
Slow	90	300	65.40	25.56	6.13	21.80	10.16	10.18	73.83
Medium	145	288	57.20	20.34	5.90	19.86	11.52	6.91	64.53
Fast	200	275	49.00	18.00	6.36	17.82	13.07	3.87	59.12
Brush 2nd coat									
Slow	125	400	65.40	18.40	4.42	16.35	7.44	7.46	54.07
Medium	200	388	57.20	14.75	4.27	14.74	8.44	5.06	47.26
Fast	275	375	49.00	13.09	4.64	13.07	9.54	2.82	43.16
Brush 3rd or additional coats									
Slow	150	550	65.40	15.33	3.69	11.89	5.87	5.88	42.66
Medium	225	525	57.20	13.11	3.78	10.90	6.95	4.17	38.91
Fast	300	500	49.00	12.00	4.22	9.80	8.07	2.39	36.48
Epoxy, 1 part, water base (material #28)									
Brush each coat									
Slow	125	400	73.20	18.40	4.42	18.30	7.81	7.83	56.76
Medium	163	388	64.10	18.10	5.21	16.52	9.96	5.98	55.77
Fast	200	375	54.90	18.00	6.36	14.64	12.09	3.58	54.67
Epoxy, 2 part system (material #29)									
Brush each coat									
Slow	100	400	120.90	23.00	5.52	30.23	11.16	11.19	81.10
Medium	138	388	105.70	21.38	6.19	27.24	13.70	8.22	76.73
Fast	175	375	90.60	20.57	7.24	24.16	16.12	4.77	72.86

Also use these formulas for concrete steps, stair treads, porches and patios. "Slow" work is based on an hourly wage of $23.00, "Medium" work on an hourly wage of $29.50, and "Fast" work on an hourly wage of $36.00. Other qualifications that apply to this table are on page 9.

	Labor SF per manhour	Material coverage SF/gallon	Material cost per gallon	Labor cost per 100 SF	Labor burden 100 SF	Material cost per 100 SF	Overhead per 100 SF	Profit per 100 SF	Total price per 100 SF
Floors, concrete, roll, interior or exterior									
Masonry (concrete) paint, water base (material #31)									
Roll 1st coat									
Slow	135	275	44.80	17.04	4.09	16.29	7.11	7.12	51.65
Medium	218	263	39.20	13.53	3.93	14.90	8.09	4.85	45.30
Fast	300	250	33.60	12.00	4.22	13.44	9.20	2.72	41.58
Roll 2nd coat									
Slow	195	350	44.80	11.79	2.84	12.80	5.21	5.22	37.86
Medium	268	325	39.20	11.01	3.18	12.06	6.56	3.94	36.75
Fast	340	300	33.60	10.59	3.73	11.20	7.91	2.34	35.77
Roll 3rd or additional coats									
Slow	210	375	44.80	10.95	2.63	11.95	4.85	4.86	35.24
Medium	300	350	39.20	9.83	2.83	11.20	5.97	3.58	33.41
Fast	390	325	33.60	9.23	3.24	10.34	7.08	2.09	31.98
Masonry (concrete) paint, oil base (material #32)									
Roll 1st coat									
Slow	135	370	65.40	17.04	4.09	17.68	7.37	7.39	53.57
Medium	218	345	57.20	13.53	3.93	16.58	8.51	5.10	47.65
Fast	300	320	49.00	12.00	4.22	15.31	9.78	2.89	44.20
Roll 2nd coat									
Slow	195	500	65.40	11.79	2.84	13.08	5.26	5.27	38.24
Medium	268	475	57.20	11.01	3.18	12.04	6.56	3.93	36.72
Fast	340	450	49.00	10.59	3.73	10.89	7.82	2.31	35.34
Roll 3rd or additional coats									
Slow	210	550	65.40	10.95	2.63	11.89	4.84	4.85	35.16
Medium	300	525	57.20	9.83	2.83	10.90	5.89	3.54	32.99
Fast	390	500	49.00	9.23	3.24	9.80	6.91	2.04	31.22
Epoxy, 1 part, water base (material #28)									
Roll each coat									
Slow	150	500	73.20	15.33	3.69	14.64	6.39	6.41	46.46
Medium	225	488	64.10	13.11	3.78	13.14	7.51	4.51	42.05
Fast	300	475	54.90	12.00	4.22	11.56	8.62	2.55	38.95
Epoxy, 2 part system (material #29)									
Roll each coat									
Slow	135	500	120.90	17.04	4.09	24.18	8.61	8.63	62.55
Medium	208	488	105.70	14.18	4.11	21.66	9.99	5.99	55.93
Fast	250	475	90.60	14.40	5.08	19.07	11.95	3.54	54.04

Also use these formulas for concrete steps, stair treads, porches and patios. "Slow" work is based on an hourly wage of $23.00, "Medium" work on an hourly wage of $29.50, and "Fast" work on an hourly wage of $36.00. Other qualifications that apply to this table are on page 9.

	Labor SF per manhour	Material coverage SF/gallon	Material cost per gallon	Labor cost per 100 SF	Labor burden 100 SF	Material cost per 100 SF	Overhead per 100 SF	Profit per 100 SF	Total price per 100 SF
Floors, concrete, spray, interior or exterior									
Masonry (concrete) paint, water base (material #31)									
Spray 1st coat									
Slow	800	175	44.80	2.88	.69	25.60	5.54	5.55	40.26
Medium	900	163	39.20	3.28	.94	24.05	7.07	4.24	39.58
Fast	1000	150	33.60	3.60	1.27	22.40	8.45	2.50	38.22
Spray 2nd coat									
Slow	900	275	44.80	2.56	.61	16.29	3.70	3.71	26.87
Medium	1000	263	39.20	2.95	.85	14.90	4.68	2.81	26.19
Fast	1100	250	33.60	3.27	1.16	13.44	5.54	1.64	25.05
Spray 3rd or additional coats									
Slow	1000	325	44.80	2.30	.55	13.78	3.16	3.17	22.96
Medium	1100	313	39.20	2.68	.78	12.52	3.99	2.40	22.37
Fast	1200	300	33.60	3.00	1.04	11.20	4.73	1.40	21.37
Masonry (concrete) paint, oil base (material #32)									
Spray 1st coat									
Slow	800	200	65.40	2.88	.69	32.70	6.89	6.91	50.07
Medium	900	188	57.20	3.28	.94	30.43	8.67	5.20	48.52
Fast	1000	175	49.00	3.60	1.27	28.00	10.19	3.01	46.07
Spray 2nd coat									
Slow	900	300	65.40	2.56	.61	21.80	4.74	4.75	34.46
Medium	1000	288	57.20	2.95	.85	19.86	5.92	3.55	33.13
Fast	1100	275	49.00	3.27	1.16	17.82	6.89	2.04	31.18
Spray 3rd or additional coats									
Slow	1000	350	65.40	2.30	.55	18.69	4.09	4.10	29.73
Medium	1100	338	57.20	2.68	.78	16.92	5.09	3.06	28.53
Fast	1200	325	49.00	3.00	1.04	15.08	5.93	1.75	26.80

Also use these formulas for concrete steps, stair treads, porches and patios. "Slow" work is based on an hourly wage of $23.00, "Medium" work on an hourly wage of $29.50, and "Fast" work on an hourly wage of $36.00. Other qualifications that apply to this table are on page 9.

	Labor SF per manhour	Material coverage SF/gallon	Material cost per gallon	Labor cost per 100 SF	Labor burden 100 SF	Material cost per 100 SF	Overhead per 100 SF	Profit per 100 SF	Total price per 100 SF
Floors, concrete, penetrating stain, interior or exterior									
Penetrating oil stain (material #13)									
Roll 1st coat									
Slow	225	450	57.90	10.22	2.44	12.87	4.85	4.86	35.24
Medium	250	425	50.70	11.80	3.41	11.93	6.79	4.07	38.00
Fast	275	400	43.40	13.09	4.64	10.85	8.85	2.62	40.05
Roll 2nd coat									
Slow	325	500	57.90	7.08	1.70	11.58	3.87	3.88	28.11
Medium	345	475	50.70	8.55	2.48	10.67	5.42	3.25	30.37
Fast	365	450	43.40	9.86	3.49	9.64	7.12	2.11	32.22
Roll 3rd and additional coats									
Slow	365	525	57.90	6.30	1.51	11.03	3.58	3.59	26.01
Medium	383	500	50.70	7.70	2.23	10.14	5.02	3.01	28.10
Fast	400	475	43.40	9.00	3.18	9.14	6.61	1.96	29.89

Also use these formulas for concrete steps, stair treads, porches and patios. "Slow" work is based on an hourly wage of $23.00, "Medium" work on an hourly wage of $29.50, and "Fast" work on an hourly wage of $36.00. Other qualifications that apply to this table are on page 9.

	Labor SF per manhour	Material coverage SF/gallon	Material cost per gallon	Labor cost per 100 SF	Labor burden 100 SF	Material cost per 100 SF	Overhead per 100 SF	Profit per 100 SF	Total price per 100 SF
Floors, wood, interior or exterior, paint grade, brush application									
Undercoat, water base (material #3)									
Brush prime coat									
Slow	275	450	36.90	8.36	2.02	8.20	3.53	3.54	25.65
Medium	300	425	32.30	9.83	2.83	7.60	5.07	3.04	28.37
Fast	325	400	27.70	11.08	3.92	6.93	6.80	2.01	30.74
Undercoat, oil base (material #4)									
Brush prime coat									
Slow	275	500	48.30	8.36	2.02	9.66	3.81	3.81	27.66
Medium	300	475	42.30	9.83	2.83	8.91	5.40	3.24	30.21
Fast	325	450	36.20	11.08	3.92	8.04	7.14	2.11	32.29
Porch & deck enamel, water base (material #26)									
Brush 1st and additional finish coats									
Slow	300	475	57.60	7.67	1.83	12.13	4.11	4.12	29.86
Medium	325	450	50.40	9.08	2.63	11.20	5.73	3.44	32.08
Fast	350	425	43.20	10.29	3.64	10.16	7.46	2.21	33.76
Porch & deck enamel, oil base (material #27)									
Brush 1st and additional finish coats									
Slow	300	550	66.60	7.67	1.83	12.11	4.11	4.12	29.84
Medium	325	525	58.30	9.08	2.63	11.10	5.70	3.42	31.93
Fast	350	500	49.90	10.29	3.64	9.98	7.41	2.19	33.51
Epoxy, 1 part, water base (material #28)									
Brush each coat									
Slow	125	450	73.20	18.40	4.42	16.27	7.43	7.44	53.96
Medium	163	425	64.10	18.10	5.21	15.08	9.60	5.76	53.75
Fast	200	400	54.90	18.00	6.36	13.73	11.80	3.49	53.38
Epoxy, 2 part system (material #29)									
Brush each coat									
Slow	100	425	120.90	23.00	5.52	28.45	10.82	10.85	78.64
Medium	138	400	105.70	21.38	6.19	26.43	13.50	8.10	75.60
Fast	175	375	90.60	20.57	7.24	24.16	16.12	4.77	72.86

"Slow" work is based on an hourly wage of $23.00, "Medium" work on an hourly wage of $29.50, and "Fast" work on an hourly wage of $36.00. Other qualifications that apply to this table are on page 9.

	Labor SF per manhour	Material coverage SF/gallon	Material cost per gallon	Labor cost per 100 SF	Labor burden 100 SF	Material cost per 100 SF	Overhead per 100 SF	Profit per 100 SF	Total price per 100 SF
Floors, wood, interior or exterior, paint grade, roll application									
Undercoat, water base (material #3)									
Roll prime coat									
Slow	400	425	36.90	5.75	1.38	8.68	3.00	3.01	21.82
Medium	438	400	32.30	6.74	1.93	8.08	4.19	2.52	23.46
Fast	475	375	27.70	7.58	2.70	7.39	5.47	1.62	24.76
Undercoat, oil base (material #4)									
Roll prime coat									
Slow	400	475	48.30	5.75	1.38	10.17	3.29	3.29	23.88
Medium	438	450	42.30	6.74	1.93	9.40	4.52	2.71	25.30
Fast	475	425	36.20	7.58	2.70	8.52	5.82	1.72	26.34
Porch & deck enamel, water base (material #26)									
Roll 1st or additional finish coats									
Slow	425	475	57.60	5.41	1.29	12.13	3.58	3.59	26.00
Medium	463	450	50.40	6.37	1.84	11.20	4.85	2.91	27.17
Fast	500	425	43.20	7.20	2.54	10.16	6.17	1.82	27.89
Porch & deck enamel, oil base (material #27)									
Roll 1st or additional finish coats									
Slow	425	525	66.60	5.41	1.29	12.69	3.69	3.69	26.77
Medium	463	500	58.30	6.37	1.84	11.66	4.97	2.98	27.82
Fast	500	475	49.90	7.20	2.54	10.51	6.28	1.86	28.39
Epoxy, 1 part, water base (material #28)									
Brush each coat									
Slow	200	425	73.20	11.50	2.76	17.22	5.98	5.99	43.45
Medium	250	400	64.10	11.80	3.41	16.03	7.81	4.69	43.74
Fast	300	375	54.90	12.00	4.22	14.64	9.57	2.83	43.26
Epoxy, 2 part system (material #29)									
Brush each coat									
Slow	175	400	120.90	13.14	3.14	30.23	8.84	8.86	64.21
Medium	225	375	105.70	13.11	3.78	28.19	11.27	6.76	63.11
Fast	275	350	90.60	13.09	4.64	25.89	13.52	4.00	61.14

"Slow" work is based on an hourly wage of $23.00, "Medium" work on an hourly wage of $29.50, and "Fast" work on an hourly wage of $36.00. Other qualifications that apply to this table are on page 9.

	Labor SF per manhour	Material coverage SF/gallon	Material cost per gallon	Labor cost per 100 SF	Labor burden 100 SF	Material cost per 100 SF	Overhead per 100 SF	Profit per 100 SF	Total price per 100 SF
Floors, wood, interior or exterior, stain grade									
Wiping stain, varnish, oil base (material #30a)									
Stain, brush 1st coat, wipe & fill									
Slow	225	500	57.10	10.22	2.44	11.42	4.58	4.59	33.25
Medium	250	475	50.00	11.80	3.41	10.53	6.44	3.86	36.04
Fast	275	450	42.90	13.09	4.64	9.53	8.44	2.50	38.20
Stain, brush 2nd coat, wipe & fill									
Slow	400	525	57.10	5.75	1.38	10.88	3.42	3.43	24.86
Medium	425	500	50.00	6.94	2.00	10.00	4.74	2.84	26.52
Fast	450	475	42.90	8.00	2.81	9.03	6.15	1.82	27.81
Stain, brush 3rd or additional coats, wipe & fill									
Slow	425	550	57.10	5.41	1.29	10.38	3.25	3.25	23.58
Medium	450	525	50.00	6.56	1.88	9.52	4.50	2.70	25.16
Fast	475	500	42.90	7.58	2.70	8.58	5.84	1.73	26.43
Sanding sealer, varnish (material #30b)									
Maple or pine, brush 1 coat									
Slow	375	475	63.80	6.13	1.48	13.43	4.00	4.00	29.04
Medium	400	450	55.80	7.38	2.13	12.40	5.48	3.29	30.68
Fast	425	425	47.80	8.47	2.98	11.25	7.04	2.08	31.82
Maple or pine, brush 2nd or additional coats									
Slow	425	550	63.80	5.41	1.29	11.60	3.48	3.49	25.27
Medium	450	525	55.80	6.56	1.88	10.63	4.77	2.86	26.70
Fast	475	500	47.80	7.58	2.70	9.56	6.14	1.82	27.80
Oak, brush 1 coat									
Slow	400	525	63.80	5.75	1.38	12.15	3.66	3.67	26.61
Medium	425	500	55.80	6.94	2.00	11.16	5.03	3.02	28.15
Fast	450	475	47.80	8.00	2.81	10.06	6.47	1.91	29.25
Oak, brush 2nd or additional coats									
Slow	500	625	63.80	4.60	1.10	10.21	3.02	3.03	21.96
Medium	525	600	55.80	5.62	1.61	9.30	4.14	2.48	23.15
Fast	550	575	47.80	6.55	2.32	8.31	5.32	1.57	24.07
Shellac, clear (material #12)									
Brush 1st coat									
Slow	275	475	70.40	8.36	2.02	14.82	4.79	4.80	34.79
Medium	300	450	61.60	9.83	2.83	13.69	6.59	3.95	36.89
Fast	325	425	52.80	11.08	3.92	12.42	8.50	2.51	38.43
Brush 2nd or additional coats									
Slow	400	500	70.40	5.75	1.38	14.08	4.03	4.04	29.28
Medium	425	475	61.60	6.94	2.00	12.97	5.48	3.29	30.68
Fast	450	450	52.80	8.00	2.81	11.73	6.99	2.07	31.60

	Labor SF per manhour	Material coverage SF/gallon	Material cost per gallon	Labor cost per 100 SF	Labor burden 100 SF	Material cost per 100 SF	Overhead per 100 SF	Profit per 100 SF	Total price per 100 SF
Varnish, gloss or flat (material #30c)									
Brush 1st coat									
Slow	275	475	81.00	8.36	2.02	17.05	5.21	5.22	37.86
Medium	300	450	70.80	9.83	2.83	15.73	7.10	4.26	39.75
Fast	325	425	60.70	11.08	3.92	14.28	9.07	2.68	41.03
Brush 2nd or additional coats									
Slow	350	600	81.00	6.57	1.59	13.50	4.11	4.12	29.89
Medium	375	575	70.80	7.87	2.28	12.31	5.61	3.37	31.44
Fast	400	550	60.70	9.00	3.18	11.04	7.20	2.13	32.55
Penetrating stain wax & wipe (material #14)									
Stain, brush 1st coat & wipe									
Slow	200	550	50.20	11.50	2.76	9.13	4.44	4.45	32.28
Medium	250	525	43.90	11.80	3.41	8.36	5.89	3.54	33.00
Fast	300	500	37.60	12.00	4.22	7.52	7.37	2.18	33.29
Stain, brush 2nd or additional coats & wipe									
Slow	250	600	50.20	9.20	2.21	8.37	3.76	3.77	27.31
Medium	300	575	43.90	9.83	2.83	7.63	5.08	3.05	28.42
Fast	350	550	37.60	10.29	3.64	6.84	6.44	1.90	29.11
Wax & polish (material #15)									
Hand apply 1 coat									
Slow	175	1000	18.60	13.14	3.14	1.86	3.45	3.46	25.05
Medium	200	950	16.30	14.75	4.27	1.72	5.18	3.11	29.03
Fast	225	900	14.00	16.00	5.63	1.56	7.20	2.13	32.52
Buffing with machine									
Slow	400	--	--	5.75	1.38	--	1.35	1.36	9.84
Medium	450	--	--	6.56	1.88	--	2.12	1.27	11.83
Fast	500	--	--	7.20	2.54	--	3.02	.89	13.65

"Slow" work is based on an hourly wage of $23.00, "Medium" work on an hourly wage of $29.50, and "Fast" work on an hourly wage of $36.00. Other qualifications that apply to this table are on page 9.

	Manhours per door	Gallons per Door	Material cost per gallon	Labor cost per door	Labor burden door	Material cost per door	Overhead per door	Profit per door	Total price per door

Garage door backs, seal coat, spray one coat

Sanding sealer, lacquer (material #11b)									
1 car garage, 8' x 7'									
Slow	0.30	0.40	50.90	6.90	1.66	20.36	5.49	5.51	39.92
Medium	0.25	0.50	44.60	7.38	2.13	22.30	7.95	4.77	44.53
Fast	0.20	0.60	38.20	7.20	2.54	22.92	10.12	2.99	45.77
2 car garage, 16' x 7'									
Slow	0.40	0.80	50.90	9.20	2.21	40.72	9.90	9.92	71.95
Medium	0.35	0.90	44.60	10.33	2.98	40.14	13.37	8.02	74.84
Fast	0.30	1.00	38.20	10.80	3.81	38.20	16.37	4.84	74.02
3 car garage, 16' x 7' + 8' x 7'									
Slow	0.60	1.00	50.90	13.80	3.31	50.90	12.92	12.95	93.88
Medium	0.55	1.10	44.60	16.23	4.69	49.06	17.50	10.50	97.98
Fast	0.50	1.20	38.20	18.00	6.36	45.84	21.76	6.44	98.40

Use the figures for Siding when estimating the cost of painting garage door fronts. These figures assume a one-car garage door measures 7' x 8' and a two-car garage door measures 7' x 16'. A three-car garage has one single and one double door. Government funded projects (FHA, VA, HUD) usually require sealing the garage door back on new construction projects. The doors are usually sprayed along with the cabinet sealer coat (as used in this table) or stained along with the exterior trim. "Slow" work is based on an hourly wage of $23.00, "Medium" work on an hourly wage of $29.50, and "Fast" work on an hourly wage of $36.00. Other qualifications that apply to this table are on page 9.

	Labor LF per manhour	Material coverage LF/gallon	Material cost per gallon	Labor cost per 100 LF	Labor burden 100 LF	Material cost per 100 LF	Overhead per 100 LF	Profit per 100 LF	Total price per 100 LF
Gutters and downspouts (galvanized), brush application									
Gutters									
Metal prime, rust inhibitor, clean metal (material #35)									
Brush prime coat									
Slow	80	400	59.70	28.75	6.90	14.93	9.61	9.63	69.82
Medium	90	375	52.30	32.78	9.47	13.95	14.05	8.43	78.68
Fast	100	350	44.80	36.00	12.71	12.80	19.07	5.64	86.22
Metal prime, rust inhibitor, rusty metal (material #36)									
Brush prime coat									
Slow	80	400	80.10	28.75	6.90	20.03	10.58	10.60	76.86
Medium	90	375	70.10	32.78	9.47	18.69	15.24	9.14	85.32
Fast	100	350	60.10	36.00	12.71	17.17	20.42	6.04	92.34
Metal finish - synthetic enamel (off white), gloss (material #37)									
Brush 1st finish coat									
Slow	100	425	63.60	23.00	5.52	14.96	8.26	8.28	60.02
Medium	110	400	55.70	26.82	7.75	13.93	12.13	7.28	67.91
Fast	120	375	47.70	30.00	10.58	12.72	16.53	4.89	74.72
Brush 2nd or additional finish coats									
Slow	120	450	63.60	19.17	4.59	14.13	7.20	7.22	52.31
Medium	130	425	55.70	22.69	6.56	13.11	10.59	6.35	59.30
Fast	140	400	47.70	25.71	9.07	11.93	14.48	4.28	65.47
Metal finish - synthetic enamel (colors except orange/red), gloss (material #38)									
Brush 1st finish coat									
Slow	100	425	66.30	23.00	5.52	15.60	8.38	8.40	60.90
Medium	110	400	58.00	26.82	7.75	14.50	12.27	7.36	68.70
Fast	120	375	49.70	30.00	10.58	13.25	16.69	4.94	75.46
Brush 2nd or additional finish coats									
Slow	120	450	66.30	19.17	4.59	14.73	7.32	7.33	53.14
Medium	130	425	58.00	22.69	6.56	13.65	10.73	6.44	60.07
Fast	140	400	49.70	25.71	9.07	12.43	14.64	4.33	66.18
Downspouts									
Metal prime, rust inhibitor, clean metal (material #35)									
Brush prime coat									
Slow	30	250	59.70	76.67	18.39	23.88	22.60	22.65	164.19
Medium	35	225	52.30	84.29	24.36	23.24	32.97	19.78	184.64
Fast	40	200	44.80	90.00	31.78	22.40	44.69	13.22	202.09
Metal prime, rust inhibitor, rusty metal (material #36)									
Brush prime coat									
Slow	30	250	80.10	76.67	18.39	32.04	24.15	24.20	175.45
Medium	35	225	70.10	84.29	24.36	31.16	34.95	20.97	195.73
Fast	40	200	60.10	90.00	31.78	30.05	47.06	13.92	212.81

	Labor LF per manhour	Material coverage LF/gallon	Material cost per gallon	Labor cost per 100 LF	Labor burden 100 LF	Material cost per 100 LF	Overhead per 100 LF	Profit per 100 LF	Total price per 100 LF
Metal finish - synthetic enamel (off white), gloss (material #37)									
Brush 1st finish coat									
Slow	50	275	63.60	46.00	11.04	23.13	15.23	15.26	110.66
Medium	60	250	55.70	49.17	14.23	22.28	21.42	12.85	119.95
Fast	70	225	47.70	51.43	18.18	21.20	28.14	8.32	127.27
Brush 2nd or additional finish coats									
Slow	70	300	63.60	32.86	7.90	21.20	11.77	11.80	85.53
Medium	80	275	55.70	36.88	10.66	20.25	16.95	10.17	94.91
Fast	90	250	47.70	40.00	14.12	19.08	22.69	6.71	102.60
Metal finish - synthetic enamel (colors except orange/red), gloss (material #38)									
Brush 1st finish coat									
Slow	50	275	66.30	46.00	11.04	24.11	15.42	15.45	112.02
Medium	60	250	58.00	49.17	14.23	23.20	21.65	12.99	121.24
Fast	70	225	49.70	51.43	18.18	22.09	28.42	8.41	128.53
Brush 2nd or additional finish coats									
Slow	70	300	66.30	32.86	7.90	22.10	11.94	11.97	86.77
Medium	80	275	58.00	36.88	10.66	21.09	17.16	10.29	96.08
Fast	90	250	49.70	40.00	14.12	19.88	22.94	6.79	103.73

NOTE: Oil base material is recommended for any metal surface. Although water base material is often used, it may cause oxidation, corrosion and rust. These figures assume that all exposed surfaces of 5" gutters and 4" downspouts are painted. For ornamental gutters and downspouts, multiply the linear feet by 1.5 before using these figures. "Slow" work is based on an hourly wage of $23.00, "Medium" work on an hourly wage of $29.50, and "Fast" work on an hourly wage of $36.00. Other qualifications that apply to this table are on page 9.

High Time Difficulty Factors

Painting takes longer and may require more material when heights above the floor exceed 8 feet. The additional time and material for working at these heights and using a roller pole or a wand on a spray gun, climbing up and down a ladder or scaffolding is applied by using one of the factors listed below. The wall area above 8 feet is typically referred to as the "Clip." To apply the high time difficulty factor, measure the surface above 8 feet which is to be painted and multiply that figure by the appropriate factor. This measurement can be listed on a separate line of your take-off and the appropriate price can be applied for a total.

For labor calculations only:
- Add 30% to the area for heights between 8 and 13 feet (multiply by 1.3)
- Add 60% to the area for heights from 13 to 17 feet (multiply by 1.6)
- Add 90% to the area for heights from 17 to 19 feet (multiply by 1.9)
- Add 120% to the area for heights from 19 to 21 feet (multiply by 2.2)

EXAMPLE: A 17 x 14 living room has a vaulted ceiling 13 feet high. Your take-off sheet might look like this:

Walls to 8 feet: 136 + 112 + 136 + 112 = 496 SF

Clip: [(5 x 14) / 2] x 2 + (5 x 17) = 70 + 85 = 155 SF

area of two triangles + rectangular area

155 SF x 1.3 (high time difficulty factor) = 202 SF

Then multiply each SF total by the appropriate price per square foot.

Mail box structures, wood, apartment type

Measure the length of each board to be painted and use the manhours and material given for Trellis or Plant-on trim or Siding.

	Labor SF per manhour	Material coverage SF/gallon	Material cost per gallon	Labor cost per 100 SF	Labor burden 100 SF	Material cost per 100 SF	Overhead per 100 SF	Profit per 100 SF	Total price per 100 SF
Masonry, anti-graffiti stain eliminator on smooth or rough surface									
Water base primer and sealer (material #39)									
Roll & brush each coat									
Slow	350	400	49.40	6.57	1.59	12.35	3.90	3.90	28.31
Medium	375	375	43.20	7.87	2.28	11.52	5.42	3.25	30.34
Fast	400	350	37.00	9.00	3.18	10.57	7.05	2.09	31.89
Oil base primer and sealer (material #40)									
Roll & brush each coat									
Slow	350	375	58.00	6.57	1.59	15.47	4.49	4.50	32.62
Medium	375	350	50.80	7.87	2.28	14.51	6.16	3.70	34.52
Fast	400	325	43.50	9.00	3.18	13.38	7.92	2.34	35.82
Polyurethane 2 part system (material #41)									
Roll & brush each coat									
Slow	300	375	187.30	7.67	1.83	49.95	11.30	11.32	82.07
Medium	325	350	163.90	9.08	2.63	46.83	14.63	8.78	81.95
Fast	350	325	140.50	10.29	3.64	43.23	17.72	5.24	80.12

Use these figures for new brick, used brick, or Concrete Masonry Units (CMU) where the block surfaces are either smooth or rough, porous or unfilled, with joints struck to average depth. The more porous the surface, the rougher the texture, the more time and material will be required. "Slow" work is based on an hourly wage of $23.00, "Medium" work on an hourly wage of $29.50, and "Fast" work on an hourly wage of $36.00. Other qualifications that apply to this table are on page 9.

	Labor SF per manhour	Material coverage SF/gallon	Material cost per gallon	Labor cost per 100 SF	Labor burden 100 SF	Material cost per 100 SF	Overhead per 100 SF	Profit per 100 SF	Total price per 100 SF
Masonry, block filler									
Brush 1 coat (material #33)									
Slow	95	75	37.10	24.21	5.82	49.47	15.10	15.13	109.73
Medium	125	65	32.50	23.60	6.82	50.00	20.11	12.06	112.59
Fast	155	55	27.80	23.23	8.19	50.55	25.41	7.52	114.90
Roll 1 coat (material #33)									
Slow	190	70	37.10	12.11	2.89	53.00	12.92	12.95	93.87
Medium	215	60	32.50	13.72	3.96	54.17	17.97	10.78	100.60
Fast	240	50	27.80	15.00	5.31	55.60	23.53	6.96	106.40
Spray 1 coat (material #33)									
Slow	425	65	37.10	5.41	1.29	57.08	12.12	12.15	88.05
Medium	525	55	32.50	5.62	1.61	59.09	16.58	9.95	92.85
Fast	625	45	27.80	5.76	2.03	61.78	21.57	6.38	97.52

Use these figures for using block filler on rough or porous masonry with joints struck to average depth. "Slow" work is based on an hourly wage of $23.00, "Medium" work on an hourly wage of $29.50, and "Fast" work on an hourly wage of $36.00. Other qualifications that apply to this table are on page 9.

	Labor SF per manhour	Material coverage SF/gallon	Material cost per gallon	Labor cost per 100 SF	Labor burden 100 SF	Material cost per 100 SF	Overhead per 100 SF	Profit per 100 SF	Total price per 100 SF
Masonry, brick, new, smooth-surface, brush									
Masonry paint, water base, flat or gloss (material #31)									
Brush 1st coat									
Slow	200	300	44.80	11.50	2.76	14.93	5.55	5.56	40.30
Medium	225	275	39.20	13.11	3.78	14.25	7.79	4.67	43.60
Fast	250	250	33.60	14.40	5.08	13.44	10.21	3.02	46.15
Brush 2nd or additional coats									
Slow	250	325	44.80	9.20	2.21	13.78	4.79	4.80	34.78
Medium	275	300	39.20	10.73	3.11	13.07	6.73	4.04	37.68
Fast	300	275	33.60	12.00	4.22	12.22	8.82	2.61	39.87
Masonry paint, oil base (material #32)									
Brush 1st coat									
Slow	200	350	65.40	11.50	2.76	18.69	6.26	6.27	45.48
Medium	225	325	57.20	13.11	3.78	17.60	8.63	5.18	48.30
Fast	250	300	49.00	14.40	5.08	16.33	11.10	3.28	50.19
Brush 2nd or additional coats									
Slow	250	400	65.40	9.20	2.21	16.35	5.27	5.28	38.31
Medium	275	363	57.20	10.73	3.11	15.76	7.40	4.44	41.44
Fast	300	325	49.00	12.00	4.22	15.08	9.71	2.87	43.88

Use these figures for new smooth-surface brick with joints struck to average depth. "Slow" work is based on an hourly wage of $23.00, "Medium" work on an hourly wage of $29.50, and "Fast" work on an hourly wage of $36.00. Other qualifications that apply to this table are on page 9.

	Labor SF per manhour	Material coverage SF/gallon	Material cost per gallon	Labor cost per 100 SF	Labor burden 100 SF	Material cost per 100 SF	Overhead per 100 SF	Profit per 100 SF	Total price per 100 SF
Masonry, brick, new, smooth-surface, roll									
Masonry paint, water base, flat or gloss (material #31)									
Roll 1st coat									
Slow	325	250	44.80	7.08	1.70	17.92	5.07	5.08	36.85
Medium	350	213	39.20	8.43	2.45	18.40	7.32	4.39	40.99
Fast	375	175	33.60	9.60	3.41	19.20	9.98	2.95	45.14
Roll 2nd or additional coats									
Slow	375	275	44.80	6.13	1.48	16.29	4.54	4.55	32.99
Medium	400	250	39.20	7.38	2.13	15.68	6.30	3.78	35.27
Fast	425	225	33.60	8.47	2.98	14.93	8.18	2.42	36.98
Masonry paint, oil base (material #32)									
Roll 1st coat									
Slow	325	325	65.40	7.08	1.70	20.12	5.49	5.50	39.89
Medium	350	288	57.20	8.43	2.45	19.86	7.68	4.61	43.03
Fast	375	250	49.00	9.60	3.41	19.60	10.10	2.99	45.70
Roll 2nd or additional coats									
Slow	375	350	65.40	6.13	1.48	18.69	5.00	5.01	36.31
Medium	400	313	57.20	7.38	2.13	18.27	6.95	4.17	38.90
Fast	425	275	49.00	8.47	2.98	17.82	9.08	2.69	41.04
Waterproofing, clear hydro sealer, oil base (material #34)									
Roll 1st coat									
Slow	200	175	40.60	11.50	2.76	23.20	7.12	7.13	51.71
Medium	225	150	35.50	13.11	3.78	23.67	10.14	6.09	56.79
Fast	250	125	30.40	14.40	5.08	24.32	13.58	4.02	61.40
Roll 2nd or additional coats									
Slow	225	200	40.60	10.22	2.44	20.30	6.26	6.28	45.50
Medium	250	190	35.50	11.80	3.41	18.68	8.47	5.08	47.44
Fast	275	180	30.40	13.09	4.64	16.89	10.73	3.17	48.52

Use these figures for new smooth-surface brick with joints struck to average depth. "Slow" work is based on an hourly wage of $23.00, "Medium" work on an hourly wage of $29.50, and "Fast" work on an hourly wage of $36.00. Other qualifications that apply to this table are on page 9.

	Labor SF per manhour	Material coverage SF/gallon	Material cost per gallon	Labor cost per 100 SF	Labor burden 100 SF	Material cost per 100 SF	Overhead per 100 SF	Profit per 100 SF	Total price per 100 SF
Masonry, brick, new, smooth-surface, spray									
Masonry paint, water base, flat or gloss (material #31)									
Spray 1st coat									
Slow	650	250	44.80	3.54	.85	17.92	4.24	4.25	30.80
Medium	750	225	39.20	3.93	1.13	17.42	5.62	3.37	31.47
Fast	850	200	33.60	4.24	1.51	16.80	6.99	2.07	31.61
Spray 2nd or additional coats									
Slow	750	275	44.80	3.07	.72	16.29	3.82	3.83	27.73
Medium	825	238	39.20	3.58	1.02	16.47	5.27	3.16	29.50
Fast	900	250	33.60	4.00	1.41	13.44	5.84	1.73	26.42
Masonry paint, oil base (material #32)									
Spray 1st coat									
Slow	650	275	65.40	3.54	.85	23.78	5.35	5.36	38.88
Medium	750	250	57.20	3.93	1.13	22.88	6.99	4.19	39.12
Fast	850	225	49.00	4.24	1.51	21.78	8.53	2.52	38.58
Spray 2nd or additional coats									
Slow	750	300	65.40	3.07	.72	21.80	4.87	4.88	35.34
Medium	825	288	57.20	3.58	1.02	19.86	6.12	3.67	34.25
Fast	900	275	49.00	4.00	1.41	17.82	7.20	2.13	32.56
Waterproofing, clear hydro sealer, oil base (material #34)									
Spray 1st coat									
Slow	700	120	40.60	3.29	.79	33.83	7.20	7.22	52.33
Medium	800	100	35.50	3.69	1.06	35.50	10.07	6.04	56.36
Fast	900	80	30.40	4.00	1.41	38.00	13.46	3.98	60.85
Spray 2nd or additional coats									
Slow	800	150	40.60	2.88	.69	27.07	5.82	5.83	42.29
Medium	900	138	35.50	3.28	.94	25.72	7.49	4.49	41.92
Fast	1000	125	30.40	3.60	1.27	24.32	9.05	2.68	40.92

Use these figures for new smooth-surface brick with joints struck to average depth. "Slow" work is based on an hourly wage of $23.00, "Medium" work on an hourly wage of $29.50, and "Fast" work on an hourly wage of $36.00. Other qualifications that apply to this table are on page 9.

	Labor SF per manhour	Material coverage SF/gallon	Material cost per gallon	Labor cost per 100 SF	Labor burden 100 SF	Material cost per 100 SF	Overhead per 100 SF	Profit per 100 SF	Total price per 100 SF
Masonry, brick, used, rough surface, brush									
Masonry paint, water base, flat or gloss (material #31)									
Brush 1st coat									
Slow	150	300	44.80	15.33	3.69	14.93	6.45	6.46	46.86
Medium	175	275	39.20	16.86	4.86	14.25	9.00	5.40	50.37
Fast	200	250	33.60	18.00	6.36	13.44	11.71	3.47	52.98
Brush 2nd or additional coats									
Slow	200	375	44.80	11.50	2.76	11.95	4.98	4.99	36.18
Medium	225	350	39.20	13.11	3.78	11.20	7.03	4.22	39.34
Fast	250	325	33.60	14.40	5.08	10.34	9.24	2.73	41.79
Masonry paint, oil base (material #32)									
Brush 1st coat									
Slow	150	325	65.40	15.33	3.69	20.12	7.43	7.45	54.02
Medium	175	300	57.20	16.86	4.86	19.07	10.20	6.12	57.11
Fast	200	275	49.00	18.00	6.36	17.82	13.07	3.87	59.12
Brush 2nd or additional coats									
Slow	200	400	65.40	11.50	2.76	16.35	5.82	5.83	42.26
Medium	225	375	57.20	13.11	3.78	15.25	8.04	4.82	45.00
Fast	250	350	49.00	14.40	5.08	14.00	10.38	3.07	46.93

Use these figures for dry pressed used brick, clay brick tile, or adobe block with joints struck to average depth. "Slow" work is based on an hourly wage of $23.00, "Medium" work on an hourly wage of $29.50, and "Fast" work on an hourly wage of $36.00. Other qualifications that apply to this table are on page 9.

	Labor SF per manhour	Material coverage SF/gallon	Material cost per gallon	Labor cost per 100 SF	Labor burden 100 SF	Material cost per 100 SF	Overhead per 100 SF	Profit per 100 SF	Total price per 100 SF
Masonry, brick, used, rough surface, roll									
Masonry paint, water base, flat or gloss (material #31)									
Roll 1st coat									
Slow	300	250	44.80	7.67	1.83	17.92	5.21	5.22	37.85
Medium	325	225	39.20	9.08	2.63	17.42	7.28	4.37	40.78
Fast	350	200	33.60	10.29	3.64	16.80	9.52	2.82	43.07
Roll 2nd or additional coats									
Slow	350	325	44.80	6.57	1.59	13.78	4.17	4.18	30.29
Medium	375	300	39.20	7.87	2.28	13.07	5.80	3.48	32.50
Fast	400	275	33.60	9.00	3.18	12.22	7.56	2.24	34.20
Masonry paint, oil base (material #32)									
Roll 1st coat									
Slow	300	275	65.40	7.67	1.83	23.78	6.33	6.34	45.95
Medium	325	250	57.20	9.08	2.63	22.88	8.65	5.19	48.43
Fast	350	225	49.00	10.29	3.64	21.78	11.07	3.27	50.05
Roll 2nd or additional coats									
Slow	350	350	65.40	6.57	1.59	18.69	5.10	5.11	37.06
Medium	375	325	57.20	7.87	2.28	17.60	6.94	4.16	38.85
Fast	400	300	49.00	9.00	3.18	16.33	8.84	2.61	39.96
Waterproofing, clear hydro sealer, oil base (material #34)									
Roll 1st coat									
Slow	150	125	40.60	15.33	3.69	32.48	9.78	9.80	71.08
Medium	175	113	35.50	16.86	4.86	31.42	13.29	7.97	74.40
Fast	200	100	30.40	18.00	6.36	30.40	16.97	5.02	76.75
Roll 2nd or additional coats									
Slow	175	150	40.60	13.14	3.14	27.07	8.24	8.26	59.85
Medium	200	138	35.50	14.75	4.27	25.72	11.18	6.71	62.63
Fast	225	125	30.40	16.00	5.63	24.32	14.25	4.22	64.42

Use these figures for dry pressed used brick, clay brick tile, or adobe block with joints struck to average depth. "Slow" work is based on an hourly wage of $23.00, "Medium" work on an hourly wage of $29.50, and "Fast" work on an hourly wage of $36.00. Other qualifications that apply to this table are on page 9.

	Labor SF per manhour	Material coverage SF/gallon	Material cost per gallon	Labor cost per 100 SF	Labor burden 100 SF	Material cost per 100 SF	Overhead per 100 SF	Profit per 100 SF	Total price per 100 SF
Masonry, brick, used, rough surface, spray									
Masonry paint, water base, flat or gloss (material #31)									
Spray 1st coat									
Slow	600	200	44.80	3.83	.93	22.40	5.16	5.17	37.49
Medium	700	175	39.20	4.21	1.23	22.40	6.96	4.17	38.97
Fast	800	150	33.60	4.50	1.59	22.40	8.83	2.61	39.93
Spray 2nd or additional coats									
Slow	700	225	44.80	3.29	.79	19.91	4.56	4.57	33.12
Medium	800	213	39.20	3.69	1.06	18.40	5.79	3.47	32.41
Fast	900	200	33.60	4.00	1.41	16.80	6.89	2.04	31.14
Masonry paint, oil base (material #32)									
Spray 1st coat									
Slow	600	225	65.40	3.83	.93	29.07	6.43	6.44	46.70
Medium	700	200	57.20	4.21	1.23	28.60	8.51	5.10	47.65
Fast	800	175	49.00	4.50	1.59	28.00	10.57	3.13	47.79
Spray 2nd or additional coats									
Slow	700	250	65.40	3.29	.79	26.16	5.75	5.76	41.75
Medium	800	238	57.20	3.69	1.06	24.03	7.20	4.32	40.30
Fast	900	225	49.00	4.00	1.41	21.78	8.43	2.49	38.11
Waterproofing, clear hydro sealer, oil base (material #34)									
Spray 1st coat									
Slow	600	80	40.60	3.83	.93	50.75	10.55	10.57	76.63
Medium	700	75	35.50	4.21	1.23	47.33	13.19	7.91	73.87
Fast	800	70	30.40	4.50	1.59	43.43	15.35	4.54	69.41
Spray 2nd coat									
Slow	800	100	40.60	2.88	.69	40.60	8.39	8.41	60.97
Medium	900	90	35.50	3.28	.94	39.44	10.92	6.55	61.13
Fast	1000	80	30.40	3.60	1.27	38.00	13.29	3.93	60.09

Use these figures for dry pressed used brick, clay brick tile, or adobe block with joints struck to average depth. "Slow" work is based on an hourly wage of $23.00, "Medium" work on an hourly wage of $29.50, and "Fast" work on an hourly wage of $36.00. Other qualifications that apply to this table are on page 9.

	Labor SF per manhour	Material coverage SF/gallon	Material cost per gallon	Labor cost per 100 SF	Labor burden 100 SF	Material cost per 100 SF	Overhead per 100 SF	Profit per 100 SF	Total price per 100 SF

Masonry, Concrete Masonry Units (CMU), rough, porous surface, brush

	Labor SF per manhour	Material coverage SF/gallon	Material cost per gallon	Labor cost per 100 SF	Labor burden 100 SF	Material cost per 100 SF	Overhead per 100 SF	Profit per 100 SF	Total price per 100 SF
Masonry paint, water base, flat or gloss (material #31)									
Brush 1st coat									
Slow	110	100	44.80	20.91	5.01	44.80	13.44	13.47	97.63
Medium	130	88	39.20	22.69	6.56	44.55	18.45	11.07	103.32
Fast	150	75	33.60	24.00	8.49	44.80	23.95	7.09	108.33
Brush 2nd or additional coats									
Slow	185	180	44.80	12.43	3.00	24.89	7.66	7.67	55.65
Medium	210	168	39.20	14.05	4.05	23.33	10.36	6.22	58.01
Fast	230	155	33.60	15.65	5.54	21.68	13.28	3.93	60.08
Masonry paint, oil base (material #32)									
Brush 1st coat									
Slow	110	130	65.40	20.91	5.01	50.31	14.49	14.52	105.24
Medium	130	120	57.20	22.69	6.56	47.67	19.23	11.54	107.69
Fast	150	110	49.00	24.00	8.49	44.55	23.88	7.06	107.98
Brush 2nd or additional coats									
Slow	185	200	65.40	12.43	3.00	32.70	9.14	9.16	66.43
Medium	208	180	57.20	14.18	4.11	31.78	12.52	7.51	70.10
Fast	230	160	49.00	15.65	5.54	30.63	16.06	4.75	72.63
Epoxy coating, 2 part system clear (material #51)									
Brush 1st coat									
Slow	95	110	160.60	24.21	5.82	146.00	33.44	33.51	242.98
Medium	115	98	140.50	25.65	7.44	143.37	44.11	26.46	247.03
Fast	135	85	120.50	26.67	9.42	141.76	55.13	16.31	249.29
Brush 2nd or additional coats									
Slow	165	200	160.60	13.94	3.34	80.30	18.54	18.58	134.70
Medium	190	188	140.50	15.53	4.47	74.73	23.69	14.21	132.63
Fast	210	175	120.50	17.14	6.05	68.86	28.54	8.44	129.03
Waterproofing, clear hydro sealer, oil base (material #34)									
Brush 1st coat									
Slow	125	90	40.60	18.40	4.42	45.11	12.91	12.93	93.77
Medium	150	80	35.50	19.67	5.70	44.38	17.43	10.46	97.64
Fast	175	70	30.40	20.57	7.24	43.43	22.09	6.53	99.86
Brush 2nd or additional coats									
Slow	230	130	40.60	10.00	2.41	31.23	8.29	8.31	60.24
Medium	275	110	35.50	10.73	3.11	32.27	11.53	6.92	64.56
Fast	295	90	30.40	12.20	4.31	33.78	15.59	4.61	70.49

Use these figures for Concrete Masonry Units (CMU) such as split face, fluted, or slump block, whose surfaces are rough, porous or unfilled, with joints struck to average depth. The more porous the surface, the rougher the texture, the more time and material will be required. For heavy waterproofing applications, see Masonry under Industrial Painting Operations. "Slow" work is based on an hourly wage of $23.00, "Medium" on an hourly wage of $29.50, and "Fast" work on an hourly wage of $36.00. Other qualifications that apply to this table are on page 9.

	Labor SF per manhour	Material coverage SF/gallon	Material cost per gallon	Labor cost per 100 SF	Labor burden 100 SF	Material cost per 100 SF	Overhead per 100 SF	Profit per 100 SF	Total price per 100 SF
Masonry, Concrete Masonry Units (CMU), rough, porous surface, roll									
Masonry paint, water base, flat or gloss (material #31)									
Roll 1st coat									
Slow	245	90	44.80	9.39	2.25	49.78	11.67	11.69	84.78
Medium	300	78	39.20	9.83	2.83	50.26	15.73	9.44	88.09
Fast	350	65	33.60	10.29	3.64	51.69	20.34	6.02	91.98
Roll 2nd or additional coats									
Slow	275	160	44.80	8.36	2.02	28.00	7.29	7.31	52.98
Medium	325	143	39.20	9.08	2.63	27.41	9.78	5.87	54.77
Fast	420	125	33.60	8.57	3.02	26.88	11.93	3.53	53.93
Masonry paint, oil base (material #32)									
Roll 1st coat									
Slow	245	110	65.40	9.39	2.25	59.45	13.51	13.54	98.14
Medium	300	98	57.20	9.83	2.83	58.37	17.76	10.66	99.45
Fast	350	85	49.00	10.29	3.64	57.65	22.19	6.56	100.33
Roll 2nd or additional coats									
Slow	275	185	65.40	8.36	2.02	35.35	8.69	8.71	63.13
Medium	325	170	57.20	9.08	2.63	33.65	11.34	6.80	63.50
Fast	420	155	49.00	8.57	3.02	31.61	13.40	3.96	60.56
Epoxy coating, 2 part system, clear (material #51)									
Roll 1st coat									
Slow	220	100	160.60	10.45	2.53	160.60	32.98	33.05	239.61
Medium	275	88	140.50	10.73	3.11	159.66	43.37	26.02	242.89
Fast	325	75	120.50	11.08	3.92	160.67	54.45	16.11	246.23
Roll 2nd or additional coats									
Slow	250	175	160.60	9.20	2.21	91.77	19.60	19.64	142.42
Medium	300	160	140.50	9.83	2.83	87.81	25.12	15.07	140.66
Fast	395	145	120.50	9.11	3.21	83.10	29.58	8.75	133.75
Waterproofing, clear hydro sealer, oil base (material #34)									
Roll 1st coat									
Slow	170	110	40.60	13.53	3.24	36.91	10.20	10.22	74.10
Medium	200	98	35.50	14.75	4.27	36.22	13.81	8.28	77.33
Fast	245	85	30.40	14.69	5.18	35.76	17.25	5.10	77.98
Roll 2nd or additional coats									
Slow	275	175	40.60	8.36	2.02	23.20	6.38	6.39	46.35
Medium	300	145	35.50	9.83	2.83	24.48	9.29	5.57	52.00
Fast	325	115	30.40	11.08	3.92	26.43	12.84	3.80	58.07

Use these figures for Concrete Masonry Units (CMU) such as split face, fluted, or slump block, whose surfaces are rough, porous or unfilled, with joints struck to average depth. The more porous the surface, the rougher the texture, the more time and material will be required. For heavy waterproofing applications, see Masonry under Industrial Painting Operations. "Slow" work is based on an hourly wage of $23.00, "Medium" on an hourly wage of $29.50, and "Fast" work on an hourly wage of $36.00. Other qualifications that apply to this table are on page 9.

	Labor SF per manhour	Material coverage SF/gallon	Material cost per gallon	Labor cost per 100 SF	Labor burden 100 SF	Material cost per 100 SF	Overhead per 100 SF	Profit per 100 SF	Total price per 100 SF

Masonry, Concrete Masonry Units (CMU), rough, porous surface, spray

	Labor SF per manhour	Material coverage SF/gallon	Material cost per gallon	Labor cost per 100 SF	Labor burden 100 SF	Material cost per 100 SF	Overhead per 100 SF	Profit per 100 SF	Total price per 100 SF
Masonry paint, water base, flat or gloss (material #31)									
Spray 1st coat									
Slow	600	100	44.80	3.83	.93	44.80	9.41	9.43	68.40
Medium	700	78	39.20	4.21	1.23	50.26	13.92	8.35	77.97
Fast	800	55	33.60	4.50	1.59	61.09	20.83	6.16	94.17
Spray 2nd or additional coats									
Slow	700	155	44.80	3.29	.79	28.90	6.27	6.28	45.53
Medium	800	133	39.20	3.69	1.06	29.47	8.56	5.13	47.91
Fast	900	110	33.60	4.00	1.41	30.55	11.15	3.30	50.41
Masonry paint, oil base (material #32)									
Spray 1st coat									
Slow	600	100	65.40	3.83	.93	65.40	13.33	13.36	96.85
Medium	700	83	57.20	4.21	1.23	68.92	18.59	11.15	104.10
Fast	800	65	49.00	4.50	1.59	75.38	25.26	7.47	114.20
Spray 2nd or additional coats									
Slow	700	160	65.40	3.29	.79	40.88	8.54	8.56	62.06
Medium	800	143	57.20	3.69	1.06	40.00	11.19	6.71	62.65
Fast	900	125	49.00	4.00	1.41	39.20	13.83	4.09	62.53
Epoxy coating, 2 part system, clear (material #51)									
Spray 1st coat									
Slow	500	85	160.60	4.60	1.10	188.94	36.98	37.06	268.68
Medium	600	68	140.50	4.92	1.43	206.62	53.24	31.94	298.15
Fast	700	50	120.50	5.14	1.83	241.00	76.86	22.74	347.57
Spray 2nd or additional coats									
Slow	600	145	160.60	3.83	.93	110.76	21.95	21.99	159.46
Medium	700	130	140.50	4.21	1.23	108.08	28.38	17.03	158.93
Fast	800	115	120.50	4.50	1.59	104.78	34.37	10.17	155.41
Waterproofing, clear hydro sealer, oil base (material #34)									
Spray 1st coat									
Slow	500	60	40.60	4.60	1.10	67.67	13.94	13.97	101.28
Medium	700	50	35.50	4.21	1.23	71.00	19.11	11.46	107.01
Fast	900	40	30.40	4.00	1.41	76.00	25.24	7.47	114.12
Spray 2nd or additional coats									
Slow	600	90	40.60	3.83	.93	45.11	9.47	9.49	68.83
Medium	800	75	35.50	3.69	1.06	47.33	13.02	7.81	72.91
Fast	1000	60	30.40	3.60	1.27	50.67	17.22	5.09	77.85

Use these figures for Concrete Masonry Units (CMU) such as split face, fluted, or slump block, whose surfaces are rough, porous or unfilled, with joints struck to average depth. The more porous the surface, the rougher the texture, the more time and material will be required. For heavy waterproofing applications, see Masonry under Industrial Painting Operations. "Slow" work is based on an hourly wage of $23.00, "Medium" on an hourly wage of $29.50, and "Fast" work on an hourly wage of $36.00. Other qualifications that apply to this table are on page 9.

	Labor SF per manhour	Material coverage SF/gallon	Material cost per gallon	Labor cost per 100 SF	Labor burden 100 SF	Material cost per 100 SF	Overhead per 100 SF	Profit per 100 SF	Total price per 100 SF
Masonry, Concrete Masonry Units (CMU), smooth-surface, brush									
Masonry paint, water base, flat or gloss (material #31)									
Brush 1st coat									
Slow	140	310	44.80	16.43	3.93	14.45	6.62	6.63	48.06
Medium	190	275	39.20	15.53	4.47	14.25	8.57	5.14	47.96
Fast	230	240	33.60	15.65	5.54	14.00	10.90	3.22	49.31
Brush 2nd or additional coats									
Slow	170	410	44.80	13.53	3.24	10.93	5.26	5.28	38.24
Medium	250	370	39.20	11.80	3.41	10.59	6.45	3.87	36.12
Fast	325	340	33.60	11.08	3.92	9.88	7.71	2.28	34.87
Masonry paint, oil base (material #32)									
Brush 1st coat									
Slow	140	350	65.40	16.43	3.93	18.69	7.42	7.44	53.91
Medium	190	320	57.20	15.53	4.47	17.88	9.48	5.69	53.05
Fast	230	290	49.00	15.65	5.54	16.90	11.80	3.49	53.38
Brush 2nd or additional coats									
Slow	170	450	65.40	13.53	3.24	14.53	5.95	5.96	43.21
Medium	250	420	57.20	11.80	3.41	13.62	7.21	4.32	40.36
Fast	325	390	49.00	11.08	3.92	12.56	8.54	2.53	38.63
Epoxy coating, 2 part system clear (material #51)									
Brush 1st coat									
Slow	120	325	160.60	19.17	4.59	49.42	13.91	13.94	101.03
Medium	160	295	140.50	18.44	5.33	47.63	17.85	10.71	99.96
Fast	200	265	120.50	18.00	6.36	45.47	21.64	6.40	97.87
Brush 2nd or additional coats									
Slow	150	425	160.60	15.33	3.69	37.79	10.79	10.81	78.41
Medium	225	395	140.50	13.11	3.78	35.57	13.12	7.87	73.45
Fast	300	365	120.50	12.00	4.22	33.01	15.27	4.52	69.02
Waterproofing, clear hydro sealer, oil base (material #34)									
Brush 1st coat									
Slow	150	100	40.60	15.33	3.69	40.60	11.33	11.35	82.30
Medium	175	88	35.50	16.86	4.86	40.34	15.52	9.31	86.89
Fast	200	75	30.40	18.00	6.36	40.53	20.11	5.95	90.95
Brush 2nd or additional coats									
Slow	230	140	40.60	10.00	2.41	29.00	7.87	7.88	57.16
Medium	275	120	35.50	10.73	3.11	29.58	10.85	6.51	60.78
Fast	295	100	30.40	12.20	4.31	30.40	14.54	4.30	65.75

Use these figures for Concrete Masonry Units (CMU) where the block surfaces are smooth with joints struck to average depth. For heavy waterproofing applications, see Masonry under Industrial Painting Operations. "Slow" work is based on an hourly wage of $23.00, "Medium" work on an hourly wage of $29.50, and "Fast" work on an hourly wage of $36.00. Other qualifications that apply to this table are on page 9.

	Labor SF per manhour	Material coverage SF/gallon	Material cost per gallon	Labor cost per 100 SF	Labor burden 100 SF	Material cost per 100 SF	Overhead per 100 SF	Profit per 100 SF	Total price per 100 SF

Masonry, Concrete Masonry Units (CMU), smooth-surface, roll

Masonry paint, water base, flat or gloss (material #31)									
Roll 1st coat									
Slow	300	240	44.80	7.67	1.83	18.67	5.35	5.36	38.88
Medium	350	228	39.20	8.43	2.45	17.19	7.02	4.21	39.30
Fast	390	215	33.60	9.23	3.24	15.63	8.72	2.58	39.40
Roll 2nd or additional coats									
Slow	350	325	44.80	6.57	1.59	13.78	4.17	4.18	30.29
Medium	400	313	39.20	7.38	2.13	12.52	5.51	3.30	30.84
Fast	450	300	33.60	8.00	2.81	11.20	6.83	2.02	30.86
Masonry paint, oil base (material #32)									
Roll 1st coat									
Slow	300	285	65.40	7.67	1.83	22.95	6.17	6.18	44.80
Medium	350	273	57.20	8.43	2.45	20.95	7.96	4.77	44.56
Fast	390	260	49.00	9.23	3.24	18.85	9.72	2.87	43.91
Roll 2nd or additional coats									
Slow	350	375	65.40	6.57	1.59	17.44	4.86	4.87	35.33
Medium	400	358	57.20	7.38	2.13	15.98	6.37	3.82	35.68
Fast	450	340	49.00	8.00	2.81	14.41	7.82	2.31	35.35
Epoxy coating, 2 part system, clear (material #51)									
Roll 1st coat									
Slow	275	265	160.60	8.36	2.02	60.60	13.48	13.51	97.97
Medium	325	253	140.50	9.08	2.63	55.53	16.81	10.08	94.13
Fast	375	240	120.50	9.60	3.41	50.21	19.59	5.80	88.61
Roll 2nd or additional coats									
Slow	325	355	160.60	7.08	1.70	45.24	10.26	10.28	74.56
Medium	375	340	140.50	7.87	2.28	41.32	12.87	7.72	72.06
Fast	425	325	120.50	8.47	2.98	37.08	15.05	4.45	68.03
Waterproofing, clear hydro sealer, oil base (material #34)									
Roll 1st coat									
Slow	170	110	40.60	13.53	3.24	36.91	10.20	10.22	74.10
Medium	200	95	35.50	14.75	4.27	37.37	14.10	8.46	78.95
Fast	245	80	30.40	14.69	5.18	38.00	17.94	5.31	81.12
Roll 2nd or additional coats									
Slow	275	150	40.60	8.36	2.02	27.07	7.11	7.13	51.69
Medium	300	125	35.50	9.83	2.83	28.40	10.27	6.16	57.49
Fast	325	100	30.40	11.08	3.92	30.40	14.07	4.16	63.63

Use these figures for Concrete Masonry Units (CMU) where the block surfaces are smooth with joints struck to average depth. For heavy waterproofing applications, see Masonry under Industrial Painting Operations. "Slow" work is based on an hourly wage of $23.00, "Medium" work on an hourly wage of $29.50, and "Fast" work on an hourly wage of $36.00. Other qualifications that apply to this table are on page 9.

	Labor SF per manhour	Material coverage SF/gallon	Material cost per gallon	Labor cost per 100 SF	Labor burden 100 SF	Material cost per 100 SF	Overhead per 100 SF	Profit per 100 SF	Total price per 100 SF
Masonry, Concrete Masonry Units (CMU), smooth-surface, spray									
Masonry paint, water base, flat or gloss (material #31)									
Spray 1st coat									
Slow	725	240	44.80	3.17	.77	18.67	4.29	4.30	31.20
Medium	788	215	39.20	3.74	1.09	18.23	5.76	3.46	32.28
Fast	850	190	33.60	4.24	1.51	17.68	7.26	2.15	32.84
Spray 2nd or additional coats									
Slow	800	320	44.80	2.88	.69	14.00	3.34	3.35	24.26
Medium	950	295	39.20	3.11	.88	13.29	4.33	2.60	24.21
Fast	1100	270	33.60	3.27	1.16	12.44	5.23	1.55	23.65
Masonry paint, oil base (material #32)									
Spray 1st coat									
Slow	725	280	65.40	3.17	.77	23.36	5.19	5.20	37.69
Medium	788	255	57.20	3.74	1.09	22.43	6.81	4.09	38.16
Fast	850	230	49.00	4.24	1.51	21.30	8.38	2.48	37.91
Spray 2nd or additional coats									
Slow	800	345	65.40	2.88	.69	18.96	4.28	4.29	31.10
Medium	950	328	57.20	3.11	.88	17.44	5.36	3.22	30.01
Fast	1100	310	49.00	3.27	1.16	15.81	6.27	1.86	28.37
Epoxy coating, 2 part system clear (material #51)									
Spray 1st coat									
Slow	675	255	160.60	3.41	.81	62.98	12.77	12.80	92.77
Medium	738	238	140.50	4.00	1.17	59.03	16.05	9.63	89.88
Fast	800	220	120.50	4.50	1.59	54.77	18.87	5.58	85.31
Spray 2nd or additional coats									
Slow	750	325	160.60	3.07	.72	49.42	10.11	10.13	73.45
Medium	900	308	140.50	3.28	.94	45.62	12.46	7.48	69.78
Fast	1050	290	120.50	3.43	1.20	41.55	14.32	4.24	64.74
Waterproofing, clear hydro sealer, oil base (material #34)									
Spray 1st coat									
Slow	500	75	40.60	4.60	1.10	54.13	11.37	11.39	82.59
Medium	750	63	35.50	3.93	1.13	56.35	15.36	9.21	85.98
Fast	900	50	30.40	4.00	1.41	60.80	20.53	6.07	92.81
Spray 2nd or additional coats									
Slow	600	100	40.60	3.83	.93	40.60	8.62	8.64	62.62
Medium	800	88	35.50	3.69	1.06	40.34	11.28	6.77	63.14
Fast	1000	75	30.40	3.60	1.27	40.53	14.07	4.16	63.63

Use these figures for Concrete Masonry Units (CMU) where the block surfaces are smooth with joints struck to average depth. For heavy waterproofing applications, see Masonry under Industrial Painting Operations. "Slow" work is based on an hourly wage of $23.00, "Medium" work on an hourly wage of $29.50, and "Fast" work on an hourly wage of $36.00. Other qualifications that apply to this table are on page 9.

	Labor SF per manhour	Material coverage SF/gallon	Material cost per gallon	Labor cost per 100 SF	Labor burden 100 SF	Material cost per 100 SF	Overhead per 100 SF	Profit per 100 SF	Total price per 100 SF
Masonry, stone, marble or granite									
Waterproof, clear hydro sealer, oil base (material #34)									
Spray 1st coat									
Slow	600	220	40.60	3.83	.93	18.45	4.41	4.42	32.04
Medium	700	200	35.50	4.21	1.23	17.75	5.80	3.48	32.47
Fast	800	180	30.40	4.50	1.59	16.89	7.12	2.11	32.21
Spray 2nd coat									
Slow	700	225	40.60	3.29	.79	18.04	4.20	4.21	30.53
Medium	800	213	35.50	3.69	1.06	16.67	5.36	3.21	29.99
Fast	900	200	30.40	4.00	1.41	15.20	6.39	1.89	28.89

Use these figures for the cost to apply waterproof sealer on stone, marble or granite surfaces. "Slow" work is based on an hourly wage of $23.00, "Medium" work on an hourly wage of $29.50, and "Fast" work on an hourly wage of $36.00. Other qualifications that apply to this table are on page 9.

	Labor LF per manhour	Material coverage LF/gallon	Material cost per gallon	Labor cost per 100 LF	Labor burden 100 LF	Material cost per 100 LF	Overhead per 100 LF	Profit per 100 LF	Total price per 100 LF
Molding, interior or exterior, paint grade, smooth-surface									
Undercoat, water or oil base (material #3 or #4)									
Brush prime coat									
Slow	135	700	42.60	17.04	4.09	6.09	5.17	5.18	37.57
Medium	205	600	37.30	14.39	4.17	6.22	6.19	3.72	34.69
Fast	275	500	31.95	13.09	4.64	6.39	7.47	2.21	33.80
Split coat (1/2 undercoat + 1/2 enamel), water or oil base (material #3 and #9, or material #4 and #10)									
Brush 1st or additional finish coats									
Slow	125	750	50.00	18.40	4.42	6.67	5.60	5.61	40.70
Medium	135	675	43.78	21.85	6.33	6.49	8.66	5.20	48.53
Fast	145	600	37.50	24.83	8.78	6.25	12.35	3.65	55.86
Enamel, interior, water or oil base (material #9 or #10)									
Brush 1st finish coat									
Slow	125	750	57.40	18.40	4.42	7.65	5.79	5.80	42.06
Medium	160	675	50.25	18.44	5.33	7.44	7.80	4.68	43.69
Fast	200	600	43.05	18.00	6.36	7.18	9.77	2.89	44.20
Brush 2nd or additional finish coats									
Slow	115	750	57.40	20.00	4.81	7.65	6.17	6.18	44.81
Medium	150	675	50.25	19.67	5.70	7.44	8.20	4.92	45.93
Fast	175	600	43.05	20.57	7.24	7.18	10.85	3.21	49.05

	Labor LF per manhour	Material coverage LF/gallon	Material cost per gallon	Labor cost per 100 LF	Labor burden 100 LF	Material cost per 100 LF	Overhead per 100 LF	Profit per 100 LF	Total price per 100 LF
Enamel, exterior, water or oil base (material #24 or #25)									
Brush 1st finish coat									
Slow	125	750	58.65	18.40	4.42	7.82	5.82	5.83	42.29
Medium	160	675	51.30	18.44	5.33	7.60	7.84	4.71	43.92
Fast	200	600	44.00	18.00	6.36	7.33	9.82	2.91	44.42
Brush 2nd or additional finish coats									
Slow	115	750	58.65	20.00	4.81	7.82	6.20	6.21	45.04
Medium	150	675	51.30	19.67	5.70	7.60	8.24	4.94	46.15
Fast	175	600	44.00	20.57	7.24	7.33	10.90	3.22	49.26
Stipple finish									
Slow	80	--	--	28.75	6.90	--	6.77	6.79	49.21
Medium	90	--	--	32.78	9.47	--	10.56	6.34	59.15
Fast	100	--	--	36.00	12.71	--	15.10	4.47	68.28
Glazing or mottling over enamel (material #16)									
Glaze & wipe, brush 1 coat									
Slow	55	950	49.10	41.82	10.03	5.17	10.84	10.86	78.72
Medium	65	900	43.00	45.38	13.11	4.78	15.82	9.49	88.58
Fast	75	850	36.90	48.00	16.93	4.34	21.48	6.35	97.10

Consider all trim to be at least 12" wide (even if it's much less than 12" wide) when calculating the area to be painted. Trim painted the same color as the wall or ceiling behind it may take no more time than painting the wall or ceiling itself. Use the slow rate when cutting-in is required to paint molding that's a different color from the surface behind the molding. "Slow" work is based on an hourly wage of $23.00, "Medium" work on an hourly wage of $29.50, and "Fast" work on an hourly wage of $36.00. Other qualifications that apply to this table are on page 9.

	Labor LF per manhour	Material coverage LF/gallon	Material cost per gallon	Labor cost per 100 LF	Labor burden 100 LF	Material cost per 100 LF	Overhead per 100 LF	Profit per 100 LF	Total price per 100 LF
Molding, interior, stain grade, smooth surface									
Stain, seal and 2 coat lacquer system (7 step process)									
STEP 1: Sand & putty;									
Slow	150	--	--	15.33	3.69	--	3.61	3.62	26.25
Medium	175	--	--	16.86	4.86	--	5.43	3.26	30.41
Fast	200	--	--	18.00	6.36	--	7.55	2.23	34.14
STEP 2 & 3: Wiping stain (material #11a) & wipe									
Brush 1 coat & wipe									
Slow	150	550	61.60	15.33	3.69	11.20	5.74	5.75	41.71
Medium	175	525	53.90	16.86	4.86	10.27	8.00	4.80	44.79
Fast	200	500	46.20	18.00	6.36	9.24	10.41	3.08	47.09
Spray 1 coat & wipe									
Slow	400	250	61.60	5.75	1.38	24.64	6.04	6.05	43.86
Medium	425	225	53.90	6.94	2.00	23.96	8.23	4.94	46.07
Fast	450	200	46.20	8.00	2.81	23.10	10.52	3.11	47.54
STEP 4 & 5: Sanding sealer (material #11b) & light sand									
Brush 1 coat									
Slow	260	575	50.90	8.85	2.13	8.85	3.77	3.77	27.37
Medium	280	563	44.60	10.54	3.04	7.92	5.38	3.23	30.11
Fast	300	550	38.20	12.00	4.22	6.95	7.19	2.13	32.49
Spray 1 coat									
Slow	450	250	50.90	5.11	1.22	20.36	5.07	5.08	36.84
Medium	475	225	44.60	6.21	1.81	19.82	6.96	4.17	38.97
Fast	500	200	38.20	7.20	2.54	19.10	8.94	2.64	40.42
STEP 6 & 7: Lacquer (material #11c), 2 coats									
Brush 1st coat									
Slow	200	300	53.50	11.50	2.76	17.83	6.10	6.11	44.30
Medium	275	288	46.80	10.73	3.11	16.25	7.52	4.51	42.12
Fast	350	275	40.10	10.29	3.64	14.58	8.84	2.61	39.96
Brush 2nd coat									
Slow	225	375	53.50	10.22	2.44	14.27	5.12	5.13	37.18
Medium	300	350	46.80	9.83	2.83	13.37	6.51	3.91	36.45
Fast	375	325	40.10	9.60	3.41	12.34	7.85	2.32	35.52
Spray 1st coat									
Slow	250	200	53.50	9.20	2.21	26.75	7.25	7.27	52.68
Medium	350	188	46.80	8.43	2.45	24.89	8.94	5.36	50.07
Fast	450	175	40.10	8.00	2.81	22.91	10.46	3.09	47.27
Spray 2nd coat									
Slow	300	250	53.50	7.67	1.83	21.40	5.87	5.88	42.65
Medium	388	225	46.80	7.60	2.21	20.80	7.65	4.59	42.85
Fast	475	200	40.10	7.58	2.70	20.05	9.40	2.78	42.51

	Labor LF per manhour	Material coverage LF/gallon	Material cost per gallon	Labor cost per 100 LF	Labor burden 100 LF	Material cost per 100 LF	Overhead per 100 LF	Profit per 100 LF	Total price per 100 LF
Complete 7 step process, stain, seal & 2 coat lacquer system (material #11)									
Brush all coats									
Slow	50	60	54.90	46.00	11.04	91.50	28.22	28.28	205.04
Medium	75	50	48.00	39.33	11.36	96.00	36.68	22.01	205.38
Fast	100	40	41.20	36.00	12.71	103.00	47.03	13.91	212.65
Spray all coats									
Slow	100	40	54.90	23.00	5.52	137.25	31.50	31.56	228.83
Medium	150	30	48.00	19.67	5.70	160.00	46.34	27.80	259.51
Fast	200	20	41.20	18.00	6.36	206.00	71.41	21.12	322.89

These figures are based on linear feet for all molding up to 12" wide. For estimating purposes, consider all molding and trim to be at least 12" wide (even if it's much less than 12") when calculating the area to be finished. The spray figures are based on finishing large quantities of molding in an area set up for spray painting before it's installed. Use the brush figures for small quantities. Use the fast brush rate for molding that's finished before it's installed. If the molding is attached and has to be masked off, use the slow brush figures. Masking time is not included. See the Preparation Operations tables for masking rates. Trim stained the same color as the surface behind it will take no more time than staining the wall itself. "Slow" work is based on an hourly wage of $23.00, "Medium" work on an hourly wage of $29.50, and "Fast" work on an hourly wage of $36.00. Other qualifications that apply to this table are on page 9.

	Labor LF per manhour	Material coverage LF/gallon	Material cost per gallon	Labor cost per 100 LF	Labor burden 100 LF	Material cost per 100 LF	Overhead per 100 LF	Profit per 100 LF	Total price per 100 LF
Molding, interior or exterior, stain grade, smooth surface									
Stain, fill and shellac or varnish									
Wiping stain (material #30a) & fill									
Brush each coat									
Slow	80	550	57.10	28.75	6.90	10.38	8.75	8.76	63.54
Medium	130	525	50.00	22.69	6.56	9.52	9.69	5.82	54.28
Fast	180	500	42.90	20.00	7.08	8.58	11.05	3.27	49.98
Shellac, clear (material #12)									
Brush each coat									
Slow	180	550	70.40	12.78	3.08	12.80	5.44	5.45	39.55
Medium	230	525	61.60	12.83	3.71	11.73	7.07	4.24	39.58
Fast	280	500	52.80	12.86	4.53	10.56	8.67	2.56	39.18
Varnish, gloss (material #30c)									
Brush each coat									
Slow	115	550	81.00	20.00	4.81	14.73	7.51	7.53	54.58
Medium	150	525	70.80	19.67	5.70	13.49	9.71	5.83	54.40
Fast	210	500	60.70	17.14	6.05	12.14	10.95	3.24	49.52
Varnish, flat (material #30c)									
Brush each coat									
Slow	125	550	81.00	18.40	4.42	14.73	7.13	7.15	51.83
Medium	160	525	70.80	18.44	5.33	13.49	9.32	5.59	52.17
Fast	220	500	60.70	16.36	5.80	12.14	10.63	3.14	48.07
Penetrating stain wax (material #14) & wipe									
Stain, brush each coat & wipe									
Slow	225	500	50.20	10.22	2.44	10.04	4.31	4.32	31.33
Medium	275	450	43.90	10.73	3.11	9.76	5.90	3.54	33.04
Fast	325	400	37.60	11.08	3.92	9.40	7.56	2.24	34.20
Polish added coats of wax									
Slow	150	--	--	15.33	3.69	--	3.61	3.62	26.25
Medium	175	--	--	16.86	4.86	--	5.43	3.26	30.41
Fast	200	--	--	18.00	6.36	--	7.55	2.23	34.14

	Labor LF per manhour	Material coverage LF/gallon	Material cost per gallon	Labor cost per 100 LF	Labor burden 100 LF	Material cost per 100 LF	Overhead per 100 LF	Profit per 100 LF	Total price per 100 LF
Steel wool buff by hand									
Slow	100	--	--	23.00	5.52	--	5.42	5.43	39.37
Medium	110	--	--	26.82	7.75	--	8.64	5.19	48.40
Fast	125	--	--	28.80	10.17	--	12.08	3.57	54.62
Wax application & polish, hand apply 1 coat									
Slow	100	--	--	23.00	5.52	--	5.42	5.43	39.37
Medium	110	--	--	26.82	7.75	--	8.64	5.19	48.40
Fast	125	--	--	28.80	10.17	--	12.08	3.57	54.62

These figures are based on linear feet for all molding up to 12" wide. For estimating purposes, consider all molding and trim to be at least 12" wide (even if it's much less than 12") when calculating the area to be finished. The spray figures are based on finishing large quantities of molding in an area set up for spray painting before it's installed. Use the brush figures for small quantities. Use the fast brush rate for molding that's finished before it's installed. If the molding is attached and has to be masked off, use the slow brush figures. Masking time is not included. See the Preparation Operations tables for masking rates. Trim stained the same color as the surface behind it will take no more time than staining the wall itself. "Slow" work is based on an hourly wage of $23.00, "Medium" work on an hourly wage of $29.50, and "Fast" work on an hourly wage of $36.00. Other qualifications that apply to this table are on page 9.

	Boxed eaves One color	Boxed eaves Two color	Exposed rafters One color	Exposed rafters Two color
Overhang difficulty factors, eaves, cornice				
One story, repaint				
Standard	1.5	2.0	2.0	2.5
Ornamental	--	--	2.5	3.0
One story, new construction	--	1.5	1.5	2.0
Two story, with scaffolding, repaint				
Standard	1.5	2.0	2.0	2.5
Ornamental	--	--	2.5	3.0
Two story, without scaffolding, repaint				
Standard	3.0	3.5	3.5	4.0
Ornamental	--	--	4.0	4.5
Two story, new construction				
With scaffolding	--	1.5	1.5	2.0
No scaffolding	2.0	2.5	2.5	3.0

Before using the figures in the tables for overhangs, apply these difficulty factors to the surface area (length times width) to be painted. Multiply the factor in this table by the overall surface area of the overhang. This allows for slower work on high eaves and the extra time needed to paint rafter tails. This table adjusts for the kind of eaves, the eave height, the number of colors used, and whether scaffolding is available and erected on site. Boxed eaves have plywood covering the rafter tails.

	Labor SF per manhour	Material coverage SF/gallon	Material cost per gallon	Labor cost per 100 SF	Labor burden 100 SF	Material cost per 100 SF	Overhead per 100 SF	Profit per 100 SF	Total price per 100 SF
Overhang at carports, large continuous surface areas									
Solid body or semi-transparent stain, water or oil base (material #18 or #19 or #20 or #21)									
Spray 1st coat									
Slow	550	95	54.50	4.18	1.01	57.37	11.88	11.91	86.35
Medium	600	90	47.70	4.92	1.43	53.00	14.84	8.90	83.09
Fast	650	85	40.88	5.54	1.96	48.09	17.23	5.10	77.92
Spray 2nd coat									
Slow	600	175	54.50	3.83	.93	31.14	6.82	6.83	49.55
Medium	650	163	47.70	4.54	1.32	29.26	8.78	5.27	49.17
Fast	700	150	40.88	5.14	1.83	27.25	10.60	3.14	47.96
Spray 3rd or additional coats									
Slow	650	225	54.50	3.54	.85	24.22	5.44	5.45	39.50
Medium	700	213	47.70	4.21	1.23	22.39	6.96	4.17	38.96
Fast	750	200	40.88	4.80	1.68	20.44	8.35	2.47	37.74

"Slow" work is based on an hourly wage of $23.00, "Medium" work on an hourly wage of $29.50, and "Fast" work on an hourly wage of $36.00. Other qualifications that apply to this table are on page 9.

	Labor SF per manhour	Material coverage SF/gallon	Material cost per gallon	Labor cost per 100 SF	Labor burden 100 SF	Material cost per 100 SF	Overhead per 100 SF	Profit per 100 SF	Total price per 100 SF
Overhang at eaves or rake, widths up to 2.5 feet									
Solid body stain, water or oil base (material #18 or #19)									
Roll & Brush 1st coat									
Slow	95	200	54.90	24.21	5.82	27.45	10.92	10.94	79.34
Medium	125	185	48.05	23.60	6.82	25.97	14.10	8.46	78.95
Fast	150	170	41.20	24.00	8.49	24.24	17.58	5.20	79.51
Roll & Brush 2nd coat									
Slow	140	260	54.90	16.43	3.93	21.12	7.88	7.90	57.26
Medium	185	240	48.05	15.95	4.62	20.02	10.15	6.09	56.83
Fast	220	220	41.20	16.36	5.80	18.73	12.67	3.75	57.31
Roll & Brush 3rd or additional coats									
Slow	170	295	54.90	13.53	3.24	18.61	6.72	6.74	48.84
Medium	225	270	48.05	13.11	3.78	17.80	8.68	5.21	48.58
Fast	275	245	41.20	13.09	4.64	16.82	10.70	3.17	48.42
Solid body stain or semi-transparent stain, water or oil base (material #18 or #19 or #20 or #21)									
Spray 1st coat									
Slow	300	150	54.50	7.67	1.83	36.33	8.71	8.73	63.27
Medium	350	125	47.70	8.43	2.45	38.16	12.26	7.35	68.65
Fast	400	100	40.88	9.00	3.18	40.88	16.45	4.87	74.38
Spray 2nd coat									
Slow	375	175	54.50	6.13	1.48	31.14	7.36	7.38	53.49
Medium	450	163	47.70	6.56	1.88	29.26	9.43	5.66	52.79
Fast	540	150	40.88	6.67	2.34	27.25	11.24	3.33	50.83
Spray 3rd or additional coats									
Slow	450	225	54.50	5.11	1.22	24.22	5.81	5.82	42.18
Medium	525	213	47.70	5.62	1.61	22.39	7.41	4.44	41.47
Fast	600	200	40.88	6.00	2.13	20.44	8.85	2.62	40.04

Use this table after multiplying the overall area (length times width) by the difficulty factor listed under Overhang, difficulty factor. Remember to ADD preparation time for masking, caulking, sanding, waterblasting, etc. "Slow" work is based on an hourly wage of $23.00, "Medium" work on an hourly wage of $29.50, and "Fast" work on an hourly wage of $36.00. Other qualifications that apply to this table are on page 9.

	Labor SF per manhour	Material coverage SF/gallon	Material cost per gallon	Labor cost per 100 SF	Labor burden 100 SF	Material cost per 100 SF	Overhead per 100 SF	Profit per 100 SF	Total price per 100 SF
Overhang at entries or decks, widths greater than 2.5 feet									
Solid body stain, water or oil base (material #18 or #19)									
Roll & Brush 1st coat									
Slow	125	225	54.90	18.40	4.42	24.40	8.97	8.99	65.18
Medium	180	210	48.05	16.39	4.75	22.88	11.00	6.60	61.62
Fast	225	195	41.20	16.00	5.63	21.13	13.26	3.92	59.94
Roll & Brush 2nd coat									
Slow	170	295	54.90	13.53	3.24	18.61	6.72	6.74	48.84
Medium	225	275	48.05	13.11	3.78	17.47	8.59	5.16	48.11
Fast	275	255	41.20	13.09	4.64	16.16	10.50	3.11	47.50
Roll & Brush 3rd or additional coats									
Slow	210	340	54.90	10.95	2.63	16.15	5.65	5.66	41.04
Medium	280	315	48.05	10.54	3.04	15.25	7.21	4.33	40.37
Fast	345	290	41.20	10.43	3.70	14.21	8.78	2.60	39.72
Solid body stain or semi-transparent stain, water or oil base (material #18 or #19 or #20 or #21)									
Spray 1st coat									
Slow	400	150	54.50	5.75	1.38	36.33	8.26	8.28	60.00
Medium	463	125	47.70	6.37	1.84	38.16	11.59	6.96	64.92
Fast	525	100	40.88	6.86	2.39	40.88	15.55	4.60	70.28
Spray 2nd coat									
Slow	450	175	54.50	5.11	1.22	31.14	7.12	7.14	51.73
Medium	550	163	47.70	5.36	1.56	29.26	9.04	5.43	50.65
Fast	650	150	40.88	5.54	1.96	27.25	10.77	3.19	48.71
Spray 3rd or additional coats									
Slow	550	225	54.50	4.18	1.01	24.22	5.59	5.60	40.60
Medium	625	213	47.70	4.72	1.36	22.39	7.12	4.27	39.86
Fast	700	200	40.88	5.14	1.83	20.44	8.49	2.51	38.41

Use this table after multiplying the overall area (length times width) by the difficulty factor listed under Overhang, difficulty factor. Remember to ADD preparation time for masking, caulking, sanding, waterblasting, etc. "Slow" work is based on an hourly wage of $23.00, "Medium" work on an hourly wage of $29.50, and "Fast" work on an hourly wage of $36.00. Other qualifications that apply to this table are on page 9.

	Labor LF per manhour	Material coverage LF/gallon	Material cost per gallon	Labor cost per 100 LF	Labor burden 100 LF	Material cost per 100 LF	Overhead per 100 LF	Profit per 100 LF	Total price per 100 LF
Pass-through shelves, wood top & wrought iron support									
Metal primer, rust inhibitor, clean metal (material #35)									
Brush 1st coat									
Slow	30	125	59.70	76.67	18.39	47.76	27.14	27.20	197.16
Medium	35	113	52.30	84.29	24.36	46.28	38.73	23.24	216.90
Fast	40	100	44.80	90.00	31.78	44.80	51.64	15.27	233.49
Metal finish - synthetic enamel (colors except orange/red) (material #38)									
Brush 1st coat									
Slow	30	125	66.30	76.67	18.39	53.04	28.14	28.20	204.44
Medium	35	113	58.00	84.29	24.36	51.33	40.00	24.00	223.98
Fast	40	100	49.70	90.00	31.78	49.70	53.16	15.72	240.36

Use these figures to estimate pass-through shelves which are approximately 12" wide and 3'0" long with a wood top and wrought iron supports. The rule-of-thumb minimum time and material is .2 hours and $1.00 for material per shelf. A two-coat system using oil based material is recommended for any metal surface. Although water based material is often used, it may cause oxidation, corrosion and rust. One coat of oil based, solid body stain is often used on exterior metal, but it may crack, peel or chip without the proper prime coat application. "Slow" work is based on an hourly wage of $23.00, "Medium" work on an hourly wage of $29.50, and "Fast" work on an hourly wage of $36.00. Other qualifications that apply to this table are on page 9.

	Labor LF per manhour	Material coverage LF/gallon	Material cost per gallon	Labor cost per 100 LF	Labor burden 100 LF	Material cost per 100 LF	Overhead per 100 LF	Profit per 100 LF	Total price per 100 LF
Plant-on trim, exterior, 2" x 2" to 2" x 4" wood									
2" x 2" to 2" x 4" rough sawn or resawn wood									
Solid body or semi-transparent stain, water base (material #18 or #20)									
Roll & brush 1st coat									
Slow	80	310	49.40	28.75	6.90	15.94	9.80	9.82	71.21
Medium	100	275	43.25	29.50	8.53	15.73	13.44	8.06	75.26
Fast	120	240	37.05	30.00	10.58	15.44	17.37	5.14	78.53
Roll & brush 2nd coat									
Slow	90	350	49.40	25.56	6.13	14.11	8.70	8.72	63.22
Medium	120	325	43.25	24.58	7.10	13.31	11.25	6.75	62.99
Fast	150	300	37.05	24.00	8.49	12.35	13.89	4.11	62.84
Roll & brush 3rd or additional coats									
Slow	100	375	49.40	23.00	5.52	13.17	7.92	7.94	57.55
Medium	140	350	43.25	21.07	6.08	12.36	9.88	5.93	55.32
Fast	180	325	37.05	20.00	7.08	11.40	11.92	3.53	53.93
Solid body or semi-transparent stain, oil base (material #19 or #21)									
Roll & brush 1st coat									
Slow	95	400	59.60	24.21	5.82	14.90	8.53	8.55	62.01
Medium	115	365	52.15	25.65	7.44	14.29	11.84	7.10	66.32
Fast	135	330	44.70	26.67	9.42	13.55	15.39	4.55	69.58

	Labor LF per manhour	Material coverage LF/gallon	Material cost per gallon	Labor cost per 100 LF	Labor burden 100 LF	Material cost per 100 LF	Overhead per 100 LF	Profit per 100 LF	Total price per 100 LF
Roll & brush 2nd coat									
Slow	105	475	59.60	21.90	5.25	12.55	7.54	7.56	54.80
Medium	135	438	52.15	21.85	6.33	11.91	10.02	6.01	56.12
Fast	165	400	44.70	21.82	7.70	11.18	12.62	3.73	57.05
Roll & brush 3rd or additional coats									
Slow	115	500	59.60	20.00	4.81	11.92	6.98	6.99	50.70
Medium	155	463	52.15	19.03	5.50	11.26	8.95	5.37	50.11
Fast	195	425	44.70	18.46	6.53	10.52	11.01	3.26	49.78
Varnish, flat or gloss (material #30c)									
Roll & brush 1st coat									
Slow	70	270	81.00	32.86	7.90	30.00	13.44	13.47	97.67
Medium	90	255	70.80	32.78	9.47	27.76	17.50	10.50	98.01
Fast	110	240	60.70	32.73	11.55	25.29	21.57	6.38	97.52
Roll & brush 2nd coat									
Slow	80	330	81.00	28.75	6.90	24.55	11.44	11.46	83.10
Medium	110	315	70.80	26.82	7.75	22.48	14.26	8.56	79.87
Fast	140	300	60.70	25.71	9.07	20.23	17.06	5.05	77.12
Roll & brush 3rd or additional coats									
Slow	90	350	81.00	25.56	6.13	23.14	10.42	10.44	75.69
Medium	130	335	70.80	22.69	6.56	21.13	12.60	7.56	70.54
Fast	170	320	60.70	21.18	7.46	18.97	14.77	4.37	66.75

Don't add additional time for plant-on trim if it's painted with the same coating as the adjacent siding. For heights above 8 feet, use the High Time Difficulty Factors on page 139. Use slow rates when cutting-in or masking adjacent surfaces. ADD preparation time for masking adjacent surfaces or protecting windows. Use fast rates when plant-on trim is finished before it's installed or on new construction projects where a prime coat can be sprayed prior to stucco color coat application. "Slow" work is based on an hourly wage of $23.00, "Medium" work on an hourly wage of $29.50, and "Fast" work on an hourly wage of $36.00. Other qualifications that apply to this table are on page 9.

	Labor LF per manhour	Material coverage LF/gallon	Material cost per gallon	Labor cost per 100 LF	Labor burden 100 LF	Material cost per 100 LF	Overhead per 100 LF	Profit per 100 LF	Total price per 100 LF

Plant-on trim, exterior, 2" x 6" to 2" x 8" wood

2" x 6" to 2" x 8" rough sawn or resawn wood

	Labor LF per manhour	Material coverage LF/gallon	Material cost per gallon	Labor cost per 100 LF	Labor burden 100 LF	Material cost per 100 LF	Overhead per 100 LF	Profit per 100 LF	Total price per 100 LF
Solid body or semi-transparent stain, water base (material #18 or #20)									
Roll & brush 1st coat									
Slow	70	210	49.40	32.86	7.90	23.52	12.21	12.24	88.73
Medium	85	175	43.25	34.71	10.01	24.71	17.36	10.42	97.21
Fast	100	140	37.05	36.00	12.71	26.46	23.30	6.89	105.36
Roll & brush 2nd coat									
Slow	80	250	49.40	28.75	6.90	19.76	10.53	10.55	76.49
Medium	110	225	43.25	26.82	7.75	19.22	13.45	8.07	75.31
Fast	140	200	37.05	25.71	9.07	18.53	16.53	4.89	74.73
Roll & brush 3rd or additional coats									
Slow	90	275	49.40	25.56	6.13	17.96	9.43	9.45	68.53
Medium	130	250	43.25	22.69	6.56	17.30	11.64	6.98	65.17
Fast	170	225	37.05	21.18	7.46	16.47	13.99	4.14	63.24
Solid body or semi-transparent stain, oil base (material #19 or #21)									
Roll & brush 1st coat									
Slow	85	300	59.60	27.06	6.48	19.87	10.15	10.17	73.73
Medium	110	265	52.15	26.82	7.75	19.68	13.56	8.14	75.95
Fast	135	230	44.70	26.67	9.42	19.43	17.21	5.09	77.82
Roll & brush 2nd coat									
Slow	95	375	59.60	24.21	5.82	15.89	8.72	8.74	63.38
Medium	125	338	52.15	23.60	6.82	15.43	11.46	6.88	64.19
Fast	155	300	44.70	23.23	8.19	14.90	14.36	4.25	64.93
Roll & brush 3rd or additional coats									
Slow	105	400	59.60	21.90	5.25	14.90	7.99	8.01	58.05
Medium	145	363	52.15	20.34	5.90	14.37	10.15	6.09	56.85
Fast	185	325	44.70	19.46	6.89	13.75	12.42	3.68	56.20
Varnish, flat or gloss (material #30c)									
Roll & brush 1st coat									
Slow	60	170	81.00	38.33	9.21	47.65	18.08	18.12	131.39
Medium	70	155	70.80	42.14	12.20	45.68	25.00	15.00	140.02
Fast	90	140	60.70	40.00	14.12	43.36	30.22	8.94	136.64

	Labor LF per manhour	Material coverage LF/gallon	Material cost per gallon	Labor cost per 100 LF	Labor burden 100 LF	Material cost per 100 LF	Overhead per 100 LF	Profit per 100 LF	Total price per 100 LF
Roll & brush 2nd coat									
Slow	70	240	81.00	32.86	7.90	33.75	14.16	14.19	102.86
Medium	100	220	70.80	29.50	8.53	32.18	17.55	10.53	98.29
Fast	130	200	60.70	27.69	9.77	30.35	21.02	6.22	95.05
Roll & brush 3rd or additional coats									
Slow	80	260	81.00	28.75	6.90	31.15	12.69	12.72	92.21
Medium	115	240	70.80	25.65	7.44	29.50	15.64	9.38	87.61
Fast	150	220	60.70	24.00	8.49	27.59	18.62	5.51	84.21

Don't add additional time for plant-on trim if it's painted with the same coating as the adjacent siding. For heights above 8 feet, use the High Time Difficulty Factors on page 139. Use slow rates when cutting-in or masking adjacent surfaces. Add preparation time for masking adjacent surfaces or protecting windows. Use fast rates when plant-on trim is finished before it's installed or on new construction projects where a prime coat can be sprayed prior to stucco color coat application. "Slow" work is based on an hourly wage of $23.00, "Medium" work on an hourly wage of $29.50, and "Fast" work on an hourly wage of $36.00. Other qualifications that apply to this table are on page 9.

	Labor LF per manhour	Material coverage LF/gallon	Material cost per gallon	Labor cost per 100 LF	Labor burden 100 LF	Material cost per 100 LF	Overhead per 100 LF	Profit per 100 LF	Total price per 100 LF
Plant-on trim, exterior, 2" x 10" to 2" x 12" wood									
2" x 10" to 2" x 12" rough sawn or resawn wood									
Solid body or semi-transparent stain, water base (material #18 or #20)									
Roll & brush 1st coat									
Slow	60	150	49.40	38.33	9.21	32.93	15.29	15.32	111.08
Medium	75	115	43.25	39.33	11.36	37.61	22.08	13.25	123.63
Fast	90	80	37.05	40.00	14.12	46.31	31.13	9.21	140.77
Roll & brush 2nd coat									
Slow	70	190	49.40	32.86	7.90	26.00	12.68	12.71	92.15
Medium	110	165	43.25	26.82	7.75	26.21	15.20	9.12	85.10
Fast	130	140	37.05	27.69	9.77	26.46	19.82	5.86	89.60
Roll & brush 3rd or additional coats									
Slow	80	215	49.40	28.75	6.90	22.98	11.14	11.16	80.93
Medium	120	190	43.25	24.58	7.10	22.76	13.61	8.17	76.22
Fast	160	165	37.05	22.50	7.94	22.45	16.40	4.85	74.14
Solid body or semi-transparent stain, oil base (material #19 or #21)									
Roll & brush 1st coat									
Slow	70	290	59.60	32.86	7.90	20.55	11.65	11.67	84.63
Medium	85	255	52.15	34.71	10.01	20.45	16.30	9.78	91.25
Fast	100	220	44.70	36.00	12.71	20.32	21.40	6.33	96.76
Roll & brush 2nd coat									
Slow	80	315	59.60	28.75	6.90	18.92	10.37	10.39	75.33
Medium	110	278	52.15	26.82	7.75	18.76	13.33	8.00	74.66
Fast	140	240	44.70	25.71	9.07	18.63	16.56	4.90	74.87
Roll & brush 3rd or additional coats									
Slow	90	340	59.60	25.56	6.13	17.53	9.35	9.37	67.94
Medium	130	303	52.15	22.69	6.56	17.21	11.62	6.97	65.05
Fast	170	265	44.70	21.18	7.46	16.87	14.11	4.17	63.79
Varnish, flat or gloss (material #30c)									
Roll & brush 1st coat									
Slow	50	120	81.00	46.00	11.04	67.50	23.66	23.71	171.91
Medium	65	100	70.80	45.38	13.11	70.80	32.32	19.39	181.00
Fast	80	80	60.70	45.00	15.89	75.88	42.40	12.54	191.71

	Labor LF per manhour	Material coverage LF/gallon	Material cost per gallon	Labor cost per 100 LF	Labor burden 100 LF	Material cost per 100 LF	Overhead per 100 LF	Profit per 100 LF	Total price per 100 LF
Roll & brush 2nd coat									
Slow	60	190	81.00	38.33	9.21	42.63	17.13	17.17	124.47
Medium	90	165	70.80	32.78	9.47	42.91	21.29	12.77	119.22
Fast	120	140	60.70	30.00	10.58	43.36	26.02	7.70	117.66
Roll & brush 3rd or additional coats									
Slow	70	200	81.00	32.86	7.90	40.50	15.44	15.47	112.17
Medium	110	175	70.80	26.82	7.75	40.46	18.76	11.25	105.04
Fast	150	150	60.70	24.00	8.49	40.47	22.61	6.69	102.26

Don't add additional time for plant-on trim if it's painted with the same coating as the adjacent siding. For heights above 8 feet, use the High Time Difficulty Factors on page 139. Use slow rates when cutting-in or masking adjacent surfaces. Add preparation time for masking adjacent surfaces or protecting windows. Use fast rates when plant-on trim is finished before it's installed or on new construction projects where a prime coat can be sprayed prior to stucco color coat application. "Slow" work is based on an hourly wage of $23.00, "Medium" work on an hourly wage of $29.50, and "Fast" work on an hourly wage of $36.00. Other qualifications that apply to this table are on page 9.

	Labor SF per manhour	Material coverage SF/gallon	Material cost per gallon	Labor cost per 100 SF	Labor burden 100 SF	Material cost per 100 SF	Overhead per 100 SF	Profit per 100 SF	Total price per 100 SF
Plaster or stucco, exterior, medium texture, brush application									
Masonry paint, water base, flat or gloss (material #31)									
Brush 1st coat									
Slow	100	225	44.80	23.00	5.52	19.91	9.20	9.22	66.85
Medium	120	213	39.20	24.58	7.10	18.40	12.52	7.51	70.11
Fast	140	200	33.60	25.71	9.07	16.80	15.99	4.73	72.30
Brush 2nd coat									
Slow	150	250	44.80	15.33	3.69	17.92	7.02	7.03	50.99
Medium	163	230	39.20	18.10	5.21	17.04	10.09	6.06	56.50
Fast	175	210	33.60	20.57	7.24	16.00	13.59	4.02	61.42
Brush 3rd or additional coats									
Slow	160	270	44.80	14.38	3.45	16.59	6.54	6.55	47.51
Medium	173	245	39.20	17.05	4.93	16.00	9.50	5.70	53.18
Fast	185	220	33.60	19.46	6.89	15.27	12.90	3.82	58.34
Masonry paint, oil base (material #32)									
Brush 1st coat									
Slow	80	265	65.40	28.75	6.90	24.68	11.46	11.49	83.28
Medium	100	250	57.20	29.50	8.53	22.88	15.23	9.14	85.28
Fast	120	235	49.00	30.00	10.58	20.85	19.05	5.63	86.11
Brush 2nd coat									
Slow	145	300	65.40	15.86	3.82	21.80	7.88	7.90	57.26
Medium	165	275	57.20	17.88	5.17	20.80	10.96	6.58	61.39
Fast	185	250	49.00	19.46	6.89	19.60	14.24	4.21	64.40
Brush 3rd or additional coats									
Slow	155	350	65.40	14.84	3.56	18.69	7.05	7.06	51.20
Medium	175	325	57.20	16.86	4.86	17.60	9.83	5.90	55.05
Fast	195	300	49.00	18.46	6.53	16.33	12.81	3.79	57.92

For heights above 8 feet, use the High Time Difficulty Factors on page 139. For oil base paint and clear hydro sealer, I recommend spraying. For painting interior plaster, see Walls, plaster. "Slow" work is based on an hourly wage of $23.00, "Medium" work on an hourly wage of $29.50, and "Fast" work on an hourly wage of $36.00. Other qualifications that apply to this table are on page 9.

	Labor SF per manhour	Material coverage SF/gallon	Material cost per gallon	Labor cost per 100 SF	Labor burden 100 SF	Material cost per 100 SF	Overhead per 100 SF	Profit per 100 SF	Total price per 100 SF
Plaster or stucco, exterior, medium texture, roll application									
Masonry paint, water base, flat or gloss (material #31)									
Roll 1st coat									
Slow	245	200	44.80	9.39	2.25	22.40	6.47	6.48	46.99
Medium	273	175	39.20	10.81	3.11	22.40	9.08	5.45	50.85
Fast	300	150	33.60	12.00	4.22	22.40	11.98	3.54	54.14
Roll 2nd coat									
Slow	300	225	44.80	7.67	1.83	19.91	5.59	5.60	40.60
Medium	320	200	39.20	9.22	2.68	19.60	7.87	4.72	44.09
Fast	340	175	33.60	10.59	3.73	19.20	10.39	3.07	46.98
Roll 3rd or additional coats									
Slow	320	250	44.80	7.19	1.74	17.92	5.10	5.11	37.06
Medium	340	225	39.20	8.68	2.50	17.42	7.15	4.29	40.04
Fast	360	200	33.60	10.00	3.54	16.80	9.40	2.78	42.52
Masonry paint, oil base (material #32)									
Roll 1st coat									
Slow	200	250	65.40	11.50	2.76	26.16	7.68	7.70	55.80
Medium	240	225	57.20	12.29	3.57	25.42	10.32	6.19	57.79
Fast	280	200	49.00	12.86	4.53	24.50	12.99	3.84	58.72
Roll 2nd coat									
Slow	220	275	65.40	10.45	2.53	23.78	6.98	7.00	50.74
Medium	265	250	57.20	11.13	3.21	22.88	9.31	5.58	52.11
Fast	305	225	49.00	11.80	4.18	21.78	11.70	3.46	52.92
Roll 3rd or additional coats									
Slow	235	300	65.40	9.79	2.36	21.80	6.45	6.46	46.86
Medium	285	275	57.20	10.35	3.00	20.80	8.54	5.12	47.81
Fast	335	250	49.00	10.75	3.81	19.60	10.58	3.13	47.87

For heights above 8 feet, use the High Time Difficulty Factors on page 139. For oil base paint and clear hydro sealer, I recommend spraying. For painting interior plaster, see Walls, plaster. "Slow" work is based on an hourly wage of $23.00, "Medium" work on an hourly wage of $29.50, and "Fast" work on an hourly wage of $36.00. Other qualifications that apply to this table are on page 9.

	Labor SF per manhour	Material coverage SF/gallon	Material cost per gallon	Labor cost per 100 SF	Labor burden 100 SF	Material cost per 100 SF	Overhead per 100 SF	Profit per 100 SF	Total price per 100 SF
Plaster or stucco, exterior, medium texture, spray application									
Masonry paint, water base, flat or gloss (material #31)									
Spray prime coat									
Slow	600	150	44.80	3.83	.93	29.87	6.58	6.59	47.80
Medium	675	120	39.20	4.37	1.26	32.67	9.58	5.75	53.63
Fast	750	90	33.60	4.80	1.68	37.33	13.58	4.02	61.41
Spray 2nd coat									
Slow	700	175	44.80	3.29	.79	25.60	5.64	5.65	40.97
Medium	800	150	39.20	3.69	1.06	26.13	7.72	4.63	43.23
Fast	900	125	33.60	4.00	1.41	26.88	10.01	2.96	45.26
Spray 3rd or additional coats									
Slow	750	200	44.80	3.07	.72	22.40	4.98	4.99	36.16
Medium	850	168	39.20	3.47	1.02	23.33	6.95	4.17	38.94
Fast	950	135	33.60	3.79	1.32	24.89	9.31	2.75	42.06
Masonry paint, oil base (material #32)									
Spray prime coat									
Slow	550	200	65.40	4.18	1.01	32.70	7.20	7.21	52.30
Medium	600	145	57.20	4.92	1.43	39.45	11.45	6.87	64.12
Fast	650	90	49.00	5.54	1.96	54.44	19.20	5.68	86.82
Spray 2nd coat									
Slow	650	225	65.40	3.54	.85	29.07	6.36	6.37	46.19
Medium	700	175	57.20	4.21	1.23	32.69	9.53	5.72	53.38
Fast	750	125	49.00	4.80	1.68	39.20	14.16	4.19	64.03
Spray 3rd or additional coats									
Slow	700	250	65.40	3.29	.79	26.16	5.75	5.76	41.75
Medium	750	193	57.20	3.93	1.13	29.64	8.68	5.21	48.59
Fast	800	135	49.00	4.50	1.59	36.30	13.14	3.89	59.42

For heights above 8 feet, use the High Time Difficulty Factors on page 139. For oil base paint and clear hydro sealer, I recommend spraying. For painting interior plaster, see Walls, plaster. "Slow" work is based on an hourly wage of $23.00, "Medium" work on an hourly wage of $29.50, and "Fast" work on an hourly wage of $36.00. Other qualifications that apply to this table are on page 9.

	Labor SF per manhour	Material coverage SF/gallon	Material cost per gallon	Labor cost per 100 SF	Labor burden 100 SF	Material cost per 100 SF	Overhead per 100 SF	Profit per 100 SF	Total price per 100 SF
Plaster or stucco, exterior, medium texture, waterproofing									
Waterproofing, clear hydro sealer (material #34)									
Brush 1st coat									
Slow	125	175	40.60	18.40	4.42	23.20	8.74	8.76	63.52
Medium	150	163	35.50	19.67	5.70	21.78	11.78	7.07	66.00
Fast	175	150	30.40	20.57	7.24	20.27	14.91	4.41	67.40
Brush 2nd or additional coats									
Slow	175	200	40.60	13.14	3.14	20.30	6.95	6.97	50.50
Medium	200	188	35.50	14.75	4.27	18.88	9.47	5.68	53.05
Fast	225	175	30.40	16.00	5.63	17.37	12.10	3.58	54.68
Roll 1st coat									
Slow	325	150	40.60	7.08	1.70	27.07	6.81	6.83	49.49
Medium	363	138	35.50	8.13	2.33	25.72	9.05	5.43	50.66
Fast	400	125	30.40	9.00	3.18	24.32	11.32	3.35	51.17
Roll 2nd or additional coats									
Slow	400	175	40.60	5.75	1.38	23.20	5.76	5.77	41.86
Medium	425	163	35.50	6.94	2.00	21.78	7.68	4.61	43.01
Fast	450	150	30.40	8.00	2.81	20.27	9.64	2.85	43.57
Spray 1st coat									
Slow	650	125	40.60	3.54	.85	32.48	7.01	7.02	50.90
Medium	700	113	35.50	4.21	1.23	31.42	9.21	5.53	51.60
Fast	750	100	30.40	4.80	1.68	30.40	11.44	3.38	51.70
Spray 2nd or additional coats									
Slow	750	150	40.60	3.07	.72	27.07	5.87	5.88	42.61
Medium	825	138	35.50	3.58	1.02	25.72	7.58	4.55	42.45
Fast	900	125	30.40	4.00	1.41	24.32	9.22	2.73	41.68

For heights above 8 feet, use the High Time Difficulty Factors on page 139. For oil base paint and clear hydro sealer, I recommend spraying. For painting interior plaster, see Walls, plaster. "Slow" work is based on an hourly wage of $23.00, "Medium" work on an hourly wage of $29.50, and "Fast" work on an hourly wage of $36.00. Other qualifications that apply to this table are on page 9.

Plaster, interior: see Walls

	Labor SF per manhour	Material coverage SF/gallon	Material cost per gallon	Labor cost per 100 SF	Labor burden 100 SF	Material cost per 100 SF	Overhead per 100 SF	Profit per 100 SF	Total price per 100 SF
Plaster or stucco, interior/exterior, medium texture, anti-graffiti stain eliminator									
Water base primer and sealer (material #39)									
Roll & brush each coat									
Slow	350	400	49.40	6.57	1.59	12.35	3.90	3.90	28.31
Medium	375	375	43.20	7.87	2.28	11.52	5.42	3.25	30.34
Fast	400	350	37.00	9.00	3.18	10.57	7.05	2.09	31.89
Oil base primer and sealer (material #40)									
Roll & brush each coat									
Slow	350	375	58.00	6.57	1.59	15.47	4.49	4.50	32.62
Medium	375	350	50.80	7.87	2.28	14.51	6.16	3.70	34.52
Fast	400	325	43.50	9.00	3.18	13.38	7.92	2.34	35.82
Polyurethane 2 part system (material #41)									
Roll & brush each coat									
Slow	300	375	187.30	7.67	1.83	49.95	11.30	11.32	82.07
Medium	325	350	163.90	9.08	2.63	46.83	14.63	8.78	81.95
Fast	350	325	140.50	10.29	3.64	43.23	17.72	5.24	80.12

For heights above 8 feet, use the High Time Difficulty Factors on page 139. For oil base paint and clear hydro sealer, I recommend spraying. For painting interior plaster, see Walls, plaster. "Slow" work is based on an hourly wage of $23.00, "Medium" work on an hourly wage of $29.50, and "Fast" work on an hourly wage of $36.00. Other qualifications that apply to this table are on page 9.

	Labor LF per manhour	Material coverage LF/gallon	Material cost per gallon	Labor cost per 100 LF	Labor burden 100 LF	Material cost per 100 LF	Overhead per 100 LF	Profit per 100 LF	Total price per 100 LF
Pot shelves, 12" to 18" wide									
Solid body or semi-transparent stain, water or oil base (material #18 or #19 or #20 or #21)									
Roll & brush each coat									
Slow	50	50	54.50	46.00	11.04	109.00	31.55	31.61	229.20
Medium	68	45	47.70	43.38	12.56	106.00	40.48	24.29	226.71
Fast	75	40	40.88	48.00	16.93	102.20	51.81	15.33	234.27

These figures are based on painting all sides of exterior or interior pot shelves. "Slow" work is based on an hourly wage of $23.00, "Medium" work on an hourly wage of $29.50, and "Fast" work on an hourly wage of $36.00. Other qualifications that apply to this table are on page 9.

	Labor LF per manhour	Material coverage LF/gallon	Material cost per gallon	Labor cost per 100 LF	Labor burden 100 LF	Material cost per 100 LF	Overhead per 100 LF	Profit per 100 LF	Total price per 100 LF
Railing, exterior, rough sawn or resawn wood									
Solid body or semi-transparent stain, water or oil base (material #18 or #19 or #20 or #21)									
Roll & brush each coat									
Slow	16	20	54.50	143.75	34.50	272.50	85.64	85.82	622.21
Medium	18	18	47.70	163.89	47.40	265.00	119.06	71.44	666.79
Fast	20	15	40.88	180.00	63.55	272.53	159.98	47.32	723.38

Use these costs for finishing railing that's 36" to 42" high and with 2" x 2" verticals spaced 4" to 6" on center. These figures include painting the rail cap, baluster, newels and spindles. "Slow" work is based on an hourly wage of $23.00, "Medium" work on an hourly wage of $29.50, and "Fast" work on an hourly wage of $36.00. Other qualifications that apply to this table are on page 9.

	Labor LF per manhour	Material coverage LF/gallon	Material cost per gallon	Labor cost per 100 LF	Labor burden 100 LF	Material cost per 100 LF	Overhead per 100 LF	Profit per 100 LF	Total price per 100 LF
Railing, exterior, stain grade, decorative wood									
STEP 1: Sand & putty									
Slow	50	--	--	46.00	11.04	--	10.84	10.86	78.74
Medium	60	--	--	49.17	14.23	--	15.85	9.51	88.76
Fast	70	--	--	51.43	18.18	--	21.57	6.38	97.56
STEP 2 & 3: Stain (material #30a) & wipe									
Brush & wipe, 1 coat									
Slow	25	60	57.10	92.00	22.08	95.17	39.76	39.84	288.85
Medium	30	55	50.00	98.33	28.42	90.91	54.42	32.65	304.73
Fast	35	50	42.90	102.86	36.30	85.80	69.74	20.63	315.33
Spray & wipe, 1 coat									
Slow	75	35	57.10	30.67	7.35	163.14	38.22	38.30	277.68
Medium	85	30	50.00	34.71	10.01	166.67	52.85	31.71	295.95
Fast	95	25	42.90	37.89	13.40	171.60	69.09	20.44	312.42
STEP 4 & 5: Sanding sealer (material #30b) & light sand									
Brush 1 coat									
Slow	45	65	63.80	51.11	12.26	98.15	30.69	30.76	222.97
Medium	50	60	55.80	59.00	17.06	93.00	42.26	25.36	236.68
Fast	55	55	47.80	65.45	23.10	86.91	54.39	16.09	245.94
Spray 1 coat									
Slow	125	35	63.80	18.40	4.42	182.29	38.97	39.05	283.13
Medium	138	30	55.80	21.38	6.19	186.00	53.39	32.03	298.99
Fast	150	25	47.80	24.00	8.49	191.20	69.34	20.51	313.54

	Labor LF per manhour	Material coverage LF/gallon	Material cost per gallon	Labor cost per 100 LF	Labor burden 100 LF	Material cost per 100 LF	Overhead per 100 LF	Profit per 100 LF	Total price per 100 LF
STEP 6 & 7: Varnish 2 coats (material #30c)									
Varnish, flat or gloss (material #30c)									
Brush 1st coat									
Slow	22	100	81.00	104.55	25.07	81.00	40.02	40.11	290.75
Medium	24	90	70.80	122.92	35.55	78.67	59.28	35.57	331.99
Fast	26	80	60.70	138.46	48.88	75.88	81.60	24.14	368.96
Brush 2nd or additional coats									
Slow	30	120	81.00	76.67	18.39	67.50	30.89	30.95	224.40
Medium	32	110	70.80	92.19	26.65	64.36	45.80	27.48	256.48
Fast	34	100	60.70	105.88	37.38	60.70	63.23	18.70	285.89
Spray 1st coat									
Slow	65	65	81.00	35.38	8.48	124.62	32.01	32.08	232.57
Medium	70	60	70.80	42.14	12.20	118.00	43.08	25.85	241.27
Fast	75	55	60.70	48.00	16.93	110.36	54.34	16.07	245.70
Spray 2nd or additional coats									
Slow	65	65	81.00	35.38	8.48	124.62	32.01	32.08	232.57
Medium	70	60	70.80	42.14	12.20	118.00	43.08	25.85	241.27
Fast	75	55	60.70	48.00	16.93	110.36	54.34	16.07	245.70
STEP 8: Steel wool, hand buff									
Slow	100	--	--	23.00	5.52	--	5.42	5.43	39.37
Medium	113	--	--	26.11	7.55	--	8.42	5.05	47.13
Fast	125	--	--	28.80	10.17	--	12.08	3.57	54.62
STEP 9: Wax & polish by hand									
Slow	100	--	--	23.00	5.52	--	5.42	5.43	39.37
Medium	113	--	--	26.11	7.55	--	8.42	5.05	47.13
Fast	125	--	--	28.80	10.17	--	12.08	3.57	54.62
Complete 9 step process (material #30)									
Brush stain, brush sanding sealer, brush varnish									
Slow	6	40	67.30	383.33	92.01	168.25	122.28	122.54	888.41
Medium	7	35	58.90	421.43	121.87	168.29	177.88	106.73	996.20
Fast	8	30	50.50	450.00	158.88	168.33	240.93	71.27	1089.41
Spray stain, spray sanding sealer, brush varnish									
Slow	11	24	67.30	209.09	50.19	280.42	102.54	102.76	745.00
Medium	14	27	58.90	210.71	60.94	218.15	122.44	73.46	685.70
Fast	17	30	50.50	211.76	74.75	168.33	141.00	41.71	637.55

Use these figures to estimate the cost of applying a natural finish on stain grade railing that's from 36" to 42" high and with spindles spaced at 4" to 6" on center. These figures include painting the rail cap, baluster, newels and spindles. "Slow" work is based on an hourly wage of $23.00, "Medium" work on an hourly wage of $29.50, and "Fast" work on an hourly wage of $36.00. Other qualifications that apply to this table are on page 9.

	Labor LF per manhour	Material coverage LF/gallon	Material cost per gallon	Labor cost per 100 LF	Labor burden 100 LF	Material cost per 100 LF	Overhead per 100 LF	Profit per 100 LF	Total price per 100 LF
Railing, interior, handrail, decorative wood									
Paint grade									
Undercoat or enamel, water or oil base (material #3 or #4 or #9 or #10)									
Brush each coat									
Slow	30	120	50.00	76.67	18.39	41.67	25.98	26.04	188.75
Medium	35	110	43.78	84.29	24.36	39.80	37.11	22.27	207.83
Fast	40	100	37.50	90.00	31.78	37.50	49.37	14.60	223.25
Stain grade, 7 step process									
Stain, seal and 2 coat lacquer system (material #11)									
Spray all coats									
Slow	55	100	54.90	41.82	10.03	54.90	20.28	20.33	147.36
Medium	60	88	48.00	49.17	14.23	54.55	29.48	17.69	165.12
Fast	65	75	41.20	55.38	19.54	54.93	40.26	11.91	182.02

Use these costs for finishing decorative wood, wall mounted handrail. "Slow" work is based on an hourly wage of $23.00, "Medium" work on an hourly wage of $29.50, and "Fast" work on an hourly wage of $36.00. Other qualifications that apply to this table are on page 9.

	Labor LF per manhour	Material coverage LF/gallon	Material cost per gallon	Labor cost per 100 LF	Labor burden 100 LF	Material cost per 100 LF	Overhead per 100 LF	Profit per 100 LF	Total price per 100 LF
Railing, interior, stain grade, decorative wood									
Stain, seal and 2 coat lacquer system (7 step process)									
STEP 1: Sand & putty									
Slow	50	--	--	46.00	11.04	--	10.84	10.86	78.74
Medium	60	--	--	49.17	14.23	--	15.85	9.51	88.76
Fast	70	--	--	51.43	18.18	--	21.57	6.38	97.56
STEP 2 & 3: Stain (material #11a) & wipe									
Brush & wipe, 1 coat									
Slow	25	60	61.60	92.00	22.08	102.67	41.18	41.27	299.20
Medium	30	55	53.90	98.33	28.42	98.00	56.19	33.71	314.65
Fast	35	50	46.20	102.86	36.30	92.40	71.79	21.24	324.59
Spray & wipe, 1 coat									
Slow	75	35	61.60	30.67	7.35	176.00	40.67	40.75	295.44
Medium	85	30	53.90	34.71	10.01	179.67	56.10	33.66	314.15
Fast	95	25	46.20	37.89	13.40	184.80	73.18	21.65	330.92

	Labor LF per manhour	Material coverage LF/gallon	Material cost per gallon	Labor cost per 100 LF	Labor burden 100 LF	Material cost per 100 LF	Overhead per 100 LF	Profit per 100 LF	Total price per 100 LF
STEP 4 & 5: Sanding sealer (material #11b) & light sand									
Brush 1 coat									
Slow	45	65	50.90	51.11	12.26	78.31	26.92	26.98	195.58
Medium	50	60	44.60	59.00	17.06	74.33	37.60	22.56	210.55
Fast	55	55	38.20	65.45	23.10	69.45	48.98	14.49	221.47
Spray 1 coat									
Slow	125	35	50.90	18.40	4.42	145.43	31.97	32.04	232.26
Medium	138	30	44.60	21.38	6.19	148.67	44.06	26.43	246.73
Fast	150	25	38.20	24.00	8.49	152.80	57.43	16.99	259.71
STEP 6 & 7: Lacquer (material #11c), 2 coats									
Brush 1st coat									
Slow	40	65	53.50	57.50	13.80	82.31	29.19	29.25	212.05
Medium	50	60	46.80	59.00	17.06	78.00	38.51	23.11	215.68
Fast	60	55	40.10	60.00	21.20	72.91	47.77	14.13	216.01
Brush 2nd coat									
Slow	45	70	53.50	51.11	12.26	76.43	26.56	26.62	192.98
Medium	55	65	46.80	53.64	15.50	72.00	35.29	21.17	197.60
Fast	65	60	40.10	55.38	19.54	66.83	43.95	13.00	198.70
Spray 1st coat									
Slow	75	55	53.50	30.67	7.35	97.27	25.71	25.76	186.76
Medium	85	50	46.80	34.71	10.01	93.60	34.59	20.75	193.66
Fast	95	45	40.10	37.89	13.40	89.11	43.52	12.87	196.79
Spray 2nd coat									
Slow	85	60	53.50	27.06	6.48	89.17	23.32	23.37	169.40
Medium	95	55	46.80	31.05	9.00	85.09	31.28	18.77	175.19
Fast	105	50	40.10	34.29	12.08	80.20	39.24	11.61	177.42
Complete stain, seal & 2 coat lacquer system (material #11)									
Brush all coats									
Slow	8	30	54.90	287.50	69.00	183.00	102.51	102.72	744.73
Medium	10	25	48.00	295.00	85.30	192.00	143.07	85.84	801.21
Fast	12	20	41.20	300.00	105.90	206.00	189.69	56.11	857.70
Spray all coats									
Slow	16	20	54.90	143.75	34.50	274.50	86.02	86.20	624.97
Medium	20	15	48.00	147.50	42.65	320.00	127.53	76.52	714.20
Fast	24	10	41.20	150.00	52.97	412.00	190.63	56.39	861.99

Use these costs for applying stain, sanding sealer and lacquer to interior railings. Typical railing is 36" to 42" height with spindles spaced at 4" to 6" on center. These costs include finishing the rail cap, baluster, newels and spindles. The typical application is one coat of stain, one coat of sanding sealer, sand, putty, and two coats of lacquer. For rough sawn wood railing with 2" x 2" spindles spaced at 4" to 6" on center, see the tables for Exterior wood railing. "Slow" work is based on an hourly wage of $23.00, "Medium" work on an hourly wage of $29.50, and "Fast" work on an hourly wage of $36.00. Other qualifications that apply to this table are on page 9.

	Labor LF per manhour	Material coverage LF/gallon	Material cost per gallon	Labor cost per 100 LF	Labor burden 100 LF	Material cost per 100 LF	Overhead per 100 LF	Profit per 100 LF	Total price per 100 LF
Railing, interior, wood, paint grade, brush application									
Undercoat, water or oil base (material #3 or #4)									
Brush 1 coat									
Slow	11	50	42.60	209.09	50.19	85.20	65.45	65.59	475.52
Medium	14	45	37.30	210.71	60.94	82.89	88.63	53.18	496.35
Fast	17	40	31.95	211.76	74.75	79.88	113.58	33.60	513.57
Split coat (1/2 undercoat + 1/2 enamel), water or oil base (material #3 and #9, or material #4 and #10)									
Brush 1 coat									
Slow	15	70	50.00	153.33	36.81	71.43	49.70	49.80	361.07
Medium	18	65	43.78	163.89	47.40	67.35	69.65	41.79	390.08
Fast	21	60	37.50	171.43	60.53	62.50	91.28	27.00	412.74
Enamel, water or oil base (material #9 or #10)									
Brush 1st finish coat									
Slow	13	60	57.40	176.92	42.46	95.67	59.86	59.99	434.90
Medium	16	55	50.25	184.38	53.31	91.36	82.26	49.35	460.66
Fast	18	50	43.05	200.00	70.63	86.10	110.58	32.71	500.02
Brush 2nd or additional finish coats									
Slow	15	70	57.40	153.33	36.81	82.00	51.70	51.81	375.65
Medium	18	65	50.25	163.89	47.40	77.31	72.14	43.28	404.02
Fast	21	60	43.05	171.43	60.53	71.75	94.14	27.85	425.70

Use these costs for applying undercoat or enamel to interior railings. Railing is based on a 36" to 42" height and with spindles spaced at 4" to 6" on center. These costs include painting the rail cap, baluster, newels and spindles. For rough sawn wood railing with 2" x 2" spindles spaced at 6" on center, see the tables for Exterior wood railing. "Slow" work is based on an hourly wage of $23.00, "Medium" work on an hourly wage of $29.50, and "Fast" work on an hourly wage of $36.00. Other qualifications that apply to this table are on page 9.

	Labor LF per manhour	Material coverage LF/gallon	Material cost per gallon	Labor cost per 100 LF	Labor burden 100 LF	Material cost per 100 LF	Overhead per 100 LF	Profit per 100 LF	Total price per 100 LF
Railing, interior, wood, paint grade, spray application									
Undercoat, water or oil base (material #3 or #4)									
Spray 1 coat									
Slow	50	40	42.60	46.00	11.04	106.50	31.07	31.14	225.75
Medium	60	35	37.30	49.17	14.23	106.57	42.49	25.49	237.95
Fast	70	30	31.95	51.43	18.18	106.50	54.58	16.15	246.84
Split coat (1/2 undercoat + 1/2 enamel), water or oil base (material #3 and #9, or material #4 and #10)									
Spray 1 coat									
Slow	65	50	50.00	35.38	8.48	100.00	27.34	27.39	198.59
Medium	75	45	43.78	39.33	11.36	97.29	37.00	22.20	207.18
Fast	85	40	37.50	42.35	14.93	93.75	46.83	13.85	211.71
Enamel, water or oil base (material #9 or #10)									
Spray 1st finish coat									
Slow	55	45	57.40	41.82	10.03	127.56	34.09	34.16	247.66
Medium	65	40	50.25	45.38	13.11	125.63	46.03	27.62	257.77
Fast	75	35	43.05	48.00	16.93	123.00	58.26	17.23	263.42
Spray 2nd or additional finish coats									
Slow	65	50	57.40	35.38	8.48	114.80	30.15	30.21	219.02
Medium	75	45	50.25	39.33	11.36	111.67	40.59	24.36	227.31
Fast	85	40	43.05	42.35	14.93	107.63	51.13	15.12	231.16

Use these costs for applying undercoat or enamel to interior railings. Railing is based on a 36" to 42" height and with spindles spaced at 4" to 6" on center. These costs include painting the rail cap, baluster, newels and spindles. For rough sawn wood railing with 2" x 2" spindles spaced at 6" on center, see the tables for Exterior wood railing. "Slow" work is based on an hourly wage of $23.00, "Medium" work on an hourly wage of $29.50, and "Fast" work on an hourly wage of $36.00. Other qualifications that apply to this table are on page 9.

	Labor LF per manhour	Material coverage LF/gallon	Material cost per gallon	Labor cost per 100 LF	Labor burden 100 LF	Material cost per 100 LF	Overhead per 100 LF	Profit per 100 LF	Total price per 100 LF
Railing, wrought iron, 36" to 42" high bars with wood cap									
Metal primer, rust inhibitor, clean metal (material #35)									
Brush 1 coat									
Slow	15	90	59.70	153.33	36.81	66.33	48.73	48.83	354.03
Medium	18	85	52.30	163.89	47.40	61.53	68.20	40.92	381.94
Fast	20	80	44.80	180.00	63.55	56.00	92.86	27.47	419.88
Metal primer, rust inhibitor, rusty metal (material #36)									
Brush 1 coat									
Slow	15	90	80.10	153.33	36.81	89.00	53.03	53.15	385.32
Medium	18	85	70.10	163.89	47.40	82.47	73.43	44.06	411.25
Fast	20	80	60.10	180.00	63.55	75.13	98.79	29.22	446.69
Metal finish, synthetic enamel, off white, gloss, interior or exterior - (material #37)									
Brush 1st finish coat									
Slow	20	110	63.60	115.00	27.60	57.82	38.08	38.16	276.66
Medium	23	100	55.70	128.26	37.09	55.70	55.26	33.15	309.46
Fast	26	90	47.70	138.46	48.88	53.00	74.51	22.04	336.89
Brush 2nd or additional finish coats									
Slow	30	125	63.60	76.67	18.39	50.88	27.73	27.79	201.46
Medium	35	120	55.70	84.29	24.36	46.42	38.77	23.26	217.10
Fast	40	115	47.70	90.00	31.78	41.48	50.61	14.97	228.84
Metal finish, synthetic enamel, colors (except orange/red), gloss, interior or exterior - (material #38)									
Brush 1st finish coat									
Slow	20	110	66.30	115.00	27.60	60.27	38.55	38.63	280.05
Medium	23	100	58.00	128.26	37.09	58.00	55.83	33.50	312.68
Fast	26	90	49.70	138.46	48.88	55.22	75.19	22.24	339.99
Brush 2nd or additional finish coats									
Slow	30	125	66.30	76.67	18.39	53.04	28.14	28.20	204.44
Medium	35	120	58.00	84.29	24.36	48.33	39.25	23.55	219.78
Fast	40	115	49.70	90.00	31.78	43.22	51.15	15.13	231.28

Use these figures for painting prefabricated preprimed wrought iron railing which is 36" to 42" high with 1/2" square vertical bars at 4" to 6" on center with a stain grade wood cap supported by a 1/2" by 1-1/2" top rail, and 1" square support posts at 6' to 10' on center and with a 1/2" by 1-1/2" bottom rail, unless otherwise noted. The metal finish figures include only minor touchup of pre-primed steel or wrought iron prefabricated railings. Note: A two coat system, prime and finish, using oil base material is recommended for any metal surface. Although water base material is often used, it may cause oxidation, corrosion and rust. Using one coat of oil base paint on exterior metal may result in cracking, peeling, or chipping without the proper prime coat application. If off white or another light colored finish paint is specified, make sure the prime coat is a light color also, or more than one finish coat will be necessary. "Slow" work is based on an hourly wage of $23.00, "Medium" work on an hourly wage of $29.50, and "Fast" work on an hourly wage of $36.00. Other qualifications that apply to this table are on page 9.

	Labor LF per manhour	Material coverage LF/gallon	Material cost per gallon	Labor cost per 100 LF	Labor burden 100 LF	Material cost per 100 LF	Overhead per 100 LF	Profit per 100 LF	Total price per 100 LF
Railing, wrought iron, 36" to 42" high bars with wrought iron cap									
Metal primer, rust inhibitor, clean metal (material #35)									
Brush 1 coat									
Slow	20	110	59.70	115.00	27.60	54.27	37.41	37.48	271.76
Medium	25	105	52.30	118.00	34.12	49.81	50.48	30.29	282.70
Fast	30	100	44.80	120.00	42.35	44.80	64.22	19.00	290.37
Metal primer, rust inhibitor, rusty metal (material #36)									
Brush 1 coat									
Slow	20	110	80.10	115.00	27.60	72.82	40.93	41.02	297.37
Medium	25	105	70.10	118.00	34.12	66.76	54.72	32.83	306.43
Fast	30	100	60.10	120.00	42.35	60.10	68.96	20.40	311.81
Metal finish, synthetic enamel, off white, gloss, interior or exterior - (material #37)									
Brush 1st finish coat									
Slow	25	120	63.60	92.00	22.08	53.00	31.75	31.81	230.64
Medium	30	115	55.70	98.33	28.42	48.43	43.80	26.28	245.26
Fast	35	110	47.70	102.86	36.30	43.36	56.58	16.74	255.84
Brush 2nd or additional finish coats									
Slow	35	135	63.60	65.71	15.77	47.11	24.43	24.48	177.50
Medium	40	130	55.70	73.75	21.33	42.85	34.48	20.69	193.10
Fast	45	125	47.70	80.00	28.23	38.16	45.38	13.42	205.19
Metal finish, synthetic enamel, colors (except orange/red), gloss, interior or exterior - (material #38)									
Brush 1st finish coat									
Slow	25	120	66.30	92.00	22.08	55.25	32.17	32.24	233.74
Medium	30	115	58.00	98.33	28.42	50.43	44.30	26.58	248.06
Fast	35	110	49.70	102.86	36.30	45.18	57.15	16.91	258.40
Brush 2nd or additional finish coats									
Slow	35	135	66.30	65.71	15.77	49.11	24.81	24.86	180.26
Medium	40	130	58.00	73.75	21.33	44.62	34.92	20.95	195.57
Fast	45	125	49.70	80.00	28.23	39.76	45.88	13.57	207.44

Use these figures for painting prefabricated preprimed wrought iron railing which is 36" to 42" high with 1/2" square vertical bars at 4" to 6" on center and a 1/2" by 1-1/2" wrought iron cap with 1" square support posts at 6' to 10' on center, and a 1/2" by 1-1/2" bottom rail, unless otherwise noted. The metal finish figures include only minor touchup of pre-primed steel or wrought iron prefabricated railings. Note: A two coat system, prime and finish, using oil base material is recommended for any metal surface. Although water base material is often used, it may cause oxidation, corrosion and rust. Using one coat of oil base paint on exterior metal may result in cracking, peeling, or chipping without the proper prime coat application. If off white or another light colored finish paint is specified, make sure the prime coat is also a light color, or more than one finish coat will be necessary. "Slow" work is based on an hourly wage of $23.00, "Medium" work on an hourly wage of $29.50, and "Fast" work on an hourly wage of $36.00. Other qualifications that apply to this table are on page 9.

	Labor LF per manhour	Material coverage LF/gallon	Material cost per gallon	Labor cost per 100 LF	Labor burden 100 LF	Material cost per 100 LF	Overhead per 100 LF	Profit per 100 LF	Total price per 100 LF
Railing, wrought iron, 60" to 72" high bars with wrought iron cap									
Metal primer, rust inhibitor, clean metal (material #35)									
Brush 1 coat									
Slow	10	90	59.70	230.00	55.20	66.33	66.79	66.93	485.25
Medium	15	85	52.30	196.67	56.88	61.53	78.76	47.26	441.10
Fast	20	80	44.80	180.00	63.55	56.00	92.86	27.47	419.88
Metal primer, rust inhibitor, rusty metal (material #36)									
Brush 1 coat									
Slow	10	90	80.10	230.00	55.20	89.00	71.10	71.25	516.55
Medium	15	85	70.10	196.67	56.88	82.47	84.00	50.40	470.42
Fast	20	80	60.10	180.00	63.55	75.13	98.79	29.22	446.69
Metal finish, synthetic enamel, off white, gloss, interior or exterior - (material #37)									
Brush 1st finish coat									
Slow	15	120	63.60	153.33	36.81	53.00	46.19	46.29	335.62
Medium	20	115	55.70	147.50	42.65	48.43	59.64	35.78	334.00
Fast	25	110	47.70	144.00	50.84	43.36	73.84	21.84	333.88
Brush 2nd or additional finish coats									
Slow	25	135	63.60	92.00	22.08	47.11	30.63	30.69	222.51
Medium	30	130	55.70	98.33	28.42	42.85	42.40	25.44	237.44
Fast	35	125	47.70	102.86	36.30	38.16	54.97	16.26	248.55
Metal finish, synthetic enamel, colors (except orange/red), gloss, interior or exterior - (material #38)									
Brush 1st finish coat									
Slow	15	120	66.30	153.33	36.81	55.25	46.62	46.72	338.73
Medium	20	115	58.00	147.50	42.65	50.43	60.14	36.08	336.80
Fast	25	110	49.70	144.00	50.84	45.18	74.40	22.01	336.43
Brush 2nd or additional finish coats									
Slow	25	135	66.30	92.00	22.08	49.11	31.01	31.07	225.27
Medium	30	130	58.00	98.33	28.42	44.62	42.84	25.71	239.92
Fast	35	125	49.70	102.86	36.30	39.76	55.47	16.41	250.80

Use these figures for painting prefabricated preprimed wrought iron railing which is 60" to 72" high with 1/2" square vertical bars at 4" to 6" on center with 1" square support posts at 6' to 10' on center and with a 1/2" by 1-1/2" bottom rail, unless otherwise noted. The metal finish figures include only minor touchup of pre-primed steel or wrought iron prefabricated railings. Note: A two coat system, prime and finish, using oil base material is recommended for any metal surface. Although water base material is often used, it may cause oxidation, corrosion and rust. Using one coat of oil base paint on exterior metal may result in cracking, peeling, or chipping without the proper prime coat application. If off white or another light colored finish paint is specified, make sure the prime coat is also light color, or more than one finish coat will be necessary. "Slow" work is based on an hourly wage of $23.00, "Medium" work on an hourly wage of $29.50, and "Fast" work on an hourly wage of $36.00. Other qualifications that apply to this table are on page 9.

	SF floor per manhour	Material SF floor per can	Material cost per can	Labor per 100 SF floor	Labor burden 100 SF	Material per 100 SF floor	Overhead per 100 SF floor	Profit per 100 SF floor	Total per 100 SF floor
Registers, HVAC, per 100 square feet of floor area									
Repaint jobs, spray cans (material #17)									
Spray 1 coat									
Slow	900	700	12.00	2.56	.61	1.71	.93	.93	6.74
Medium	950	650	10.50	3.11	.88	1.62	1.41	.84	7.86
Fast	1000	600	9.00	3.60	1.27	1.50	1.97	.58	8.92
New construction projects, spray cans (material #17)									
Spray 1 coat									
Slow	2500	800	12.00	.92	.22	1.50	.50	.50	3.64
Medium	2750	750	10.50	1.07	.30	1.40	.70	.42	3.89
Fast	3000	700	9.00	1.20	.41	1.29	.90	.27	4.07

These costs assume HVAC registers are painted with spray cans (bombs) to match the adjacent walls. Costs are based on square footage of the floor area of the building. These rates include time to remove, paint and replace the HVAC registers. Use the square feet of floor area divided by these rates to find manhours and the number of spray bombs needed to paint all the heat registers in a building. Rule of Thumb: 2 minutes per 100 square feet of floor is for new construction projects. "Slow" work is based on an hourly wage of $23.00, "Medium" work on an hourly wage of $29.50, and "Fast" work on an hourly wage of $36.00. Other qualifications that apply to this table are on page 9.

	Manhours per 1500 SF roof	Material gallons 1500 SF	Material cost per gallon	Labor per 1500 SF roof	Labor burden 1500 SF	Material per 1500 SF roof	Overhead per 1500 SF roof	Profit per 1500 SF roof	Total per 1500 SF roof
Roof jacks, per 1500 square feet of roof area									
Metal primer, clean metal (material #35)									
1 story building, brush prime coat									
Slow	0.40	0.30	59.70	9.20	2.21	17.91	5.57	5.58	40.47
Medium	0.35	0.33	52.30	10.33	2.98	17.26	7.65	4.59	42.81
Fast	0.30	0.35	44.80	10.80	3.81	15.68	9.39	2.78	42.46
2 story building, brush prime coat									
Slow	0.50	0.40	59.70	11.50	2.76	23.88	7.25	7.26	52.65
Medium	0.45	0.43	52.30	13.28	3.83	22.49	9.90	5.94	55.44
Fast	0.40	0.45	44.80	14.40	5.08	20.16	12.29	3.64	55.57
Metal primer, rusty metal (material #36)									
1 story building, brush prime coat									
Slow	0.40	0.30	80.10	9.20	2.21	24.03	6.73	6.75	48.92
Medium	0.35	0.33	70.10	10.33	2.98	23.13	9.11	5.47	51.02
Fast	0.30	0.35	60.10	10.80	3.81	21.04	11.05	3.27	49.97
2 story building, brush prime coat									
Slow	0.50	0.40	80.10	11.50	2.76	32.04	8.80	8.82	63.92
Medium	0.45	0.43	70.10	13.28	3.83	30.14	11.82	7.09	66.16
Fast	0.40	0.45	60.10	14.40	5.08	27.05	14.42	4.27	65.22
Metal finish, synthetic enamel - off white, gloss, interior or exterior (material #37)									
1 story building, brush each coat									
Slow	0.30	0.20	63.60	6.90	1.66	12.72	4.04	4.05	29.37
Medium	0.25	0.25	55.70	7.38	2.12	13.93	5.86	3.52	32.81
Fast	0.20	0.30	47.70	7.20	2.54	14.31	7.46	2.21	33.72
2 story building, brush each coat									
Slow	0.40	0.30	63.60	9.20	2.21	19.08	5.79	5.80	42.08
Medium	0.35	0.33	55.70	10.33	2.98	18.38	7.93	4.76	44.38
Fast	0.30	0.35	47.70	10.80	3.81	16.70	9.71	2.87	43.89

	Manhours per 1500 SF roof	Material gallons 1500 SF	Material cost per gallon	Labor per 1500 SF roof	Labor burden 1500 SF	Material per 1500 SF roof	Overhead per 1500 SF roof	Profit per 1500 SF roof	Total per 1500 SF roof
Metal finish, synthetic enamel, colors (except orange/red), gloss, interior or exterior - (material #38)									
1 story building, brush each coat									
Slow	0.30	0.20	66.30	6.90	1.66	13.26	4.15	4.16	30.13
Medium	0.25	0.25	58.00	7.38	2.12	14.50	6.00	3.60	33.60
Fast	0.20	0.30	49.70	7.20	2.54	14.91	7.64	2.26	34.55
2 story building, brush each coat									
Slow	0.40	0.30	66.30	9.20	2.21	19.89	5.95	5.96	43.21
Medium	0.35	0.33	58.00	10.33	2.98	19.14	8.12	4.87	45.44
Fast	0.30	0.35	49.70	10.80	3.81	17.40	9.92	2.94	44.87

Production rates and coverage figures are minimum values based on 1 or 2 story roof areas of up to 1500 square feet. For example, to apply metal primer on clean metal roof jacks on a 3000 SF one-story building at a medium rate, use two times the cost of $42.99 or $85.98. This figure includes ladder time. See the paragraphs below on Roof pitch difficulty factors and Roof area conversion factors to adjust for roof slope and type. Note: A two coat system, prime and finish, using oil base material is recommended for any metal surface. Although water base material is often used, it may cause oxidation, corrosion and rust. One coat of oil base solid body stain is often used on exterior metal but it may crack, peel or chip without the proper prime coat application. "Slow" work is based on an hourly wage of $23.00, "Medium" work on an hourly wage of $29.50, and "Fast" work on an hourly wage of $36.00. Other qualifications that apply to this table are on page 9.

Roof area conversion factors

For an arched roof, multiply the building length by the building width, then multiply by 1.5.
For a gambrel roof, multiply the building length by the building width, then multiply by 1.33.

Roof pitch difficulty factors

It's harder to paint on a sloped surface than on a flat surface. The steeper the slope, the more difficult the work. Roof slope is usually measured in inches of rise per inch of horizontal run. For example, a 3 in 12 pitch means the roof rises 3 inches for each 12 inches of run, measuring horizontally. Use the difficulty factors that follow when estimating the time needed to paint on a sloping roof.

On a flat roof, or roof with a pitch of less than 3 in 12, calculate the roof area without modification.
If the pitch is 3 in 12, multiply the surface area by 1.1.
If the pitch is 4 in 12, multiply the surface area by 1.2.
If the pitch is 6 in 12, multiply the surface area by 1.3.

	Labor SF per manhour	Material coverage SF/gallon	Material cost per gallon	Labor cost per 100 SF	Labor burden 100 SF	Material cost per 100 SF	Overhead per 100 SF	Profit per 100 SF	Total price per 100 SF
Roofing, composition shingles, brush application									
Solid body stain, water base (material #18)									
Brush 1st coat									
Slow	45	220	48.90	51.11	12.26	22.23	16.27	16.30	118.17
Medium	60	200	42.80	49.17	14.23	21.40	21.20	12.72	118.72
Fast	80	180	36.70	45.00	15.89	20.39	25.20	7.45	113.93
Brush 2nd coat									
Slow	65	330	48.90	35.38	8.48	14.82	11.15	11.17	81.00
Medium	80	210	42.80	36.88	10.66	20.38	16.98	10.19	95.09
Fast	100	290	36.70	36.00	12.71	12.66	19.02	5.63	86.02
Brush 3rd or additional coats									
Slow	85	405	48.90	27.06	6.48	12.07	8.67	8.69	62.97
Medium	100	385	42.80	29.50	8.53	11.12	12.29	7.37	68.81
Fast	120	365	36.70	30.00	10.58	10.05	15.70	4.64	70.97
Solid body stain, oil base (material #19)									
Brush 1st coat									
Slow	45	270	60.90	51.11	12.26	22.56	16.33	16.36	118.62
Medium	60	250	53.30	49.17	14.23	21.32	21.18	12.71	118.61
Fast	80	230	45.70	45.00	15.89	19.87	25.04	7.41	113.21
Brush 2nd coat									
Slow	65	360	60.90	35.38	8.48	16.92	11.55	11.57	83.90
Medium	80	345	53.30	36.88	10.66	15.45	15.75	9.45	88.19
Fast	100	330	45.70	36.00	12.71	13.85	19.39	5.74	87.69
Brush 3rd or additional coats									
Slow	85	425	60.90	27.06	6.48	14.33	9.10	9.12	66.09
Medium	100	405	53.30	29.50	8.53	13.16	12.80	7.68	71.67
Fast	120	385	45.70	30.00	10.58	11.87	16.26	4.81	73.52

Use these figures for repaint jobs only. Some older composition shingles may contain asbestos. It has been established that asbestos fibers are a known carcinogen (cancer causing) and it is likely that no new construction projects will specify materials or products which contain asbestos. Furthermore, roofing materials for a new construction project would not need painting. Roofing and siding products usually contain very little asbestos and are typically non-friable (hand pressure can not crumble, pulverize or reduce to a powder when dry). There is danger when asbestos is being removed because of exposure to airborne particulate matter. Apparently, there is little danger when painting asbestos roofing or siding, but it is a good idea to have your painters wear respirators or particle masks for their safety. Coverage figures are based on shingles or shakes with average moisture content. See the paragraphs on Roof pitch difficulty factors and Roof area conversion factors to adjust for roof slope and type. "Slow" work is based on an hourly wage of $23.00, "Medium" work on an hourly wage of $29.50, and "Fast" work on an hourly wage of $36.00. Other qualifications that apply to this table are on page 9.

	Labor SF per manhour	Material coverage SF/gallon	Material cost per gallon	Labor cost per 100 SF	Labor burden 100 SF	Material cost per 100 SF	Overhead per 100 SF	Profit per 100 SF	Total price per 100 SF
Roofing, composition shingles, roll application									
Solid body stain, water base (material #18)									
Roll 1st coat									
Slow	150	190	48.90	15.33	3.69	25.74	8.50	8.52	61.78
Medium	170	180	42.80	17.35	5.01	23.78	11.54	6.92	64.60
Fast	200	170	36.70	18.00	6.36	21.59	14.24	4.21	64.40
Roll 2nd coat									
Slow	250	300	48.90	9.20	2.21	16.30	5.26	5.28	38.25
Medium	305	290	42.80	9.67	2.80	14.76	6.81	4.08	38.12
Fast	360	280	36.70	10.00	3.54	13.11	8.26	2.44	37.35
Roll 3rd or additional coats									
Slow	360	385	48.90	6.39	1.54	12.70	3.92	3.93	28.48
Medium	385	375	42.80	7.66	2.23	11.41	5.32	3.19	29.81
Fast	420	365	36.70	8.57	3.02	10.05	6.71	1.99	30.34
Solid body stain, oil base (material #19)									
Roll 1st coat									
Slow	150	220	60.90	15.33	3.69	27.68	8.87	8.89	64.46
Medium	170	205	53.30	17.35	5.01	26.00	12.09	7.25	67.70
Fast	200	190	45.70	18.00	6.36	24.05	15.00	4.44	67.85
Roll 2nd coat									
Slow	250	330	60.90	9.20	2.21	18.45	5.67	5.68	41.21
Medium	305	320	53.30	9.67	2.80	16.66	7.28	4.37	40.78
Fast	360	310	45.70	10.00	3.54	14.74	8.76	2.59	39.63
Roll 3rd or additional coats									
Slow	360	405	60.90	6.39	1.54	15.04	4.36	4.37	31.70
Medium	385	395	53.30	7.66	2.23	13.49	5.84	3.50	32.72
Fast	420	385	45.70	8.57	3.02	11.87	7.28	2.15	32.89

	Labor SF per manhour	Material coverage SF/gallon	Material cost per gallon	Labor cost per 100 SF	Labor burden 100 SF	Material cost per 100 SF	Overhead per 100 SF	Profit per 100 SF	Total price per 100 SF
Waterproofing, clear hydro sealer (material #34)									
Roll 1st coat									
Slow	100	300	40.60	23.00	5.52	13.53	7.99	8.01	58.05
Medium	200	275	35.50	14.75	4.27	12.91	7.98	4.79	44.70
Fast	300	250	30.40	12.00	4.22	12.16	8.80	2.60	39.78
Roll 2nd or additional coats									
Slow	150	350	40.60	15.33	3.69	11.60	5.82	5.83	42.27
Medium	250	325	35.50	11.80	3.41	10.92	6.53	3.92	36.58
Fast	350	300	30.40	10.29	3.64	10.13	7.46	2.21	33.73

Use these figures for repaint jobs only. Some older composition shingles may contain asbestos. It has been established that asbestos fibers are a known carcinogen (cancer causing) and it is likely that no new construction projects will specify materials or products which contain asbestos. Furthermore, roofing materials for a new construction project would not need painting. Roofing and siding products usually contain very little asbestos and are typically non-friable (hand pressure can not crumble, pulverize or reduce to a powder when dry). There is danger when asbestos is being removed because of exposure to airborne particulate matter. Apparently, there is little danger when painting asbestos roofing or siding, but it is a good idea to have your painters wear respirators or particle masks for their safety. Coverage figures are based on shingles or shakes with average moisture content. See the paragraphs on Roof pitch difficulty factors and Roof area conversion factors to adjust for roof slope and type. "Slow" work is based on an hourly wage of $23.00, "Medium" work on an hourly wage of $29.50, and "Fast" work on an hourly wage of $36.00. Other qualifications that apply to this table are on page 9.

	Labor SF per manhour	Material coverage SF/gallon	Material cost per gallon	Labor cost per 100 SF	Labor burden 100 SF	Material cost per 100 SF	Overhead per 100 SF	Profit per 100 SF	Total price per 100 SF
Roofing, composition shingles, spray application									
Solid body stain, water base (material #18)									
Spray 1st coat									
Slow	325	200	48.90	7.08	1.70	24.45	6.31	6.33	45.87
Medium	350	180	42.80	8.43	2.45	23.78	8.66	5.20	48.52
Fast	375	160	36.70	9.60	3.41	22.94	11.14	3.29	50.38
Spray 2nd coat									
Slow	425	290	48.90	5.41	1.29	16.86	4.48	4.49	32.53
Medium	450	280	42.80	6.56	1.88	15.29	5.94	3.56	33.23
Fast	475	270	36.70	7.58	2.70	13.59	7.39	2.19	33.45
Spray 3rd or additional coats									
Slow	500	365	48.90	4.60	1.10	13.40	3.63	3.64	26.37
Medium	538	355	42.80	5.48	1.59	12.06	4.78	2.87	26.78
Fast	575	345	36.70	6.26	2.22	10.64	5.92	1.75	26.79
Solid body stain, oil base (material #19)									
Spray 1st coat									
Slow	325	230	60.90	7.08	1.70	26.48	6.70	6.71	48.67
Medium	350	205	53.30	8.43	2.45	26.00	9.22	5.53	51.63
Fast	375	180	45.70	9.60	3.41	25.39	11.90	3.52	53.82
Spray 2nd coat									
Slow	425	310	60.90	5.41	1.29	19.65	5.01	5.02	36.38
Medium	450	300	53.30	6.56	1.88	17.77	6.56	3.93	36.70
Fast	475	290	45.70	7.58	2.70	15.76	8.07	2.39	36.50
Spray 3rd or additional coats									
Slow	500	380	60.90	4.60	1.10	16.03	4.13	4.14	30.00
Medium	538	370	53.30	5.48	1.59	14.41	5.37	3.22	30.07
Fast	575	360	45.70	6.26	2.22	12.69	6.56	1.94	29.67

	Labor SF per manhour	Material coverage SF/gallon	Material cost per gallon	Labor cost per 100 SF	Labor burden 100 SF	Material cost per 100 SF	Overhead per 100 SF	Profit per 100 SF	Total price per 100 SF
Waterproofing, clear hydro sealer (material #34)									
Spray 1st coat									
Slow	550	100	40.60	4.18	1.01	40.60	8.70	8.72	63.21
Medium	600	88	35.50	4.92	1.43	40.34	11.67	7.00	65.36
Fast	650	75	30.40	5.54	1.96	40.53	14.89	4.40	67.32
Spray 2nd or additional coats									
Slow	600	150	40.60	3.83	.93	27.07	6.05	6.06	43.94
Medium	650	138	35.50	4.54	1.32	25.72	7.89	4.74	44.21
Fast	700	125	30.40	5.14	1.83	24.32	9.69	2.87	43.85

Use these figures for repaint jobs only. Some older composition shingles may contain asbestos. It has been established that asbestos fibers are a known carcinogen (cancer causing) and it is likely that no new construction projects will specify materials or products which contain asbestos. Furthermore, roofing materials for a new construction project would not need painting. Roofing and siding products usually contain very little asbestos and are typically non-friable (hand pressure can not crumble, pulverize or reduce to a powder when dry). There is danger when asbestos is being removed because of exposure to airborne particulate matter. Apparently, there is little danger when painting asbestos roofing or siding, but it is a good idea to have your painters wear respirators or particle masks for their safety. Coverage figures are based on shingles or shakes with average moisture content. See the paragraphs on Roof pitch difficulty factors and Roof area conversion factors to adjust for roof slope and type. "Slow" work is based on an hourly wage of $23.00, "Medium" work on an hourly wage of $29.50, and "Fast" work on an hourly wage of $36.00. Other qualifications that apply to this table are on page 9.

	Labor SF per manhour	Material coverage SF/gallon	Material cost per gallon	Labor cost per 100 SF	Labor burden 100 SF	Material cost per 100 SF	Overhead per 100 SF	Profit per 100 SF	Total price per 100 SF
Roofing, wood shingles or shakes, brush application									
Solid body stain, water base (material #18)									
Brush 1st coat									
Slow	100	240	48.90	23.00	5.52	20.38	9.29	9.31	67.50
Medium	155	228	42.80	19.03	5.50	18.77	10.83	6.50	60.63
Fast	210	215	36.70	17.14	6.05	17.07	12.48	3.69	56.43
Brush 2nd or additional coats									
Slow	150	290	48.90	15.33	3.69	16.86	6.82	6.83	49.53
Medium	195	278	42.80	15.13	4.38	15.40	8.73	5.24	48.88
Fast	240	265	36.70	15.00	5.31	13.85	10.59	3.13	47.88
Solid body stain, oil base (material #19)									
Brush 1st coat									
Slow	100	160	60.90	23.00	5.52	38.06	12.65	12.68	91.91
Medium	155	150	53.30	19.03	5.50	35.53	15.02	9.01	84.09
Fast	210	140	45.70	17.14	6.05	32.64	17.31	5.12	78.26
Brush 2nd or additional coats									
Slow	150	260	60.90	15.33	3.69	23.42	8.06	8.08	58.58
Medium	195	250	53.30	15.13	4.38	21.32	10.21	6.12	57.16
Fast	240	240	45.70	15.00	5.31	19.04	12.20	3.61	55.16
Semi-transparent stain, water base (material #20)									
Brush 1st coat									
Slow	120	260	49.90	19.17	4.59	19.19	8.16	8.18	59.29
Medium	175	248	43.70	16.86	4.86	17.62	9.84	5.90	55.08
Fast	230	235	37.40	15.65	5.54	15.91	11.49	3.40	51.99
Brush 2nd or additional coats									
Slow	160	300	49.90	14.38	3.45	16.63	6.55	6.56	47.57
Medium	205	288	43.70	14.39	4.17	15.17	8.43	5.06	47.22
Fast	250	275	37.40	14.40	5.08	13.60	10.25	3.03	46.36
Semi-transparent stain, oil base (material #21)									
Brush 1st coat									
Slow	120	180	58.30	19.17	4.59	32.39	10.67	10.69	77.51
Medium	175	170	51.00	16.86	4.86	30.00	12.93	7.76	72.41
Fast	230	160	43.70	15.65	5.54	27.31	15.03	4.45	67.98
Brush 2nd or additional coats									
Slow	160	280	58.30	14.38	3.45	20.82	7.34	7.36	53.35
Medium	205	270	51.00	14.39	4.17	18.89	9.36	5.62	52.43
Fast	250	260	43.70	14.40	5.08	16.81	11.25	3.33	50.87

	Labor SF per manhour	Material coverage SF/gallon	Material cost per gallon	Labor cost per 100 SF	Labor burden 100 SF	Material cost per 100 SF	Overhead per 100 SF	Profit per 100 SF	Total price per 100 SF
Penetrating oil stain (material #13)									
Brush 1st coat									
Slow	100	160	57.90	23.00	5.52	36.19	12.29	12.32	89.32
Medium	155	150	50.70	19.03	5.50	33.80	14.58	8.75	81.66
Fast	210	140	43.40	17.14	6.05	31.00	16.80	4.97	75.96
Brush 2nd or additional coats									
Slow	150	205	57.90	15.33	3.69	28.24	8.98	9.00	65.24
Medium	195	195	50.70	15.13	4.38	26.00	11.38	6.83	63.72
Fast	240	185	43.40	15.00	5.31	23.46	13.57	4.01	61.35

Coverage figures are based on shingles or shakes with average moisture content. See the paragraphs on Roof pitch difficulty factors and Roof area conversion factors to adjust for roof slope and type. "Slow" work is based on an hourly wage of $23.00, "Medium" work on an hourly wage of $29.50, and "Fast" work on an hourly wage of $36.00. Other qualifications that apply to this table are on page 9.

	Labor SF per manhour	Material coverage SF/gallon	Material cost per gallon	Labor cost per 100 SF	Labor burden 100 SF	Material cost per 100 SF	Overhead per 100 SF	Profit per 100 SF	Total price per 100 SF
Roofing, wood shingles or shakes, roll application									
Solid body stain, water base (material #18)									
Roll 1st coat									
Slow	210	225	48.90	10.95	2.63	21.73	6.71	6.72	48.74
Medium	258	213	42.80	11.43	3.33	20.09	8.71	5.22	48.78
Fast	305	200	36.70	11.80	4.18	18.35	10.64	3.15	48.12
Roll 2nd or additional coats									
Slow	250	275	48.90	9.20	2.21	17.78	5.55	5.56	40.30
Medium	300	263	42.80	9.83	2.83	16.27	7.24	4.34	40.51
Fast	350	250	36.70	10.29	3.64	14.68	8.87	2.62	40.10
Solid body stain, oil base (material #19)									
Roll 1st coat									
Slow	210	150	60.90	10.95	2.63	40.60	10.29	10.32	74.79
Medium	255	140	53.30	11.57	3.34	38.07	13.25	7.95	74.18
Fast	305	130	45.70	11.80	4.18	35.15	15.85	4.69	71.67
Roll 2nd or additional coats									
Slow	250	245	60.90	9.20	2.21	24.86	6.89	6.91	50.07
Medium	300	235	53.30	9.83	2.83	22.68	8.84	5.30	49.48
Fast	350	225	45.70	10.29	3.64	20.31	10.61	3.14	47.99
Semi-transparent stain, water base (material #20)									
Roll 1st coat									
Slow	240	250	49.90	9.58	2.31	19.96	6.05	6.06	43.96
Medium	288	238	43.70	10.24	2.96	18.36	7.89	4.73	44.18
Fast	335	225	37.40	10.75	3.81	16.62	9.66	2.86	43.70
Roll 2nd or additional coats									
Slow	270	290	49.90	8.52	2.03	17.21	5.28	5.29	38.33
Medium	320	278	43.70	9.22	2.68	15.72	6.90	4.14	38.66
Fast	370	255	37.40	9.73	3.42	14.67	8.63	2.55	39.00
Semi-transparent stain, oil base (material #21)									
Roll 1st coat									
Slow	240	175	58.30	9.58	2.31	33.31	8.59	8.60	62.39
Medium	288	165	51.00	10.24	2.96	30.91	11.03	6.62	61.76
Fast	335	155	43.70	10.75	3.81	28.19	13.25	3.92	59.92
Roll 2nd or additional coats									
Slow	270	260	58.30	8.52	2.03	22.42	6.27	6.28	45.52
Medium	320	250	51.00	9.22	2.68	20.40	8.07	4.84	45.21
Fast	370	240	43.70	9.73	3.42	18.21	9.72	2.88	43.96

	Labor SF per manhour	Material coverage SF/gallon	Material cost per gallon	Labor cost per 100 SF	Labor burden 100 SF	Material cost per 100 SF	Overhead per 100 SF	Profit per 100 SF	Total price per 100 SF
Penetrating oil stain (material #13)									
Roll 1st coat									
Slow	210	200	57.90	10.95	2.63	28.95	8.08	8.10	58.71
Medium	258	190	50.70	11.43	3.33	26.68	10.35	6.21	58.00
Fast	305	180	43.40	11.80	4.18	24.11	12.42	3.68	56.19
Roll 2nd or additional coats									
Slow	250	295	57.90	9.20	2.21	19.63	5.90	5.91	42.85
Medium	300	285	50.70	9.83	2.83	17.79	7.62	4.57	42.64
Fast	350	275	43.40	10.29	3.64	15.78	9.21	2.72	41.64
Waterproofing, clear hydro sealer (material #34)									
Roll 1st coat									
Slow	80	225	40.60	28.75	6.90	18.04	10.20	10.22	74.11
Medium	160	195	35.50	18.44	5.33	18.21	10.50	6.30	58.78
Fast	230	165	30.40	15.65	5.54	18.42	12.27	3.63	55.51
Roll 2nd or additional coats									
Slow	125	150	40.60	18.40	4.42	27.07	9.48	9.50	68.87
Medium	190	138	35.50	15.53	4.47	25.72	11.44	6.86	64.02
Fast	255	125	30.40	14.12	4.97	24.32	13.46	3.98	60.85

Coverage figures are based on shingles or shakes with average moisture content. See the paragraphs on Roof pitch difficulty factors and Roof area conversion factors to adjust for roof slope and type. "Slow" work is based on an hourly wage of $23.00, "Medium" work on an hourly wage of $29.50, and "Fast" work on an hourly wage of $36.00. Other qualifications that apply to this table are on page 9.

	Labor SF per manhour	Material coverage SF/gallon	Material cost per gallon	Labor cost per 100 SF	Labor burden 100 SF	Material cost per 100 SF	Overhead per 100 SF	Profit per 100 SF	Total price per 100 SF
Roofing, wood shingles or shakes, spray application									
Solid body stain, water base (material #18)									
Spray 1st coat									
Slow	600	230	48.90	3.83	.93	21.26	4.94	4.95	35.91
Medium	700	220	42.80	4.21	1.23	19.45	6.22	3.73	34.84
Fast	800	200	36.70	4.50	1.59	18.35	7.58	2.24	34.26
Spray 2nd or additional coats									
Slow	700	250	48.90	3.29	.79	19.56	4.49	4.50	32.63
Medium	800	235	42.80	3.69	1.06	18.21	5.74	3.45	32.15
Fast	900	220	36.70	4.00	1.41	16.68	6.85	2.03	30.97
Solid body stain, oil base (material #19)									
Spray 1st coat									
Slow	600	170	60.90	3.83	.93	35.82	7.71	7.72	56.01
Medium	700	150	53.30	4.21	1.23	35.53	10.24	6.14	57.35
Fast	800	130	45.70	4.50	1.59	35.15	12.78	3.78	57.80
Spray 2nd or additional coats									
Slow	700	230	60.90	3.29	.79	26.48	5.81	5.82	42.19
Medium	800	215	53.30	3.69	1.06	24.79	7.39	4.43	41.36
Fast	900	200	45.70	4.00	1.41	22.85	8.76	2.59	39.61
Semi-transparent stain, water base (material #20)									
Spray 1st coat									
Slow	650	265	49.90	3.54	.85	18.83	4.41	4.42	32.05
Medium	750	253	43.70	3.93	1.13	17.27	5.59	3.35	31.27
Fast	850	240	37.40	4.24	1.51	15.58	6.61	1.96	29.90
Spray 2nd or additional coats									
Slow	750	305	49.90	3.07	.72	16.36	3.83	3.84	27.82
Medium	850	293	43.70	3.47	1.02	14.91	4.85	2.91	27.16
Fast	950	270	37.40	3.79	1.32	13.85	5.88	1.74	26.58
Semi-transparent stain, oil base (material #21)									
Spray 1st coat									
Slow	650	190	58.30	3.54	.85	30.68	6.66	6.68	48.41
Medium	750	180	51.00	3.93	1.13	28.33	8.35	5.01	46.75
Fast	850	170	43.70	4.24	1.51	25.71	9.75	2.88	44.09
Spray 2nd or additional coats									
Slow	750	275	58.30	3.07	.72	21.20	4.75	4.76	34.50
Medium	850	265	51.00	3.47	1.02	19.25	5.93	3.56	33.23
Fast	950	255	43.70	3.79	1.32	17.14	6.90	2.04	31.19

	Labor SF per manhour	Material coverage SF/gallon	Material cost per gallon	Labor cost per 100 SF	Labor burden 100 SF	Material cost per 100 SF	Overhead per 100 SF	Profit per 100 SF	Total price per 100 SF
Penetrating oil stain (material #13)									
Spray 1st coat									
Slow	600	200	57.90	3.83	.93	28.95	6.40	6.42	46.53
Medium	700	170	50.70	4.21	1.23	29.82	8.81	5.29	49.36
Fast	800	140	43.40	4.50	1.59	31.00	11.50	3.40	51.99
Spray 2nd or additional coats									
Slow	700	300	57.90	3.29	.79	19.30	4.44	4.45	32.27
Medium	800	260	50.70	3.69	1.06	19.50	6.07	3.64	33.96
Fast	900	240	43.40	4.00	1.41	18.08	7.28	2.15	32.92
Waterproofing, clear hydro sealer (material #34)									
Spray 1st coat									
Slow	450	75	40.60	5.11	1.22	54.13	11.49	11.51	83.46
Medium	475	63	35.50	6.21	1.81	56.35	16.09	9.65	90.11
Fast	500	50	30.40	7.20	2.54	60.80	21.87	6.47	98.88
Spray 2nd or additional coats									
Slow	500	150	40.60	4.60	1.10	27.07	6.23	6.24	45.24
Medium	525	138	35.50	5.62	1.61	25.72	8.24	4.94	46.13
Fast	550	125	30.40	6.55	2.32	24.32	10.29	3.04	46.52

Coverage figures are based on shingles or shakes with average moisture content. See the paragraphs on Roof pitch difficulty factors and Roof area conversion factors to adjust for roof slope and type. "Slow" work is based on an hourly wage of $23.00, "Medium" work on an hourly wage of $29.50, and "Fast" work on an hourly wage of $36.00. Other qualifications that apply to this table are on page 9.

	Labor LF per manhour	Material coverage LF/gallon	Material cost per gallon	Labor cost per 100 LF	Labor burden 100 LF	Material cost per 100 LF	Overhead per 100 LF	Profit per 100 LF	Total price per 100 LF
Sheet metal cap or flashing									
3" to 8" wide:									
Metal primer, clean metal (material #35)									
Brush prime coat									
Slow	215	525	59.70	10.70	2.56	11.37	4.68	4.69	34.00
Medium	228	500	52.30	12.94	3.76	10.46	6.79	4.07	38.02
Fast	240	475	44.80	15.00	5.31	9.43	9.22	2.73	41.69
Metal primer, rusty metal (material #36)									
Brush prime coat									
Slow	215	525	80.10	10.70	2.56	15.26	5.42	5.43	39.37
Medium	228	500	70.10	12.94	3.76	14.02	7.68	4.61	43.01
Fast	240	475	60.10	15.00	5.31	12.65	10.21	3.02	46.19
Metal finish, synthetic enamel - off white, gloss, interior or exterior (material #37)									
Brush 1st or additional finish coats									
Slow	275	550	63.60	8.36	2.02	11.56	4.17	4.18	30.29
Medium	288	525	55.70	10.24	2.96	10.61	5.95	3.57	33.33
Fast	300	500	47.70	12.00	4.22	9.54	7.99	2.36	36.11
Metal finish, synthetic enamel, colors (except orange/red), gloss, interior or exterior (material #38)									
Brush 1st or additional finish coats									
Slow	275	550	66.30	8.36	2.02	12.05	4.26	4.27	30.96
Medium	288	525	58.00	10.24	2.96	11.05	6.06	3.64	33.95
Fast	300	500	49.70	12.00	4.22	9.94	8.12	2.40	36.68

This table is based on a two coat system, prime and finish, using oil base material which is recommended for any metal surface. Although water base material is often used, it may cause oxidation, corrosion and rust. One coat of oil base solid body stain is often used on exterior metal but it may crack, peel or chip without the proper prime coat application. If off white or another light colored finish paint is specified, make sure the prime coat is also a light color, or more than one finish coat will be necessary. "Slow" work is based on an hourly wage of $23.00, "Medium" work on an hourly wage of $29.50, and "Fast" work on an hourly wage of $36.00. Other qualifications that apply to this table are on page 9.

	Labor LF per manhour	Material coverage LF/gallon	Material cost per gallon	Labor cost per 100 LF	Labor burden 100 LF	Material cost per 100 LF	Overhead per 100 LF	Profit per 100 LF	Total price per 100 LF
Sheet metal cap or flashing									
8" to 12" wide:									
Metal primer, clean metal (material #35)									
Brush prime coat									
Slow	200	500	59.70	11.50	2.76	11.94	4.98	4.99	36.17
Medium	215	475	52.30	13.72	3.96	11.01	7.18	4.31	40.18
Fast	230	450	44.80	15.65	5.54	9.96	9.65	2.85	43.65
Metal primer, rusty metal (material #36)									
Brush prime coat									
Slow	200	500	80.10	11.50	2.76	16.02	5.75	5.76	41.79
Medium	215	475	70.10	13.72	3.96	14.76	8.11	4.87	45.42
Fast	230	450	60.10	15.65	5.54	13.36	10.70	3.17	48.42
Metal finish, synthetic enamel - off white, gloss, interior or exterior (material #37)									
Brush 1st or additional finish coats									
Slow	250	525	63.60	9.20	2.21	12.11	4.47	4.48	32.47
Medium	265	500	55.70	11.13	3.21	11.14	6.37	3.82	35.67
Fast	280	475	47.70	12.86	4.53	10.04	8.51	2.52	38.46
Metal finish, synthetic enamel, colors (except orange/red), gloss, interior or exterior (material #38)									
Brush 1st or additional finish coats									
Slow	250	525	66.30	9.20	2.21	12.63	4.57	4.58	33.19
Medium	265	500	58.00	11.13	3.21	11.60	6.49	3.89	36.32
Fast	280	475	49.70	12.86	4.53	10.46	8.64	2.56	39.05

This table is based on a two coat system, prime and finish, using oil base material which is recommended for any metal surface. Although water base material is often used, it may cause oxidation, corrosion and rust. One coat of oil base solid body stain is often used on exterior metal but it may crack, peel or chip without the proper prime coat application. If off white or another light colored finish paint is specified, make sure the prime coat is also a light color, or more than one finish coat will be necessary. "Slow" work is based on an hourly wage of $23.00, "Medium" work on an hourly wage of $29.50, and "Fast" work on an hourly wage of $36.00. Other qualifications that apply to this table are on page 9.

	Labor LF per manhour	Material coverage LF/gallon	Material cost per gallon	Labor cost per 100 LF	Labor burden 100 LF	Material cost per 100 LF	Overhead per 100 LF	Profit per 100 LF	Total price per 100 LF
Sheet metal, diverters or gravel stop									
Up to 3" wide:									
Metal primer, clean metal (material #35)									
Brush prime coat									
Slow	230	600	59.70	10.00	2.41	9.95	4.25	4.26	30.87
Medium	240	550	52.30	12.29	3.57	9.51	6.34	3.80	35.51
Fast	250	500	44.80	14.40	5.08	8.96	8.82	2.61	39.87
Metal primer, rusty metal (material #36)									
Brush prime coat									
Slow	230	600	80.10	10.00	2.41	13.35	4.89	4.90	35.55
Medium	240	550	70.10	12.29	3.57	12.75	7.15	4.29	40.05
Fast	250	500	60.10	14.40	5.08	12.02	9.77	2.89	44.16
Metal finish, synthetic enamel - off white, gloss, interior or exterior (material #37)									
Brush 1st or additional finish coats									
Slow	300	650	63.60	7.67	1.83	9.78	3.67	3.67	26.62
Medium	313	600	55.70	9.42	2.71	9.28	5.36	3.21	29.98
Fast	325	550	47.70	11.08	3.92	8.67	7.33	2.17	33.17
Metal finish, synthetic enamel, colors (except orange/red), gloss, interior or exterior (material #38)									
Brush 1st or additional finish coats									
Slow	300	650	66.30	7.67	1.83	10.20	3.74	3.75	27.19
Medium	313	600	58.00	9.42	2.71	9.67	5.45	3.27	30.52
Fast	325	550	49.70	11.08	3.92	9.04	7.45	2.20	33.69

This table is based on a two coat system, prime and finish, using oil base material which is recommended for any metal surface. Although water base material is often used, it may cause oxidation, corrosion and rust. One coat of oil base solid body stain is often used on exterior metal but it may crack, peel or chip without the proper prime coat application. If off white or another light colored finish paint is specified, make sure the prime coat is also a light color, or more than one finish coat will be necessary. "Slow" work is based on an hourly wage of $23.00, "Medium" work on an hourly wage of $29.50, and "Fast" work on an hourly wage of $36.00. Other qualifications that apply to this table are on page 9.

	Manhours per home	Material gallons per home	Material cost per gallon	Labor cost per home	Labor burden per home	Material cost per home	Overhead per home	Profit per home	Total price per home
Sheet metal vents & flashing									
Metal primer, clean metal (material #35)									
1 story home, brush prime coat									
Slow	0.7	0.40	59.70	16.10	3.86	23.88	8.33	8.35	60.52
Medium	0.6	0.45	52.30	17.70	5.12	23.54	11.59	6.95	64.90
Fast	0.5	0.50	44.80	18.00	6.36	22.40	14.49	4.29	65.54
2 story home, brush prime coat									
Slow	0.9	0.50	59.70	20.70	4.97	29.85	10.55	10.57	76.64
Medium	0.8	0.55	52.30	23.60	6.82	28.77	14.80	8.88	82.87
Fast	0.7	0.60	44.80	25.20	8.90	26.88	18.90	5.59	85.47
Multiple units with attached roofs (per unit or 900 SF of roof area), brush 1st or additional finish coats									
Slow	0.6	0.40	59.70	13.80	3.31	23.88	7.79	7.80	56.58
Medium	0.5	0.45	52.30	14.75	4.27	23.54	10.64	6.38	59.58
Fast	0.4	0.50	44.80	14.40	5.08	22.40	12.98	3.84	58.70
Metal primer, rusty metal (material #36)									
1 story home, brush prime coat									
Slow	0.7	0.40	80.10	16.10	3.86	32.04	9.88	9.90	71.78
Medium	0.6	0.45	70.10	17.70	5.12	31.55	13.59	8.16	76.12
Fast	0.5	0.50	60.10	18.00	6.36	30.05	16.86	4.99	76.26
2 story home, brush prime coat									
Slow	0.9	0.50	80.10	20.70	4.97	40.05	12.49	12.51	90.72
Medium	0.8	0.55	70.10	23.60	6.82	38.56	17.25	10.35	96.58
Fast	0.7	0.60	60.10	25.20	8.90	36.06	21.75	6.43	98.34
Multiple units with attached roofs (per unit or 900 SF of roof area), brush 1st or additional finish coats									
Slow	0.6	0.40	80.10	13.80	3.31	32.04	9.34	9.36	67.85
Medium	0.5	0.45	70.10	14.75	4.27	31.55	12.64	7.58	70.79
Fast	0.4	0.50	60.10	14.40	5.08	30.05	15.35	4.54	69.42
Metal finish, synthetic enamel - off white, gloss, interior or exterior (material #37)									
1 story home, brush 1st or additional finish coats									
Slow	0.7	0.40	63.60	16.10	3.86	25.44	8.63	8.64	62.67
Medium	0.6	0.45	55.70	17.70	5.12	25.07	11.97	7.18	67.04
Fast	0.5	0.50	47.70	18.00	6.36	23.85	14.94	4.42	67.57
2 story home, brush 1st or additional finish coats									
Slow	0.9	0.50	63.60	20.70	4.97	31.80	10.92	10.94	79.33
Medium	0.8	0.55	55.70	23.60	6.82	30.64	15.27	9.16	85.49
Fast	0.7	0.60	47.70	25.20	8.90	28.62	19.44	5.75	87.91

	Manhours per home	Material gallons per home	Material cost per gallon	Labor cost per home	Labor burden per home	Material cost per home	Overhead per home	Profit per home	Total price per home
Multiple units with attached roofs (per unit or 900 SF of roof area), brush 1st or additional finish coats									
Slow	0.6	0.40	63.60	13.80	3.31	25.44	8.08	8.10	58.73
Medium	0.5	0.45	55.70	14.75	4.27	25.07	11.02	6.61	61.72
Fast	0.4	0.50	47.70	14.40	5.08	23.85	13.43	3.97	60.73
Metal finish, synthetic enamel - colors (except orange/red), gloss, interior or exterior (material #38)									
1 story home, brush 1st or additional finish coats									
Slow	0.7	0.40	66.30	16.10	3.86	26.52	8.83	8.85	64.16
Medium	0.6	0.45	58.00	17.70	5.12	26.10	12.23	7.34	68.49
Fast	0.5	0.50	49.70	18.00	6.36	24.85	15.25	4.51	68.97
2 story home, brush 1st or additional finish coats									
Slow	0.9	0.50	66.30	20.70	4.97	33.15	11.18	11.20	81.20
Medium	0.8	0.55	58.00	23.60	6.82	31.90	15.58	9.35	87.25
Fast	0.7	0.60	49.70	25.20	8.90	29.82	19.82	5.86	89.60
Multiple units with attached roofs (per unit or 900 SF of roof area), brush 1st or additional finish coats									
Slow	0.6	0.40	66.30	13.80	3.31	26.52	8.29	8.31	60.23
Medium	0.5	0.45	58.00	14.75	4.27	26.10	11.28	6.77	63.17
Fast	0.4	0.50	49.70	14.40	5.08	24.85	13.74	4.06	62.13

Use this table with metal prime and finish paints only. This table shows the time needed to paint sheet metal vents on an average home or attached dwelling unit. Use it to estimate the costs for residential units without having to take off each vent or piece of flashing. Use the "attached units" section based on 900 square feet of roof area to estimate commercial buildings. Note: A two coat system, prime and finish, using oil base material is recommended for any metal surface. Although water base material is often used, it may cause oxidation, corrosion and rust. One coat of oil base solid body stain is often used on exterior metal but it may crack, peel or chip without the proper prime coat application. If off white or another light colored finish paint is specified, make sure the prime coat is also a light color, or more than one finish coat will be necessary. "Slow" work is based on an hourly wage of $23.00, "Medium" work on an hourly wage of $29.50, and "Fast" work on an hourly wage of $36.00. Other qualifications that apply to this table are on page 9.

	Shutters per manhour	Shutters per gallon	Material cost per gallon	Labor cost per shutter	Labor burden per shutter	Material cost per shutter	Overhead per shutter	Profit per shutter	Total price per shutter
Shutters or blinds, 2' x 4' average size									
Brush each coat									
Undercoat, water or oil base (material #3 or #4)									
Slow	2.5	12	42.60	9.20	2.21	3.55	2.84	2.85	20.65
Medium	3.0	11	37.30	9.83	2.83	3.39	4.02	2.41	22.48
Fast	3.5	10	31.95	10.29	3.64	3.20	5.31	1.57	24.01
Split coat (1/2 undercoat + 1/2 enamel), water or oil base (material #3 and #24, or material #4 and #25)									
Slow	3.0	15	50.63	7.67	1.83	3.38	2.45	2.45	17.78
Medium	3.5	14	44.30	8.43	2.45	3.16	3.51	2.10	19.65
Fast	4.0	13	37.98	9.00	3.18	2.92	4.68	1.38	21.16
Exterior enamel, water or oil base (material #24 or #25)									
Slow	2.0	15	58.65	11.50	2.76	3.91	3.45	3.46	25.08
Medium	2.5	14	51.30	11.80	3.41	3.66	4.72	2.83	26.42
Fast	3.0	13	44.00	12.00	4.22	3.38	6.08	1.80	27.48
Spray each coat									
Undercoat, water or oil base (material #3 or #4)									
Slow	8	10	42.60	2.88	.69	4.26	1.49	1.49	10.81
Medium	9	9	37.30	3.28	.94	4.14	2.09	1.26	11.71
Fast	10	8	31.95	3.60	1.27	3.99	2.75	.81	12.42
Split coat (1/2 undercoat + 1/2 enamel), water or oil base (material #3 and #24, or material #4 and #25)									
Slow	8	12	50.63	2.88	.69	4.22	1.48	1.48	10.75
Medium	10	11	44.30	2.95	.85	4.03	1.96	1.17	10.96
Fast	12	10	37.98	3.00	1.04	3.80	2.44	.72	11.00
Exterior enamel, water or oil base (material #24 or #25)									
Slow	7	12	58.65	3.29	.79	4.89	1.70	1.71	12.38
Medium	8	11	51.30	3.69	1.06	4.66	2.36	1.41	13.18
Fast	9	10	44.00	4.00	1.41	4.40	3.04	.90	13.75

Use these figures to estimate the costs to paint all six sides of solid face, paint grade, interior or exterior, "false" plant-on type shutters or blinds. Costs are based on the number of single-panel 2' x 4' false (solid) shutters or blinds that can be painted in one hour (manhour). For real louvered shutters, multiply the quantity of shutters by a difficulty factor of 1.5 and then use this table. "Slow" work is based on an hourly wage of $23.00, "Medium" work on an hourly wage of $29.50, and "Fast" work on an hourly wage of $36.00. Other qualifications that apply to this table are on page 9.

	Labor SF per manhour	Material coverage SF/gallon	Material cost per gallon	Labor cost per 100 SF	Labor burden 100 SF	Material cost per 100 SF	Overhead per 100 SF	Profit per 100 SF	Total price per 100 SF
Siding, aluminum									
Metal primer, clean metal (material #35)									
Brush prime coat									
Slow	215	440	59.70	10.70	2.56	13.57	5.10	5.11	37.04
Medium	235	420	52.30	12.55	3.65	12.45	7.16	4.29	40.10
Fast	255	400	44.80	14.12	4.97	11.20	9.39	2.78	42.46
Metal primer, rusty metal (material #36)									
Brush prime coat									
Slow	215	440	80.10	10.70	2.56	18.20	5.98	5.99	43.43
Medium	235	420	70.10	12.55	3.65	16.69	8.22	4.93	46.04
Fast	255	400	60.10	14.12	4.97	15.03	10.58	3.13	47.83
Metal finish, synthetic enamel - off white, gloss, interior or exterior (material #37)									
Brush 1st or additional finish coats									
Slow	265	480	63.60	8.68	2.07	13.25	4.56	4.57	33.13
Medium	285	465	55.70	10.35	3.00	11.98	6.33	3.80	35.46
Fast	305	450	47.70	11.80	4.18	10.60	8.24	2.44	37.26
Metal finish, synthetic enamel - colors (except orange/red), gloss, interior or exterior (material #38)									
Brush 1st or additional finish coats									
Slow	265	480	66.30	8.68	2.07	13.81	4.67	4.68	33.91
Medium	285	465	58.00	10.35	3.00	12.47	6.45	3.87	36.14
Fast	305	450	49.70	11.80	4.18	11.04	8.37	2.48	37.87

Don't deduct for openings under 100 square feet. For heights above 8 feet, use the High Time Difficulty Factors on page 139. Note: A two coat system, prime and finish, using oil base material is recommended for any metal surface. Although water base material is often used, it may cause oxidation, corrosion and rust. One coat of oil base solid body stain is often used on exterior metal but it may crack, peel or chip without the proper prime coat application. If off white or another light colored finish paint is specified, make sure the prime coat is also a light color, or more than one finish coat will be necessary. "Slow" work is based on an hourly wage of $23.00, "Medium" work on an hourly wage of $29.50, and "Fast" work on an hourly wage of $36.00. Other qualifications that apply to this table are on page 9.

	Labor SF per manhour	Material coverage SF/gallon	Material cost per gallon	Labor cost per 100 SF	Labor burden 100 SF	Material cost per 100 SF	Overhead per 100 SF	Profit per 100 SF	Total price per 100 SF
Siding, composition shingle, brush application									
Solid body stain, water base (material #18)									
Brush 1st coat									
Slow	65	240	48.90	35.38	8.48	20.38	12.21	12.23	88.68
Medium	85	220	42.80	34.71	10.01	19.45	16.05	9.63	89.85
Fast	105	200	36.70	34.29	12.08	18.35	20.07	5.94	90.73
Brush 2nd coat									
Slow	90	350	48.90	25.56	6.13	13.97	8.68	8.69	63.03
Medium	110	335	42.80	26.82	7.75	12.78	11.84	7.10	66.29
Fast	130	320	36.70	27.69	9.77	11.47	15.17	4.49	68.59
Brush 3rd or additional coats									
Slow	110	425	48.90	20.91	5.01	11.51	7.11	7.13	51.67
Medium	130	405	42.80	22.69	6.56	10.57	9.96	5.97	55.75
Fast	150	385	36.70	24.00	8.49	9.53	13.02	3.85	58.89
Solid body stain, oil base (material #19)									
Brush 1st coat									
Slow	65	290	60.90	35.38	8.48	21.00	12.33	12.35	89.54
Medium	85	270	53.30	34.71	10.01	19.74	16.12	9.67	90.25
Fast	105	250	45.70	34.29	12.08	18.28	20.05	5.93	90.63
Brush 2nd coat									
Slow	90	390	60.90	25.56	6.13	15.62	8.99	9.01	65.31
Medium	110	375	53.30	26.82	7.75	14.21	12.20	7.32	68.30
Fast	130	360	45.70	27.69	9.77	12.69	15.55	4.60	70.30
Brush 3rd or additional coats									
Slow	110	445	60.90	20.91	5.01	13.69	7.53	7.54	54.68
Medium	130	425	53.30	22.69	6.56	12.54	10.45	6.27	58.51
Fast	150	405	45.70	24.00	8.49	11.28	13.56	4.01	61.34

Use these figures for repaint jobs only. Don't deduct for openings under 100 square feet. For heights above 8 feet, use the High Time Difficulty Factors on page 139. It has been established that asbestos fibers are known carcinogens (cancer causing) and it is likely that materials or products which contain asbestos will not be specified on any new construction projects in the future. Siding and roofing products usually contain very little asbestos and are typically non-friable (hand pressure can not crumble, pulverize or reduce to powder when dry). There is danger when asbestos is being removed because of exposure to airborne particulate matter. Apparently, there is little danger when painting asbestos siding or roofing, but it is a good idea to have your painters wear respirators or particle masks for their safety. "Slow" work is based on an hourly wage of $23.00, "Medium" work an hourly wage of $29.50, and "Fast" work on an hourly wage of $36.00. Other qualifications that apply to this table are on page 9.

	Labor SF per manhour	Material coverage SF/gallon	Material cost per gallon	Labor cost per 100 SF	Labor burden 100 SF	Material cost per 100 SF	Overhead per 100 SF	Profit per 100 SF	Total price per 100 SF
Siding, composition shingle, roll application									
Solid body stain, water base (material #18)									
Roll 1st coat									
Slow	140	230	48.90	16.43	3.93	21.26	7.91	7.93	57.46
Medium	160	215	42.80	18.44	5.33	19.91	10.92	6.55	61.15
Fast	180	200	36.70	20.00	7.08	18.35	14.08	4.16	63.67
Roll 2nd coat									
Slow	190	330	48.90	12.11	2.89	14.82	5.67	5.68	41.17
Medium	210	315	42.80	14.05	4.05	13.59	7.93	4.76	44.38
Fast	230	300	36.70	15.65	5.54	12.23	10.35	3.06	46.83
Roll 3rd or additional coats									
Slow	250	395	48.90	9.20	2.21	12.38	4.52	4.53	32.84
Medium	280	385	42.80	10.54	3.04	11.12	6.18	3.71	34.59
Fast	300	375	36.70	12.00	4.22	9.79	8.07	2.39	36.47
Solid body stain, oil base (material #19)									
Roll 1st coat									
Slow	140	260	60.90	16.43	3.93	23.42	8.32	8.34	60.44
Medium	160	240	53.30	18.44	5.33	22.21	11.50	6.90	64.38
Fast	180	220	45.70	20.00	7.08	20.77	14.83	4.39	67.07
Roll 2nd coat									
Slow	190	360	60.90	12.11	2.89	16.92	6.07	6.08	44.07
Medium	210	345	53.30	14.05	4.05	15.45	8.39	5.03	46.97
Fast	230	330	45.70	15.65	5.54	13.85	10.86	3.21	49.11
Roll 3rd or additional coats									
Slow	250	420	60.90	9.20	2.21	14.50	4.92	4.93	35.76
Medium	280	408	53.30	10.54	3.04	13.06	6.66	4.00	37.30
Fast	300	395	45.70	12.00	4.22	11.57	8.62	2.55	38.96

	Labor SF per manhour	Material coverage SF/gallon	Material cost per gallon	Labor cost per 100 SF	Labor burden 100 SF	Material cost per 100 SF	Overhead per 100 SF	Profit per 100 SF	Total price per 100 SF
Waterproofing, clear hydro seal, oil base (material #34)									
Roll 1st coat									
Slow	75	275	40.60	30.67	7.35	14.76	10.03	10.05	72.86
Medium	150	250	35.50	19.67	5.70	14.20	9.89	5.93	55.39
Fast	225	225	30.40	16.00	5.63	13.51	10.90	3.22	49.26
Roll 2nd or additional coats									
Slow	125	325	40.60	18.40	4.42	12.49	6.71	6.72	48.74
Medium	200	300	35.50	14.75	4.27	11.83	7.71	4.63	43.19
Fast	275	275	30.40	13.09	4.64	11.05	8.92	2.64	40.34

Use these figures for repaint jobs only. Don't deduct for openings under 100 square feet. For heights above 8 feet, use the High Time Difficulty Factors on page 139. It has been established that asbestos fibers are known carcinogens (cancer causing) and it is likely that materials or products which contain asbestos will not be specified on any new construction projects in the future. Siding and roofing products usually contain very little asbestos and are typically non-friable (hand pressure can not crumble, pulverize or reduce to powder when dry). There is danger when asbestos is being removed because of exposure to airborne particulate matter. Apparently, there is little danger when painting asbestos siding or roofing, but it is a good idea to have your painters wear respirators or particle masks for their safety. "Slow" work is based on an hourly wage of $23.00, "Medium" work an hourly wage of $29.50, and "Fast" work on an hourly wage of $36.00. Other qualifications that apply to this table are on page 9.

	Labor SF per manhour	Material coverage SF/gallon	Material cost per gallon	Labor cost per 100 SF	Labor burden 100 SF	Material cost per 100 SF	Overhead per 100 SF	Profit per 100 SF	Total price per 100 SF
Siding, composition shingle, spray application									
Solid body stain, water base (material #18)									
Spray 1st coat									
Slow	325	160	48.90	7.08	1.70	30.56	7.47	7.49	54.30
Medium	350	150	42.80	8.43	2.45	28.53	9.85	5.91	55.17
Fast	375	140	36.70	9.60	3.41	26.21	12.15	3.59	54.96
Spray 2nd coat									
Slow	350	240	48.90	6.57	1.59	20.38	5.42	5.43	39.39
Medium	375	228	42.80	7.87	2.28	18.77	7.23	4.34	40.49
Fast	400	215	36.70	9.00	3.18	17.07	9.07	2.68	41.00
Spray 3rd or additional coats									
Slow	425	300	48.90	5.41	1.29	16.30	4.37	4.38	31.75
Medium	450	288	42.80	6.56	1.88	14.86	5.83	3.50	32.63
Fast	475	275	36.70	7.58	2.70	13.35	7.32	2.17	33.12
Solid body stain, oil base (material #19)									
Spray 1st coat									
Slow	325	190	60.90	7.08	1.70	32.05	7.76	7.77	56.36
Medium	350	180	53.30	8.43	2.45	29.61	10.12	6.07	56.68
Fast	375	170	45.70	9.60	3.41	26.88	12.36	3.66	55.91
Spray 2nd coat									
Slow	350	270	60.90	6.57	1.59	22.56	5.83	5.85	42.40
Medium	375	258	53.30	7.87	2.28	20.66	7.70	4.62	43.13
Fast	400	245	45.70	9.00	3.18	18.65	9.56	2.83	43.22
Spray 3rd or additional coats									
Slow	425	320	60.90	5.41	1.29	19.03	4.89	4.90	35.52
Medium	450	308	53.30	6.56	1.88	17.31	6.44	3.87	36.06
Fast	475	295	45.70	7.58	2.70	15.49	7.98	2.36	36.11

	Labor SF per manhour	Material coverage SF/gallon	Material cost per gallon	Labor cost per 100 SF	Labor burden 100 SF	Material cost per 100 SF	Overhead per 100 SF	Profit per 100 SF	Total price per 100 SF
Waterproofing, clear hydro seal, oil base (material #34)									
Spray 1st coat									
Slow	475	100	40.60	4.84	1.18	40.60	8.85	8.87	64.34
Medium	525	88	35.50	5.62	1.61	40.34	11.90	7.14	66.61
Fast	575	75	30.40	6.26	2.22	40.53	15.19	4.49	68.69
Spray 2nd or additional coats									
Slow	500	150	40.60	4.60	1.10	27.07	6.23	6.24	45.24
Medium	550	125	35.50	5.36	1.56	28.40	8.83	5.30	49.45
Fast	600	100	30.40	6.00	2.13	30.40	11.94	3.53	54.00

Use these figures for repaint jobs only. Don't deduct for openings under 100 square feet. For heights above 8 feet, use the High Time Difficulty Factors on page 139. It has been established that asbestos fibers are known carcinogens (cancer causing) and it is likely that materials or products which contain asbestos will not be specified on any new construction projects in the future. Siding and roofing products usually contain very little asbestos and are typically non-friable (hand pressure can not crumble, pulverize or reduce to powder when dry). There is danger when asbestos is being removed because of exposure to airborne particulate matter. Apparently, there is little danger when painting asbestos siding or roofing, but it is a good idea to have your painters wear respirators or particle masks for their safety. "Slow" work is based on an hourly wage of $23.00, "Medium" work an hourly wage of $29.50, and "Fast" work on an hourly wage of $36.00. Other qualifications that apply to this table are on page 9.

	Labor SF per manhour	Material coverage SF/gallon	Material cost per gallon	Labor cost per 100 SF	Labor burden 100 SF	Material cost per 100 SF	Overhead per 100 SF	Profit per 100 SF	Total price per 100 SF
Siding, rough sawn or resawn wood, brush									
Solid body or semi-transparent stain, water base (material #18 or #20)									
Brush 1st coat									
Slow	100	250	49.40	23.00	5.52	19.76	9.17	9.19	66.64
Medium	135	238	43.25	21.85	6.33	18.17	11.58	6.95	64.88
Fast	170	225	37.05	21.18	7.46	16.47	13.99	4.14	63.24
Brush 2nd coat									
Slow	135	300	49.40	17.04	4.09	16.47	7.14	7.16	51.90
Medium	168	288	43.25	17.56	5.07	15.02	9.41	5.65	52.71
Fast	200	275	37.05	18.00	6.36	13.47	11.72	3.47	53.02
Brush 3rd or additional coats									
Slow	150	335	49.40	15.33	3.69	14.75	6.41	6.43	46.61
Medium	183	323	43.25	16.12	4.64	13.39	8.54	5.13	47.82
Fast	215	310	37.05	16.74	5.91	11.95	10.73	3.17	48.50
Solid body or semi-transparent stain, oil base (material #19 or #21)									
Brush 1st coat									
Slow	100	275	59.60	23.00	5.52	21.67	9.54	9.56	69.29
Medium	135	250	52.15	21.85	6.33	20.86	12.26	7.35	68.65
Fast	170	225	44.70	21.18	7.46	19.87	15.04	4.45	68.00
Brush 2nd coat									
Slow	135	350	59.60	17.04	4.09	17.03	7.25	7.27	52.68
Medium	168	325	52.15	17.56	5.07	16.05	9.67	5.80	54.15
Fast	200	300	44.70	18.00	6.36	14.90	12.17	3.60	55.03
Brush 3rd or additional coats									
Slow	150	400	59.60	15.33	3.69	14.90	6.44	6.46	46.82
Medium	183	375	52.15	16.12	4.64	13.91	8.67	5.20	48.54
Fast	215	350	44.70	16.74	5.91	12.77	10.98	3.25	49.65
Penetrating oil stain (material #13)									
Brush 1st coat									
Slow	100	230	57.90	23.00	5.52	25.17	10.20	10.22	74.11
Medium	135	210	50.70	21.85	6.33	24.14	13.08	7.85	73.25
Fast	170	190	43.40	21.18	7.46	22.84	15.97	4.72	72.17

	Labor SF per manhour	Material coverage SF/gallon	Material cost per gallon	Labor cost per 100 SF	Labor burden 100 SF	Material cost per 100 SF	Overhead per 100 SF	Profit per 100 SF	Total price per 100 SF
Brush 2nd coat									
Slow	135	275	57.90	17.04	4.09	21.05	8.01	8.03	58.22
Medium	168	255	50.70	17.56	5.07	19.88	10.63	6.38	59.52
Fast	200	235	43.40	18.00	6.36	18.47	13.27	3.93	60.03
Brush 3rd or additional coats									
Slow	150	350	57.90	15.33	3.69	16.54	6.75	6.77	49.08
Medium	183	330	50.70	16.12	4.64	15.36	9.04	5.42	50.58
Fast	215	310	43.40	16.74	5.91	14.00	11.36	3.36	51.37

Use this table to estimate the cost of painting shingle, shake, resawn or rough sawn wood or plywood siding with average moisture content. Don't deduct for openings under 100 square feet. For heights above 8 feet, use the High Time Difficulty Factors on page 139. For wood or composition drop siding with exposed bevel edges, multiply the surface area by 1.12 to allow for the extra time and material needed to paint the underside of each board. "Slow" work is based on an hourly wage of $23.00, "Medium" work on an hourly wage of $29.50, and "Fast" work on an hourly wage of $36.00. Other qualifications that apply to this table are on page 9.

	Labor SF per manhour	Material coverage SF/gallon	Material cost per gallon	Labor cost per 100 SF	Labor burden 100 SF	Material cost per 100 SF	Overhead per 100 SF	Profit per 100 SF	Total price per 100 SF
Siding, rough sawn or resawn wood, roll									
Solid body or semi-transparent stain, water base (material #18 or #20)									
Roll 1st coat									
Slow	150	235	49.40	15.33	3.69	21.02	7.61	7.62	55.27
Medium	200	213	43.25	14.75	4.27	20.31	9.83	5.90	55.06
Fast	250	210	37.05	14.40	5.08	17.64	11.51	3.40	52.03
Roll 2nd coat									
Slow	200	285	49.40	11.50	2.76	17.33	6.00	6.01	43.60
Medium	250	273	43.25	11.80	3.41	15.84	7.76	4.66	43.47
Fast	300	260	37.05	12.00	4.22	14.25	9.45	2.80	42.72
Roll 3rd or additional coats									
Slow	225	335	49.40	10.22	2.44	14.75	5.21	5.22	37.84
Medium	288	323	43.25	10.24	2.96	13.39	6.65	3.99	37.23
Fast	350	310	37.05	10.29	3.64	11.95	8.02	2.37	36.27
Solid body or semi-transparent stain, oil base (material #19 or #21)									
Roll 1st coat									
Slow	150	250	59.60	15.33	3.69	23.84	8.14	8.16	59.16
Medium	225	225	52.15	13.11	3.78	23.18	10.02	6.01	56.10
Fast	275	200	44.70	13.09	4.64	22.35	12.42	3.67	56.17
Roll 2nd coat									
Slow	200	330	59.60	11.50	2.76	18.06	6.14	6.15	44.61
Medium	275	305	52.15	10.73	3.11	17.10	7.73	4.64	43.31
Fast	350	280	44.70	10.29	3.64	15.96	9.26	2.74	41.89
Roll 3rd or additional coats									
Slow	260	415	59.60	8.85	2.13	14.36	4.81	4.82	34.97
Medium	335	390	52.15	8.81	2.56	13.37	6.18	3.71	34.63
Fast	410	365	44.70	8.78	3.11	12.25	7.48	2.21	33.83
Penetrating oil stain (material #13)									
Roll 1st coat									
Slow	150	150	57.90	15.33	3.69	38.60	10.95	10.97	79.54
Medium	225	125	50.70	13.11	3.78	40.56	14.37	8.62	80.44
Fast	275	100	43.40	13.09	4.64	43.40	18.94	5.60	85.67

	Labor SF per manhour	Material coverage SF/gallon	Material cost per gallon	Labor cost per 100 SF	Labor burden 100 SF	Material cost per 100 SF	Overhead per 100 SF	Profit per 100 SF	Total price per 100 SF
Roll 2nd coat									
Slow	200	280	57.90	11.50	2.76	20.68	6.64	6.65	48.23
Medium	275	240	50.70	10.73	3.11	21.13	8.74	5.24	48.95
Fast	350	200	43.40	10.29	3.64	21.70	11.04	3.27	49.94
Roll 3rd or additional coats									
Slow	260	365	57.90	8.85	2.13	15.86	5.10	5.11	37.05
Medium	335	330	50.70	8.81	2.56	15.36	6.68	4.01	37.42
Fast	410	295	43.40	8.78	3.11	14.71	8.24	2.44	37.28

Use this table to estimate the cost of painting shingle, shake, resawn or rough sawn wood or plywood siding with average moisture content. Don't deduct for openings under 100 square feet. For heights above 8 feet, use the High Time Difficulty Factors on page 139. For wood or composition drop siding with exposed bevel edges, multiply the surface area by 1.12 to allow for the extra time and material needed to paint the underside of each board. "Slow" work is based on an hourly wage of $23.00, "Medium" work on an hourly wage of $29.50, and "Fast" work on an hourly wage of $36.00. Other qualifications that apply to this table are on page 9.

	Labor SF per manhour	Material coverage SF/gallon	Material cost per gallon	Labor cost per 100 SF	Labor burden 100 SF	Material cost per 100 SF	Overhead per 100 SF	Profit per 100 SF	Total price per 100 SF
Siding, rough sawn or resawn wood, spray									
Solid body or semi-transparent stain, water base (material #18 or #20)									
Spray 1st coat									
Slow	400	140	49.40	5.75	1.38	35.29	8.06	8.08	58.56
Medium	500	115	43.25	5.90	1.71	37.61	11.31	6.78	63.31
Fast	600	90	37.05	6.00	2.13	41.17	15.28	4.52	69.10
Spray 2nd coat									
Slow	500	150	49.40	4.60	1.10	32.93	7.34	7.36	53.33
Medium	600	125	43.25	4.92	1.43	34.60	10.24	6.14	57.33
Fast	700	100	37.05	5.14	1.83	37.05	13.64	4.03	61.69
Spray 3rd or additional coats									
Slow	550	200	49.40	4.18	1.01	24.70	5.68	5.69	41.26
Medium	650	170	43.25	4.54	1.32	25.44	7.82	4.69	43.81
Fast	750	140	37.05	4.80	1.68	26.46	10.21	3.02	46.17
Solid body or semi-transparent stain, oil base (material #19 or #21)									
Spray 1st coat									
Slow	400	170	59.60	5.75	1.38	35.06	8.02	8.03	58.24
Medium	500	150	52.15	5.90	1.71	34.77	10.60	6.36	59.34
Fast	600	130	44.70	6.00	2.13	34.38	13.18	3.90	59.59
Spray 2nd coat									
Slow	450	255	59.60	5.11	1.22	23.37	5.64	5.66	41.00
Medium	550	235	52.15	5.36	1.56	22.19	7.28	4.37	40.76
Fast	650	215	44.70	5.54	1.96	20.79	8.77	2.59	39.65
Spray 3rd or additional coats									
Slow	550	355	59.60	4.18	1.01	16.79	4.17	4.18	30.33
Medium	650	335	52.15	4.54	1.32	15.57	5.36	3.21	30.00
Fast	750	315	44.70	4.80	1.68	14.19	6.41	1.90	28.98
Penetrating oil stain (material #13)									
Spray 1st coat									
Slow	400	200	57.90	5.75	1.38	28.95	6.86	6.87	49.81
Medium	500	180	50.70	5.90	1.71	28.17	8.95	5.37	50.10
Fast	600	160	43.40	6.00	2.13	27.13	10.93	3.23	49.42
Spray 2nd coat									
Slow	450	290	57.90	5.11	1.22	19.97	5.00	5.01	36.31
Medium	550	245	50.70	5.36	1.56	20.69	6.90	4.14	38.65
Fast	650	200	43.40	5.54	1.96	21.70	9.05	2.68	40.93
Spray 3rd or additional coats									
Slow	550	390	57.90	4.18	1.01	14.85	3.81	3.81	27.66
Medium	650	360	50.70	4.54	1.32	14.08	4.98	2.99	27.91
Fast	750	330	43.40	4.80	1.68	13.15	6.09	1.80	27.52

	Labor SF per manhour	Material coverage SF/gallon	Material cost per gallon	Labor cost per 100 SF	Labor burden 100 SF	Material cost per 100 SF	Overhead per 100 SF	Profit per 100 SF	Total price per 100 SF
Waterproofing, clear hydro seal, oil base (material #34)									
Spray 1st coat									
Slow	500	150	40.60	4.60	1.10	27.07	6.23	4.68	43.68
Medium	575	113	35.50	5.13	1.49	31.42	9.51	5.70	53.25
Fast	650	75	30.40	5.54	1.96	40.53	14.89	4.40	67.32
Spray 2nd coat									
Slow	575	175	40.60	4.00	.96	23.20	5.35	5.36	38.87
Medium	675	150	35.50	4.37	1.26	23.67	7.33	4.40	41.03
Fast	775	125	30.40	4.65	1.63	24.32	9.49	2.81	42.90
Spray 3rd or additional coats									
Slow	650	200	40.60	3.54	.85	20.30	4.69	4.70	34.08
Medium	750	175	35.50	3.93	1.13	20.29	6.34	3.80	35.49
Fast	850	150	30.40	4.24	1.51	20.27	8.06	2.38	36.46

Use this table to estimate the cost of painting shingle, shake, resawn or rough sawn wood or plywood siding with average moisture content. Don't deduct for openings under 100 square feet. For heights above 8 feet, use the High Time Difficulty Factors on page 139. For wood or composition drop siding with exposed beveled edges, multiply the surface area by 1.12 to allow for the extra time and material needed to paint the underside of each board. "Slow" work is based on an hourly wage of $23.00, "Medium" work on an hourly wage of $29.50, and "Fast" work on an hourly wage of $36.00. Other qualifications that apply to this table are on page 9.

	Labor SF per manhour	Material coverage SF/gallon	Material cost per gallon	Labor cost per 100 SF	Labor burden 100 SF	Material cost per 100 SF	Overhead per 100 SF	Profit per 100 SF	Total price per 100 SF
Siding, smooth wood, brush									
Solid body or semi-transparent stain, water base (material #18 or #20)									
Brush 1st coat									
Slow	100	275	49.40	23.00	5.52	17.96	8.83	8.85	64.16
Medium	125	363	43.25	23.60	6.82	11.91	10.58	6.35	59.26
Fast	150	250	37.05	24.00	8.49	14.82	14.66	4.34	66.31
Brush 2nd coat									
Slow	135	350	49.40	17.04	4.09	14.11	6.70	6.71	48.65
Medium	168	325	43.25	17.56	5.07	13.31	8.99	5.39	50.32
Fast	200	300	37.05	18.00	6.36	12.35	11.38	3.37	51.46
Brush 3rd or additional coats									
Slow	150	425	49.40	15.33	3.69	11.62	5.82	5.83	42.29
Medium	188	400	43.25	15.69	4.54	10.81	7.76	4.65	43.45
Fast	215	375	37.05	16.74	5.91	9.88	10.08	2.98	45.59
Solid body or semi-transparent stain, oil base (material #19 or #21)									
Brush 1st coat									
Slow	100	400	59.60	23.00	5.52	14.90	8.25	8.27	59.94
Medium	125	363	52.15	23.60	6.82	14.37	11.20	6.72	62.71
Fast	150	325	44.70	24.00	8.49	13.75	14.33	4.24	64.81
Brush 2nd coat									
Slow	135	450	59.60	17.04	4.09	13.24	6.53	6.54	47.44
Medium	168	408	52.15	17.56	5.07	12.78	8.85	5.31	49.57
Fast	200	375	44.70	18.00	6.36	11.92	11.24	3.33	50.85
Brush 3rd or additional coats									
Slow	150	525	59.60	15.33	3.69	11.35	5.77	5.78	41.92
Medium	188	438	52.15	15.69	4.54	11.91	8.03	4.82	44.99
Fast	215	450	44.70	16.74	5.91	9.93	10.10	2.99	45.67
Penetrating oil stain (material #13)									
Brush 1st coat									
Slow	100	315	57.90	23.00	5.52	18.38	8.91	8.93	64.74
Medium	125	303	50.70	23.60	6.82	16.73	11.79	7.07	66.01
Fast	150	290	43.40	24.00	8.49	14.97	14.71	4.35	66.52

	Labor SF per manhour	Material coverage SF/gallon	Material cost per gallon	Labor cost per 100 SF	Labor burden 100 SF	Material cost per 100 SF	Overhead per 100 SF	Profit per 100 SF	Total price per 100 SF
Brush 2nd coat									
Slow	135	355	57.90	17.04	4.09	16.31	7.11	7.13	51.68
Medium	168	343	50.70	17.56	5.07	14.78	9.35	5.61	52.37
Fast	200	330	43.40	18.00	6.36	13.15	11.63	3.44	52.58
Brush 3rd or additional coats									
Slow	150	430	57.90	15.33	3.69	13.47	6.17	6.18	44.84
Medium	188	418	50.70	15.69	4.54	12.13	8.09	4.85	45.30
Fast	215	405	43.40	16.74	5.91	10.72	10.34	3.06	46.77

Use this table for butt or tongue and groove siding, joint lap, drop, beveled or board and batten siding in redwood, plywood, fir, hemlock or pine. Don't deduct for openings under 100 square feet. For heights above 8 feet, use the High Time Difficulty Factors on page 139. For wood or composition drop siding with exposed bevel edges, multiply the surface area by 1.12 to allow for the extra time and material needed to paint the underside of each board. "Slow" work is based on an hourly wage of $23.00, "Medium" work on an hourly wage of $29.50, and "Fast" work on an hourly wage of $36.00. Other qualifications that apply to this table are on page 9.

	Labor SF per manhour	Material coverage SF/gallon	Material cost per gallon	Labor cost per 100 SF	Labor burden 100 SF	Material cost per 100 SF	Overhead per 100 SF	Profit per 100 SF	Total price per 100 SF
Siding, smooth wood, roll									
Solid body or semi-transparent stain, water base (material #18 or #20)									
Roll 1st coat									
Slow	175	250	49.40	13.14	3.14	19.76	6.85	6.86	49.75
Medium	225	238	43.25	13.11	3.78	18.17	8.77	5.26	49.09
Fast	275	225	37.05	13.09	4.64	16.47	10.60	3.13	47.93
Roll 2nd coat									
Slow	225	300	49.40	10.22	2.44	16.47	5.54	5.55	40.22
Medium	275	288	43.25	10.73	3.11	15.02	7.21	4.33	40.40
Fast	325	275	37.05	11.08	3.92	13.47	8.82	2.61	39.90
Roll 3rd or additional coats									
Slow	260	375	49.40	8.85	2.13	13.17	4.59	4.60	33.34
Medium	335	338	43.25	8.81	2.56	12.80	6.04	3.62	33.83
Fast	410	300	37.05	8.78	3.11	12.35	7.51	2.22	33.97
Solid body or semi-transparent stain, oil base (material #19 or #21)									
Roll 1st coat									
Slow	175	350	59.60	13.14	3.14	17.03	6.33	6.34	45.98
Medium	225	325	52.15	13.11	3.78	16.05	8.24	4.94	46.12
Fast	275	300	44.70	13.09	4.64	14.90	10.11	2.99	45.73
Roll 2nd coat									
Slow	225	400	59.60	10.22	2.44	14.90	5.24	5.25	38.05
Medium	275	375	52.15	10.73	3.11	13.91	6.94	4.16	38.85
Fast	325	350	44.70	11.08	3.92	12.77	8.61	2.55	38.93
Roll 3rd or additional coats									
Slow	260	425	59.60	8.85	2.13	14.02	4.75	4.76	34.51
Medium	335	413	52.15	8.81	2.56	12.63	6.00	3.60	33.60
Fast	410	400	44.70	8.78	3.11	11.18	7.15	2.11	32.33
Penetrating oil stain (material #13)									
Roll 1st coat									
Slow	175	200	57.90	13.14	3.14	28.95	8.60	8.61	62.44
Medium	225	150	50.70	13.11	3.78	33.80	12.68	7.61	70.98
Fast	275	100	43.40	13.09	4.64	43.40	18.94	5.60	85.67

	Labor SF per manhour	Material coverage SF/gallon	Material cost per gallon	Labor cost per 100 SF	Labor burden 100 SF	Material cost per 100 SF	Overhead per 100 SF	Profit per 100 SF	Total price per 100 SF
Roll 2nd coat									
Slow	225	250	57.90	10.22	2.44	23.16	6.81	6.82	49.45
Medium	275	200	50.70	10.73	3.11	25.35	9.80	5.88	54.87
Fast	325	150	43.40	11.08	3.92	28.93	13.62	4.03	61.58
Roll 3rd or additional coats									
Slow	260	300	57.90	8.85	2.13	19.30	5.75	5.76	41.79
Medium	335	250	50.70	8.81	2.56	20.28	7.91	4.75	44.31
Fast	410	200	43.40	8.78	3.11	21.70	10.41	3.08	47.08

Use this table for butt or tongue and groove siding, joint lap, drop, beveled or board and batten siding in redwood, plywood, fir, hemlock or pine. Don't deduct for openings under 100 square feet. For heights above 8 feet, use the High Time Difficulty Factors on page 139. For wood or composition drop siding with exposed beveled edges, multiply the surface area by 1.12 to allow for the extra time and material needed to paint the underside of each board. "Slow" work is based on an hourly wage of $23.00, "Medium" work on an hourly wage of $29.50, and "Fast" work on an hourly wage of $36.00. Other qualifications that apply to this table are on page 9.

	Labor SF per manhour	Material coverage SF/gallon	Material cost per gallon	Labor cost per 100 SF	Labor burden 100 SF	Material cost per 100 SF	Overhead per 100 SF	Profit per 100 SF	Total price per 100 SF
Siding, smooth wood, spray									
Solid body or semi-transparent stain, water base (material #18 or #20)									
Spray 1st coat									
Slow	450	150	49.40	5.11	1.22	32.93	7.46	7.48	54.20
Medium	550	125	43.25	5.36	1.56	34.60	10.38	6.23	58.13
Fast	650	100	37.05	5.54	1.96	37.05	13.81	4.09	62.45
Spray 2nd coat									
Slow	550	250	49.40	4.18	1.01	19.76	4.74	4.75	34.44
Medium	650	225	43.25	4.54	1.32	19.22	6.27	3.76	35.11
Fast	750	200	37.05	4.80	1.68	18.53	7.76	2.29	35.06
Spray 3rd or additional coats									
Slow	650	350	49.40	3.54	.85	14.11	3.52	3.52	25.54
Medium	750	325	43.25	3.93	1.13	13.31	4.60	2.76	25.73
Fast	850	300	37.05	4.24	1.51	12.35	5.61	1.66	25.37
Solid body or semi-transparent stain, oil base (material #19 or #21)									
Spray 1st coat									
Slow	450	170	59.60	5.11	1.22	35.06	7.87	7.88	57.14
Medium	550	150	52.15	5.36	1.56	34.77	10.42	6.25	58.36
Fast	650	130	44.70	5.54	1.96	34.38	12.98	3.84	58.70
Spray 2nd coat									
Slow	550	300	59.60	4.18	1.01	19.87	4.76	4.77	34.59
Medium	650	273	52.15	4.54	1.32	19.10	6.24	3.74	34.94
Fast	750	245	44.70	4.80	1.68	18.24	7.67	2.27	34.66
Spray 3rd or additional coats									
Slow	650	400	59.60	3.54	.85	14.90	3.67	3.67	26.63
Medium	750	373	52.15	3.93	1.13	13.98	4.76	2.86	26.66
Fast	850	345	44.70	4.24	1.51	12.96	5.80	1.72	26.23
Penetrating oil stain (material #13)									
Spray 1st coat									
Slow	450	150	57.90	5.11	1.22	38.60	8.54	8.56	62.03
Medium	550	113	50.70	5.36	1.56	44.87	12.95	7.77	72.51
Fast	650	75	43.40	5.54	1.96	57.87	20.26	5.99	91.62
Spray 2nd coat									
Slow	550	225	57.90	4.18	1.01	25.73	5.87	5.88	42.67
Medium	650	188	50.70	4.54	1.32	26.97	8.21	4.92	45.96
Fast	750	150	43.40	4.80	1.68	28.93	10.98	3.25	49.64
Spray 3rd or additional coats									
Slow	650	250	57.90	3.54	.85	23.16	5.23	5.24	38.02
Medium	750	225	50.70	3.93	1.13	22.53	6.90	4.14	38.63
Fast	850	200	43.40	4.24	1.51	21.70	8.51	2.52	38.48

	Labor SF per manhour	Material coverage SF/gallon	Material cost per gallon	Labor cost per 100 SF	Labor burden 100 SF	Material cost per 100 SF	Overhead per 100 SF	Profit per 100 SF	Total price per 100 SF
Waterproofing, clear hydro seal, oil base (material #34)									
Spray 1st coat									
Slow	550	250	40.60	4.18	1.01	16.24	4.07	3.06	28.56
Medium	650	200	35.50	4.54	1.32	17.75	5.90	3.54	33.05
Fast	750	150	30.40	4.80	1.68	20.27	8.30	2.45	37.50
Spray 2nd coat									
Slow	650	300	40.60	3.54	.85	13.53	3.40	3.41	24.73
Medium	750	250	35.50	3.93	1.13	14.20	4.82	2.89	26.97
Fast	850	200	30.40	4.24	1.51	15.20	6.49	1.92	29.36
Spray 3rd or additional coats									
Slow	700	325	40.60	3.29	.79	12.49	3.15	3.16	22.88
Medium	800	275	35.50	3.69	1.06	12.91	4.42	2.65	24.73
Fast	900	225	30.40	4.00	1.41	13.51	5.87	1.74	26.53

Use this table for butt or tongue and groove siding, joint lap, drop, beveled or board and batten siding in redwood, plywood, fir, hemlock or pine. Don't deduct for openings under 100 square feet. For heights above 8 feet, use the High Time Difficulty Factors on page 139. For wood or composition drop siding with exposed bevel edges, multiply the surface area by 1.12 to allow for the extra time and material needed to paint the underside of each board. "Slow" work is based on an hourly wage of $23.00, "Medium" work on an hourly wage of $29.50, and "Fast" work on an hourly wage of $36.00. Other qualifications that apply to this table are on page 9.

Stair steps, interior or exterior, wood

To estimate the cost to paint or stain stairs, find the surface area. Then use the tables for wood siding. To find the surface area of each tread and riser, multiply the length by the width. To find the tread length, add the run, the rise, and the tread nosing. For example, a tread with a 12" run, an 8" rise, and 1" nosing, has a 23" surface area (measured one side). For estimating purposes, figure any length from 14" to 26" as 2 feet. Use the actual width of the tread if the stringers are calculated separately. If the tread in the example is 3 feet wide and you use 2 feet for the length, the surface area is 6 feet. If there are 15 treads, the total top surface area is 90 square feet.

If you're calculating the area to paint the stair treads and stringers in one operation, add 2 feet to the actual tread width to include the stringers. That would make the effective width of the tread in the example 5 feet. Then multiply 5 feet by 2 feet to find the area of each tread, 10 square feet. For 15 treads, the total surface area is 150 square feet.

	Labor LF per manhour	Material coverage LF/gallon	Material cost per gallon	Labor cost per 100 LF	Labor burden 100 LF	Material cost per 100 LF	Overhead per 100 LF	Profit per 100 LF	Total price per 100 LF

Stair stringers, exterior, metal, shapes up to 14" wide, each side

	Labor LF per manhour	Material coverage LF/gallon	Material cost per gallon	Labor cost per 100 LF	Labor burden 100 LF	Material cost per 100 LF	Overhead per 100 LF	Profit per 100 LF	Total price per 100 LF
Metal primer - rust inhibitor, clean metal (material #35)									
Roll & brush prime coat									
Slow	50	120	59.70	46.00	11.04	49.75	20.29	20.33	147.41
Medium	55	115	52.30	53.64	15.50	45.48	28.66	17.19	160.47
Fast	60	110	44.80	60.00	21.20	40.73	37.79	11.18	170.90
Metal primer - rust inhibitor, rusty metal (material #36)									
Roll & brush prime coat									
Slow	50	120	80.10	46.00	11.04	66.75	23.52	23.57	170.88
Medium	55	115	70.10	53.64	15.50	60.96	32.53	19.52	182.15
Fast	60	110	60.10	60.00	21.20	54.64	42.10	12.45	190.39
Metal finish - synthetic enamel, off white (material #37)									
Roll & brush 1st or additional finish coats									
Slow	50	135	63.60	46.00	11.04	47.11	19.79	19.83	143.77
Medium	55	130	55.70	53.64	15.50	42.85	28.00	16.80	156.79
Fast	60	125	47.70	60.00	21.20	38.16	37.00	10.94	167.30
Metal finish - synthetic enamel, colors - except orange/red (material #38)									
Roll & brush 1st or additional finish coats									
Slow	50	135	66.30	46.00	11.04	49.11	20.17	20.21	146.53
Medium	55	130	58.00	53.64	15.50	44.62	28.44	17.06	159.26
Fast	60	125	49.70	60.00	21.20	39.76	37.49	11.09	169.54

Use these figures to paint each side of installed stair stringers. Measurements are based on linear feet of each stringer. Note: A two coat system, prime and finish, using oil base material is recommended for any metal surface. Although water base material is often used, it may cause oxidation, corrosion and rust. Using one coat of oil base paint on exterior metal may result in cracking, peeling or chipping without the proper prime coat application. Pre-primed steel or wrought iron generally requires only one coat to cover. The metal finish figures include minor touchup to the prime coat. If off white or other light colored finish paint is specified, make sure the prime coat is also a light color, or more than one finish coat will be necessary. "Slow" work is based on an hourly wage of $23.00, "Medium" work on an hourly wage of $29.50, and "Fast" work on an hourly wage of $36.00. Other qualifications that apply to this table are on page 9.

Stair stringers, exterior, rough sawn wood up to 4" x 12"

	Labor LF per manhour	Material coverage LF/gallon	Material cost per gallon	Labor cost per 100 LF	Labor burden 100 LF	Material cost per 100 LF	Overhead per 100 LF	Profit per 100 LF	Total price per 100 LF
Solid body or semi-transparent stain, water or oil base (material #18, #19, #20 or #21)									
Roll & brush each coat									
Slow	40	70	54.50	57.50	13.80	77.86	28.34	28.40	205.90
Medium	45	65	47.70	65.56	18.94	73.38	39.47	23.68	221.03
Fast	50	60	40.88	72.00	25.42	68.13	51.32	15.18	232.05

Measurements are based on the linear feet of each stringer. "Slow" work is based on an hourly wage of $23.00, "Medium" work on an hourly wage of $29.50, and "Fast" work on an hourly wage of $36.00. Other qualifications that apply to this table are on page 9.

	Labor LF per manhour	Material coverage LF/gallon	Material cost per gallon	Labor cost per 100 LF	Labor burden 100 LF	Material cost per 100 LF	Overhead per 100 LF	Profit per 100 LF	Total price per 100 LF
Stair stringers, interior, metal, shapes up to 14" wide, each side									
Metal primer - rust inhibitor, clean metal (material #35)									
Roll & brush prime coat									
Slow	45	130	59.70	51.11	12.26	45.92	20.77	20.81	150.87
Medium	50	125	52.30	59.00	17.06	41.84	29.47	17.68	165.05
Fast	55	120	44.80	65.45	23.10	37.33	39.02	11.54	176.44
Metal primer - rust inhibitor, rusty metal (material #36)									
Roll & brush prime coat									
Slow	45	130	80.10	51.11	12.26	61.62	23.75	23.80	172.54
Medium	50	125	70.10	59.00	17.06	56.08	33.03	19.82	184.99
Fast	55	120	60.10	65.45	23.10	50.08	42.98	12.71	194.32
Metal finish - synthetic enamel, off white (material #37)									
Roll & brush 1st or additional finish coats									
Slow	45	145	63.60	51.11	12.26	43.86	20.38	20.42	148.03
Medium	50	140	55.70	59.00	17.06	39.79	28.96	17.38	162.19
Fast	55	135	47.70	65.45	23.10	35.33	38.40	11.36	173.64
Metal finish - synthetic enamel, colors - except orange/red (material #38)									
Roll & brush 1st or additional finish coats									
Slow	45	145	66.30	51.11	12.26	45.72	20.73	20.77	150.59
Medium	50	140	58.00	59.00	17.06	41.43	29.37	17.62	164.48
Fast	55	135	49.70	65.45	23.10	36.81	38.86	11.50	175.72

Use these figures to paint each side of installed stair stringers. Measurements are based on linear feet of each stringer. Note: A two coat system, prime and finish, using oil base material is recommended for any metal surface. Although water base material is often used, it may cause oxidation, corrosion and rust. Using one coat of oil base paint on exterior metal may result in cracking, peeling or chipping without the proper prime coat application. Pre-primed steel or wrought iron generally requires only one coat to cover. The metal finish figures include minor touchup to the prime coat. If off white or other light colored finish paint is specified, make sure the prime coat is also a light color, or more than one finish coat will be necessary. "Slow" work is based on an hourly wage of $23.00, "Medium" work on an hourly wage of $29.50, and "Fast" work on an hourly wage of $36.00. Other qualifications that apply to this table are on page 9.

	Labor LF per manhour	Material coverage LF/gallon	Material cost per gallon	Labor cost per 100 LF	Labor burden 100 LF	Material cost per 100 LF	Overhead per 100 LF	Profit per 100 LF	Total price per 100 LF

Stair stringers, interior, rough sawn wood up to 4" x 12", each side

Solid body or semi-transparent stain, water or oil base (material #18, #19, #20 or #21)

Roll & brush each coat									
Slow	35	60	54.50	65.71	15.77	90.83	32.74	32.81	237.86
Medium	40	55	47.70	73.75	21.33	86.73	45.45	27.27	254.53
Fast	45	50	40.88	80.00	28.23	81.76	58.90	17.42	266.31

Measurements are based on the linear feet of each side of each stringer. "Slow" work is based on an hourly wage of $23.00, "Medium" work on an hourly wage of $29.50, and "Fast" work on an hourly wage of $36.00. Other qualifications that apply to this table are on page 9.

Stucco: see Plaster and stucco

Touchup, brush as required

	Percentage of interior manhours	**Percentage of interior material costs**
Interior & exterior		
Slow	20.0%	1.0%
Medium	18.0%	1.0%
Fast	15.0%	1.0%
Interior only		
Slow	10.0%	0.5%
Medium	9.0%	0.5%
Fast	7.5%	0.5%

	Percentage of exterior manhours	**Percentage of exterior material costs**
Exterior only		
Slow	10.0%	0.5%
Medium	9.0%	0.5%
Fast	7.5%	0.5%

Touchup will be required on nearly all repaint jobs. Using these percentages is an easy but accurate way to calculate touchup costs. When painting both interiors and exteriors, use the appropriate percentage of interior hours only. When painting either the interior or the exterior of a building, use the appropriate figures as indicated for touchup. Multiply the percentages above times the total manhours and material costs as indicated to allow enough time for production and customer service touchup. The skill of your paint crews and the type of job will determine the time and material needed. To calculate an accurate percentage for your company, use your actual time and material costs for touchup on previous projects (historical costs) and convert this figure into a percentage of the total job cost.

	Labor LF per manhour	Material coverage LF/gallon	Material cost per gallon	Labor cost per 100 LF	Labor burden 100 LF	Material cost per 100 LF	Overhead per 100 LF	Profit per 100 LF	Total price per 100 LF

Trellis or lattice, roll and brush

2" x 2" to 2" x 6", roll & brush all sides, each coat

Solid body or semi-transparent stain, water or oil base (material #18, #19, #20 or #21)

Slow	120	130	54.50	19.17	4.59	41.92	12.48	12.51	90.67
Medium	125	120	47.70	23.60	6.82	39.75	17.54	10.53	98.24
Fast	130	110	40.88	27.69	9.77	37.16	23.13	6.84	104.59

2" x 8" to 4" x 12", roll & brush all sides, each coat

Solid body or semi-transparent stain, water or oil base (material #18, #19, #20 or #21)

Slow	100	100	54.50	23.00	5.52	54.50	15.77	15.81	114.60
Medium	110	90	47.70	26.82	7.75	53.00	21.89	13.14	122.60
Fast	120	80	40.88	30.00	10.58	51.10	28.42	8.41	128.51

Measurements are based on accumulated total linear feet of each trellis or lattice member. These figures are based on roll and brush staining of all four sides and the ends of each member per coat. "Slow" work is based on an hourly wage of $23.00, "Medium" work on an hourly wage of $29.50, and "Fast" work on an hourly wage of $36.00. Other qualifications that apply to this table are on page 9.

	SF surface area per manhour	SF surface area per gallon	Material cost per gallon	Labor cost per 100 SF	Labor burden 100 SF	Material cost per 100 SF	Overhead per 100 SF	Profit per 100 SF	Total price per 100 SF

Trellis or lattice, spray

2" x 2" at 3" on center with 2" x 8" supports, spray all sides, each coat

Solid body or semi-transparent stain, water or oil base (material #18, #19, #20 or #21)

Slow	50	60	54.50	46.00	11.04	90.83	28.10	28.16	204.13
Medium	55	55	47.70	53.64	15.50	86.73	38.97	23.38	218.22
Fast	60	50	40.88	60.00	21.20	81.76	50.51	14.94	228.41

Measurements are based on the square feet of the surface area footprint of the trellis or lattice structure. (The footprint is the surface area seen from the plan or overhead view.) These figures are based on staining all four sides of each member per coat. "Slow" work is based on an hourly wage of $23.00, "Medium" work on an hourly wage of $29.50, and "Fast" work on an hourly wage of $36.00. Other qualifications that apply to this table are on page 9.

	Labor LF per manhour	Material coverage LF/gallon	Material cost per gallon	Labor cost per 100 LF	Labor burden 100 LF	Material cost per 100 LF	Overhead per 100 LF	Profit per 100 LF	Total price per 100 LF
Valances for light fixtures, 2" x 8"									
Solid body or semi-transparent stain, water or oil base (material #18, #19, #20 or #21)									
Brush each coat									
Slow	30	100	54.50	76.67	18.39	54.50	28.42	28.48	206.46
Medium	35	95	47.70	84.29	24.36	50.21	39.72	23.83	222.41
Fast	40	90	40.88	90.00	31.78	45.42	51.83	15.33	234.36

Rough sawn or resawn 2" x 8" wood valances are commonly found in baths and kitchens surrounding light fixtures or supporting plastic cracked-ice diffusers. Measurements are based on the linear feet of the valance. "Slow" work is based on an hourly wage of $23.00, "Medium" work on an hourly wage of $29.50, and "Fast" work on an hourly wage of $36.00. Other qualifications that apply to this table are on page 9.

Walls, Concrete tilt-up: See Industrial, Institutional and Heavy Commercial Painting Costs, page 412

	Labor SF per manhour	Material coverage SF/gallon	Material cost per gallon	Labor cost per 100 SF	Labor burden 100 SF	Material cost per 100 SF	Overhead per 100 SF	Profit per 100 SF	Total price per 100 SF

Walls, gypsum drywall, anti-graffiti stain eliminator, per 100 SF of wall area

Water base primer and sealer (material #39)									
Roll & brush each coat									
Slow	375	450	49.40	6.13	1.48	10.98	3.53	3.54	25.66
Medium	400	425	43.20	7.38	2.13	10.16	4.92	2.95	27.54
Fast	425	400	37.00	8.47	2.98	9.25	6.42	1.90	29.02
Oil base primer and sealer (material #40)									
Roll & brush each coat									
Slow	375	400	58.00	6.13	1.48	14.50	4.20	4.21	30.52
Medium	400	388	50.80	7.38	2.13	13.09	5.65	3.39	31.64
Fast	425	375	43.50	8.47	2.98	11.60	7.15	2.11	32.31
Polyurethane 2 part system (material #41)									
Roll & brush each coat									
Slow	325	400	187.30	7.08	1.70	46.83	10.57	10.59	76.77
Medium	350	375	163.90	8.43	2.45	43.71	13.65	8.19	76.43
Fast	375	350	140.50	9.60	3.41	40.14	16.47	4.87	74.49

Measurements are based on the square feet of wall coated. Do not deduct for openings less than 100 square feet. These figures assume paint products are being applied over a smooth finish. For heights above 8 feet, use the High Time Difficulty Factors on page 139. "Slow" work is based on an hourly wage of $23.00, "Medium" work on an hourly wage of $29.50, and "Fast" work on an hourly wage of $36.00. Other qualifications that apply to this table are on page 9.

	Labor SF per manhour	Material coverage SF/gallon	Material cost per gallon	Labor cost per 100 SF	Labor burden 100 SF	Material cost per 100 SF	Overhead per 100 SF	Profit per 100 SF	Total price per 100 SF
Walls, gypsum drywall, orange peel or knock-down, brush, per 100 SF of wall area									
Flat latex, water base (material #5)									
Brush 1st coat									
Slow	150	300	36.00	15.33	3.69	12.00	5.89	5.90	42.81
Medium	175	288	31.50	16.86	4.86	10.94	8.17	4.90	45.73
Fast	200	275	27.00	18.00	6.36	9.82	10.59	3.13	47.90
Brush 2nd coat									
Slow	175	350	36.00	13.14	3.14	10.29	5.05	5.06	36.68
Medium	200	338	31.50	14.75	4.27	9.32	7.08	4.25	39.67
Fast	225	325	27.00	16.00	5.63	8.31	9.29	2.75	41.98
Brush 3rd or additional coats									
Slow	200	400	36.00	11.50	2.76	9.00	4.42	4.43	32.11
Medium	225	375	31.50	13.11	3.78	8.40	6.33	3.80	35.42
Fast	250	350	27.00	14.40	5.08	7.71	8.43	2.49	38.11
Sealer (drywall), water base (material #1)									
Brush prime coat									
Slow	125	300	36.00	18.40	4.42	12.00	6.62	6.63	48.07
Medium	163	288	31.50	18.10	5.21	10.94	8.57	5.14	47.96
Fast	200	275	27.00	18.00	6.36	9.82	10.59	3.13	47.90
Sealer (drywall), oil base (material #2)									
Brush prime coat									
Slow	125	250	46.20	18.40	4.42	18.48	7.85	7.86	57.01
Medium	163	238	40.50	18.10	5.21	17.02	10.09	6.05	56.47
Fast	200	225	34.70	18.00	6.36	15.42	12.33	3.65	55.76
Enamel, water base (material #9)									
Brush 1st finish coat									
Slow	100	300	53.30	23.00	5.52	17.77	8.80	8.81	63.90
Medium	150	288	46.70	19.67	5.70	16.22	10.39	6.24	58.22
Fast	200	275	40.00	18.00	6.36	14.55	12.06	3.57	54.54
Brush 2nd or additional finish coats									
Slow	125	350	53.30	18.40	4.42	15.23	7.23	7.24	52.52
Medium	163	325	46.70	18.10	5.21	14.37	9.43	5.66	52.77
Fast	200	300	40.00	18.00	6.36	13.33	11.68	3.46	52.83
Enamel, oil base (material #10)									
Brush 1st finish coat									
Slow	100	325	61.50	23.00	5.52	18.92	9.01	9.03	65.48
Medium	150	300	53.80	19.67	5.70	17.93	10.82	6.49	60.61
Fast	200	275	46.10	18.00	6.36	16.76	12.74	3.77	57.63

	Labor SF per manhour	Material coverage SF/gallon	Material cost per gallon	Labor cost per 100 SF	Labor burden 100 SF	Material cost per 100 SF	Overhead per 100 SF	Profit per 100 SF	Total price per 100 SF
Brush 2nd or additional finish coats									
Slow	125	350	61.50	18.40	4.42	17.57	7.67	7.69	55.75
Medium	163	325	53.80	18.10	5.21	16.55	9.97	5.98	55.81
Fast	200	300	46.10	18.00	6.36	15.37	12.31	3.64	55.68
Epoxy coating, 2 part system - white (material #52)									
Brush 1st coat									
Slow	175	350	157.00	13.14	3.14	44.86	11.62	11.64	84.40
Medium	200	325	137.40	14.75	4.27	42.28	15.32	9.19	85.81
Fast	225	300	117.70	16.00	5.63	39.23	18.87	5.58	85.31
Brush 2nd or additional coats									
Slow	200	375	157.00	11.50	2.76	41.87	10.66	10.69	77.48
Medium	225	350	137.40	13.11	3.78	39.26	14.04	8.42	78.61
Fast	250	325	117.70	14.40	5.08	36.22	17.27	5.11	78.08

Measurements are based on the square feet of wall coated. Do not deduct for openings less than 100 square feet. These figures assume paint products are being applied over a smooth finish. For heights above 8 feet, use the High Time Difficulty Factors on page 139. These figures include brushing-in at corners when all walls are the same color, and at ceilings that are the same color or finished with acoustic spray-on texture. ADD for cutting-in at ceilings if they're a different color than the walls, or at corners where walls in the same room are painted different colors. Do not include cutting-in time for ceilings unless you're only painting the ceilings, not the walls. "Slow" work is based on an hourly wage of $23.00, "Medium" work on an hourly wage of $29.50, and "Fast" work on an hourly wage of $36.00. Other qualifications that apply to this table are on page 9.

	Labor SF per manhour	Material coverage SF/gallon	Material cost per gallon	Labor cost per 100 SF	Labor burden 100 SF	Material cost per 100 SF	Overhead per 100 SF	Profit per 100 SF	Total price per 100 SF
Walls, gypsum drywall, orange peel or knock-down, roll, per 100 SF of wall area									
Flat latex, water base (material #5)									
Roll 1st coat									
Slow	400	300	36.00	5.75	1.38	12.00	3.63	3.64	26.40
Medium	538	275	31.50	5.48	1.59	11.45	4.63	2.78	25.93
Fast	675	250	27.00	5.33	1.88	10.80	5.58	1.65	25.24
Roll 2nd coat									
Slow	500	325	36.00	4.60	1.10	11.08	3.19	3.20	23.17
Medium	600	313	31.50	4.92	1.43	10.06	4.10	2.46	22.97
Fast	700	300	27.00	5.14	1.83	9.00	4.94	1.46	22.37
Roll 3rd or additional coats									
Slow	550	350	36.00	4.18	1.01	10.29	2.94	2.95	21.37
Medium	650	338	31.50	4.54	1.32	9.32	3.79	2.28	21.25
Fast	750	325	27.00	4.80	1.68	8.31	4.59	1.36	20.74
Sealer (drywall), water base (material #1)									
Roll prime coat									
Slow	325	275	36.00	7.08	1.70	13.09	4.16	4.16	30.19
Medium	500	263	31.50	5.90	1.71	11.98	4.90	2.94	27.43
Fast	675	250	27.00	5.33	1.88	10.80	5.58	1.65	25.24
Sealer (drywall), oil base (material #2)									
Roll prime coat									
Slow	325	275	46.20	7.08	1.70	16.80	4.86	4.87	35.31
Medium	500	263	40.50	5.90	1.71	15.40	5.75	3.45	32.21
Fast	675	250	34.70	5.33	1.88	13.88	6.54	1.93	29.56
Enamel, water base (material #9)									
Roll 1st finish coat									
Slow	300	285	53.30	7.67	1.83	18.70	5.36	5.37	38.93
Medium	450	263	46.70	6.56	1.88	17.76	6.56	3.93	36.69
Fast	600	240	40.00	6.00	2.13	16.67	7.68	2.27	34.75
Roll 2nd finish coat									
Slow	325	300	53.30	7.08	1.70	17.77	5.04	5.05	36.64
Medium	475	288	46.70	6.21	1.81	16.22	6.06	3.63	33.93
Fast	625	275	40.00	5.76	2.03	14.55	6.93	2.05	31.32
Enamel, oil base (material #10)									
Roll 1st finish coat									
Slow	300	250	61.50	7.67	1.83	24.60	6.48	6.49	47.07
Medium	450	238	53.80	6.56	1.88	22.61	7.77	4.66	43.48
Fast	600	225	46.10	6.00	2.13	20.49	8.87	2.62	40.11

	Labor SF per manhour	Material coverage SF/gallon	Material cost per gallon	Labor cost per 100 SF	Labor burden 100 SF	Material cost per 100 SF	Overhead per 100 SF	Profit per 100 SF	Total price per 100 SF
Roll 2nd finish coat									
Slow	325	275	61.50	7.08	1.70	22.36	5.92	5.93	42.99
Medium	475	263	53.80	6.21	1.81	20.46	7.12	4.27	39.87
Fast	625	250	46.10	5.76	2.03	18.44	8.13	2.41	36.77
Epoxy coating, 2 part system - white (material #52)									
Roll 1st coat									
Slow	325	300	157.00	7.08	1.70	52.33	11.61	11.64	84.36
Medium	488	288	137.40	6.05	1.75	47.71	13.88	8.33	77.72
Fast	700	275	117.70	5.14	1.83	42.80	15.42	4.56	69.75
Roll 2nd or additional coats									
Slow	400	325	157.00	5.75	1.38	48.31	10.53	10.56	76.53
Medium	575	313	137.40	5.13	1.49	43.90	12.63	7.58	70.73
Fast	750	300	117.70	4.80	1.68	39.23	14.17	4.19	64.07

Measurements are based on the square feet of wall coated. Do not deduct for openings less than 100 square feet. These figures assume paint products are being applied over an orange peel or knock-down texture finish. For heights above 8 feet, use the High Time Difficulty Factors on page 139. These figures include brushing-in at corners when all walls are the same color, and at ceilings that are the same color or finished with acoustic spray-on texture. ADD for cutting-in at ceilings if they're a different color than the walls, or at corners where walls in the same room are painted different colors. Do not include cutting-in time for ceilings unless you're only painting the ceilings, not the walls. "Slow" work is based on an hourly wage of $23.00, "Medium" work on an hourly wage of $29.50, and "Fast" work on an hourly wage of $36.00. Other qualifications that apply to this table are on page 9.

	SF of floor area per manhour	SF of floor area per gallon	Material cost per gallon	Labor cost per 100 SF	Labor burden 100 SF	Material cost per 100 SF	Overhead per 100 SF	Profit per 100 SF	Total price per 100 SF
Walls, gypsum drywall, orange peel or knock-down, roll, per 100 SF of floor area									
Flat latex, water base (material #5)									
Roll 1st coat on *walls only*									
Slow	250	175	36.00	9.20	2.21	20.57	6.08	6.09	44.15
Medium	325	158	31.50	9.08	2.63	19.94	7.91	4.75	44.31
Fast	400	140	27.00	9.00	3.18	19.29	9.76	2.89	44.12
Roll 2nd coat on *walls only*									
Slow	300	200	36.00	7.67	1.83	18.00	5.23	5.24	37.97
Medium	400	188	31.50	7.38	2.13	16.76	6.57	3.94	36.78
Fast	500	175	27.00	7.20	2.54	15.43	7.80	2.31	35.28
Sealer (drywall), water base (material #1) on *walls and ceilings*									
Roll prime coat									
Slow	100	100	36.00	23.00	5.52	36.00	12.26	12.28	89.06
Medium	170	88	31.50	17.35	5.01	35.80	14.54	8.72	81.42
Fast	240	75	27.00	15.00	5.31	36.00	17.45	5.16	78.92
Sealer (drywall), oil base (material #2) on *walls and ceilings*									
Roll prime coat									
Slow	100	100	46.20	23.00	5.52	46.20	14.20	14.23	103.15
Medium	170	88	40.50	17.35	5.01	46.02	17.10	10.26	95.74
Fast	240	75	34.70	15.00	5.31	46.27	20.64	6.10	93.32
Enamel, water base (material #9) on *walls and ceilings*									
Roll 1st finish coat									
Slow	70	100	53.30	32.86	7.90	53.30	17.87	17.91	129.84
Medium	100	90	46.70	29.50	8.53	51.89	22.48	13.49	125.89
Fast	135	80	40.00	26.67	9.42	50.00	26.68	7.89	120.66
Roll 2nd finish coat									
Slow	125	150	53.30	18.40	4.42	35.53	11.09	11.11	80.55
Medium	175	125	46.70	16.86	4.86	37.36	14.77	8.86	82.71
Fast	225	100	40.00	16.00	5.63	40.00	19.11	5.65	86.39

	SF of floor area per manhour	SF of floor area per gallon	Material cost per gallon	Labor cost per 100 SF	Labor burden 100 SF	Material cost per 100 SF	Overhead per 100 SF	Profit per 100 SF	Total price per 100 SF
Enamel, oil base (material #10) on walls and ceilings									
Roll 1st finish coat									
Slow	70	100	61.50	32.86	7.90	61.50	19.43	19.47	141.16
Medium	100	90	53.80	29.50	8.53	59.78	24.45	14.67	136.93
Fast	135	80	46.10	26.67	9.42	57.63	29.05	8.59	131.36
Roll 2nd finish coat									
Slow	125	150	61.50	18.40	4.42	41.00	12.13	12.15	88.10
Medium	175	125	53.80	16.86	4.86	43.04	16.19	9.72	90.67
Fast	225	100	46.10	16.00	5.63	46.10	21.00	6.21	94.94

Measurements for these costs are based on square feet of floor area. The flat wall figures are for painting walls only but the Sealer and Enamel figures are for painting walls and ceilings in wet areas, i.e. kitchens, baths, utility areas, etc. The floor area measurements are from outside wall to outside wall or from the edge of the concrete slab or from the outside edge of an interior wall. This method of figuring the costs to paint the walls and ceilings is not as accurate as measuring the actual surface area of the wall or ceiling area directly, but it is much less time consuming. For heights above 8 feet, use the High Time Difficulty Factors on page 139. These figures include brushing-in at corners when all walls are the same color, and at ceilings that are the same color or finished with acoustic spray-on texture. ADD for cutting-in at ceilings if they're a different color than the walls, or at corners where walls in the same room are painted different colors. Do not include cutting-in time for ceilings unless you're only painting the ceilings, not the walls. "Slow" work is based on an hourly wage of $23.00, "Medium" work on an hourly wage of $29.50, and "Fast" work on an hourly wage of $36.00. Other qualifications that apply to this table are on page 9.

	Labor SF per manhour	Material coverage SF/gallon	Material cost per gallon	Labor cost per 100 SF	Labor burden 100 SF	Material cost per 100 SF	Overhead per 100 SF	Profit per 100 SF	Total price per 100 SF
Walls, gypsum drywall, orange peel or knock-down, spray, per 100 SF of wall area									
Flat latex, water base (material #5)									
Spray 1st coat									
Slow	700	250	36.00	3.29	.79	14.40	3.51	3.52	25.51
Medium	800	225	31.50	3.69	1.06	14.00	4.69	2.81	26.25
Fast	900	200	27.00	4.00	1.41	13.50	5.86	1.73	26.50
Spray 2nd coat									
Slow	800	300	36.00	2.88	.69	12.00	2.96	2.96	21.49
Medium	900	275	31.50	3.28	.94	11.45	3.92	2.35	21.94
Fast	1000	250	27.00	3.60	1.27	10.80	4.86	1.44	21.97
Spray 3rd or additional coats									
Slow	850	325	36.00	2.71	.66	11.08	2.74	2.75	19.94
Medium	950	300	31.50	3.11	.88	10.50	3.63	2.18	20.30
Fast	1050	275	27.00	3.43	1.20	9.82	4.48	1.33	20.26
Sealer (drywall), water base (material #1)									
Spray prime coat									
Slow	575	250	36.00	4.00	.96	14.40	3.68	3.69	26.73
Medium	738	225	31.50	4.00	1.17	14.00	4.79	2.87	26.83
Fast	900	200	27.00	4.00	1.41	13.50	5.86	1.73	26.50
Sealer (drywall), oil base (material #2)									
Spray prime coat									
Slow	575	250	46.20	4.00	.96	18.48	4.45	4.46	32.35
Medium	738	225	40.50	4.00	1.17	18.00	5.79	3.47	32.43
Fast	900	200	34.70	4.00	1.41	17.35	7.06	2.09	31.91
Enamel, water base (material #9)									
Spray 1st finish coat									
Slow	500	250	53.30	4.60	1.10	21.32	5.13	5.14	37.29
Medium	675	238	46.70	4.37	1.26	19.62	6.31	3.79	35.35
Fast	850	225	40.00	4.24	1.51	17.78	7.29	2.16	32.98
Spray 2nd finish coat									
Slow	525	275	53.30	4.38	1.04	19.38	4.71	4.72	34.23
Medium	700	263	46.70	4.21	1.23	17.76	5.80	3.48	32.48
Fast	875	250	40.00	4.11	1.44	16.00	6.68	1.98	30.21
Spray 3rd or additional finish coats									
Slow	575	300	53.30	4.00	.96	17.77	4.32	4.33	31.38
Medium	775	275	46.70	3.81	1.10	16.98	5.47	3.28	30.64
Fast	925	250	40.00	3.89	1.37	16.00	6.59	1.95	29.80

	Labor SF per manhour	Material coverage SF/gallon	Material cost per gallon	Labor cost per 100 SF	Labor burden 100 SF	Material cost per 100 SF	Overhead per 100 SF	Profit per 100 SF	Total price per 100 SF
Enamel, oil base (material #10)									
Spray 1st finish coat									
Slow	500	250	61.50	4.60	1.10	24.60	5.76	5.77	41.83
Medium	675	225	53.80	4.37	1.26	23.91	7.39	4.43	41.36
Fast	850	200	46.10	4.24	1.51	23.05	8.92	2.64	40.36
Spray 2nd finish coat									
Slow	525	275	61.50	4.38	1.04	22.36	5.28	5.29	38.35
Medium	700	250	53.80	4.21	1.23	21.52	6.74	4.04	37.74
Fast	875	225	46.10	4.11	1.44	20.49	8.08	2.39	36.51
Spray 3rd or additional finish coat									
Slow	575	300	61.50	4.00	.96	20.50	4.84	4.85	35.15
Medium	775	275	53.80	3.81	1.10	19.56	6.12	3.67	34.26
Fast	925	250	46.10	3.89	1.37	18.44	7.35	2.17	33.22

Measurements are based on the square feet of wall coated. Do not deduct for openings less than 100 square feet. These figures assume paint products are being applied over a smooth finish. For heights above 8 feet, use the High Time Difficulty Factors on page 139. These figures include spraying at corners when all walls are the same color, and at ceiling-to-wall intersection when ceilings are the same color. ADD for cutting-in at ceilings and protecting adjacent surfaces from overspray if they're a different color than the walls, or at corners where walls in the same room are painted different colors. Do not include cutting-in time for ceilings unless you're only painting the ceilings, not the walls. "Slow" work is based on an hourly wage of $23.00, "Medium" work on an hourly wage of $29.50, and "Fast" work on hourly wage of $36.00. Other qualifications that apply to this table are on page 9.

	Labor SF per manhour	Material coverage SF/gallon	Material cost per gallon	Labor cost per 100 SF	Labor burden 100 SF	Material cost per 100 SF	Overhead per 100 SF	Profit per 100 SF	Total price per 100 SF
Walls, gypsum drywall, skip trowel or sand finish, brush, per 100 SF of wall area									
Flat latex, water base (material #5)									
Brush 1st coat									
Slow	175	325	36.00	13.14	3.14	11.08	5.20	5.21	37.77
Medium	200	313	31.50	14.75	4.27	10.06	7.27	4.36	40.71
Fast	225	300	27.00	16.00	5.63	9.00	9.50	2.81	42.94
Brush 2nd coat									
Slow	200	400	36.00	11.50	2.76	9.00	4.42	4.43	32.11
Medium	225	375	31.50	13.11	3.78	8.40	6.33	3.80	35.42
Fast	250	350	27.00	14.40	5.08	7.71	8.43	2.49	38.11
Brush 3rd or additional coats									
Slow	225	425	36.00	10.22	2.44	8.47	4.02	4.03	29.18
Medium	250	400	31.50	11.80	3.41	7.88	5.77	3.46	32.32
Fast	275	375	27.00	13.09	4.64	7.20	7.72	2.28	34.93
Sealer (drywall), water base (material #1)									
Brush prime coat									
Slow	140	325	36.00	16.43	3.93	11.08	5.98	5.99	43.41
Medium	183	313	31.50	16.12	4.64	10.06	7.71	4.63	43.16
Fast	225	300	27.00	16.00	5.63	9.00	9.50	2.81	42.94
Sealer (drywall), oil base (material #2)									
Brush prime coat									
Slow	140	350	46.20	16.43	3.93	13.20	6.38	6.39	46.33
Medium	183	338	40.50	16.12	4.64	11.98	8.19	4.91	45.84
Fast	225	325	34.70	16.00	5.63	10.68	10.02	2.96	45.29
Enamel, water base (material #9)									
Brush 1st finish coat									
Slow	125	350	53.30	18.40	4.42	15.23	7.23	7.24	52.52
Medium	175	325	46.70	16.86	4.86	14.37	9.03	5.42	50.54
Fast	225	300	40.00	16.00	5.63	13.33	10.84	3.21	49.01
Brush 2nd or additional finish coats									
Slow	140	350	53.30	16.43	3.93	15.23	6.76	6.78	49.13
Medium	185	325	46.70	15.95	4.62	14.37	8.73	5.24	48.91
Fast	235	300	40.00	15.32	5.43	13.33	10.56	3.12	47.76
Enamel, oil base (material #10)									
Brush 1st finish coat									
Slow	125	350	61.50	18.40	4.42	17.57	7.67	7.69	55.75
Medium	175	325	53.80	16.86	4.86	16.55	9.57	5.74	53.58
Fast	225	300	46.10	16.00	5.63	15.37	11.48	3.40	51.88

	Labor SF per manhour	Material coverage SF/gallon	Material cost per gallon	Labor cost per 100 SF	Labor burden 100 SF	Material cost per 100 SF	Overhead per 100 SF	Profit per 100 SF	Total price per 100 SF
Brush 2nd or additional finish coats									
Slow	140	350	61.50	16.43	3.93	17.57	7.21	7.22	52.36
Medium	185	338	53.80	15.95	4.62	15.92	9.12	5.47	51.08
Fast	235	325	46.10	15.32	5.43	14.18	10.82	3.20	48.95
Epoxy coating, 2 part system - white (material #52)									
Brush 1st coat									
Slow	200	375	157.00	11.50	2.76	41.87	10.66	10.69	77.48
Medium	225	350	137.40	13.11	3.78	39.26	14.04	8.42	78.61
Fast	250	325	117.70	14.40	5.08	36.22	17.27	5.11	78.08
Brush 2nd or additional coats									
Slow	225	400	157.00	10.22	2.44	39.25	9.86	9.88	71.65
Medium	250	375	137.40	11.80	3.41	36.64	12.96	7.78	72.59
Fast	275	350	117.70	13.09	4.64	33.63	15.92	4.71	71.99

Measurements are based on the square feet of wall coated. Do not deduct for openings less than 100 square feet. These figures assume paint products are being applied over a smooth finish. For heights above 8 feet, use the High Time Difficulty Factors on page 139. These figures include brushing-in at corners when all walls are the same color, and at ceilings that are the same color or finished with acoustic spray-on texture. ADD for cutting-in at ceilings if they're a different color than the walls, or at corners where walls in the same room are painted different colors. Do not include cutting-in time for ceilings unless you're only painting the ceilings, not the walls. "Slow" work is based on an hourly wage of $23.00, "Medium" work on an hourly wage of $29.50, and "Fast" work on an hourly wage of $36.00. Other qualifications that apply to this table are on page 9.

	Labor SF per manhour	Material coverage SF/gallon	Material cost per gallon	Labor cost per 100 SF	Labor burden 100 SF	Material cost per 100 SF	Overhead per 100 SF	Profit per 100 SF	Total price per 100 SF
Walls, gypsum drywall, skip trowel or sand finish, roll, per 100 SF of wall area									
Flat latex, water base (material #5)									
Roll 1st coat									
Slow	275	325	36.00	8.36	2.02	11.08	4.08	4.08	29.62
Medium	488	300	31.50	6.05	1.75	10.50	4.58	2.75	25.63
Fast	700	275	27.00	5.14	1.83	9.82	5.20	1.54	23.53
Roll 2nd coat									
Slow	350	350	36.00	6.57	1.59	10.29	3.50	3.51	25.46
Medium	538	338	31.50	5.48	1.59	9.32	4.10	2.46	22.95
Fast	725	325	27.00	4.97	1.75	8.31	4.66	1.38	21.07
Roll 3rd or additional coats									
Slow	425	350	36.00	5.41	1.29	10.29	3.23	3.24	23.46
Medium	600	338	31.50	4.92	1.43	9.32	3.92	2.35	21.94
Fast	775	325	27.00	4.65	1.63	8.31	4.53	1.34	20.46
Sealer (drywall), water base (material #1)									
Roll prime coat									
Slow	225	325	36.00	10.22	2.44	11.08	4.51	4.52	32.77
Medium	463	300	31.50	6.37	1.84	10.50	4.68	2.81	26.20
Fast	700	275	27.00	5.14	1.83	9.82	5.20	1.54	23.53
Sealer (drywall), oil base (material #2)									
Roll prime coat									
Slow	225	300	46.20	10.22	2.44	15.40	5.33	5.34	38.73
Medium	463	275	40.50	6.37	1.84	14.73	5.74	3.44	32.12
Fast	700	250	34.70	5.14	1.83	13.88	6.46	1.91	29.22
Enamel, water base (material #9)									
Roll 1st finish coat									
Slow	225	300	53.30	10.22	2.44	17.77	5.78	5.80	42.01
Medium	400	288	46.70	7.38	2.13	16.22	6.43	3.86	36.02
Fast	600	275	40.00	6.00	2.13	14.55	7.03	2.08	31.79
Roll 2nd or additional finish coats									
Slow	275	300	53.30	8.36	2.02	17.77	5.35	5.36	38.86
Medium	450	288	46.70	6.56	1.88	16.22	6.17	3.70	34.53
Fast	650	275	40.00	5.54	1.96	14.55	6.84	2.02	30.91
Enamel, oil base (material #10)									
Roll 1st finish coat									
Slow	225	275	61.50	10.22	2.44	22.36	6.66	6.67	48.35
Medium	400	263	53.80	7.38	2.13	20.46	7.49	4.50	41.96
Fast	600	250	46.10	6.00	2.13	18.44	8.23	2.44	37.24

	Labor SF per manhour	Material coverage SF/gallon	Material cost per gallon	Labor cost per 100 SF	Labor burden 100 SF	Material cost per 100 SF	Overhead per 100 SF	Profit per 100 SF	Total price per 100 SF
Roll 2nd or additional finish coats									
Slow	275	300	61.50	8.36	2.02	20.50	5.87	5.88	42.63
Medium	450	288	53.80	6.56	1.88	18.68	6.79	4.07	37.98
Fast	650	275	46.10	5.54	1.96	16.76	7.52	2.22	34.00
Epoxy coating, 2 part system - white (material #52)									
Roll 1st coat									
Slow	350	350	157.00	6.57	1.59	44.86	10.07	10.09	73.18
Medium	550	325	137.40	5.36	1.56	42.28	12.30	7.38	68.88
Fast	725	300	117.70	4.97	1.75	39.23	14.24	4.21	64.40
Roll 2nd or additional coats									
Slow	425	375	157.00	5.41	1.29	41.87	9.23	9.25	67.05
Medium	600	350	137.40	4.92	1.43	39.26	11.40	6.84	63.85
Fast	775	325	117.70	4.65	1.63	36.22	13.18	3.90	59.58

Measurements are based on the square feet of wall coated. Do not deduct for openings less than 100 square feet. These figures assume paint products are being applied over a smooth finish. For heights above 8 feet, use the High Time Difficulty Factors on page 139. These figures include brushing-in at corners when all walls are the same color, and at ceilings that are the same color or finished with acoustic spray-on texture. ADD for cutting-in at ceilings if they're a different color than the walls, or at corners where walls in the same room are painted different colors. Do not include cutting-in time for ceilings unless you're only painting the ceilings, not the walls. "Slow" work is based on an hourly wage of $23.00, "Medium" work on an hourly wage of $29.50, and "Fast" work on an hourly wage of $36.00. Other qualifications that apply to this table are on page 9.

	Labor SF per manhour	Material coverage SF/gallon	Material cost per gallon	Labor cost per 100 SF	Labor burden 100 SF	Material cost per 100 SF	Overhead per 100 SF	Profit per 100 SF	Total price per 100 SF
Walls, gypsum drywall, skip trowel or sand finish, spray, per 100 SF of wall area									
Flat latex, water base (material #5)									
Spray 1st coat									
Slow	700	275	36.00	3.29	.79	13.09	3.26	3.27	23.70
Medium	800	250	31.50	3.69	1.06	12.60	4.34	2.60	24.29
Fast	900	225	27.00	4.00	1.41	12.00	5.40	1.60	24.41
Spray 2nd coat									
Slow	800	325	36.00	2.88	.69	11.08	2.78	2.79	20.22
Medium	900	300	31.50	3.28	.94	10.50	3.68	2.21	20.61
Fast	1000	275	27.00	3.60	1.27	9.82	4.55	1.35	20.59
Spray 3rd or additional coats									
Slow	850	325	36.00	2.71	.66	11.08	2.74	2.75	19.94
Medium	950	313	31.50	3.11	.88	10.06	3.52	2.11	19.68
Fast	1050	300	27.00	3.43	1.20	9.00	4.23	1.25	19.11
Sealer (drywall), water base (material #1)									
Spray prime coat									
Slow	575	275	36.00	4.00	.96	13.09	3.43	3.44	24.92
Medium	738	250	31.50	4.00	1.17	12.60	4.44	2.66	24.87
Fast	900	225	27.00	4.00	1.41	12.00	5.40	1.60	24.41
Sealer (drywall), oil base (material #2)									
Spray prime coat									
Slow	575	275	46.20	4.00	.96	16.80	4.13	4.14	30.03
Medium	738	250	40.50	4.00	1.17	16.20	5.34	3.20	29.91
Fast	900	225	34.70	4.00	1.41	15.42	6.46	1.91	29.20
Enamel, water base (material #9)									
Spray 1st finish coat									
Slow	500	275	53.30	4.60	1.10	19.38	4.77	4.78	34.63
Medium	675	250	46.70	4.37	1.26	18.68	6.08	3.65	34.04
Fast	850	225	40.00	4.24	1.51	17.78	7.29	2.16	32.98
Spray 2nd or additional finish coats									
Slow	525	275	53.30	4.38	1.04	19.38	4.71	4.72	34.23
Medium	700	263	46.70	4.21	1.23	17.76	5.80	3.48	32.48
Fast	900	250	40.00	4.00	1.41	16.00	6.64	1.96	30.01

	Labor SF per manhour	Material coverage SF/gallon	Material cost per gallon	Labor cost per 100 SF	Labor burden 100 SF	Material cost per 100 SF	Overhead per 100 SF	Profit per 100 SF	Total price per 100 SF
Enamel, oil base (material #10)									
Spray 1st finish coat									
Slow	500	250	61.50	4.60	1.10	24.60	5.76	5.77	41.83
Medium	675	238	53.80	4.37	1.26	22.61	7.06	4.24	39.54
Fast	850	225	46.10	4.24	1.51	20.49	8.13	2.41	36.78
Spray 2nd or additional finish coats									
Slow	525	275	61.50	4.38	1.04	22.36	5.28	5.29	38.35
Medium	700	263	53.80	4.21	1.23	20.46	6.47	3.88	36.25
Fast	900	250	46.10	4.00	1.41	18.44	7.39	2.19	33.43

Measurements are based on the square feet of wall coated. Do not deduct for openings less than 100 square feet. These figures assume paint products are being applied over a smooth finish. For heights above 8 feet, use the High Time Difficulty Factors on page 139. These figures include spraying at corners when all walls are the same color, and at ceiling-to-wall intersection when ceilings are the same color. ADD for cutting-in at ceilings and protecting adjacent surfaces from overspray if they're a different color than the walls, or at corners where walls in the same room are painted different colors. Do not include cutting-in time for ceilings unless you're only painting the ceilings, not the walls. "Slow" work is based on an hourly wage of $23.00, "Medium" work on an hourly wage of $29.50, and "Fast" work on an hourly wage of $36.00. Other qualifications that apply to this table are on page 9.

	Labor SF per manhour	Material coverage SF/gallon	Material cost per gallon	Labor cost per 100 SF	Labor burden 100 SF	Material cost per 100 SF	Overhead per 100 SF	Profit per 100 SF	Total price per 100 SF
Walls, gypsum drywall, smooth-wall finish, brush, per 100 SF of wall area									
Flat latex, water base (material #5)									
Brush 1st coat									
Slow	175	325	36.00	13.14	3.14	11.08	5.20	5.21	37.77
Medium	200	313	31.50	14.75	4.27	10.06	7.27	4.36	40.71
Fast	225	300	27.00	16.00	5.63	9.00	9.50	2.81	42.94
Brush 2nd coat									
Slow	225	400	36.00	10.22	2.44	9.00	4.12	4.13	29.91
Medium	250	375	31.50	11.80	3.41	8.40	5.90	3.54	33.05
Fast	275	350	27.00	13.09	4.64	7.71	7.88	2.33	35.65
Brush 3rd or additional coats									
Slow	250	425	36.00	9.20	2.21	8.47	3.78	3.79	27.45
Medium	275	400	31.50	10.73	3.11	7.88	5.43	3.26	30.41
Fast	300	375	27.00	12.00	4.22	7.20	7.27	2.15	32.84
Sealer (drywall), water base (material #1)									
Brush prime coat									
Slow	150	325	36.00	15.33	3.69	11.08	5.72	5.73	41.55
Medium	188	313	31.50	15.69	4.54	10.06	7.57	4.54	42.40
Fast	225	300	27.00	16.00	5.63	9.00	9.50	2.81	42.94
Sealer (drywall), oil base (material #2)									
Brush prime coat									
Slow	150	350	46.20	15.33	3.69	13.20	6.12	6.13	44.47
Medium	188	338	40.50	15.69	4.54	11.98	8.05	4.83	45.09
Fast	225	325	34.70	16.00	5.63	10.68	10.02	2.96	45.29
Enamel, water base (material #9)									
Brush 1st finish coat									
Slow	140	350	53.30	16.43	3.93	15.23	6.76	6.78	49.13
Medium	185	325	46.70	15.95	4.62	14.37	8.73	5.24	48.91
Fast	235	300	40.00	15.32	5.43	13.33	10.56	3.12	47.76
Brush 2nd or additional finish coats									
Slow	150	350	53.30	15.33	3.69	15.23	6.51	6.52	47.28
Medium	200	333	46.70	14.75	4.27	14.02	8.26	4.95	46.25
Fast	250	315	40.00	14.40	5.08	12.70	9.98	2.95	45.11
Enamel, oil base (material #10)									
Brush 1st finish coat									
Slow	140	350	61.50	16.43	3.93	17.57	7.21	7.22	52.36
Medium	185	338	53.80	15.95	4.62	15.92	9.12	5.47	51.08
Fast	235	325	46.10	15.32	5.43	14.18	10.82	3.20	48.95

	Labor SF per manhour	Material coverage SF/gallon	Material cost per gallon	Labor cost per 100 SF	Labor burden 100 SF	Material cost per 100 SF	Overhead per 100 SF	Profit per 100 SF	Total price per 100 SF
Brush 2nd or additional finish coats									
Slow	150	360	61.50	15.33	3.69	17.08	6.86	6.87	49.83
Medium	200	348	53.80	14.75	4.27	15.46	8.62	5.17	48.27
Fast	250	335	46.10	14.40	5.08	13.76	10.30	3.05	46.59
Stipple finish									
Slow	225	--	--	10.22	2.44	--	2.41	2.41	17.48
Medium	250	--	--	11.80	3.41	--	3.80	2.28	21.29
Fast	275	--	--	13.09	4.64	--	5.49	1.62	24.84
Epoxy coating, 2 part system - white (material #52)									
Brush 1st coat									
Slow	225	425	157.00	10.22	2.44	36.94	9.43	9.45	68.48
Medium	250	400	137.40	11.80	3.41	34.35	12.39	7.43	69.38
Fast	275	375	117.70	13.09	4.64	31.39	15.22	4.50	68.84
Brush 2nd or additional coats									
Slow	250	450	157.00	9.20	2.21	34.89	8.80	8.82	63.92
Medium	275	425	137.40	10.73	3.11	32.33	11.54	6.92	64.63
Fast	300	400	117.70	12.00	4.22	29.43	14.16	4.19	64.00

Measurements are based on the square feet of wall coated. Do not deduct for openings less than 100 square feet. These figures assume paint products are being applied over a smooth finish. For heights above 8 feet, use the High Time Difficulty Factors on page 139. These figures include brushing-in at corners when all walls are the same color, and at ceilings that are the same color or finished with acoustic spray-on texture. ADD for cutting-in at ceilings if they're a different color than the walls, or at corners where walls in the same room are painted different colors. Do not include cutting-in time for ceilings unless you're only painting the ceilings, not the walls. "Slow" work is based on an hourly wage of $23.00, "Medium" work on an hourly wage of $29.50, and "Fast" work on an hourly wage of $36.00. Other qualifications that apply to this table are on page 9.

	Labor SF per manhour	Material coverage SF/gallon	Material cost per gallon	Labor cost per 100 SF	Labor burden 100 SF	Material cost per 100 SF	Overhead per 100 SF	Profit per 100 SF	Total price per 100 SF
Walls, gypsum drywall, smooth-wall finish, roll, per 100 SF of wall area									
Flat latex, water base (material #5)									
Roll 1st coat									
Slow	300	325	36.00	7.67	1.83	11.08	3.91	3.92	28.41
Medium	513	313	31.50	5.75	1.67	10.06	4.37	2.62	24.47
Fast	725	300	27.00	4.97	1.75	9.00	4.87	1.44	22.03
Roll 2nd coat									
Slow	375	375	36.00	6.13	1.48	9.60	3.27	3.28	23.76
Medium	563	363	31.50	5.24	1.53	8.68	3.86	2.31	21.62
Fast	750	350	27.00	4.80	1.68	7.71	4.40	1.30	19.89
Roll 3rd or additional coats									
Slow	450	400	36.00	5.11	1.22	9.00	2.91	2.92	21.16
Medium	625	388	31.50	4.72	1.36	8.12	3.55	2.13	19.88
Fast	800	375	27.00	4.50	1.59	7.20	4.12	1.22	18.63
Sealer (drywall), water base (material #1)									
Roll prime coat									
Slow	245	350	36.00	9.39	2.25	10.29	4.17	4.18	30.28
Medium	485	325	31.50	6.08	1.75	9.69	4.38	2.63	24.53
Fast	725	300	27.00	4.97	1.75	9.00	4.87	1.44	22.03
Sealer (drywall), oil base (material #2)									
Roll prime coat									
Slow	245	325	46.20	9.39	2.25	14.22	4.91	4.92	35.69
Medium	485	300	40.50	6.08	1.75	13.50	5.34	3.20	29.87
Fast	725	275	34.70	4.97	1.75	12.62	6.00	1.77	27.11
Enamel, water base (material #9)									
Roll 1st finish coat									
Slow	235	325	53.30	9.79	2.36	16.40	5.42	5.43	39.40
Medium	438	313	46.70	6.74	1.93	14.92	5.90	3.54	33.03
Fast	640	300	40.00	5.63	1.97	13.33	6.49	1.92	29.34
Roll 2nd or additional finish coats									
Slow	280	350	53.30	8.21	1.97	15.23	4.83	4.84	35.08
Medium	465	338	46.70	6.34	1.84	13.82	5.50	3.30	30.80
Fast	680	325	40.00	5.29	1.87	12.31	6.04	1.79	27.30
Enamel, oil base (material #10)									
Roll 1st finish coat									
Slow	235	300	61.50	9.79	2.36	20.50	6.20	6.21	45.06
Medium	438	288	53.80	6.74	1.93	18.68	6.84	4.11	38.30
Fast	640	275	46.10	5.63	1.97	16.76	7.56	2.24	34.16

	Labor SF per manhour	Material coverage SF/gallon	Material cost per gallon	Labor cost per 100 SF	Labor burden 100 SF	Material cost per 100 SF	Overhead per 100 SF	Profit per 100 SF	Total price per 100 SF
Roll 2nd or additional finish coats									
Slow	280	325	61.50	8.21	1.97	18.92	5.53	5.54	40.17
Medium	465	313	53.80	6.34	1.84	17.19	6.34	3.80	35.51
Fast	680	300	46.10	5.29	1.87	15.37	6.98	2.07	31.58
Epoxy coating, 2 part system - white (material #52)									
Roll 1st coat									
Slow	375	400	157.00	6.13	1.48	39.25	8.90	8.92	64.68
Medium	550	375	137.40	5.36	1.56	36.64	10.89	6.53	60.98
Fast	750	350	117.70	4.80	1.68	33.63	12.44	3.68	56.23
Roll 2nd or additional coats									
Slow	450	425	157.00	5.11	1.22	36.94	8.22	8.24	59.73
Medium	625	400	137.40	4.72	1.36	34.35	10.11	6.06	56.60
Fast	800	375	117.70	4.50	1.59	31.39	11.62	3.44	52.54

Measurements are based on the square feet of wall coated. Do not deduct for openings less than 100 square feet. These figures assume paint products are being applied over a smooth finish. For heights above 8 feet, use the High Time Difficulty Factors on page 139. These figures include brushing-in at corners when all walls are the same color, and at ceilings that are the same color or finished with acoustic spray-on texture. ADD for cutting-in at ceilings if they're a different color than the walls, or at corners where walls in the same room are painted different colors. Do not include cutting-in time for ceilings unless you're only painting the ceilings, not the walls. "Slow" work is based on an hourly wage of $23.00, "Medium" work on an hourly wage of $29.50, and "Fast" work on an hourly wage of $36.00. Other qualifications that apply to this table are on page 9.

	Labor SF per manhour	Material coverage SF/gallon	Material cost per gallon	Labor cost per 100 SF	Labor burden 100 SF	Material cost per 100 SF	Overhead per 100 SF	Profit per 100 SF	Total price per 100 SF
Walls, gypsum drywall, smooth-wall finish, spray, per 100 SF of wall area									
Flat latex, water base (material #5)									
Spray 1st coat									
Slow	750	300	36.00	3.07	.72	12.00	3.00	3.01	21.80
Medium	850	275	31.50	3.47	1.02	11.45	3.98	2.39	22.31
Fast	950	250	27.00	3.79	1.32	10.80	4.94	1.46	22.31
Spray 2nd coat									
Slow	850	350	36.00	2.71	.66	10.29	2.59	2.60	18.85
Medium	950	325	31.50	3.11	.88	9.69	3.43	2.06	19.17
Fast	1050	300	27.00	3.43	1.20	9.00	4.23	1.25	19.11
Spray 3rd or additional coats									
Slow	950	375	36.00	2.42	.57	9.60	2.39	2.40	17.38
Medium	1050	350	31.50	2.81	.80	9.00	3.16	1.89	17.66
Fast	1150	325	27.00	3.13	1.11	8.31	3.89	1.15	17.59
Sealer (drywall), water base (material #1)									
Spray prime coat									
Slow	600	300	36.00	3.83	.93	12.00	3.18	3.19	23.13
Medium	775	275	31.50	3.81	1.10	11.45	4.09	2.45	22.90
Fast	950	250	27.00	3.79	1.32	10.80	4.94	1.46	22.31
Sealer (drywall), oil base (material #2)									
Spray prime coat									
Slow	600	275	46.20	3.83	.93	16.80	4.09	4.10	29.75
Medium	775	250	40.50	3.81	1.10	16.20	5.28	3.17	29.56
Fast	950	225	34.70	3.79	1.32	15.42	6.37	1.88	28.78
Enamel, water base (material #9)									
Spray 1st finish coat									
Slow	525	300	53.30	4.38	1.04	17.77	4.41	4.42	32.02
Medium	713	275	46.70	4.14	1.18	16.98	5.58	3.35	31.23
Fast	900	250	40.00	4.00	1.41	16.00	6.64	1.96	30.01
Spray 2nd or additional finish coats									
Slow	600	300	53.30	3.83	.93	17.77	4.28	4.29	31.10
Medium	788	288	46.70	3.74	1.09	16.22	5.26	3.16	29.47
Fast	975	275	40.00	3.69	1.33	14.55	6.06	1.79	27.42

	Labor SF per manhour	Material coverage SF/gallon	Material cost per gallon	Labor cost per 100 SF	Labor burden 100 SF	Material cost per 100 SF	Overhead per 100 SF	Profit per 100 SF	Total price per 100 SF
Enamel, oil base (material #10)									
Spray 1st finish coat									
Slow	525	275	61.50	4.38	1.04	22.36	5.28	5.29	38.35
Medium	713	263	53.80	4.14	1.18	20.46	6.45	3.87	36.10
Fast	900	250	46.10	4.00	1.41	18.44	7.39	2.19	33.43
Spray 2nd or additional finish coats									
Slow	600	300	61.50	3.83	.93	20.50	4.80	4.81	34.87
Medium	788	288	53.80	3.74	1.09	18.68	5.88	3.53	32.92
Fast	975	275	46.10	3.69	1.33	16.76	6.74	1.99	30.51

Measurements are based on the square feet of wall coated. Do not deduct for openings less than 100 square feet. These figures assume paint products are being applied over a smooth finish. For heights above 8 feet, use the High Time Difficulty Factors on page 139. These figures include spraying at corners when all walls are the same color, and at ceiling-to-wall intersection when ceilings are the same color. ADD for cutting-in at ceilings and protecting adjacent surfaces from overspray if they're a different color than the walls, or at corners where walls in the same room are painted different colors. Do not include cutting-in time for ceilings unless you're only painting the ceilings, not the walls. "Slow" work is based on an hourly wage of $23.00, "Medium" work on an hourly wage of $29.50, and "Fast" work on an hourly wage of $36.00. Other qualifications that apply to this table are on page 9.

Walls, plaster, exterior: see Plaster and stucco

	Labor SF per manhour	Material coverage SF/gallon	Material cost per gallon	Labor cost per 100 SF	Labor burden 100 SF	Material cost per 100 SF	Overhead per 100 SF	Profit per 100 SF	Total price per 100 SF
Walls, plaster, interior, medium texture, per 100 SF of wall area									
Anti-graffiti stain eliminator									
Water base primer and sealer (material #39)									
Roll & brush each coat									
Slow	375	425	49.40	6.13	1.48	11.62	3.65	3.66	26.54
Medium	400	400	43.20	7.38	2.13	10.80	5.08	3.05	28.44
Fast	425	375	37.00	8.47	2.98	9.87	6.61	1.96	29.89
Oil base primer and sealer (material #40)									
Roll & brush each coat									
Slow	375	400	58.00	6.13	1.48	14.50	4.20	4.21	30.52
Medium	400	375	50.80	7.38	2.13	13.55	5.77	3.46	32.29
Fast	425	350	43.50	8.47	2.98	12.43	7.41	2.19	33.48
Polyurethane 2 part system (material #41)									
Roll & brush each coat									
Slow	325	375	187.30	7.08	1.70	49.95	11.16	11.18	81.07
Medium	350	350	163.90	8.43	2.45	46.83	14.43	8.66	80.80
Fast	375	325	140.50	9.60	3.41	43.23	17.43	5.16	78.83

Measurements are based on the square feet of wall coated. Do not deduct for openings less than 100 square feet. These figures assume paint products are being applied over a smooth or medium texture finish. For heights above 8 feet, use the High Time Difficulty Factors on page 139. "Slow" work is based on an hourly wage of $23.00, "Medium" work on an hourly wage of $29.50, and "Fast" work on an hourly wage of $36.00. Other qualifications that apply to this table are on page 9.

	Labor SF per manhour	Material coverage SF/gallon	Material cost per gallon	Labor cost per 100 SF	Labor burden 100 SF	Material cost per 100 SF	Overhead per 100 SF	Profit per 100 SF	Total price per 100 SF
Walls, plaster, interior, medium texture, brush, per 100 SF of wall area									
Flat latex, water base (material #5)									
Brush 1st coat									
Slow	125	300	36.00	18.40	4.42	12.00	6.62	6.63	48.07
Medium	150	288	31.50	19.67	5.70	10.94	9.07	5.44	50.82
Fast	175	275	27.00	20.57	7.24	9.82	11.67	3.45	52.75
Brush 2nd coat									
Slow	150	325	36.00	15.33	3.69	11.08	5.72	5.73	41.55
Medium	168	313	31.50	17.56	5.07	10.06	8.17	4.90	45.76
Fast	185	300	27.00	19.46	6.89	9.00	10.95	3.24	49.54
Brush 3rd or additional coats									
Slow	160	350	36.00	14.38	3.45	10.29	5.34	5.35	38.81
Medium	185	338	31.50	15.95	4.62	9.32	7.47	4.48	41.84
Fast	210	325	27.00	17.14	6.05	8.31	9.77	2.89	44.16
Enamel, water base (material #9)									
Brush 1st finish coat									
Slow	100	300	53.30	23.00	5.52	17.77	8.80	8.81	63.90
Medium	125	288	46.70	23.60	6.82	16.22	11.66	7.00	65.30
Fast	150	275	40.00	24.00	8.49	14.55	14.58	4.31	65.93
Brush 2nd finish coat									
Slow	125	325	53.30	18.40	4.42	16.40	7.45	7.47	54.14
Medium	143	313	46.70	20.63	5.95	14.92	10.38	6.23	58.11
Fast	160	300	40.00	22.50	7.94	13.33	13.57	4.01	61.35
Brush 3rd or additional finish coats									
Slow	135	350	53.30	17.04	4.09	15.23	6.91	6.92	50.19
Medium	160	338	46.70	18.44	5.33	13.82	9.40	5.64	52.63
Fast	185	325	40.00	19.46	6.89	12.31	11.98	3.54	54.18
Enamel, oil base (material #10)									
Brush 1st finish coat									
Slow	100	325	61.50	23.00	5.52	18.92	9.01	9.03	65.48
Medium	125	313	53.80	23.60	6.82	17.19	11.90	7.14	66.65
Fast	150	300	46.10	24.00	8.49	15.37	14.83	4.39	67.08
Brush 2nd finish coat									
Slow	125	400	61.50	18.40	4.42	15.38	7.26	7.27	52.73
Medium	143	375	53.80	20.63	5.95	14.35	10.24	6.14	57.31
Fast	160	350	46.10	22.50	7.94	13.17	13.52	4.00	61.13

	Labor SF per manhour	Material coverage SF/gallon	Material cost per gallon	Labor cost per 100 SF	Labor burden 100 SF	Material cost per 100 SF	Overhead per 100 SF	Profit per 100 SF	Total price per 100 SF
Brush 3rd or additional finish coats									
Slow	135	425	61.50	17.04	4.09	14.47	6.76	6.78	49.14
Medium	160	400	53.80	18.44	5.33	13.45	9.31	5.58	52.11
Fast	185	375	46.10	19.46	6.89	12.29	11.97	3.54	54.15
Stipple finish									
Slow	125	--	--	18.40	4.42	--	4.34	4.35	31.51
Medium	143	--	--	20.63	5.95	--	6.65	3.99	37.22
Fast	160	--	--	22.50	7.94	--	9.44	2.79	42.67
Epoxy coating, 2 part system - white (material #52)									
Brush 1st coat									
Slow	150	400	157.00	15.33	3.69	39.25	11.07	11.09	80.43
Medium	165	388	137.40	17.88	5.17	35.41	14.62	8.77	81.85
Fast	185	375	117.70	19.46	6.89	31.39	17.89	5.29	80.92
Brush 2nd or additional coats									
Slow	160	425	157.00	14.38	3.45	36.94	10.41	10.43	75.61
Medium	185	413	137.40	15.95	4.62	33.27	13.46	8.07	75.37
Fast	210	400	117.70	17.14	6.05	29.43	16.31	4.83	73.76
Glazing & mottling over enamel (material #16)									
Brush each coat									
Slow	50	900	49.10	46.00	11.04	5.46	11.88	11.90	86.28
Medium	65	800	43.00	45.38	13.11	5.38	15.97	9.58	89.42
Fast	80	700	36.90	45.00	15.89	5.27	20.51	6.07	92.74
Stipple									
Slow	100	--	--	23.00	5.52	--	5.42	5.43	39.37
Medium	113	--	--	26.11	7.55	--	8.42	5.05	47.13
Fast	125	--	--	28.80	10.17	--	12.08	3.57	54.62

Measurements are based on the square feet of wall coated. Do not deduct for openings less than 100 square feet. These figures assume paint products are being applied over a smooth or medium texture finish. For heights above 8 feet, use the High Time Difficulty Factors on page 139. These figures include brushing-in at corners when all walls are the same color, and at ceilings that are the same color or finished with acoustic spray-on texture. ADD for cutting-in at ceilings if they're a different color than the walls, or at corners where walls in the same room are painted different colors. Do not include cutting-in time for ceilings unless you're only painting the ceilings, not the walls. "Slow" work is based on an hourly wage of $23.00, "Medium" work on an hourly wage of $29.50, and "Fast" work on an hourly wage of $36.00. Other qualifications that apply to this table are on page 9.

	Labor SF per manhour	Material coverage SF/gallon	Material cost per gallon	Labor cost per 100 SF	Labor burden 100 SF	Material cost per 100 SF	Overhead per 100 SF	Profit per 100 SF	Total price per 100 SF
Walls, plaster, interior, medium texture, roll, per 100 SF of wall area									
Flat latex, water base (material #5)									
Roll 1st coat									
Slow	225	250	36.00	10.22	2.44	14.40	5.14	5.15	37.35
Medium	438	238	31.50	6.74	1.93	13.24	5.48	3.29	30.68
Fast	650	225	27.00	5.54	1.96	12.00	6.05	1.79	27.34
Roll 2nd coat									
Slow	250	300	36.00	9.20	2.21	12.00	4.45	4.46	32.32
Medium	463	288	31.50	6.37	1.84	10.94	4.79	2.87	26.81
Fast	675	275	27.00	5.33	1.88	9.82	5.28	1.56	23.87
Roll 3rd or additional coats									
Slow	275	325	36.00	8.36	2.02	11.08	4.08	4.08	29.62
Medium	500	313	31.50	5.90	1.71	10.06	4.42	2.65	24.74
Fast	725	300	27.00	4.97	1.75	9.00	4.87	1.44	22.03
Enamel, water base (material #9)									
Roll 1st finish coat									
Slow	200	250	53.30	11.50	2.76	21.32	6.76	6.77	49.11
Medium	413	238	46.70	7.14	2.06	19.62	7.21	4.32	40.35
Fast	625	225	40.00	5.76	2.03	17.78	7.93	2.35	35.85
Roll 2nd finish coat									
Slow	225	300	53.30	10.22	2.44	17.77	5.78	5.80	42.01
Medium	438	288	46.70	6.74	1.93	16.22	6.23	3.74	34.86
Fast	650	275	40.00	5.54	1.96	14.55	6.84	2.02	30.91
Roll 3rd or additional finish coats									
Slow	250	325	53.30	9.20	2.21	16.40	5.28	5.29	38.38
Medium	475	313	46.70	6.21	1.81	14.92	5.73	3.44	32.11
Fast	700	300	40.00	5.14	1.83	13.33	6.29	1.86	28.45
Enamel, oil base (material #10)									
Roll 1st finish coat									
Slow	200	275	61.50	11.50	2.76	22.36	6.96	6.97	50.55
Medium	413	263	53.80	7.14	2.06	20.46	7.42	4.45	41.53
Fast	625	250	46.10	5.76	2.03	18.44	8.13	2.41	36.77
Roll 2nd finish coat									
Slow	225	350	61.50	10.22	2.44	17.57	5.75	5.76	41.74
Medium	438	325	53.80	6.74	1.93	16.55	6.31	3.79	35.32
Fast	650	300	46.10	5.54	1.96	15.37	7.09	2.10	32.06

	Labor SF per manhour	Material coverage SF/gallon	Material cost per gallon	Labor cost per 100 SF	Labor burden 100 SF	Material cost per 100 SF	Overhead per 100 SF	Profit per 100 SF	Total price per 100 SF
Roll 3rd or additional finish coats									
Slow	250	375	61.50	9.20	2.21	16.40	5.28	5.29	38.38
Medium	475	350	53.80	6.21	1.81	15.37	5.84	3.51	32.74
Fast	700	325	46.10	5.14	1.83	14.18	6.55	1.94	29.64
Epoxy coating, 2 part system - white (material #52)									
Roll 1st coat									
Slow	250	350	157.00	9.20	2.21	44.86	10.69	10.71	77.67
Medium	463	335	137.40	6.37	1.84	41.01	12.31	7.38	68.91
Fast	675	320	117.70	5.33	1.88	36.78	13.64	4.03	61.66
Roll 2nd or additional coats									
Slow	275	400	157.00	8.36	2.02	39.25	9.43	9.45	68.51
Medium	500	375	137.40	5.90	1.71	36.64	11.06	6.64	61.95
Fast	725	350	117.70	4.97	1.75	33.63	12.51	3.70	56.56

Measurements are based on the square feet of wall coated. Do not deduct for openings less than 100 square feet. These figures assume paint products are being applied over a smooth or medium texture finish. For heights above 8 feet, use the High Time Difficulty Factors on page 139. These figures include brushing-in at corners when all walls are the same color, and at ceilings that are the same color or finished with acoustic spray-on texture. ADD for cutting-in at ceilings if they're a different color than the walls, or at corners where walls in the same room are painted different colors. Do not include cutting-in time for ceilings unless you're only painting the ceilings, not the walls. "Slow" work is based on an hourly wage of $23.00, "Medium" work on an hourly wage of $29.50, and "Fast" work on an hourly wage of $36.00. Other qualifications that apply to this table are on page 9.

	Labor SF per manhour	Material coverage SF/gallon	Material cost per gallon	Labor cost per 100 SF	Labor burden 100 SF	Material cost per 100 SF	Overhead per 100 SF	Profit per 100 SF	Total price per 100 SF
Walls, plaster, interior, medium texture, spray, per 100 SF of wall area									
Flat latex, water base (material #5)									
Spray 1st coat									
Slow	475	350	36.00	4.84	1.18	10.29	3.10	3.10	22.51
Medium	600	313	31.50	4.92	1.43	10.06	4.10	2.46	22.97
Fast	725	275	27.00	4.97	1.75	9.82	5.13	1.52	23.19
Spray 2nd coat									
Slow	525	400	36.00	4.38	1.04	9.00	2.74	2.75	19.91
Medium	675	350	31.50	4.37	1.26	9.00	3.66	2.19	20.48
Fast	825	300	27.00	4.36	1.53	9.00	4.62	1.37	20.88
Spray 3rd or additional coats									
Slow	575	450	36.00	4.00	.96	8.00	2.46	2.47	17.89
Medium	750	388	31.50	3.93	1.13	8.12	3.30	1.98	18.46
Fast	925	325	27.00	3.89	1.37	8.31	4.21	1.24	19.02
Enamel, water base (material #9)									
Spray 1st finish coat									
Slow	450	350	53.30	5.11	1.22	15.23	4.10	4.11	29.77
Medium	575	313	46.70	5.13	1.49	14.92	5.38	3.23	30.15
Fast	700	275	40.00	5.14	1.83	14.55	6.67	1.97	30.16
Spray 2nd finish coat									
Slow	500	400	53.30	4.60	1.10	13.33	3.62	3.62	26.27
Medium	650	350	46.70	4.54	1.32	13.34	4.80	2.88	26.88
Fast	800	300	40.00	4.50	1.59	13.33	6.02	1.78	27.22
Spray 3rd or additional finish coats									
Slow	550	450	53.30	4.18	1.01	11.84	3.23	3.24	23.50
Medium	750	388	46.70	3.93	1.13	12.04	4.28	2.57	23.95
Fast	900	325	40.00	4.00	1.41	12.31	5.49	1.62	24.83
Enamel, oil base (material #10)									
Spray 1st finish coat									
Slow	450	400	61.50	5.11	1.22	15.38	4.13	4.14	29.98
Medium	575	363	53.80	5.13	1.49	14.82	5.36	3.21	30.01
Fast	700	325	46.10	5.14	1.83	14.18	6.55	1.94	29.64
Spray 2nd finish coat									
Slow	500	425	61.50	4.60	1.10	14.47	3.83	3.84	27.84
Medium	650	388	53.80	4.54	1.32	13.87	4.93	2.96	27.62
Fast	800	350	46.10	4.50	1.59	13.17	5.97	1.77	27.00

	Labor SF per manhour	Material coverage SF/gallon	Material cost per gallon	Labor cost per 100 SF	Labor burden 100 SF	Material cost per 100 SF	Overhead per 100 SF	Profit per 100 SF	Total price per 100 SF
Spray 3rd or additional finish coats									
Slow	550	450	61.50	4.18	1.01	13.67	3.58	3.59	26.03
Medium	725	400	53.80	4.07	1.18	13.45	4.68	2.81	26.19
Fast	900	375	46.10	4.00	1.41	12.29	5.49	1.62	24.81
Epoxy coating, 2 part system - white (material #52)									
Spray 1st coat									
Slow	525	325	157.00	4.38	1.04	48.31	10.21	10.23	74.17
Medium	675	300	137.40	4.37	1.26	45.80	12.86	7.71	72.00
Fast	825	275	117.70	4.36	1.53	42.80	15.10	4.47	68.26
Spray 2nd or additional coats									
Slow	575	350	157.00	4.00	.96	44.86	9.47	9.49	68.78
Medium	725	325	137.40	4.07	1.18	42.28	11.88	7.13	66.54
Fast	875	300	117.70	4.11	1.44	39.23	13.88	4.11	62.77

Measurements are based on the square feet of wall coated. Do not deduct for openings less than 100 square feet. These figures assume paint products are being applied over a smooth or medium texture finish. For heights above 8 feet, use the High Time Difficulty Factors on page 139. These figures include spraying at corners when all walls are the same color, and at ceiling-to-wall intersection when ceilings are the same color. ADD for cutting-in at ceilings and protecting adjacent surfaces from overspray if they're a different color than the walls, or at corners where walls in the same room are painted different colors. "Slow" work is based on an hourly wage of $23.00, "Medium" work on an hourly wage of $29.50, and "Fast" work on an hourly wage of $36.00. Other qualifications that apply to this table are on page 9.

	Labor SF per manhour	Material coverage SF/gallon	Material cost per gallon	Labor cost per 100 SF	Labor burden 100 SF	Material cost per 100 SF	Overhead per 100 SF	Profit per 100 SF	Total price per 100 SF
Walls, plaster, interior, rough texture, per 100 SF of wall area									
Anti-graffiti stain eliminator									
Water base primer and sealer (material #39)									
Roll & brush each coat									
Slow	350	400	49.40	6.57	1.59	12.35	3.90	3.90	28.31
Medium	375	375	43.20	7.87	2.28	11.52	5.42	3.25	30.34
Fast	400	350	37.00	9.00	3.18	10.57	7.05	2.09	31.89
Oil base primer and sealer (material #40)									
Roll & brush each coat									
Slow	350	375	58.00	6.57	1.59	15.47	4.49	4.50	32.62
Medium	375	350	50.80	7.87	2.28	14.51	6.16	3.70	34.52
Fast	400	325	43.50	9.00	3.18	13.38	7.92	2.34	35.82
Polyurethane 2 part system (material #41)									
Roll & brush each coat									
Slow	300	350	187.30	7.67	1.83	53.51	11.97	12.00	86.98
Medium	325	325	163.90	9.08	2.63	50.43	15.53	9.32	86.99
Fast	350	300	140.50	10.29	3.64	46.83	18.83	5.57	85.16

Measurements are based on the square feet of wall coated. Do not deduct for openings less than 100 square feet. These figures assume paint products are being applied over a rough finish, sand finish, or orange peel texture finish. For heights above 8 feet, use the High Time Difficulty Factors on page 139. "Slow" work is based on an hourly wage of $23.00, "Medium" work on an hourly wage of $29.50, and "Fast" work on an hourly wage of $36.00. Other qualifications that apply to this table are on page 9.

	Labor SF per manhour	Material coverage SF/gallon	Material cost per gallon	Labor cost per 100 SF	Labor burden 100 SF	Material cost per 100 SF	Overhead per 100 SF	Profit per 100 SF	Total price per 100 SF

Walls, plaster, interior, rough texture, brush, per 100 SF of wall area

	Labor SF per manhour	Material coverage SF/gallon	Material cost per gallon	Labor cost per 100 SF	Labor burden 100 SF	Material cost per 100 SF	Overhead per 100 SF	Profit per 100 SF	Total price per 100 SF
Flat latex, water base (material #5)									
Brush 1st coat									
Slow	115	300	36.00	20.00	4.81	12.00	6.99	7.01	50.81
Medium	140	275	31.50	21.07	6.08	11.45	9.65	5.79	54.04
Fast	165	250	27.00	21.82	7.70	10.80	12.50	3.70	56.52
Brush 2nd coat									
Slow	125	325	36.00	18.40	4.42	11.08	6.44	6.45	46.79
Medium	153	300	31.50	19.28	5.59	10.50	8.84	5.30	49.51
Fast	180	275	27.00	20.00	7.08	9.82	11.43	3.38	51.71
Brush 3rd or additional coats									
Slow	135	350	36.00	17.04	4.09	10.29	5.97	5.98	43.37
Medium	168	325	31.50	17.56	5.07	9.69	8.08	4.85	45.25
Fast	200	300	27.00	18.00	6.36	9.00	10.34	3.06	46.76
Enamel, water base (material #9)									
Brush 1st finish coat									
Slow	100	300	53.30	23.00	5.52	17.77	8.80	8.81	63.90
Medium	125	275	46.70	23.60	6.82	16.98	11.85	7.11	66.36
Fast	150	250	40.00	24.00	8.49	16.00	15.03	4.45	67.97
Brush 2nd finish coat									
Slow	115	325	53.30	20.00	4.81	16.40	7.83	7.84	56.88
Medium	140	300	46.70	21.07	6.08	15.57	10.68	6.41	59.81
Fast	165	275	40.00	21.82	7.70	14.55	13.66	4.04	61.77
Brush 3rd or additional finish coats									
Slow	125	350	53.30	18.40	4.42	15.23	7.23	7.24	52.52
Medium	160	325	46.70	18.44	5.33	14.37	9.54	5.72	53.40
Fast	185	300	40.00	19.46	6.89	13.33	12.29	3.64	55.61
Enamel, oil base (material #10)									
Brush 1st finish coat									
Slow	100	300	61.50	23.00	5.52	20.50	9.31	9.33	67.66
Medium	125	288	53.80	23.60	6.82	18.68	12.28	7.37	68.75
Fast	150	275	46.10	24.00	8.49	16.76	15.26	4.51	69.02
Brush 2nd finish coat									
Slow	115	375	61.50	20.00	4.81	16.40	7.83	7.84	56.88
Medium	140	350	53.80	21.07	6.08	15.37	10.63	6.38	59.53
Fast	165	325	46.10	21.82	7.70	14.18	13.55	4.01	61.26

	Labor SF per manhour	Material coverage SF/gallon	Material cost per gallon	Labor cost per 100 SF	Labor burden 100 SF	Material cost per 100 SF	Overhead per 100 SF	Profit per 100 SF	Total price per 100 SF
Brush 3rd or additional finish coats									
Slow	125	400	61.50	18.40	4.42	15.38	7.26	7.27	52.73
Medium	160	375	53.80	18.44	5.33	14.35	9.53	5.72	53.37
Fast	185	350	46.10	19.46	6.89	13.17	12.25	3.62	55.39
Epoxy coating, 2 part system - white (material #52)									
Brush 1st coat									
Slow	125	375	157.00	18.40	4.42	41.87	12.29	12.32	89.30
Medium	160	363	137.40	18.44	5.33	37.85	15.41	9.24	86.27
Fast	180	350	117.70	20.00	7.08	33.63	18.81	5.57	85.09
Brush 2nd or additional coats									
Slow	135	400	157.00	17.04	4.09	39.25	11.47	11.50	83.35
Medium	175	388	137.40	16.86	4.86	35.41	14.29	8.57	79.99
Fast	200	375	117.70	18.00	6.36	31.39	17.28	5.11	78.14
Glazing & mottling over enamel (material #16)									
Brush each coat									
Slow	40	875	49.10	57.50	13.80	5.61	14.61	14.64	106.16
Medium	50	838	43.00	59.00	17.06	5.13	20.30	12.18	113.67
Fast	60	800	36.90	60.00	21.20	4.61	26.59	7.87	120.27
Stipple									
Slow	90	--	--	25.56	6.13	--	6.02	6.03	43.74
Medium	103	--	--	28.64	8.29	--	9.23	5.54	51.70
Fast	115	--	--	31.30	11.08	--	13.13	3.88	59.39

Measurements are based on the square feet of wall coated. Do not deduct for openings less than 100 square feet. These figures assume paint products are being applied over a rough finish. For heights above 8 feet, use the High Time Difficulty Factors on page 139. These figures include brushing-in at corners when all walls are the same color, and at ceilings that are the same color or finished with acoustic spray-on texture. ADD for cutting-in at ceilings if they're a different color than the walls, or at corners where walls in the same room are painted different colors. Do not include cutting-in time for ceilings unless you're only painting the ceilings, not the walls. "Slow" work is based on an hourly wage of $23.00, "Medium" work on an hourly wage of $29.50, and "Fast" work on an hourly wage of $36.00. Other qualifications that apply to this table are on page 9.

	Labor SF per manhour	Material coverage SF/gallon	Material cost per gallon	Labor cost per 100 SF	Labor burden 100 SF	Material cost per 100 SF	Overhead per 100 SF	Profit per 100 SF	Total price per 100 SF
Walls, plaster, interior, rough texture, roll, per 100 SF of wall area									
Flat latex, water base (material #5)									
Roll 1st coat									
Slow	200	250	36.00	11.50	2.76	14.40	5.45	5.46	39.57
Medium	413	238	31.50	7.14	2.06	13.24	5.61	3.37	31.42
Fast	625	225	27.00	5.76	2.03	12.00	6.13	1.81	27.73
Roll 2nd coat									
Slow	225	300	36.00	10.22	2.44	12.00	4.69	4.70	34.05
Medium	438	288	31.50	6.74	1.93	10.94	3.73	3.74	27.08
Fast	650	275	27.00	5.54	1.96	9.82	5.37	1.59	24.28
Roll 3rd or additional coats									
Slow	250	325	36.00	9.20	2.21	11.08	4.27	4.28	31.04
Medium	463	313	31.50	6.37	1.84	10.06	4.57	2.74	25.58
Fast	675	300	27.00	5.33	1.88	9.00	5.03	1.49	22.73
Enamel, water base (material #9)									
Roll 1st finish coat									
Slow	175	250	53.30	13.14	3.14	21.32	7.15	7.16	51.91
Medium	388	238	46.70	7.60	2.21	19.62	7.36	4.41	41.20
Fast	600	225	40.00	6.00	2.13	17.78	8.03	2.38	36.32
Roll 2nd finish coat									
Slow	200	325	53.30	11.50	2.76	16.40	5.83	5.84	42.33
Medium	413	313	46.70	7.14	2.06	14.92	6.03	3.62	33.77
Fast	625	300	40.00	5.76	2.03	13.33	6.55	1.94	29.61
Roll 3rd or additional finish coats									
Slow	225	325	53.30	10.22	2.44	16.40	5.52	5.53	40.11
Medium	438	313	46.70	6.74	1.93	14.92	5.90	3.54	33.03
Fast	650	300	40.00	5.54	1.96	13.33	6.46	1.91	29.20
Enamel, oil base (material #10)									
Roll 1st finish coat									
Slow	175	300	61.50	13.14	3.14	20.50	6.99	7.00	50.77
Medium	388	288	53.80	7.60	2.21	18.68	7.12	4.27	39.88
Fast	600	275	46.10	6.00	2.13	16.76	7.71	2.28	34.88
Roll 2nd finish coat									
Slow	200	350	61.50	11.50	2.76	17.57	6.05	6.06	43.94
Medium	413	325	53.80	7.14	2.06	16.55	6.44	3.86	36.05
Fast	625	300	46.10	5.76	2.03	15.37	7.18	2.12	32.46

	Labor SF per manhour	Material coverage SF/gallon	Material cost per gallon	Labor cost per 100 SF	Labor burden 100 SF	Material cost per 100 SF	Overhead per 100 SF	Profit per 100 SF	Total price per 100 SF
Roll 3rd or additional finish coats									
Slow	225	375	61.50	10.22	2.44	16.40	5.52	5.53	40.11
Medium	438	350	53.80	6.74	1.93	15.37	6.02	3.61	33.67
Fast	650	325	46.10	5.54	1.96	14.18	6.72	1.99	30.39
Epoxy coating, 2 part system - white (material #52)									
Roll 1st coat									
Slow	225	350	157.00	10.22	2.44	44.86	10.93	10.95	79.40
Medium	438	325	137.40	6.74	1.93	42.28	12.74	7.65	71.34
Fast	650	300	117.70	5.54	1.96	39.23	14.49	4.29	65.51
Roll 2nd or additional coats									
Slow	250	400	157.00	9.20	2.21	39.25	9.63	9.65	69.94
Medium	463	375	137.40	6.37	1.84	36.64	11.21	6.73	62.79
Fast	675	350	117.70	5.33	1.88	33.63	12.66	3.75	57.25

Measurements are based on the square feet of wall coated. Do not deduct for openings less than 100 square feet. These figures assume paint products are being applied over a rough finish. For heights above 8 feet, use the High Time Difficulty Factors on page 139. These figures include brushing-in at corners when all walls are the same color, and at ceilings that are the same color or finished with acoustic spray-on texture. ADD for cutting-in at ceilings if they're a different color than the walls, or at corners where walls in the same room are painted different colors. Do not include cutting-in time for ceilings unless you're only painting the ceilings, not the walls. "Slow" work is based on an hourly wage of $23.00, "Medium" work on an hourly wage of $29.50, and "Fast" work on an hourly wage of $36.00. Other qualifications that apply to this table are on page 9.

	Labor SF per manhour	Material coverage SF/gallon	Material cost per gallon	Labor cost per 100 SF	Labor burden 100 SF	Material cost per 100 SF	Overhead per 100 SF	Profit per 100 SF	Total price per 100 SF
Walls, plaster, interior, rough texture, spray, per 100 SF of wall area									
Flat latex, water base (material #5)									
Spray 1st coat									
Slow	500	325	36.00	4.60	1.10	11.08	3.19	3.20	23.17
Medium	600	288	31.50	4.92	1.43	10.94	4.32	2.59	24.20
Fast	700	250	27.00	5.14	1.83	10.80	5.50	1.63	24.90
Spray 2nd coat									
Slow	600	400	36.00	3.83	.93	9.00	2.61	2.62	18.99
Medium	700	350	31.50	4.21	1.23	9.00	3.61	2.16	20.21
Fast	800	300	27.00	4.50	1.59	9.00	4.68	1.38	21.15
Spray 3rd or additional coats									
Slow	700	425	36.00	3.29	.79	8.47	2.38	2.39	17.32
Medium	800	375	31.50	3.69	1.06	8.40	3.29	1.97	18.41
Fast	900	325	27.00	4.00	1.41	8.31	4.25	1.26	19.23
Enamel, water base (material #9)									
Spray 1st finish coat									
Slow	450	325	53.30	5.11	1.22	16.40	4.32	4.33	31.38
Medium	550	288	46.70	5.36	1.56	16.22	5.78	3.47	32.39
Fast	650	250	40.00	5.54	1.96	16.00	7.29	2.16	32.95
Spray 2nd finish coat									
Slow	550	400	53.30	4.18	1.01	13.33	3.52	3.52	25.56
Medium	650	350	46.70	4.54	1.32	13.34	4.80	2.88	26.88
Fast	750	300	40.00	4.80	1.68	13.33	6.14	1.82	27.77
Spray 3rd or additional finish coats									
Slow	650	425	53.30	3.54	.85	12.54	3.22	3.22	23.37
Medium	750	375	46.70	3.93	1.13	12.45	4.38	2.63	24.52
Fast	850	325	40.00	4.24	1.51	12.31	5.60	1.66	25.32
Enamel, oil base (material #10)									
Spray 1st finish coat									
Slow	450	325	61.50	5.11	1.22	18.92	4.80	4.81	34.86
Medium	550	300	53.80	5.36	1.56	17.93	6.21	3.73	34.79
Fast	650	275	46.10	5.54	1.96	16.76	7.52	2.22	34.00
Spray 2nd finish coat									
Slow	550	400	61.50	4.18	1.01	15.38	3.91	3.92	28.40
Medium	650	362	53.80	4.54	1.32	14.86	5.18	3.11	29.01
Fast	750	325	46.10	4.80	1.68	14.18	6.41	1.90	28.97

	Labor SF per manhour	Material coverage SF/gallon	Material cost per gallon	Labor cost per 100 SF	Labor burden 100 SF	Material cost per 100 SF	Overhead per 100 SF	Profit per 100 SF	Total price per 100 SF
Spray 3rd or additional finish coats									
Slow	650	425	61.50	3.54	.85	14.47	3.58	3.59	26.03
Medium	750	388	53.80	3.93	1.13	13.87	4.74	2.84	26.51
Fast	850	350	46.10	4.24	1.51	13.17	5.86	1.73	26.51
Epoxy coating, 2 part system - white (material #52)									
Spray 1st coat									
Slow	525	325	157.00	4.38	1.04	48.31	10.21	10.23	74.17
Medium	663	313	137.40	4.45	1.29	43.90	12.41	7.45	69.50
Fast	800	300	117.70	4.50	1.59	39.23	14.05	4.16	63.53
Spray 2nd or additional coats									
Slow	575	375	157.00	4.00	.96	41.87	8.90	8.92	64.65
Medium	713	363	137.40	4.14	1.18	37.85	10.80	6.48	60.45
Fast	850	350	117.70	4.24	1.51	33.63	12.20	3.61	55.19

Measurements are based on the square feet of wall coated. Do not deduct for openings less than 100 square feet. These figures assume paint products are being applied over a rough finish. For heights above 8 feet, use the High Time Difficulty Factors on page 139. These figures include spraying at corners when all walls are the same color, and at ceiling-to-wall intersection when ceilings are the same color. ADD for cutting-in at ceilings and protecting adjacent surfaces from overspray if they're a different color than the walls, or at corners where walls in the same room are painted different colors. Do not include cutting-in time for ceilings unless you're only painting the ceilings, not the walls. "Slow" work is based on an hourly wage of $23.00, "Medium" work on an hourly wage of $29.50, and "Fast" work on an hourly wage of $36.00. Other qualifications that apply to this table are on page 9.

	Labor SF per manhour	Material coverage SF/gallon	Material cost per gallon	Labor cost per 100 SF	Labor burden 100 SF	Material cost per 100 SF	Overhead per 100 SF	Profit per 100 SF	Total price per 100 SF
Walls, plaster, interior, smooth finish, per 100 SF of wall area									
Anti-graffiti stain eliminator									
Water base primer and sealer (material #39)									
Roll & brush each coat									
Slow	375	425	49.40	6.13	1.48	11.62	3.65	3.66	26.54
Medium	400	400	43.20	7.38	2.13	10.80	5.08	3.05	28.44
Fast	425	375	37.00	8.47	2.98	9.87	6.61	1.96	29.89
Oil base primer and sealer (material #40)									
Roll & brush each coat									
Slow	375	400	58.00	6.13	1.48	14.50	4.20	4.21	30.52
Medium	400	375	50.80	7.38	2.13	13.55	5.77	3.46	32.29
Fast	425	350	43.50	8.47	2.98	12.43	7.41	2.19	33.48
Polyurethane 2 part system (material #41)									
Roll & brush each coat									
Slow	325	375	187.30	7.08	1.70	49.95	11.16	11.18	81.07
Medium	350	350	163.90	8.43	2.45	46.83	14.43	8.66	80.80
Fast	375	325	140.50	9.60	3.41	43.23	17.43	5.16	78.83

Measurements are based on the square feet of wall coated. Do not deduct for openings less than 100 square feet. These figures assume paint products are being applied over a smooth finish. For heights above 8 feet, use the High Time Difficulty Factors on page 139. "Slow" work is based on an hourly wage of $23.00, "Medium" work on an hourly wage of $29.50, and "Fast" work on an hourly wage of $36.00. Other qualifications that apply to this table are on page 9.

	Labor SF per manhour	Material coverage SF/gallon	Material cost per gallon	Labor cost per 100 SF	Labor burden 100 SF	Material cost per 100 SF	Overhead per 100 SF	Profit per 100 SF	Total price per 100 SF

Walls, plaster, interior, smooth finish, brush, per 100 SF of wall area

	Labor SF per manhour	Material coverage SF/gallon	Material cost per gallon	Labor cost per 100 SF	Labor burden 100 SF	Material cost per 100 SF	Overhead per 100 SF	Profit per 100 SF	Total price per 100 SF
Flat latex, water base (material #5)									
Brush 1st coat									
Slow	150	350	36.00	15.33	3.69	10.29	5.57	5.58	40.46
Medium	175	325	31.50	16.86	4.86	9.69	7.86	4.71	43.98
Fast	200	300	27.00	18.00	6.36	9.00	10.34	3.06	46.76
Brush 2nd coat									
Slow	175	375	36.00	13.14	3.14	9.60	4.92	4.93	35.73
Medium	200	350	31.50	14.75	4.27	9.00	7.00	4.20	39.22
Fast	225	325	27.00	16.00	5.63	8.31	9.29	2.75	41.98
Brush 3rd or additional coats									
Slow	200	400	36.00	11.50	2.76	9.00	4.42	4.43	32.11
Medium	225	375	31.50	13.11	3.78	8.40	6.33	3.80	35.42
Fast	250	350	27.00	14.40	5.08	7.71	8.43	2.49	38.11
Enamel, water base (material #9)									
Brush 1st finish coat									
Slow	125	350	53.30	18.40	4.42	15.23	7.23	7.24	52.52
Medium	163	325	46.70	18.10	5.21	14.37	9.43	5.66	52.77
Fast	200	300	40.00	18.00	6.36	13.33	11.68	3.46	52.83
Brush 2nd finish coat									
Slow	150	375	53.30	15.33	3.69	14.21	6.31	6.32	45.86
Medium	175	350	46.70	16.86	4.86	13.34	8.77	5.26	49.09
Fast	200	325	40.00	18.00	6.36	12.31	11.36	3.36	51.39
Brush 3rd or additional finish coats									
Slow	175	400	53.30	13.14	3.14	13.33	5.63	5.64	40.88
Medium	200	375	46.70	14.75	4.27	12.45	7.87	4.72	44.06
Fast	225	350	40.00	16.00	5.63	11.43	10.25	3.03	46.34
Enamel, oil base (material #10)									
Brush 1st finish coat									
Slow	125	400	61.50	18.40	4.42	15.38	7.26	7.27	52.73
Medium	163	375	53.80	18.10	5.21	14.35	9.42	5.65	52.73
Fast	200	350	46.10	18.00	6.36	13.17	11.63	3.44	52.60
Brush 2nd finish coat									
Slow	150	425	61.50	15.33	3.69	14.47	6.36	6.37	46.22
Medium	175	400	53.80	16.86	4.86	13.45	8.80	5.28	49.25
Fast	200	375	46.10	18.00	6.36	12.29	11.36	3.36	51.37

	Labor SF per manhour	Material coverage SF/gallon	Material cost per gallon	Labor cost per 100 SF	Labor burden 100 SF	Material cost per 100 SF	Overhead per 100 SF	Profit per 100 SF	Total price per 100 SF
Brush 3rd or additional finish coats									
Slow	175	450	61.50	13.14	3.14	13.67	5.69	5.70	41.34
Medium	200	425	53.80	14.75	4.27	12.66	7.92	4.75	44.35
Fast	225	400	46.10	16.00	5.63	11.53	10.29	3.04	46.49
Epoxy coating, 2 part system - white (material #52)									
Brush 1st coat									
Slow	175	400	157.00	13.14	3.14	39.25	10.55	10.57	76.65
Medium	200	388	137.40	14.75	4.27	35.41	13.61	8.16	76.20
Fast	225	375	117.70	16.00	5.63	31.39	16.44	4.86	74.32
Brush 2nd or additional coats									
Slow	200	425	157.00	11.50	2.76	36.94	9.73	9.75	70.68
Medium	225	413	137.40	13.11	3.78	33.27	12.54	7.53	70.23
Fast	250	400	117.70	14.40	5.08	29.43	15.16	4.48	68.55
Glazing & mottling over enamel (material #16)									
Brush each coat									
Slow	75	900	49.10	30.67	7.35	5.46	8.26	8.28	60.02
Medium	98	850	43.00	30.10	8.69	5.06	10.97	6.58	61.40
Fast	120	800	36.90	30.00	10.58	4.61	14.01	4.14	63.34
Stipple finish									
Slow	100	--	--	23.00	5.52	--	5.42	5.43	39.37
Medium	123	--	--	23.98	6.94	--	7.73	4.64	43.29
Fast	135	--	--	26.67	9.42	--	11.18	3.31	50.58

Measurements are based on the square feet of wall coated. Do not deduct for openings less than 100 square feet. These figures assume paint products are being applied over a smooth finish. For heights above 8 feet, use the High Time Difficulty Factors on page 139. These figures include brushing-in at corners when all walls are the same color, and at ceilings that are the same color or finished with acoustic spray-on texture. ADD for cutting-in at ceilings if they're a different color than the walls, or at corners where walls in the same room are painted different colors. Do not include cutting-in time for ceilings unless you're only painting the ceilings, not the walls. "Slow" work is based on an hourly wage of $23.00, "Medium" work on an hourly wage of $29.50, and "Fast" work on an hourly wage of $36.00. Other qualifications that apply to this table are on page 9.

	Labor SF per manhour	Material coverage SF/gallon	Material cost per gallon	Labor cost per 100 SF	Labor burden 100 SF	Material cost per 100 SF	Overhead per 100 SF	Profit per 100 SF	Total price per 100 SF
Walls, plaster, interior, smooth finish, roll, per 100 SF of wall area									
Flat latex, water base (material #5)									
Roll 1st coat									
Slow	260	350	36.00	8.85	2.13	10.29	4.04	4.05	29.36
Medium	430	325	31.50	6.86	2.00	9.69	4.63	2.78	25.96
Fast	640	300	27.00	5.63	1.97	9.00	5.15	1.52	23.27
Roll 2nd coat									
Slow	300	375	36.00	7.67	1.83	9.60	3.63	3.64	26.37
Medium	488	350	31.50	6.05	1.75	9.00	4.20	2.52	23.52
Fast	675	325	27.00	5.33	1.88	8.31	4.81	1.42	21.75
Roll 3rd or additional coats									
Slow	325	400	36.00	7.08	1.70	9.00	3.38	3.39	24.55
Medium	513	375	31.50	5.75	1.67	8.40	3.95	2.37	22.14
Fast	700	350	27.00	5.14	1.83	7.71	4.54	1.34	20.56
Enamel, water base (material #9)									
Roll 1st finish coat									
Slow	235	350	53.30	9.79	2.36	15.23	5.20	5.21	37.79
Medium	423	325	46.70	6.97	2.01	14.37	5.84	3.50	32.69
Fast	615	300	40.00	5.85	2.09	13.33	6.59	1.95	29.81
Roll 2nd finish coat									
Slow	275	375	53.30	8.36	2.02	14.21	4.67	4.68	33.94
Medium	453	350	46.70	6.51	1.89	13.34	5.43	3.26	30.43
Fast	630	325	40.00	5.71	2.03	12.31	6.21	1.84	28.10
Roll 3rd or additional finish coats									
Slow	300	400	53.30	7.67	1.83	13.33	4.34	4.35	31.52
Medium	475	375	46.70	6.21	1.81	12.45	5.11	3.07	28.65
Fast	650	350	40.00	5.54	1.96	11.43	5.87	1.74	26.54
Enamel, oil base (material #10)									
Roll 1st finish coat									
Slow	235	375	61.50	9.79	2.36	16.40	5.42	5.43	39.40
Medium	423	350	53.80	6.97	2.01	15.37	6.09	3.65	34.09
Fast	615	325	46.10	5.85	2.09	14.18	6.85	2.03	31.00
Roll 2nd finish coat									
Slow	275	400	61.50	8.36	2.02	15.38	4.89	4.90	35.55
Medium	453	375	53.80	6.51	1.89	14.35	5.69	3.41	31.85
Fast	630	350	46.10	5.71	2.03	13.17	6.48	1.92	29.31
Roll 3rd or additional finish coats									
Slow	300	425	61.50	7.67	1.83	14.47	4.56	4.57	33.10
Medium	475	400	53.80	6.21	1.81	13.45	5.36	3.22	30.05
Fast	650	375	46.10	5.54	1.96	12.29	6.13	1.81	27.73

	Labor SF per manhour	Material coverage SF/gallon	Material cost per gallon	Labor cost per 100 SF	Labor burden 100 SF	Material cost per 100 SF	Overhead per 100 SF	Profit per 100 SF	Total price per 100 SF
Stipple finish									
Slow	130	--	--	17.69	4.24	--	4.17	4.18	30.28
Medium	150	--	--	19.67	5.70	--	6.34	3.80	35.51
Fast	170	--	--	21.18	7.46	--	8.88	2.63	40.15
Epoxy coating, 2 part system - white (material #52)									
Roll 1st coat									
Slow	300	375	157.00	7.67	1.83	41.87	9.76	9.78	70.91
Medium	488	350	137.40	6.05	1.75	39.26	11.77	7.06	65.89
Fast	675	325	117.70	5.33	1.88	36.22	13.46	3.98	60.87
Roll 2nd or additional coats									
Slow	325	425	157.00	7.08	1.70	36.94	8.69	8.71	63.12
Medium	513	400	137.40	5.75	1.67	34.35	10.44	6.26	58.47
Fast	700	375	117.70	5.14	1.83	31.39	11.89	3.52	53.77

Measurements are based on the square feet of wall coated. Do not deduct for openings less than 100 square feet. These figures assume paint products are being applied over a smooth finish. For heights above 8 feet, use the High Time Difficulty Factors on page 139. These figures include brushing-in at corners when all walls are the same color, and at ceilings that are the same color or finished with acoustic spray-on texture. ADD for cutting-in at ceilings if they're a different color than the walls, or at corners where walls in the same room are painted different colors. Do not include cutting-in time for ceilings unless you're only painting the ceilings, not the walls. "Slow" work is based on an hourly wage of $23.00, "Medium" work on an hourly wage of $29.50, and "Fast" work on an hourly wage of $36.00. Other qualifications that apply to this table are on page 9.

	Labor SF per manhour	Material coverage SF/gallon	Material cost per gallon	Labor cost per 100 SF	Labor burden 100 SF	Material cost per 100 SF	Overhead per 100 SF	Profit per 100 SF	Total price per 100 SF
Walls, plaster, interior, smooth finish, spray, per 100 SF of wall area									
Flat latex, water base (material #5)									
Spray 1st coat									
Slow	500	375	36.00	4.60	1.10	9.60	2.91	2.91	21.12
Medium	625	338	31.50	4.72	1.36	9.32	3.85	2.31	21.56
Fast	750	300	27.00	4.80	1.68	9.00	4.80	1.42	21.70
Spray 2nd coat									
Slow	550	400	36.00	4.18	1.01	9.00	2.69	2.70	19.58
Medium	700	363	31.50	4.21	1.23	8.68	3.53	2.12	19.77
Fast	850	325	27.00	4.24	1.51	8.31	4.36	1.29	19.71
Spray 3rd or additional coats									
Slow	600	425	36.00	3.83	.93	8.47	2.51	2.52	18.26
Medium	775	388	31.50	3.81	1.10	8.12	3.26	1.95	18.24
Fast	950	350	27.00	3.79	1.32	7.71	3.98	1.18	17.98
Enamel, water base (material #9)									
Spray 1st finish coat									
Slow	475	375	53.30	4.84	1.18	14.21	3.84	3.85	27.92
Medium	600	338	46.70	4.92	1.43	13.82	5.04	3.02	28.23
Fast	725	300	40.00	4.97	1.75	13.33	6.22	1.84	28.11
Spray 2nd finish coat									
Slow	525	400	53.30	4.38	1.04	13.33	3.56	3.57	25.88
Medium	675	388	46.70	4.37	1.26	12.04	4.42	2.65	24.74
Fast	825	325	40.00	4.36	1.53	12.31	5.65	1.67	25.52
Spray 3rd or additional finish coats									
Slow	575	425	53.30	4.00	.96	12.54	3.33	3.33	24.16
Medium	775	388	46.70	3.81	1.10	12.04	4.24	2.54	23.73
Fast	925	350	40.00	3.89	1.37	11.43	5.17	1.53	23.39
Enamel, oil base (material #10)									
Spray 1st finish coat									
Slow	475	425	61.50	4.84	1.18	14.47	3.89	3.90	28.28
Medium	575	388	53.80	5.13	1.49	13.87	5.12	3.07	28.68
Fast	725	350	46.10	4.97	1.75	13.17	6.17	1.82	27.88
Spray 2nd finish coat									
Slow	525	450	61.50	4.38	1.04	13.67	3.63	3.64	26.36
Medium	675	413	53.80	4.37	1.26	13.03	4.67	2.80	26.13
Fast	825	375	46.10	4.36	1.53	12.29	5.64	1.67	25.49

	Labor SF per manhour	Material coverage SF/gallon	Material cost per gallon	Labor cost per 100 SF	Labor burden 100 SF	Material cost per 100 SF	Overhead per 100 SF	Profit per 100 SF	Total price per 100 SF
Spray 3rd or additional finish coats									
Slow	575	475	61.50	4.00	.96	12.95	3.40	3.41	24.72
Medium	750	438	53.80	3.93	1.13	12.28	4.34	2.60	24.28
Fast	925	400	46.10	3.89	1.37	11.53	5.20	1.54	23.53
Epoxy coating, 2 part system - white (material #52)									
Spray 1st coat									
Slow	550	325	157.00	4.18	1.01	48.31	10.16	10.18	73.84
Medium	700	300	137.40	4.21	1.23	45.80	12.81	7.68	71.73
Fast	850	275	117.70	4.24	1.51	42.80	15.05	4.45	68.05
Spray 2nd or additional coats									
Slow	600	375	157.00	3.83	.93	41.87	8.86	8.88	64.37
Medium	750	350	137.40	3.93	1.13	39.26	11.08	6.65	62.05
Fast	900	325	117.70	4.00	1.41	36.22	12.91	3.82	58.36

Measurements are based on the square feet of wall coated. Do not deduct for openings less than 100 square feet. These figures assume paint products are being applied over a smooth finish. For heights above 8 feet, use the High Time Difficulty Factors on page 139. These figures include spraying at corners when all walls are the same color, and at ceiling-to-wall intersection when ceilings are the same color. ADD for cutting-in at ceilings and protecting adjacent surfaces from overspray if they're a different color than the walls, or at corners where walls in the same room are painted different colors. Do not include cutting-in time for ceilings unless you're only painting the ceilings, not the walls. "Slow" work is based on an hourly wage of $23.00, "Medium" work on an hourly wage of $29.50, and "Fast" work on an hourly wage of $36.00. Other qualifications that apply to this table are on page 9.

	Labor SF per manhour	Material coverage SF/gallon	Material cost per gallon	Labor cost per 100 SF	Labor burden 100 SF	Material cost per 100 SF	Overhead per 100 SF	Profit per 100 SF	Total price per 100 SF
Walls, wood paneled, interior, paint grade, brush, per 100 SF of wall area									
Undercoat, water base (material #3)									
Brush, 1 coat									
Slow	65	300	36.90	35.38	8.48	12.30	10.67	10.69	77.52
Medium	75	288	32.30	39.33	11.36	11.22	15.48	9.29	86.68
Fast	85	275	27.70	42.35	14.93	10.07	20.88	6.18	94.41
Undercoat, oil base (material #4)									
Brush, 1 coat									
Slow	65	375	48.30	35.38	8.48	12.88	10.78	10.80	78.32
Medium	75	363	42.30	39.33	11.36	11.65	15.59	9.35	87.28
Fast	85	350	36.20	42.35	14.93	10.34	20.97	6.20	94.79
Split coat (1/2 undercoat + 1/2 enamel), water base (material #3 & #9)									
Brush, 1 coat									
Slow	55	300	45.10	41.82	10.03	15.03	12.71	12.74	92.33
Medium	65	288	39.50	45.38	13.11	13.72	18.05	10.83	101.09
Fast	75	275	33.85	48.00	16.93	12.31	23.95	7.08	108.27
Split coat (1/2 undercoat + 1/2 enamel), oil base (material #4 & #10)									
Brush, 1 coat									
Slow	55	375	54.90	41.82	10.03	14.64	12.64	12.66	91.79
Medium	65	363	48.05	45.38	13.11	13.24	17.93	10.76	100.42
Fast	75	350	41.15	48.00	16.93	11.76	23.78	7.03	107.50
Enamel, water base (material #9)									
Brush 1st finish coat									
Slow	80	350	53.30	28.75	6.90	15.23	9.67	9.69	70.24
Medium	95	338	46.70	31.05	9.00	13.82	13.46	8.08	75.41
Fast	110	325	40.00	32.73	11.55	12.31	17.54	5.19	79.32
Brush 2nd or additional finish coats									
Slow	100	375	53.30	23.00	5.52	14.21	8.12	8.14	58.99
Medium	110	363	46.70	26.82	7.75	12.87	11.86	7.12	66.42
Fast	120	350	40.00	30.00	10.58	11.43	16.13	4.77	72.91

	Labor SF per manhour	Material coverage SF/gallon	Material cost per gallon	Labor cost per 100 SF	Labor burden 100 SF	Material cost per 100 SF	Overhead per 100 SF	Profit per 100 SF	Total price per 100 SF
Enamel, oil base (material #10)									
Brush 1st finish coat									
Slow	80	400	61.50	28.75	6.90	15.38	9.70	9.72	70.45
Medium	95	388	53.80	31.05	9.00	13.87	13.47	8.08	75.47
Fast	110	375	46.10	32.73	11.55	12.29	17.54	5.19	79.30
Brush 2nd or additional finish coats									
Slow	100	425	61.50	23.00	5.52	14.47	8.17	8.19	59.35
Medium	110	413	53.80	26.82	7.75	13.03	11.90	7.14	66.64
Fast	120	400	46.10	30.00	10.58	11.53	16.16	4.78	73.05

These costs are based on painting interior tongue and groove, wood veneer or plain wainscot wood paneling. Do not deduct for openings less than 100 square feet. For heights above 8 feet, use the High Time Difficulty Factors on page 139. ADD for masking-off or cutting-in at wall-to-ceiling intersections and for protecting adjacent surfaces as necessary. "Slow" work is based on an hourly wage of $23.00, "Medium" work on an hourly wage of $29.50, and "Fast" work on an hourly wage of $36.00. Other qualifications that apply to this table are on page 9.

	Labor SF per manhour	Material coverage SF/gallon	Material cost per gallon	Labor cost per 100 SF	Labor burden 100 SF	Material cost per 100 SF	Overhead per 100 SF	Profit per 100 SF	Total price per 100 SF
Walls, wood paneled, interior, paint grade, roll, per 100 SF of wall area									
Undercoat, water base (material #3)									
Roll, 1 coat									
Slow	200	275	36.90	11.50	2.76	13.42	5.26	5.27	38.21
Medium	300	263	32.30	9.83	2.83	12.28	6.24	3.74	34.92
Fast	400	250	27.70	9.00	3.18	11.08	7.21	2.13	32.60
Undercoat, oil base (material #4)									
Roll, 1 coat									
Slow	200	350	48.30	11.50	2.76	13.80	5.33	5.34	38.73
Medium	300	325	42.30	9.83	2.83	13.02	6.42	3.85	35.95
Fast	400	300	36.20	9.00	3.18	12.07	7.52	2.22	33.99
Split coat (1/2 undercoat + 1/2 enamel), water base (material #3 & #9)									
Roll, 1 coat									
Slow	175	275	45.10	13.14	3.14	16.40	6.21	6.22	45.11
Medium	275	263	39.50	10.73	3.11	15.02	7.21	4.33	40.40
Fast	375	250	33.85	9.60	3.41	13.54	8.22	2.43	37.20
Split coat (1/2 undercoat + 1/2 enamel), oil base (material #4 & #10)									
Roll, 1 coat									
Slow	175	350	54.90	13.14	3.14	15.69	6.08	6.09	44.14
Medium	275	325	48.05	10.73	3.11	14.78	7.15	4.29	40.06
Fast	375	300	41.15	9.60	3.41	13.72	8.28	2.45	37.46
Enamel, water base (material #9)									
Roll 1st finish coat									
Slow	250	325	53.30	9.20	2.21	16.40	5.28	5.29	38.38
Medium	375	313	46.70	7.87	2.28	14.92	6.27	3.76	35.10
Fast	500	300	40.00	7.20	2.54	13.33	7.15	2.12	32.34
Roll 2nd or additional finish coats									
Slow	300	350	53.30	7.67	1.83	15.23	4.70	4.71	34.14
Medium	425	338	46.70	6.94	2.00	13.82	5.69	3.42	31.87
Fast	550	325	40.00	6.55	2.32	12.31	6.56	1.94	29.68

	Labor SF per manhour	Material coverage SF/gallon	Material cost per gallon	Labor cost per 100 SF	Labor burden 100 SF	Material cost per 100 SF	Overhead per 100 SF	Profit per 100 SF	Total price per 100 SF
Enamel, oil base (material #10)									
Roll 1st finish coat									
Slow	250	375	61.50	9.20	2.21	16.40	5.28	5.29	38.38
Medium	375	363	53.80	7.87	2.28	14.82	6.24	3.74	34.95
Fast	500	350	46.10	7.20	2.54	13.17	7.10	2.10	32.11
Roll 2nd or additional finish coats									
Slow	300	400	61.50	7.67	1.83	15.38	4.73	4.74	34.35
Medium	425	388	53.80	6.94	2.00	13.87	5.71	3.42	31.94
Fast	550	375	46.10	6.55	2.32	12.29	6.56	1.94	29.66

These costs are based on painting interior tongue and groove, wood veneer or plain wainscot wood paneling. Do not deduct for openings less than 100 square feet. For heights above 8 feet, use the High Time Difficulty Factors on page 139. ADD for masking-off or cutting-in at wall-to-ceiling intersections and for protecting adjacent surfaces as necessary. "Slow" work is based on an hourly wage of $23.00, "Medium" work on an hourly wage of $29.50, and "Fast" work on an hourly wage of $36.00. Other qualifications that apply to this table are on page 9.

	Labor SF per manhour	Material coverage SF/gallon	Material cost per gallon	Labor cost per 100 SF	Labor burden 100 SF	Material cost per 100 SF	Overhead per 100 SF	Profit per 100 SF	Total price per 100 SF
Walls, wood paneled, interior, paint grade, spray, per 100 SF of wall area									
Undercoat, water base (material #3)									
Spray, 1 coat									
Slow	350	175	36.90	6.57	1.59	21.09	5.56	5.57	40.38
Medium	425	150	32.30	6.94	2.00	21.53	7.62	4.57	42.66
Fast	500	125	27.70	7.20	2.54	22.16	9.89	2.93	44.72
Undercoat, oil base (material #4)									
Spray, 1 coat									
Slow	350	200	48.30	6.57	1.59	24.15	6.14	6.15	44.60
Medium	425	188	42.30	6.94	2.00	22.50	7.86	4.72	44.02
Fast	500	175	36.20	7.20	2.54	20.69	9.43	2.79	42.65
Split coat (1/2 undercoat + 1/2 enamel), water base (material #3 & #9)									
Spray, 1 coat									
Slow	325	175	45.10	7.08	1.70	25.77	6.56	6.58	47.69
Medium	400	150	39.50	7.38	2.13	26.33	8.96	5.38	50.18
Fast	475	125	33.85	7.58	2.70	27.08	11.58	3.42	52.36
Split coat (1/2 undercoat + 1/2 enamel), oil base (material #4 & #10)									
Spray, 1 coat									
Slow	325	200	54.90	7.08	1.70	27.45	6.88	6.90	50.01
Medium	400	188	48.05	7.38	2.13	25.56	8.77	5.26	49.10
Fast	475	175	41.15	7.58	2.70	23.51	10.47	3.10	47.36
Enamel, water base (material #9)									
Spray 1st finish coat									
Slow	500	250	53.30	4.60	1.10	21.32	5.13	5.14	37.29
Medium	550	225	46.70	5.36	1.56	20.76	6.92	4.15	38.75
Fast	600	200	40.00	6.00	2.13	20.00	8.72	2.58	39.43
Spray 2nd or additional finish coats									
Slow	600	350	53.30	3.83	.93	15.23	3.80	3.80	27.59
Medium	650	325	46.70	4.54	1.32	14.37	5.06	3.03	28.32
Fast	700	300	40.00	5.14	1.83	13.33	6.29	1.86	28.45

	Labor SF per manhour	Material coverage SF/gallon	Material cost per gallon	Labor cost per 100 SF	Labor burden 100 SF	Material cost per 100 SF	Overhead per 100 SF	Profit per 100 SF	Total price per 100 SF
Enamel, oil base (material #10)									
Spray 1st finish coat									
Slow	500	300	61.50	4.60	1.10	20.50	4.98	4.99	36.17
Medium	550	275	53.80	5.36	1.56	19.56	6.62	3.97	37.07
Fast	600	250	46.10	6.00	2.13	18.44	8.23	2.44	37.24
Spray 2nd or additional finish coats									
Slow	600	400	61.50	3.83	.93	15.38	3.82	3.83	27.79
Medium	650	375	53.80	4.54	1.32	14.35	5.05	3.03	28.29
Fast	700	350	46.10	5.14	1.83	13.17	6.24	1.85	28.23

These costs are based on painting interior tongue and groove, wood veneer or plain wainscot wood paneling. Do not deduct for openings less than 100 square feet. For heights above 8 feet, use the High Time Difficulty Factors on page 139. ADD for masking-off or cutting-in at wall-to-ceiling intersections and for protecting adjacent surfaces as necessary. "Slow" work is based on an hourly wage of $23.00, "Medium" work on an hourly wage of $29.50, and "Fast" work on an hourly wage of $36.00. Other qualifications that apply to this table are on page 9.

	Labor SF per manhour	Material coverage SF/gallon	Material cost per gallon	Labor cost per 100 SF	Labor burden 100 SF	Material cost per 100 SF	Overhead per 100 SF	Profit per 100 SF	Total price per 100 SF

Walls, wood paneled, interior, stain grade, per 100 SF of wall area

Solid body or semi-transparent stain, water or oil base (material #18 or #19 or #20 or #21)

Roll & brush each coat

	Labor SF per manhour	Material coverage SF/gallon	Material cost per gallon	Labor cost per 100 SF	Labor burden 100 SF	Material cost per 100 SF	Overhead per 100 SF	Profit per 100 SF	Total price per 100 SF
Slow	225	500	54.50	10.22	2.44	10.90	4.48	4.49	32.53
Medium	263	450	47.70	11.22	3.23	10.60	6.27	3.76	35.08
Fast	300	400	40.88	12.00	4.22	10.22	8.20	2.43	37.07

Solid body or semi-transparent stain, water or oil base (material #18 or #19 or #20 or #21)

Spray each coat

	Labor SF per manhour	Material coverage SF/gallon	Material cost per gallon	Labor cost per 100 SF	Labor burden 100 SF	Material cost per 100 SF	Overhead per 100 SF	Profit per 100 SF	Total price per 100 SF
Slow	350	300	54.50	6.57	1.59	18.17	5.00	5.01	36.34
Medium	400	250	47.70	7.38	2.13	19.08	7.15	4.29	40.03
Fast	450	200	40.88	8.00	2.81	20.44	9.69	2.87	43.81

Use these figures for quantities greater than 100 square feet. For quantities less than 100 square feet, use Fireplace siding. These costs are based on painting interior tongue and groove, wood veneer or plain wainscot wood paneling. Do not deduct for openings less than 100 square feet. For heights above 8 feet, use the High Time Difficulty Factors on page 139. ADD for masking-off or cutting-in at wall-to-ceiling intersections and for protecting adjacent surfaces as necessary. "Slow" work is based on an hourly wage of $23.00, "Medium" work on an hourly wage of $29.50, and "Fast" work on an hourly wage of $36.00. Other qualifications that apply to this table are on page 9.

	Labor SF per manhour	Material coverage SF/gallon	Material cost per gallon	Labor cost per 100 SF	Labor burden 100 SF	Material cost per 100 SF	Overhead per 100 SF	Profit per 100 SF	Total price per 100 SF
Walls, wood paneled, interior, stain grade, per 100 SF of wall area, brush									
Stain, seal and 2 coat lacquer system (7 step process)									
STEP 1: Sand & putty									
Slow	175	--	--	13.14	3.14	--	3.10	3.10	22.48
Medium	200	--	--	14.75	4.27	--	4.75	2.85	26.62
Fast	225	--	--	16.00	5.63	--	6.71	1.99	30.33
STEP 2 & 3: Wiping stain, oil base (material #11a) & wipe									
Brush 1 coat & wipe									
Slow	100	400	61.60	23.00	5.52	15.40	8.34	8.36	60.62
Medium	125	375	53.90	23.60	6.82	14.37	11.20	6.72	62.71
Fast	150	350	46.20	24.00	8.49	13.20	14.16	4.19	64.04
STEP 4 & 5: Sanding sealer (material #11b) & sand lightly									
Brush 1 coat & wipe									
Slow	200	450	50.90	11.50	2.76	11.31	4.86	4.87	35.30
Medium	220	425	44.60	13.41	3.89	10.49	6.95	4.17	38.91
Fast	240	400	38.20	15.00	5.31	9.55	9.25	2.74	41.85
STEP 6 & 7: Lacquer, 2 coats (material #11c)									
Brush 1st coat									
Slow	175	375	53.50	13.14	3.14	14.27	5.81	5.82	42.18
Medium	225	350	46.80	13.11	3.78	13.37	7.57	4.54	42.37
Fast	300	325	40.10	12.00	4.22	12.34	8.86	2.62	40.04
Brush 2nd coat									
Slow	225	400	53.50	10.22	2.44	13.38	4.95	4.96	35.95
Medium	288	388	46.80	10.24	2.96	12.06	6.32	3.79	35.37
Fast	350	375	40.10	10.29	3.64	10.69	7.63	2.26	34.51
Complete 7 step stain, seal & 2 coat lacquer system (material #11)									
Brush all coats									
Slow	40	170	54.90	57.50	13.80	32.29	19.68	19.72	142.99
Medium	45	158	48.00	65.56	18.94	30.38	28.72	17.23	160.83
Fast	50	145	41.20	72.00	25.42	28.41	39.01	11.54	176.38
Penetrating stain wax (material #14)									
Brush each coat									
Slow	300	500	50.20	7.67	1.83	10.04	3.71	3.72	26.97
Medium	350	475	43.90	8.43	2.45	9.24	5.03	3.02	28.17
Fast	400	450	37.60	9.00	3.18	8.36	6.37	1.88	28.79

These costs are based on painting interior tongue and groove, wood veneer or plain wainscot wood paneling. Do not deduct for openings less than 100 square feet. For heights above 8 feet, use the High Time Difficulty Factors on page 139. ADD for masking-off or cutting-in at wall-to-ceiling intersections and for protecting adjacent surfaces as necessary. "Slow" work is based on an hourly wage of $23.00, "Medium" work on an hourly wage of $29.50, and "Fast" work on an hourly wage of $36.00. Other qualifications that apply to this table are on page 9.

	Labor SF per manhour	Material coverage SF/gallon	Material cost per gallon	Labor cost per 100 SF	Labor burden 100 SF	Material cost per 100 SF	Overhead per 100 SF	Profit per 100 SF	Total price per 100 SF
Walls, wood paneled, interior, stain grade, per 100 SF of wall area, spray									
Stain, seal and 2 coat lacquer system (7 step process)									
STEP 1: Sand & putty									
Slow	175	--	--	13.14	3.14	--	3.10	3.10	22.48
Medium	200	--	--	14.75	4.27	--	4.75	2.85	26.62
Fast	225	--	--	16.00	5.63	--	6.71	1.99	30.33
STEP 2 & 3: Wiping stain, oil base (material #11a) & wipe									
Spray 1 coat & wipe									
Slow	350	175	61.60	6.57	1.59	35.20	8.24	8.25	59.85
Medium	425	150	53.90	6.94	2.00	35.93	11.22	6.73	62.82
Fast	500	125	46.20	7.20	2.54	36.96	14.48	4.28	65.46
STEP 4 & 5: Sanding sealer (material #11b) & sand lightly									
Spray 1 coat & wipe									
Slow	400	175	50.90	5.75	1.38	29.09	6.88	6.90	50.00
Medium	500	150	44.60	5.90	1.71	29.73	9.34	5.60	52.28
Fast	600	125	38.20	6.00	2.13	30.56	11.99	3.55	54.23
STEP 6 & 7: Lacquer, 2 coats (material #11c)									
Spray 1st coat									
Slow	450	150	53.50	5.11	1.22	35.67	7.98	8.00	57.98
Medium	550	125	46.80	5.36	1.56	37.44	11.09	6.65	62.10
Fast	650	100	40.10	5.54	1.96	40.10	14.76	4.37	66.73
Spray 2nd coat									
Slow	450	225	53.50	5.11	1.22	23.78	5.72	5.73	41.56
Medium	550	200	46.80	5.36	1.56	23.40	7.58	4.55	42.45
Fast	650	175	40.10	5.54	1.96	22.91	9.43	2.79	42.63
Complete 7 step stain, seal & 2 coat lacquer system (material #11)									
Spray all coats									
Slow	70	60	54.90	32.86	7.90	91.50	25.13	25.18	182.57
Medium	80	50	48.00	36.88	10.66	96.00	35.89	21.53	200.96
Fast	90	40	41.20	40.00	14.12	103.00	48.71	14.41	220.24

These costs are based on painting interior tongue and groove, wood veneer or plain wainscot wood paneling. Do not deduct for openings less than 100 square feet. For heights above 8 feet, use the High Time Difficulty Factors on page 139. ADD for masking-off or cutting-in at wall-to-ceiling intersections and for protecting adjacent surfaces as necessary. "Slow" work is based on an hourly wage of $23.00, "Medium" work on an hourly wage of $29.50, and "Fast" work on an hourly wage of $36.00. Other qualifications that apply to this table are on page 9.

	Frames per manhour	Frames per gallon	Material cost per gallon	Labor cost per frame	Labor burden frame	Material cost per frame	Overhead per frame	Profit per frame	Total price per frame
Window screen frames, paint grade, per frame (15 square feet)									
Undercoat, water or oil base (material #3 or #4)									
Brush 1 coat									
Slow	5	50	42.60	4.60	1.10	.85	1.24	1.25	9.04
Medium	6	45	37.30	4.92	1.43	.83	1.79	1.08	10.05
Fast	7	40	31.95	5.14	1.83	.80	2.40	.71	10.88
Split coat (1/2 undercoat + 1/2 enamel), water or oil base (material #3 or #4 or #9 or #10)									
Brush 1 coat									
Slow	8	60	50.00	2.88	.69	.83	.84	.84	6.08
Medium	9	58	43.78	3.28	.94	.75	1.25	.75	6.97
Fast	10	55	37.50	3.60	1.27	.68	1.72	.51	7.78
Enamel, water or oil base (material #9 or #10)									
Brush each finish coat									
Slow	6	55	57.40	3.83	.93	1.04	1.10	1.10	8.00
Medium	7	53	50.25	4.21	1.23	.95	1.60	.96	8.95
Fast	8	50	43.05	4.50	1.59	.86	2.15	.64	9.74

These figures will apply when painting all sides of wood window screens up to 15 square feet (length times width). Add: Preparation time for protecting adjacent surfaces with masking tape and paper as required. "Slow" work is based on an hourly wage of $23.00, "Medium" work on an hourly wage of $29.50, and "Fast" work on an hourly wage of $36.00. Other qualifications that apply to this table are on page 9.

	Labor SF per manhour	Material coverage SF/gallon	Material cost per gallon	Labor cost per 100 SF	Labor burden 100 SF	Material cost per 100 SF	Overhead per 100 SF	Profit per 100 SF	Total price per 100 SF
Window seats, wood, paint grade, per 100 square feet coated									
Undercoat, water or oil base (material #3 or #4)									
Brush 1 coat									
Slow	20	45	42.60	115.00	27.60	94.67	45.08	45.18	327.53
Medium	25	43	37.30	118.00	34.12	86.74	59.71	35.83	334.40
Fast	30	40	31.95	120.00	42.35	79.88	75.09	22.21	339.53
Split coat (1/2 undercoat + 1/2 enamel), water or oil base (material #3 or #4 or #9 or #10)									
Brush 1 coat									
Slow	30	60	50.00	76.67	18.39	83.33	33.90	33.97	246.26
Medium	35	58	43.78	84.29	24.36	75.48	46.03	27.62	257.78
Fast	40	55	37.50	90.00	31.78	68.18	58.88	17.42	266.26
Enamel, water or oil base (material #9 or #10)									
Brush each finish coat									
Slow	25	55	57.40	92.00	22.08	104.36	41.50	41.59	301.53
Medium	30	53	50.25	98.33	28.42	94.81	55.39	33.23	310.18
Fast	35	50	43.05	102.86	36.30	86.10	69.83	20.66	315.75

Measurements are based on square feet of surface area of each window seat. "Slow" work is based on an hourly wage of $23.00, "Medium" work on an hourly wage of $29.50, and "Fast" work on an hourly wage of $36.00. Other qualifications that apply to this table are on page 9.

	Labor LF per manhour	Material coverage LF/gallon	Material cost per gallon	Labor cost per 100 LF	Labor burden 100 LF	Material cost per 100 LF	Overhead per 100 LF	Profit per 100 LF	Total price per 100 LF
Window sills, wood, paint grade, per 100 linear feet coated									
Undercoat, water or oil base (material #3 or #4)									
Brush 1 coat									
Slow	40	140	42.60	57.50	13.80	30.43	19.33	19.37	140.43
Medium	50	130	37.30	59.00	17.06	28.69	26.19	15.71	146.65
Fast	60	120	31.95	60.00	21.20	26.63	33.42	9.89	151.14
Split coat (1/2 undercoat + 1/2 enamel), water or oil base (material #3 or #4 or #9 or #10)									
Brush 1 coat									
Slow	60	180	50.00	38.33	9.21	27.78	14.31	14.34	103.97
Medium	70	170	43.78	42.14	12.20	25.75	20.02	12.01	112.12
Fast	80	160	37.50	45.00	15.89	23.44	26.14	7.73	118.20
Enamel, water or oil base (material #9 or #10)									
Brush each finish coat									
Slow	50	175	57.40	46.00	11.04	32.80	17.07	17.11	124.02
Medium	60	163	50.25	49.17	14.23	30.83	23.55	14.13	131.91
Fast	70	150	43.05	51.43	18.18	28.70	30.47	9.01	137.79

Measurements are based on linear feet of each window sill. Add: Preparation time for protecting adjacent surfaces with masking tape and paper as required. "Slow" work is based on an hourly wage of $23.00, "Medium" work on an hourly wage of $29.50, and "Fast" work on an hourly wage of $36.00. Other qualifications that apply to this table are on page 9.

	Frames per manhour	Frames per gallon	Material cost per gallon	Labor cost per frame	Labor burden frame	Material cost per frame	Overhead per frame	Profit per frame	Total price per frame
Window storm sash, paint grade, per 15 square feet painted									
Undercoat, water or oil base (material #3 or #4)									
Brush 1 coat									
Slow	3	25	42.60	7.67	1.83	1.70	2.13	2.13	15.46
Medium	4	24	37.30	7.38	2.13	1.55	2.77	1.66	15.49
Fast	5	22	31.95	7.20	2.54	1.45	3.47	1.03	15.69
Split coat (1/2 undercoat + 1/2 enamel), water or oil base (material #3 or #4 or #9 or #10)									
Brush 1 coat									
Slow	5	35	50.00	4.60	1.10	1.43	1.35	1.36	9.84
Medium	6	33	43.78	4.92	1.43	1.33	1.92	1.15	10.75
Fast	7	30	37.50	5.14	1.83	1.25	2.54	.75	11.51
Enamel, water or oil base (material #9 or #10)									
Brush each finish coat									
Slow	4	30	57.40	5.75	1.38	1.91	1.72	1.72	12.48
Medium	5	28	50.25	5.90	1.71	1.79	2.35	1.41	13.16
Fast	6	25	43.05	6.00	2.13	1.72	3.05	.90	13.80

These figures will apply when painting all sides of two-lite wood storm sash measuring up to 15 square feet overall (length times width). Add: Preparation time for protecting adjacent surfaces with window protective coating or masking tape and paper as required. "Slow" work is based on an hourly wage of $23.00, "Medium" work on an hourly wage of $29.50, and "Fast" work on an hourly wage of $36.00. Other qualifications that apply to this table are on page 9.

Windows, wood, exterior or interior, *per window basis*

Windows 15 square feet or smaller

Use the figures in the following tables to estimate the costs for finishing exterior or interior wood windows on a *per window basis* where the windows are 15 square feet or less in area (length times width). For estimating windows larger than 15 square feet, use these same tables and estimate additional time and material proportionately or use the system for windows, wood, exterior or interior *square foot basis* on page 290. Both the *per window basis* and the *square foot basis* include time and material needed to paint the sash (mullions or muntins), trim, frames, jambs, sill and apron on ONE SIDE ONLY. The stain, seal and finish coat systems include one coat of stain sanding sealer, light sanding and one finish coat of either varnish for exterior or lacquer for interior. In addition, finalizing the varnish application usually includes a steel wool buff and wax application with minimum material usage. Add preparation time for sanding, putty and for protecting adjacent surfaces and protecting window panes with window protective coating (wax) or masking tape and paper as required.

For heights above 8 feet, use the High Time Difficulty Factors on page 139. "Slow" work is based on an hourly wage of $23.00, "Medium" work on an hourly wage of $29.50, and "Fast" work on an hourly wage of $36.00. Other qualifications that might apply to the following *per window basis* tables are listed on page 9.

	Manhours per window	Windows per gallon	Material cost per gallon	Labor cost per window	Labor burden window	Material cost per window	Overhead per window	Profit per window	Total price per window
Windows, wood, exterior, *per window basis,* 1, 2 and 3 panes, 15 square feet or smaller									
Undercoat, water or oil base (material #3 or #4)									
Brush 1 coat									
Slow	.25	15.0	42.60	5.75	1.38	2.84	1.89	1.90	13.76
Medium	.20	14.5	37.30	5.90	1.71	2.57	2.55	1.53	14.26
Fast	.15	14.0	31.95	5.40	1.91	2.28	2.97	.88	13.44
Split coat (1/2 undercoat + 1/2 enamel), water or oil base (material #3 or #4 or #9 or #10)									
Brush 1 coat									
Slow	.30	17.0	50.00	6.90	1.66	2.94	2.19	2.19	15.88
Medium	.25	16.5	43.78	7.38	2.13	2.65	3.04	1.82	17.02
Fast	.20	16.0	37.50	7.20	2.54	2.34	3.74	1.11	16.93
Enamel, water or oil base (material #9 or #10)									
Brush each finish coat									
Slow	.35	16.0	57.40	8.05	1.93	3.59	2.58	2.58	18.73
Medium	.30	15.5	50.25	8.85	2.56	3.24	3.66	2.20	20.51
Fast	.25	15.0	43.05	9.00	3.18	2.87	4.67	1.38	21.10
Stain, seal & 1 coat varnish system (material #30)									
Brush each coat									
Slow	.70	11.0	67.30	16.10	3.86	6.12	4.96	4.97	36.01
Medium	.65	10.0	58.90	19.18	5.54	5.89	7.65	4.59	42.85
Fast	.60	9.0	50.50	21.60	7.63	5.61	10.80	3.19	48.83
Varnish (material #30c)									
Brush additional coats of varnish									
Slow	.25	22.0	81.00	5.75	1.38	3.68	2.05	2.06	14.92
Medium	.23	20.0	70.80	6.79	1.96	3.54	3.07	1.84	17.20
Fast	.20	18.0	60.70	7.20	2.54	3.37	4.06	1.20	18.37
Buff & wax after varnish application (material - minimal)									
Steel wool buff									
Slow	.25	--	--	5.75	1.38	--	1.35	1.36	9.84
Medium	.20	--	--	5.90	1.71	--	1.90	1.14	10.65
Fast	.15	--	--	5.40	1.91	--	2.27	.67	10.25
Wax application (material - minimal)									
Slow	.25	--	--	5.75	1.38	--	1.35	1.36	9.84
Medium	.20	--	--	5.90	1.71	--	1.90	1.14	10.65
Fast	.15	--	--	5.40	1.91	--	2.27	.67	10.25

For notes on this table, see the note section on page 279 under Windows, wood, exterior or interior, *per window basis.*

	Manhours per window	Windows per gallon	Material cost per gallon	Labor cost per window	Labor burden window	Material cost per window	Overhead per window	Profit per window	Total price per window
Windows, wood, exterior, *per window basis,* 4 to 6 panes									
Undercoat, water or oil base (material #3 or #4)									
Brush 1 coat									
Slow	.35	14.0	42.60	8.05	1.93	3.04	2.47	2.48	17.97
Medium	.30	13.5	37.30	8.85	2.56	2.76	3.54	2.13	19.84
Fast	.25	13.0	31.95	9.00	3.18	2.46	4.54	1.34	20.52
Split coat (1/2 undercoat + 1/2 enamel), water or oil base (material #3 or #4 or #9 or #10)									
Brush 1 coat									
Slow	.40	16.0	50.00	9.20	2.21	3.13	2.76	2.77	20.07
Medium	.35	15.5	43.78	10.33	2.98	2.82	4.04	2.42	22.59
Fast	.30	15.0	37.50	10.80	3.81	2.50	5.30	1.57	23.98
Enamel, water or oil base (material #9 or #10)									
Brush each finish coat									
Slow	.50	15.0	57.40	11.50	2.76	3.83	3.44	3.44	24.97
Medium	.45	14.5	50.25	13.28	3.83	3.47	5.15	3.09	28.82
Fast	.40	14.0	43.05	14.40	5.08	3.08	6.99	2.07	31.62
Stain, seal & 1 coat varnish system (material #30)									
Brush each coat									
Slow	.98	10.0	67.30	22.54	5.41	6.73	6.59	6.60	47.87
Medium	.88	9.0	58.90	25.96	7.51	6.54	10.00	6.00	56.01
Fast	.78	8.0	50.50	28.08	9.91	6.31	13.73	4.06	62.09
Varnish (material #30c)									
Brush additional coats of varnish									
Slow	.33	21.0	81.00	7.59	1.82	3.86	2.52	2.53	18.32
Medium	.27	19.0	70.80	7.97	2.30	3.73	3.50	2.10	19.60
Fast	.22	17.0	60.70	7.92	2.80	3.57	4.43	1.31	20.03
Buff & wax after varnish application (material - minimal)									
Steel wool buff									
Slow	.33	--	--	7.59	1.82	--	1.79	1.79	12.99
Medium	.27	--	--	7.97	2.30	--	2.57	1.54	14.38
Fast	.22	--	--	7.92	2.80	--	3.32	.98	15.02
Wax application (material - minimal)									
Slow	.33	--	--	7.59	1.82	--	1.79	1.79	12.99
Medium	.27	--	--	7.97	2.30	--	2.57	1.54	14.38
Fast	.22	--	--	7.92	2.80	--	3.32	.98	15.02

For notes on this table, see the note section on page 279 under Windows, wood, exterior or interior, *per window basis.*

	Manhours per window	Windows per gallon	Material cost per gallon	Labor cost per window	Labor burden window	Material cost per window	Overhead per window	Profit per window	Total price per window
Windows, wood, exterior, *per window basis,* 7 to 8 panes									
Undercoat, water or oil base (material #3 or #4)									
Brush 1 coat									
Slow	.45	13	42.60	10.35	2.48	3.28	3.06	3.07	22.24
Medium	.40	13	37.30	11.80	3.41	2.87	4.52	2.71	25.31
Fast	.35	12	31.95	12.60	4.45	2.66	6.11	1.81	27.63
Split coat (1/2 undercoat + 1/2 enamel), water or oil base (material #3 or #4 or #9 or #10)									
Brush 1 coat									
Slow	.55	15	50.00	12.65	3.04	3.33	3.61	3.62	26.25
Medium	.50	15	43.78	14.75	4.27	2.92	5.48	3.29	30.71
Fast	.45	14	37.50	16.20	5.72	2.68	7.63	2.26	34.49
Enamel, water or oil base (material #9 or #10)									
Brush each finish coat									
Slow	.67	14	57.40	15.41	3.70	4.10	4.41	4.42	32.04
Medium	.62	14	50.25	18.29	5.29	3.59	6.79	4.08	38.04
Fast	.57	13	43.05	20.52	7.24	3.31	9.63	2.85	43.55
Stain, seal & 1 coat varnish system (material #30)									
Brush each coat									
Slow	1.20	9	67.30	27.60	6.62	7.48	7.92	7.94	57.56
Medium	1.10	8	58.90	32.45	9.38	7.36	12.30	7.38	68.87
Fast	1.00	7	50.50	36.00	12.71	7.21	17.34	5.13	78.39
Varnish (material #30c)									
Brush additional coats of varnish									
Slow	.38	20	81.00	8.74	2.10	4.05	2.83	2.84	20.56
Medium	.33	18	70.80	9.74	2.81	3.93	4.12	2.47	23.07
Fast	.27	16	60.70	9.72	3.43	3.79	5.25	1.55	23.74
Buff & wax after varnish application (material - minimal)									
Steel wool buff									
Slow	.38	--	--	8.74	2.10	--	2.06	2.06	14.96
Medium	.33	--	--	9.74	2.81	--	3.14	1.88	17.57
Fast	.27	--	--	9.72	3.43	--	4.08	1.21	18.44
Wax application (material - minimal)									
Slow	.38	--	--	8.74	2.10	--	2.06	2.06	14.96
Medium	.33	--	--	9.74	2.81	--	3.14	1.88	17.57
Fast	.27	--	--	9.72	3.43	--	4.08	1.21	18.44

For notes on this table, see the note section on page 279 under Windows, wood, exterior or interior, *per window basis.*

	Manhours per window	Windows per gallon	Material cost per gallon	Labor cost per window	Labor burden window	Material cost per window	Overhead per window	Profit per window	Total price per window
Windows, wood, exterior, *per window basis,* 9 to 11 panes									
Undercoat, water or oil base (material #3 or #4)									
Brush 1 coat									
Slow	.55	12	42.60	12.65	3.04	3.55	3.66	3.66	26.56
Medium	.50	12	37.30	14.75	4.27	3.11	5.53	3.32	30.98
Fast	.45	11	31.95	16.20	5.72	2.90	7.69	2.28	34.79
Split coat (1/2 undercoat + 1/2 enamel), water or oil base (material #3 or #4 or #9 or #10)									
Brush 1 coat									
Slow	.67	14	50.00	15.41	3.70	3.57	4.31	4.32	31.31
Medium	.62	14	43.78	18.29	5.29	3.13	6.68	4.01	37.40
Fast	.57	13	37.50	20.52	7.24	2.88	9.50	2.81	42.95
Enamel, water or oil base (material #9 or #10)									
Brush each finish coat									
Slow	.78	13	57.40	17.94	4.31	4.42	5.07	5.08	36.82
Medium	.73	13	50.25	21.54	6.22	3.87	7.91	4.75	44.29
Fast	.68	12	43.05	24.48	8.64	3.59	11.38	3.37	51.46
Stain, seal & 1 coat varnish system (material #30)									
Brush each coat									
Slow	1.50	7	67.30	34.50	8.28	9.61	9.95	9.97	72.31
Medium	1.40	6	58.90	41.30	11.94	9.82	15.77	9.46	88.29
Fast	1.30	5	50.50	46.80	16.52	10.10	22.76	6.73	102.91
Varnish (material #30c)									
Brush additional coats of varnish									
Slow	.45	19	81.00	10.35	2.48	4.26	3.25	3.25	23.59
Medium	.40	17	70.80	11.80	3.41	4.16	4.84	2.91	27.12
Fast	.35	15	60.70	12.60	4.45	4.05	6.54	1.93	29.57
Buff & wax after varnish application (material - minimal)									
Steel wool buff									
Slow	.45	--	--	10.35	2.48	--	2.44	2.44	17.71
Medium	.40	--	--	11.80	3.41	--	3.80	2.28	21.29
Fast	.35	--	--	12.60	4.45	--	5.29	1.56	23.90
Wax application (material - minimal)									
Slow	.45	--	--	10.35	2.48	--	2.44	2.44	17.71
Medium	.40	--	--	11.80	3.41	--	3.80	2.28	21.29
Fast	.35	--	--	12.60	4.45	--	5.29	1.56	23.90

For notes on this table, see the note section on page 279 under Windows, wood, exterior or interior, *per window basis.*

	Manhours per window	Windows per gallon	Material cost per gallon	Labor cost per window	Labor burden window	Material cost per window	Overhead per window	Profit per window	Total price per window
Windows, wood, exterior, *per window basis,* 12 pane									
Undercoat, water or oil base (material #3 or #4)									
Brush 1 coat									
Slow	.67	11	42.60	15.41	3.70	3.87	4.37	4.38	31.73
Medium	.62	11	37.30	18.29	5.29	3.39	6.74	4.05	37.76
Fast	.57	10	31.95	20.52	7.24	3.20	9.60	2.84	43.40
Split coat (1/2 undercoat + 1/2 enamel), water or oil base (material #3 or #4 or #9 or #10)									
Brush 1 coat									
Slow	.72	13	50.00	16.56	3.97	3.85	4.63	4.64	33.65
Medium	.67	13	43.78	19.77	5.71	3.37	7.21	4.33	40.39
Fast	.62	12	37.50	22.32	7.88	3.13	10.33	3.06	46.72
Enamel, water or oil base (material #9 or #10)									
Brush each finish coat									
Slow	.85	12	57.40	19.55	4.69	4.78	5.51	5.52	40.05
Medium	.80	12	50.25	23.60	6.82	4.19	8.65	5.19	48.45
Fast	.75	11	43.05	27.00	9.53	3.91	12.54	3.71	56.69
Stain, seal & 1 coat varnish system (material #30)									
Brush each coat									
Slow	1.70	6	67.30	39.10	9.38	11.22	11.34	11.37	82.41
Medium	1.60	5	58.90	47.20	13.65	11.78	18.16	10.89	101.68
Fast	1.50	4	50.50	54.00	19.07	12.63	26.56	7.86	120.12
Varnish (material #30c)									
Brush additional coats of varnish									
Slow	.50	18	81.00	11.50	2.76	4.50	3.56	3.57	25.89
Medium	.45	16	70.80	13.28	3.83	4.43	5.39	3.23	30.16
Fast	.40	14	60.70	14.40	5.08	4.34	7.38	2.18	33.38
Buff & wax after varnish application (material - minimal)									
Steel wool buff									
Slow	.50	--	--	11.50	2.76	--	2.71	2.72	19.69
Medium	.45	--	--	13.28	3.83	--	4.28	2.57	23.96
Fast	.40	--	--	14.40	5.08	--	6.04	1.79	27.31
Wax application (material - minimal)									
Slow	.50	--	--	11.50	2.76	--	2.71	2.72	19.69
Medium	.45	--	--	13.28	3.83	--	4.28	2.57	23.96
Fast	.40	--	--	14.40	5.08	--	6.04	1.79	27.31

For notes on this table, see the note section on page 279 under Windows, wood, exterior or interior, *per window basis.*

	Manhour per window	Windows per gallon	Material cost per gallon	Labor cost per window	Labor burden window	Material cost per window	Overhead per window	Profit per window	Total price per window
Windows, wood, interior, *per window basis*, 1, 2 or 3 panes, 15 square feet or smaller									
Undercoat, water or oil base (material #3 or #4)									
Brush 1 coat									
Slow	.25	15	42.60	5.75	1.38	2.84	1.89	1.90	13.76
Medium	.20	15	37.30	5.90	1.71	2.49	2.53	1.52	14.15
Fast	.15	14	31.95	5.40	1.91	2.28	2.97	.88	13.44
Split coat (1/2 undercoat + 1/2 enamel), water or oil base (material #3 or #4 or #9 or #10)									
Brush 1 coat									
Slow	.30	17	50.00	6.90	1.66	2.94	2.19	2.19	15.88
Medium	.25	17	43.78	7.38	2.13	2.58	3.02	1.81	16.92
Fast	.20	16	37.50	7.20	2.54	2.34	3.74	1.11	16.93
Enamel, water or oil base (material #9 or #10)									
Brush each finish coat									
Slow	.35	16	57.40	8.05	1.93	3.59	2.58	2.58	18.73
Medium	.30	16	50.25	8.85	2.56	3.14	3.64	2.18	20.37
Fast	.25	15	43.05	9.00	3.18	2.87	4.67	1.38	21.10
Stain, seal & 1 coat lacquer system, (material #11)									
Brush each coat									
Slow	.70	11	54.90	16.10	3.86	4.99	4.74	4.75	34.44
Medium	.65	10	48.00	19.18	5.54	4.80	7.38	4.43	41.33
Fast	.60	9	41.20	21.60	7.63	4.58	10.48	3.10	47.39
Lacquer (material #11c)									
Brush additional coats of lacquer									
Slow	.25	22	53.50	5.75	1.38	2.43	1.82	1.82	13.20
Medium	.23	20	46.80	6.79	1.96	2.34	2.77	1.66	15.52
Fast	.20	18	40.10	7.20	2.54	2.23	3.71	1.10	16.78

For notes on this table, see the note section on page 279 under Windows, wood, exterior or interior, *per window basis.*

	Manhours per window	Windows per gallon	Material cost per gallon	Labor cost per window	Labor burden window	Material cost per window	Overhead per window	Profit per window	Total price per window
Windows, wood, interior, *per window basis,* 4 to 6 panes									
Undercoat, water or oil base (material #3 or #4)									
Brush 1 coat									
Slow	.35	14	42.60	8.05	1.93	3.04	2.47	2.48	17.97
Medium	.30	14	37.30	8.85	2.56	2.66	3.52	2.11	19.70
Fast	.25	13	31.95	9.00	3.18	2.46	4.54	1.34	20.52
Split coat (1/2 undercoat + 1/2 enamel), water or oil base (material #3 or #4 or #9 or #10)									
Brush 1 coat									
Slow	.40	16	50.00	9.20	2.21	3.13	2.76	2.77	20.07
Medium	.35	16	43.78	10.33	2.98	2.74	4.02	2.41	22.48
Fast	.30	15	37.50	10.80	3.81	2.50	5.30	1.57	23.98
Enamel, water or oil base (material #9 or #10)									
Brush each finish coat									
Slow	.50	15	57.40	11.50	2.76	3.83	3.44	3.44	24.97
Medium	.45	15	50.25	13.28	3.83	3.35	5.12	3.07	28.65
Fast	.40	14	43.05	14.40	5.08	3.08	6.99	2.07	31.62
Stain, seal & 1 coat lacquer system, (material #11)									
Brush each coat									
Slow	.98	10	54.90	22.54	5.41	5.49	6.35	6.37	46.16
Medium	.88	9	48.00	25.96	7.51	5.33	9.70	5.82	54.32
Fast	.78	8	41.20	28.08	9.91	5.15	13.37	3.96	60.47
Lacquer (material #11c)									
Brush additional coats of lacquer									
Slow	.33	21	53.50	7.59	1.82	2.55	2.27	2.28	16.51
Medium	.27	19	46.80	7.97	2.30	2.46	3.18	1.91	17.82
Fast	.22	17	40.10	7.92	2.80	2.36	4.05	1.20	18.33

For notes on this table, see the note section on page 279 under Windows, wood, exterior or interior, *per window basis.*

	Manhours per window	Windows per gallon	Material cost per gallon	Labor cost per window	Labor burden window	Material cost per window	Overhead per window	Profit per window	Total price per window
Windows, wood, interior, *per window basis,* 7 to 8 panes									
Undercoat, water or oil base (material #3 or #4)									
Brush 1 coat									
Slow	.45	13	42.60	10.35	2.48	3.28	3.06	3.07	22.24
Medium	.40	13	37.30	11.80	3.41	2.87	4.52	2.71	25.31
Fast	.35	12	31.95	12.60	4.45	2.66	6.11	1.81	27.63
Split coat (1/2 undercoat + 1/2 enamel), water or oil base (material #3 or #4 or #9 or #10)									
Brush 1 coat									
Slow	.55	15	50.00	12.65	3.04	3.33	3.61	3.62	26.25
Medium	.50	15	43.78	14.75	4.27	2.92	5.48	3.29	30.71
Fast	.45	14	37.50	16.20	5.72	2.68	7.63	2.26	34.49
Enamel, water or oil base (material #9 or #10)									
Brush each finish coat									
Slow	.67	14	57.40	15.41	3.70	4.10	4.41	4.42	32.04
Medium	.62	14	50.25	18.29	5.29	3.59	6.79	4.08	38.04
Fast	.57	13	43.05	20.52	7.24	3.31	9.63	2.85	43.55
Stain, seal & 1 coat lacquer system, (material #11)									
Brush each coat									
Slow	1.20	9	54.90	27.60	6.62	6.10	7.66	7.68	55.66
Medium	1.10	8	48.00	32.45	9.38	6.00	11.96	7.17	66.96
Fast	1.00	7	41.20	36.00	12.71	5.89	16.93	5.01	76.54
Lacquer (material #11c)									
Brush additional coats of lacquer									
Slow	.38	20	53.50	8.74	2.10	2.68	2.57	2.57	18.66
Medium	.33	18	46.80	9.74	2.81	2.60	3.79	2.27	21.21
Fast	.27	16	40.10	9.72	3.43	2.51	4.85	1.44	21.95

For notes on this table, see the note section on page 279 under Windows, wood, exterior or interior, *per window basis.*

	Manhours per window	Windows per gallon	Material cost per gallon	Labor cost per window	Labor burden window	Material cost per window	Overhead per window	Profit per window	Total price per window
Windows, wood, interior, *per window basis,* 9 to 11 panes									
Undercoat, water or oil base (material #3 or #4)									
Brush 1 coat									
Slow	.55	12	42.60	12.65	3.04	3.55	3.66	3.66	26.56
Medium	.50	12	37.30	14.75	4.27	3.11	5.53	3.32	30.98
Fast	.45	11	31.95	16.20	5.72	2.90	7.69	2.28	34.79
Split coat (1/2 undercoat + 1/2 enamel), water or oil base (material #3 or #4 or #9 or #10)									
Brush 1 coat									
Slow	.67	14	50.00	15.41	3.70	3.57	4.31	4.32	31.31
Medium	.62	14	43.78	18.29	5.29	3.13	6.68	4.01	37.40
Fast	.57	13	37.50	20.52	7.24	2.88	9.50	2.81	42.95
Enamel, water or oil base (material #9 or #10)									
Brush each finish coat									
Slow	.78	13	57.40	17.94	4.31	4.42	5.07	5.08	36.82
Medium	.73	13	50.25	21.54	6.22	3.87	7.91	4.75	44.29
Fast	.68	12	43.05	24.48	8.64	3.59	11.38	3.37	51.46
Stain, seal & 1 coat lacquer system, (material #11)									
Brush each coat									
Slow	1.50	8	54.90	34.50	8.28	6.86	9.43	9.45	68.52
Medium	1.40	7	48.00	41.30	11.94	6.86	15.03	9.02	84.15
Fast	1.30	6	41.20	46.80	16.52	6.87	21.76	6.44	98.39
Lacquer (material #11c)									
Brush additional coats of lacquer									
Slow	.45	19	53.50	10.35	2.48	2.82	2.97	2.98	21.60
Medium	.40	17	46.80	11.80	3.41	2.75	4.49	2.69	25.14
Fast	.35	15	40.10	12.60	4.45	2.67	6.11	1.81	27.64

For notes on this table, see the note section on page 279 under Windows, wood, exterior or interior, *per window basis.*

	Manhours per window	Windows per gallon	Material cost per gallon	Labor cost per window	Labor burden window	Material cost per window	Overhead per window	Profit per window	Total price per window
Windows, wood, interior, *per window basis,* 12 pane									
Undercoat, water or oil base (material #3 or #4)									
Brush 1 coat									
Slow	.67	11	42.60	15.41	3.70	3.87	4.37	4.38	31.73
Medium	.62	11	37.30	18.29	5.29	3.39	6.74	4.05	37.76
Fast	.57	10	31.95	20.52	7.24	3.20	9.60	2.84	43.40
Split coat (1/2 undercoat + 1/2 enamel), water or oil base (material #3 or #4 or #9 or #10)									
Brush 1 coat									
Slow	.72	13	50.00	16.56	3.97	3.85	4.63	4.64	33.65
Medium	.67	13	43.78	19.77	5.71	3.37	7.21	4.33	40.39
Fast	.62	12	37.50	22.32	7.88	3.13	10.33	3.06	46.72
Enamel, water or oil base (material #9 or #10)									
Brush each finish coat									
Slow	.85	12	57.40	19.55	4.69	4.78	5.51	5.52	40.05
Medium	.80	12	50.25	23.60	6.82	4.19	8.65	5.19	48.45
Fast	.75	11	43.05	27.00	9.53	3.91	12.54	3.71	56.69
Stain, seal & 1 coat lacquer system, (material #11)									
Brush each coat									
Slow	1.70	7	54.90	39.10	9.38	7.84	10.70	10.72	77.74
Medium	1.60	6	48.00	47.20	13.65	8.00	17.21	10.33	96.39
Fast	1.50	5	41.20	54.00	19.07	8.24	25.20	7.46	113.97
Lacquer (material #11c)									
Brush additional coats of lacquer									
Slow	.50	18	53.50	11.50	2.76	2.97	3.27	3.28	23.78
Medium	.45	16	46.80	13.28	3.83	2.93	5.01	3.01	28.06
Fast	.40	14	40.10	14.40	5.08	2.86	6.93	2.05	31.32

For notes on this table, see the note section on page 279 under Windows, wood, exterior or interior, *per window basis.*

Windows, wood, exterior or interior, *square foot basis*

Windows larger than 15 square feet

Use the figures in the following tables to estimate exterior or interior wood windows on a *square foot basis* where the windows are larger than 15 square feet in area (length times width). For estimating windows 15 square feet or smaller, use the system for windows, wood exterior, or interior on a *per window basis*. Both the square foot basis and the per window basis include time and material needed to paint the sash (mullions or muntins), trim, frames, jambs, sill and apron on ONE SIDE ONLY. The stain, seal and finish coat systems include one coat of stain, sanding sealer, light sanding and one finish coat of either varnish for exterior or lacquer for interior. In addition, finalizing the varnish application usually includes a steel wool buff and wax application with minimum material usage. Add preparation time for sanding, putty, protecting adjacent surfaces and protecting window panes with window protective coating (wax) or masking tape and paper as required. For heights above 8 feet, use the High Time Difficulty Factors on page 139. "Slow" work is based on an hourly wage of $23.00, "Medium" work on an hourly wage of $29.50, and "Fast" work on an hourly wage of $36.00. Other qualifications that might apply to the following *per window basis* tables are listed on page 9.

Example calculation: Measure each window and add 1 foot to each dimension before calculating the area. For example, a window measuring 4'0" x 4'0" with 1 foot added to the top, bottom, right side and left side is now a 6 x 6 dimension or 36 square feet. Then, add an additional 2 square feet for each window pane, with allows time to finish the mullions, muntins and sash. The square footage calculation for this six pane window would be 36 + (2 x 6) or 48 square feet. Use this number and apply it to the appropriate manhour and material coverage figures in the table.

	Labor SF per manhour	Material coverage SF/gallon	Material cost per gallon	Labor cost per 100 SF	Labor burden 100 SF	Material cost per 100 SF	Overhead per 100 SF	Profit per 100 SF	Total price per 100 SF
Paint grade									
Undercoat, water or oil base (material #3 or #4)									
Brush 1 coat									
Slow	150	460	42.60	15.33	3.69	9.26	5.37	5.38	39.03
Medium	165	450	37.30	17.88	5.17	8.29	7.84	4.70	43.88
Fast	180	440	31.95	20.00	7.08	7.26	10.64	3.15	48.13
Split coat (1/2 undercoat + 1/2 enamel), water or oil base (material #3 or #4 or #9 or #10)									
Brush 1 coat									
Slow	120	520	50.00	19.17	4.59	9.62	6.34	6.36	46.08
Medium	135	500	43.78	21.85	6.33	8.76	9.23	5.54	51.71
Fast	150	480	37.50	24.00	8.49	7.81	12.49	3.69	56.48
Enamel, water or oil base (material #9 or #10)									
Brush each finish coat									
Slow	100	500	57.40	23.00	5.52	11.48	7.60	7.62	55.22
Medium	113	480	50.25	26.11	7.55	10.47	11.03	6.62	61.78
Fast	125	460	43.05	28.80	10.17	9.36	14.98	4.43	67.74

	Labor SF per manhour	Material coverage SF/gallon	Material cost per gallon	Labor cost per 100 SF	Labor burden 100 SF	Material cost per 100 SF	Overhead per 100 SF	Profit per 100 SF	Total price per 100 SF
Stain grade, exterior									
Stain grade: stain, seal & 1 coat varnish system (material #30)									
Brush all coats									
Slow	45	335	67.30	51.11	12.26	20.09	15.86	15.89	115.21
Medium	55	305	58.90	53.64	15.50	19.31	22.11	13.27	123.83
Fast	65	280	50.50	55.38	19.54	18.04	28.82	8.53	130.31
Brush additional coats of varnish (material #30c)									
Slow	140	675	81.00	16.43	3.93	12.00	6.15	6.16	44.67
Medium	165	620	70.80	17.88	5.17	11.42	8.62	5.17	48.26
Fast	190	565	60.70	18.95	6.67	10.74	11.28	3.34	50.98
Buff & wax after varnish application (material - minimal)									
Steel wool buff									
Slow	140	--	--	16.43	3.93	--	3.87	3.88	28.11
Medium	165	--	--	17.88	5.17	--	5.76	3.46	32.27
Fast	190	--	--	18.95	6.67	--	7.95	2.35	35.92
Wax application									
Slow	140	--	--	16.43	3.93	--	3.87	3.88	28.11
Medium	165	--	--	17.88	5.17	--	5.76	3.46	32.27
Fast	190	--	--	18.95	6.67	--	7.95	2.35	35.92
Stain grade, interior									
Stain, seal & 1 coat lacquer system (material #11)									
Brush all coats									
Slow	45	335	54.90	51.11	12.26	16.39	15.16	15.19	110.11
Medium	55	305	48.00	53.64	15.50	15.74	21.22	12.73	118.83
Fast	65	280	41.20	55.38	19.54	14.71	27.79	8.22	125.64
Brush additional coats of lacquer (material #11c)									
Slow	140	675	53.50	16.43	3.93	7.93	5.38	5.39	39.06
Medium	165	620	46.80	17.88	5.17	7.55	7.65	4.59	42.84
Fast	190	565	40.10	18.95	6.67	7.10	10.15	3.00	45.87

For notes on this table, see page 290 under Windows, wood, exterior or interior, *square foot basis.*

Window conversion factors -- a window area calculation shortcut

Panes	Manhours per SF conversion factor	Material per SF conversion factor
1, 2 or 3 panes	L x W x 2.0	L x W x 2.0
4 to 6 panes	L x W x 3.0	L x W x 2.2
7 to 8 panes	L x W x 4.0	L x W x 2.4
9 to 11 panes	L x W x 5.0	L x W x 2.6
12 panes	L x W x 6.0	L x W x 2.9

Use this table in conjunction with the Windows, exterior or interior, *square foot basis* table on page 290 for a guide to calculating window area on a "square foot basis." To convert the window area, calculate the actual square shortcut footage of a window, say, 4'0" x 4'0" or 16 square feet, and look in the conversion table under the number panes, say 6. The conversion factor of a 6-pane window is 3. Multiply the 16 square feet by 3 to equal 48 which is the number used in the square foot basis tables on pages 290 and 291. To undercoat at a slow rate, divide 48 by 150 to come up with .32 hours. Then divide 48 by 460 to result in .10 gallons to undercoat that window. Divide 1 gallon by .1 to find that you can undercoat 10 windows with 1 gallon of undercoat material.

	Labor SF per manhour	Material coverage SF/gallon	Material cost per gallon	Labor cost per 100 SF	Labor burden 100 SF	Material cost per 100 SF	Overhead per 100 SF	Profit per 100 SF	Total price per 100 SF
Wine racks, paint grade, spray, square feet of face									
Undercoat, water base (material #3)									
Spray 1 coat									
Slow	50	100	36.90	46.00	11.04	36.90	17.85	17.89	129.68
Medium	65	88	32.30	45.38	13.11	36.70	23.80	14.28	133.27
Fast	80	75	27.70	45.00	15.89	36.93	30.32	8.97	137.11
Undercoat, oil base (material #4)									
Spray 1 coat									
Slow	50	100	48.30	46.00	11.04	48.30	20.01	20.06	145.41
Medium	65	88	42.30	45.38	13.11	48.07	26.64	15.98	149.18
Fast	80	75	36.20	45.00	15.89	48.27	33.84	10.01	153.01
Split coat (1/2 undercoat + 1/2 enamel), water base (material #3 or #9)									
Spray 1 coat									
Slow	75	150	45.10	30.67	7.35	30.07	12.94	12.97	94.00
Medium	90	138	39.50	32.78	9.47	28.62	17.72	10.63	99.22
Fast	125	125	33.85	28.80	10.17	27.08	20.48	6.06	92.59
Split coat (1/2 undercoat + 1/2 enamel), oil base (material #4 or #10)									
Spray 1 coat									
Slow	75	150	54.90	30.67	7.35	36.60	14.18	14.21	103.01
Medium	90	138	48.05	32.78	9.47	34.82	19.27	11.56	107.90
Fast	125	125	41.15	28.80	10.17	32.92	22.29	6.59	100.77
Enamel, water base (material #9)									
Spray 1st finish coat									
Slow	65	125	53.30	35.38	8.48	42.64	16.44	16.47	119.41
Medium	88	113	46.70	33.52	9.68	41.33	21.14	12.68	118.35
Fast	100	100	40.00	36.00	12.71	40.00	27.50	8.13	124.34
Spray additional finish coats									
Slow	75	150	53.30	30.67	7.35	35.53	13.98	14.01	101.54
Medium	100	138	46.70	29.50	8.53	33.84	17.97	10.78	100.62
Fast	125	125	40.00	28.80	10.17	32.00	22.00	6.51	99.48
Enamel, oil base (material #10)									
Spray 1st finish coat									
Slow	65	125	61.50	35.38	8.48	49.20	17.68	17.72	128.46
Medium	88	113	53.80	33.52	9.68	47.61	22.71	13.62	127.14
Fast	100	100	46.10	36.00	12.71	46.10	29.39	8.69	132.89
Spray additional finish coats									
Slow	75	150	61.50	30.67	7.35	41.00	15.02	15.05	109.09
Medium	100	138	53.80	29.50	8.53	38.99	19.26	11.55	107.83
Fast	125	125	46.10	28.80	10.17	36.88	23.51	6.96	106.32

These figures include coating all interior and exterior surfaces and are based on overall dimensions (length times width) of the wine rack face. For heights above 8 feet, use the High Time Difficulty Factors on page 139. "Slow" work is based on an hourly wage of $23.00, "Medium" work on an hourly wage of $29.50, and "Fast" work on an hourly wage of $36.00. Other qualifications that apply to this table are on page 9.

	Labor SF per manhour	Material coverage SF/gallon	Material cost per gallon	Labor cost per 100 SF	Labor burden 100 SF	Material cost per 100 SF	Overhead per 100 SF	Profit per 100 SF	Total price per 100 SF
Wine racks, stain grade, per square feet of face									
Stain, seal and 2 coat lacquer system (7 step process)									
STEP 1: Sand & putty									
Slow	50	--	--	46.00	11.04	--	10.84	10.86	78.74
Medium	75	--	--	39.33	11.36	--	12.68	7.61	70.98
Fast	100	--	--	36.00	12.71	--	15.10	4.47	68.28
STEP 2 & 3: Wiping stain, oil base (material #11a) & wipe									
Spray 1 coat & wipe									
Slow	100	150	61.60	23.00	5.52	41.07	13.22	13.25	96.06
Medium	175	113	53.90	16.86	4.86	47.70	17.36	10.41	97.19
Fast	225	75	46.20	16.00	5.63	61.60	25.81	7.63	116.67
STEP 4: Sanding sealer (material #11b)									
Spray 1 coat & sand									
Slow	130	150	50.90	17.69	4.24	33.93	10.62	10.64	77.12
Medium	208	113	44.60	14.18	4.11	39.47	14.44	8.66	80.86
Fast	275	75	38.20	13.09	4.64	50.93	21.28	6.29	96.23
STEP 5: Sand lightly									
Slow	75	--	--	30.67	7.35	--	7.23	7.24	52.49
Medium	100	--	--	29.50	8.53	--	9.51	5.70	53.24
Fast	125	--	--	28.80	10.17	--	12.08	3.57	54.62
STEP 6 & 7: Lacquer (material #11c)									
Spray 1st coat									
Slow	100	100	53.50	23.00	5.52	53.50	15.58	15.62	113.22
Medium	200	75	46.80	14.75	4.27	62.40	20.35	12.21	113.98
Fast	300	50	40.10	12.00	4.22	80.20	29.90	8.84	135.16
Spray 2nd coat									
Slow	175	100	53.50	13.14	3.14	53.50	13.26	13.29	96.33
Medium	313	75	46.80	9.42	2.71	62.40	18.64	11.18	104.35
Fast	450	50	40.10	8.00	2.81	80.20	28.22	8.35	127.58
Complete 7 step stain, seal & 2 coat lacquer system (material #11)									
Spray all coats									
Slow	20	25	54.90	115.00	27.60	219.60	68.82	68.96	499.98
Medium	28	20	48.00	105.36	30.45	240.00	93.95	56.37	526.13
Fast	35	15	41.20	102.86	36.30	274.67	128.29	37.95	580.07

These figures include coating all interior and exterior surfaces and are based on overall dimensions (length times width) of the wine rack face. For heights above 8 feet, use the High Time Difficulty Factors on page 139. "Slow" work is based on an hourly wage of $23.00, "Medium" work on an hourly wage of $29.50, and "Fast" work on an hourly wage of $36.00. Other qualifications that apply to this table are on page 9.

Part II

Preparation Costs

	Labor LF per manhour	Material coverage LF/gallon	Material cost per gallon	Labor cost per 100 LF	Labor burden 100 LF	Material cost per 100 LF	Overhead per 100 LF	Profit per 100 LF	Total price per 100 LF
Acid wash gutters & downspouts									
Acid wash, muriatic acid (material #49)									
Brush or mitt 1 coat									
Slow	80	450	21.20	28.75	6.90	4.71	7.67	7.68	55.71
Medium	95	425	18.50	31.05	9.00	4.35	11.09	6.66	62.15
Fast	110	400	15.90	32.73	11.55	3.98	14.96	4.43	67.65

For heights above one story, use the High Time Difficulty Factors on page 139. "Slow" work is based on an hourly wage of $23.00, "Medium" work on an hourly wage of $29.50, and "Fast" work on an hourly wage of $36.00. Other qualifications that apply to this table are on page 9.

	Labor SF per manhour	Material coverage SF/gallon	Material cost per gallon	Labor cost per 100 SF	Labor burden 100 SF	Material cost per 100 SF	Overhead per 100 SF	Profit per 100 SF	Total price per 100 SF
Airblast, compressed air									
Average production									
Slow	150	--	--	15.33	3.69	--	3.61	3.62	26.25
Medium	175	--	--	16.86	4.86	--	5.43	3.26	30.41
Fast	200	--	--	18.00	6.36	--	7.55	2.23	34.14

All widths less than 12", consider as 1 square foot per linear foot. Add equipment rental costs with Overhead and Profit. For heights above 8 feet, use the High Time Difficulty Factors on page 139. "Slow" work is based on an hourly wage of $23.00, "Medium" work on an hourly wage of $29.50, and "Fast" work on an hourly wage of $36.00. Other qualifications that apply to this table are on page 9.

	Labor SF per manhour	Material coverage SF/gallon	Material cost per gallon	Labor cost per 100 SF	Labor burden 100 SF	Material cost per 100 SF	Overhead per 100 SF	Profit per 100 SF	Total price per 100 SF
Burn off paint									
Exterior:									
Exterior trim									
Slow	15	--	--	153.33	36.81	--	36.12	36.20	262.46
Medium	20	--	--	147.50	42.65	--	47.53	28.52	266.20
Fast	25	--	--	144.00	50.84	--	60.40	17.87	273.11
Plain surfaces									
Slow	30	--	--	76.67	18.39	--	18.06	18.10	131.22
Medium	40	--	--	73.75	21.33	--	23.77	14.26	133.11
Fast	50	--	--	72.00	25.42	--	30.20	8.93	136.55
Beveled wood siding									
Slow	20	--	--	115.00	27.60	--	27.09	27.15	196.84
Medium	30	--	--	98.33	28.42	--	31.69	19.01	177.45
Fast	40	--	--	90.00	31.78	--	37.75	11.17	170.70
Interior:									
Interior trim									
Slow	10	--	--	230.00	55.20	--	54.19	54.30	393.69
Medium	15	--	--	196.67	56.88	--	63.38	38.03	354.96
Fast	20	--	--	180.00	63.55	--	75.50	22.33	341.38
Plain surfaces									
Slow	15	--	--	153.33	36.81	--	36.12	36.20	262.46
Medium	25	--	--	118.00	34.12	--	38.03	22.82	212.97
Fast	35	--	--	102.86	36.30	--	43.14	12.76	195.06

All widths less than 12", consider as 1 square foot per linear foot. Note: Because surfaces and the material being removed vary widely, it's best to quote prices for burning-off existing finishes on a Time and Material or Cost Plus Fee basis at a preset hourly rate. "Slow" work is based on an hourly wage of $23.00, "Medium" work on an hourly wage of $29.50, and "Fast" work on an hourly wage of $36.00. Other qualifications that apply to this table are on page 9.

	Labor LF per manhour	Material LF/fluid oz ounce	Material cost per ounce	Labor cost per 100 LF	Labor burden 100 LF	Material cost per 100 LF	Overhead per 100 LF	Profit per 100 LF	Total price per 100 LF
Caulk									
1/8" gap (material #42)									
Slow	60	14	.55	38.33	9.21	3.93	9.78	9.80	71.05
Medium	65	13	.49	45.38	13.11	3.77	15.57	9.34	87.17
Fast	70	12	.42	51.43	18.18	3.50	22.65	6.70	102.46
1/4" gap (material #42)									
Slow	50	3.5	.55	46.00	11.04	15.71	13.82	13.85	100.42
Medium	55	3.3	.49	53.64	15.50	14.85	21.00	12.60	117.59
Fast	60	3.0	.42	60.00	21.20	14.00	29.51	8.73	133.44
3/8" gap (material #42)									
Slow	40	1.5	.55	57.50	13.80	36.67	20.51	20.56	149.04
Medium	45	1.4	.49	65.56	18.94	35.00	29.88	17.93	167.31
Fast	50	1.3	.42	72.00	25.42	32.31	40.22	11.90	181.85
1/2" gap (material #42)									
Slow	33	1.0	.55	69.70	16.72	55.00	26.87	26.93	195.22
Medium	38	0.9	.49	77.63	22.46	54.44	38.63	23.18	216.34
Fast	43	0.8	.42	83.72	29.58	52.50	51.39	15.20	232.39

Caulking that's part of normal surface preparation is included in the painting cost tables. When extra caulking is required, use this cost guide. It's based on oil or latex base, silicone or urethane caulk in 10 ounce tubes. "Slow" work is based on an hourly wage of $23.00, "Medium" work on an hourly wage of $29.50, and "Fast" work on an hourly wage of $36.00. Other qualifications that apply to this table are on page 9.

	Labor SF per manhour	Material coverage SF/gallon	Material cost per gallon	Labor cost per 100 SF	Labor burden 100 SF	Material cost per 100 SF	Overhead per 100 SF	Profit per 100 SF	Total price per 100 SF
Cut cracks									
Varnish or hard oil and repair cracks									
Slow	120	--	--	19.17	4.59	--	4.52	4.53	32.81
Medium	130	--	--	22.69	6.56	--	7.31	4.39	40.95
Fast	140	--	--	25.71	9.07	--	10.78	3.19	48.75
Gloss painted walls and fix cracks									
Slow	125	--	--	18.40	4.42	--	4.34	4.35	31.51
Medium	135	--	--	21.85	6.33	--	7.04	4.22	39.44
Fast	145	--	--	24.83	8.78	--	10.41	3.08	47.10

All widths less than 12", consider as 1 square foot per linear foot. "Slow" work is based on an hourly wage of $23.00, "Medium" work on an hourly wage of $29.50, and "Fast" work on an hourly wage of $36.00. Other qualifications that apply to this table are on page 9.

	Labor SF per manhour	Material coverage SF/gallon	Material cost per gallon	Labor cost per 100 SF	Labor burden 100 SF	Material cost per 100 SF	Overhead per 100 SF	Profit per 100 SF	Total price per 100 SF
Fill wood floors									
Fill and wipe wood floors (material #48)									
Slow	45	155	55.40	51.11	12.26	35.74	18.83	18.87	136.81
Medium	60	145	48.40	49.17	14.23	33.38	24.19	14.51	135.48
Fast	75	135	41.50	48.00	16.93	30.74	29.66	8.77	134.10

All widths less than 12", consider as 1 square foot per linear foot. "Slow" work is based on an hourly wage of $23.00, "Medium" work on an hourly wage of $29.50, and "Fast" work on an hourly wage of $36.00. Other qualifications that apply to this table are on page 9.

	Labor SF per manhour	Material coverage SF/pound	Material cost per pound	Labor cost per 100 SF	Labor burden 100 SF	Material cost per 100 SF	Overhead per 100 SF	Profit per 100 SF	Total price per 100 SF
Putty application									
Good condition, 1 coat (material #45)									
Slow	60	150	8.00	38.33	9.21	5.33	10.04	10.06	72.97
Medium	90	135	7.00	32.78	9.47	5.19	11.86	7.12	66.42
Fast	120	120	6.00	30.00	10.58	5.00	14.13	4.18	63.89
Average condition, 1 coat (material #45)									
Slow	35	90	8.00	65.71	15.77	8.89	17.17	17.21	124.75
Medium	65	75	7.00	45.38	13.11	9.33	16.96	10.17	94.95
Fast	95	60	6.00	37.89	13.40	10.00	18.99	5.62	85.90
Poor condition, 1 coat (material #45)									
Slow	15	40	8.00	153.33	36.81	20.00	39.92	40.01	290.07
Medium	30	30	7.00	98.33	28.42	23.33	37.52	22.51	210.11
Fast	45	20	6.00	80.00	28.23	30.00	42.85	12.68	193.76

These figures apply to either spackle or Swedish putty. All widths less than 12", consider as 1 square foot per linear foot. For heights above 8 feet, use the High Time Difficulty Factors on page 139. For flat trim or sash: Estimate 1 linear foot of trim as 1 square foot of surface. "Slow" work is based on an hourly wage of $23.00, "Medium" work on an hourly wage of $29.50, and "Fast" work on an hourly wage of $36.00. Other qualifications that apply to this table are on page 9.

	Labor SF per manhour	Material coverage SF/gallon	Material cost per gallon	Labor cost per 100 SF	Labor burden 100 SF	Material cost per 100 SF	Overhead per 100 SF	Profit per 100 SF	Total price per 100 SF
Sand, medium (before first coat)									
Interior flatwall areas									
Slow	275	--	--	8.36	2.02	--	2.59	2.07	15.04
Medium	300	--	--	9.83	2.83	--	3.17	1.90	17.73
Fast	325	--	--	11.08	3.92	--	4.65	1.37	21.02
Interior enamel areas									
Slow	250	--	--	9.20	2.21	--	2.17	2.17	15.75
Medium	275	--	--	10.73	3.11	--	3.46	2.07	19.37
Fast	300	--	--	12.00	4.22	--	5.03	1.49	22.74

All widths less than 12", consider as 1 square foot per linear foot. For heights above 8 feet, use the High Time Difficulty Factors on page 139. "Slow" work is based on an hourly wage of $23.00, "Medium" work on an hourly wage of $29.50, and "Fast" work on an hourly wage of $36.00. Other qualifications that apply to this table are on page 9.

	Labor SF per manhour	Material coverage SF/gallon	Material cost per gallon	Labor cost per 100 SF	Labor burden 100 SF	Material cost per 100 SF	Overhead per 100 SF	Profit per 100 SF	Total price per 100 SF
Sand & putty (before second coat)									
Interior flatwall areas									
Slow	190	--	--	12.11	2.89	--	2.85	2.86	20.71
Medium	200	--	--	14.75	4.27	--	4.75	2.85	26.62
Fast	210	--	--	17.14	6.05	--	7.19	2.13	32.51
Interior enamel areas									
Slow	110	--	--	20.91	5.01	--	4.93	4.94	35.79
Medium	125	--	--	23.60	6.82	--	7.61	4.56	42.59
Fast	140	--	--	25.71	9.07	--	10.78	3.19	48.75
Exterior siding & trim - plain									
Slow	180	--	--	12.78	3.08	--	3.01	3.02	21.89
Medium	200	--	--	14.75	4.27	--	4.75	2.85	26.62
Fast	220	--	--	16.36	5.80	--	6.86	2.03	31.05
Exterior trim only									
Slow	100	--	--	23.00	5.52	--	5.42	5.43	39.37
Medium	110	--	--	26.82	7.75	--	8.64	5.19	48.40
Fast	120	--	--	30.00	10.58	--	12.58	3.72	56.88
Bookshelves									
Slow	100	--	--	23.00	5.52	--	5.42	5.43	39.37
Medium	125	--	--	23.60	6.82	--	7.61	4.56	42.59
Fast	150	--	--	24.00	8.49	--	10.07	2.98	45.54
Cabinets									
Slow	125	--	--	18.40	4.42	--	4.34	4.35	31.51
Medium	150	--	--	19.67	5.70	--	6.34	3.80	35.51
Fast	175	--	--	20.57	7.24	--	8.63	2.55	38.99

All trim which is less than 12" wide, consider to be 12" wide. For heights above 8 feet, use the High Time Difficulty Factors on page 139. High grade work - Use the manhours equal to 1 coat of paint. For medium grade work - Use half (50%) of the manhours for 1 coat of paint. "Slow" work is based on an hourly wage of $23.00, "Medium" work on an hourly wage of $29.50, and "Fast" work on an hourly wage of $36.00. Other qualifications that apply to this table are on page 9.

Sand, light (before third coat)

	Labor SF per manhour	Material coverage SF/gallon	Material cost per gallon	Labor cost per 100 SF	Labor burden 100 SF	Material cost per 100 SF	Overhead per 100 SF	Profit per 100 SF	Total price per 100 SF
Interior flatwall areas									
Slow	335	--	--	6.87	1.66	--	1.62	1.62	11.77
Medium	345	--	--	8.55	2.48	--	2.76	1.65	15.44
Fast	355	--	--	10.14	3.60	--	4.25	1.26	19.25
Interior enamel areas									
Slow	130	--	--	17.69	4.24	--	4.17	4.18	30.28
Medium	140	--	--	21.07	6.08	--	6.79	4.07	38.01
Fast	150	--	--	24.00	8.49	--	10.07	2.98	45.54
Exterior siding & trim - plain									
Slow	250	--	--	9.20	2.21	--	2.17	2.17	15.75
Medium	275	--	--	10.73	3.11	--	3.46	2.07	19.37
Fast	300	--	--	12.00	4.22	--	5.03	1.49	22.74
Exterior trim only									
Slow	150	--	--	15.33	3.69	--	3.61	3.62	26.25
Medium	175	--	--	16.86	4.86	--	5.43	3.26	30.41
Fast	200	--	--	18.00	6.36	--	7.55	2.23	34.14
Bookshelves									
Slow	175	--	--	13.14	3.14	--	3.10	3.10	22.48
Medium	225	--	--	13.11	3.78	--	4.23	2.54	23.66
Fast	275	--	--	13.09	4.64	--	5.49	1.62	24.84
Cabinets									
Slow	200	--	--	11.50	2.76	--	2.71	2.72	19.69
Medium	250	--	--	11.80	3.41	--	3.80	2.28	21.29
Fast	300	--	--	12.00	4.22	--	5.03	1.49	22.74

All widths less than 12", consider 1 square foot per linear foot. For heights above 8 feet, use the High Time Difficulty Factors on page 139. "Slow" work is based on an hourly wage of $23.00, "Medium" work on an hourly wage of $29.50, and "Fast" work on an hourly wage of $36.00. Other qualifications that apply to this table are on page 9.

Sand, extra fine, flat surfaces, varnish

	Labor SF per manhour	Material coverage SF/gallon	Material cost per gallon	Labor cost per 100 SF	Labor burden 100 SF	Material cost per 100 SF	Overhead per 100 SF	Profit per 100 SF	Total price per 100 SF
Sand or steel wool									
Slow	50	--	--	46.00	11.04	--	10.84	10.86	78.74
Medium	88	--	--	33.52	9.68	--	10.80	6.48	60.48
Fast	125	--	--	28.80	10.17	--	12.08	3.57	54.62

All widths less than 12", consider as 1 square foot per linear foot. For heights above 8 feet, use the High Time Difficulty Factors on page 139. "Slow" work is based on an hourly wage of $23.00, "Medium" work on an hourly wage of $29.50, and "Fast" work on an hourly wage of $36.00. Other qualifications that apply to this table are on page 9.

Sandblast, general

Sandblasting production rates may vary widely. Use the following figures as a reference for estimating and to establish performance data for your company. The abrasive material used in sandblasting is usually white silica sand although slags have recently gained in popularity. Material consumption varies with several factors:

1) Type of finish required
2) Condition of the surface
3) Quality of abrasive material (sharpness, cleanliness and hardness)
4) Nozzle size
5) Equipment arrangement and placement
6) Operator skill

All material consumption values are based on three uses of a 25 to 35 mesh white silica sand abrasive at a cost of $40 to $60 per ton. (Check the current price in your area.) Note: See the Structural Steel Conversion table at Figure 23 on pages 391 through 399 for converting linear feet or tons of structural steel to square feet.

	Labor SF per manhour	Material coverage pounds/SF	Material cost per pound	Labor cost per 100 SF	Labor burden 100 SF	Material cost per 100 SF	Overhead per 100 SF	Profit per 100 SF	Total price per 100 SF
Sandblast, brush-off blast									
Surface condition basis - large projects & surface areas (material #46)									
Remove cement base paint									
Slow	150	2.0	.66	15.33	3.69	132.00	28.69	28.75	208.46
Medium	175	2.5	.58	16.86	4.86	145.00	41.68	25.01	233.41
Fast	200	3.0	.50	18.00	6.36	150.00	54.05	15.99	244.40
Remove oil or latex base paint									
Slow	100	3.0	.66	23.00	5.52	198.00	43.04	43.13	312.69
Medium	125	3.5	.58	23.60	6.82	203.00	58.36	35.01	326.79
Fast	150	4.0	.50	24.00	8.49	200.00	72.07	21.32	325.88
Surface area basis - large projects & surface areas (material #46)									
Pipe up to 12" O/D									
Slow	125	4.0	.66	18.40	4.42	264.00	54.50	54.61	395.93
Medium	150	4.5	.58	19.67	5.70	261.00	71.59	42.95	400.91
Fast	175	5.0	.50	20.57	7.24	250.00	86.13	25.48	389.42
Structural steel									
Sizes up to 2 SF/LF									
Slow	150	4.0	.66	15.33	3.69	264.00	53.77	53.88	390.67
Medium	175	4.5	.58	16.86	4.86	261.00	70.68	42.41	395.81
Fast	200	5.0	.50	18.00	6.36	250.00	85.05	25.16	384.57
Sizes from 2 to 5 SF/LF									
Slow	200	3.0	.66	11.50	2.76	198.00	40.33	40.41	293.00
Medium	225	3.5	.58	13.11	3.78	203.00	54.98	32.99	307.86
Fast	250	4.0	.50	14.40	5.08	200.00	68.04	20.13	307.65
Sizes over 5 SF/LF									
Slow	250	2.0	.66	9.20	2.21	132.00	27.25	27.31	197.97
Medium	275	3.0	.58	10.73	3.11	174.00	46.96	28.17	262.97
Fast	300	4.0	.50	12.00	4.22	200.00	67.03	19.83	303.08
Tanks and vessels									
Sizes up to 12'0" O/D									
Slow	200	3.0	.66	11.50	2.76	198.00	40.33	40.41	293.00
Medium	225	3.5	.58	13.11	3.78	203.00	54.98	32.99	307.86
Fast	250	4.0	.50	14.40	5.08	200.00	68.04	20.13	307.65
Sizes over 12'0" O/D									
Slow	250	2.0	.66	9.20	2.21	132.00	27.25	27.31	197.97
Medium	275	3.0	.58	10.73	3.11	174.00	46.96	28.17	262.97
Fast	300	4.0	.50	12.00	4.22	200.00	67.03	19.83	303.08

For heights above 8 feet, use the High Time Difficulty Factors on page 139. "Slow" work is based on an hourly wage of $23.00, "Medium" work on an hourly wage of $29.50, and "Fast" work on an hourly wage of $36.00. Other qualifications that apply to this table are on page 9.

	Labor SF per manhour	Material coverage pounds/SF	Material cost per pound	Labor cost per 100 SF	Labor burden 100 SF	Material cost per 100 SF	Overhead per 100 SF	Profit per 100 SF	Total price per 100 SF
Sandblast, commercial blast (67% white)									
Surface condition basis - large projects & surface areas (material #46)									
Loose mill scale & fine powder rust									
Slow	150	4.0	.66	15.33	3.69	264.00	53.77	53.88	390.67
Medium	175	4.5	.58	16.86	4.86	261.00	70.68	42.41	395.81
Fast	200	5.0	.50	18.00	6.36	250.00	85.05	25.16	384.57
Tight mill scale & little or no rust									
Slow	125	5.0	.66	18.40	4.42	330.00	67.04	67.18	487.04
Medium	150	5.5	.58	19.67	5.70	319.00	86.09	51.65	482.11
Fast	175	6.0	.50	20.57	7.24	300.00	101.63	30.06	459.50
Hard scale, blistered, rusty surface									
Slow	75	6.0	.66	30.67	7.35	396.00	82.47	82.64	599.13
Medium	100	7.0	.58	29.50	8.53	406.00	111.01	66.60	621.64
Fast	125	8.0	.50	28.80	10.17	400.00	136.08	40.25	615.30
Rust nodules and pitted surface									
Slow	50	8.0	.66	46.00	11.04	528.00	111.16	111.39	807.59
Medium	60	9.5	.58	49.17	14.23	551.00	153.60	92.16	860.16
Fast	70	11.0	.50	51.43	18.18	550.00	192.07	56.82	868.50
Surface area basis - large projects & surface areas (material #46)									
Pipe up to 12" O/D									
Slow	45	5.0	.66	51.11	12.26	330.00	74.74	74.90	543.01
Medium	60	6.0	.58	49.17	14.23	348.00	102.85	61.71	575.96
Fast	75	7.0	.50	48.00	16.93	350.00	128.63	38.05	581.61
Structural steel									
Sizes up to 2 SF/LF									
Slow	70	5.0	.66	32.86	7.90	330.00	70.44	70.59	511.79
Medium	85	6.0	.58	34.71	10.01	348.00	98.19	58.91	549.82
Fast	100	7.0	.50	36.00	12.71	350.00	123.60	36.56	558.87
Sizes from 2 to 5 SF/LF									
Slow	80	5.0	.66	28.75	6.90	330.00	69.47	69.62	504.74
Medium	95	5.5	.58	31.05	9.00	319.00	89.76	53.85	502.66
Fast	110	6.0	.50	32.73	11.55	300.00	106.73	31.57	482.58
Sizes over 5 SF/LF									
Slow	85	5.0	.66	27.06	6.48	330.00	69.07	69.22	501.83
Medium	100	5.5	.58	29.50	8.53	319.00	89.26	53.55	499.84
Fast	115	6.0	.50	31.30	11.08	300.00	106.13	31.39	479.90

	Labor SF per manhour	Material coverage pounds/SF	Material cost per pound	Labor cost per 100 SF	Labor burden 100 SF	Material cost per 100 SF	Overhead per 100 SF	Profit per 100 SF	Total price per 100 SF
Tanks and vessels									
Sizes up to 12'0" O/D									
Slow	80	6.0	.66	28.75	6.90	396.00	82.01	82.19	595.85
Medium	95	6.5	.58	31.05	9.00	377.00	104.26	62.55	583.86
Fast	110	7.0	.50	32.73	11.55	350.00	122.23	36.16	552.67
Sizes over 12'0" O/D									
Slow	75	6.0	.66	30.67	7.35	396.00	82.47	82.64	599.13
Medium	100	6.3	.58	29.50	8.53	365.40	100.86	60.51	564.80
Fast	125	6.5	.50	28.80	10.17	325.00	112.83	33.38	510.18

For heights above 8 feet, use the High Time Difficulty Factors on page 139. "Slow" work is based on an hourly wage of $23.00, "Medium" work on an hourly wage of $29.50, and "Fast" work on an hourly wage of $36.00. Other qualifications that apply to this table are on page 9.

	Labor SF per manhour	Material coverage pounds/SF	Material cost per pound	Labor cost per 100 SF	Labor burden 100 SF	Material cost per 100 SF	Overhead per 100 SF	Profit per 100 SF	Total price per 100 SF
Sandblast, near white blast (95% white)									
Surface condition basis - large projects & surface areas (material #46)									
Loose mill scale & fine powder rust									
Slow	125	5.0	.66	18.40	4.42	330.00	67.04	67.18	487.04
Medium	150	6.0	.58	19.67	5.70	348.00	93.34	56.00	522.71
Fast	175	7.0	.50	20.57	7.24	350.00	117.13	34.65	529.59
Tight mill scale & little or no rust									
Slow	75	7.0	.66	30.67	7.35	462.00	95.01	95.21	690.24
Medium	100	8.0	.58	29.50	8.53	464.00	125.51	75.30	702.84
Fast	125	9.0	.50	28.80	10.17	450.00	151.58	44.84	685.39
Hard scale, blistered, rusty surface									
Slow	50	9.0	.66	46.00	11.04	594.00	123.70	123.96	898.70
Medium	75	11.0	.58	39.33	11.36	638.00	172.18	103.31	964.18
Fast	100	13.0	.50	36.00	12.71	650.00	216.60	64.07	979.38
Rust nodules and pitted surface									
Slow	35	12.0	.66	65.71	15.77	792.00	165.96	166.31	1205.75
Medium	50	14.5	.58	59.00	17.06	841.00	229.26	137.56	1283.88
Fast	65	17.0	.50	55.38	19.54	850.00	286.73	84.82	1296.47
Surface area basis - large projects & surface areas (material #46)									
Pipe up to 12" O/D									
Slow	30	8.0	.66	76.67	18.39	528.00	118.38	118.63	860.07
Medium	45	9.0	.58	65.56	18.94	522.00	151.63	90.98	849.11
Fast	60	10.0	.50	60.00	21.20	500.00	180.17	53.29	814.66
Structural steel									
Sizes up to 2 SF/LF									
Slow	40	7.0	.66	57.50	13.80	462.00	101.33	101.54	736.17
Medium	55	8.5	.58	53.64	15.50	493.00	140.54	84.32	787.00
Fast	70	10.0	.50	51.43	18.18	500.00	176.57	52.23	798.41
Sizes from 2 to 5 SF/LF									
Slow	45	7.0	.66	51.11	12.26	462.00	99.82	100.03	725.22
Medium	60	8.0	.58	49.17	14.23	464.00	131.85	79.11	738.36
Fast	75	9.0	.50	48.00	16.93	450.00	159.63	47.22	721.78
Sizes over 5 SF/LF									
Slow	55	8.0	.66	41.82	10.03	528.00	110.17	110.40	800.42
Medium	70	8.5	.58	42.14	12.20	493.00	136.83	82.10	766.27
Fast	85	9.0	.50	42.35	14.93	450.00	157.26	46.52	711.06

	Labor SF per manhour	Material coverage pounds/SF	Material cost per pound	Labor cost per 100 SF	Labor burden 100 SF	Material cost per 100 SF	Overhead per 100 SF	Profit per 100 SF	Total price per 100 SF
Tanks and vessels									
Sizes up to 12'0" O/D									
Slow	65	6.0	.66	35.38	8.48	396.00	83.58	83.75	607.19
Medium	80	7.0	.58	36.88	10.66	406.00	113.39	68.03	634.96
Fast	95	8.0	.50	37.89	13.40	400.00	139.89	41.38	632.56
Sizes over 12'0" O/D									
Slow	70	6.0	.66	32.86	7.90	396.00	82.98	83.16	602.90
Medium	85	7.0	.58	34.71	10.01	406.00	112.69	67.61	631.02
Fast	100	8.0	.50	36.00	12.71	400.00	139.10	41.15	628.96

For heights above 8 feet, use the High Time Difficulty Factors on page 139. "Slow" work is based on an hourly wage of $23.00, "Medium" work on an hourly wage of $29.50, and "Fast" work on an hourly wage of $36.00. Other qualifications that apply to this table are on page 9.

	Labor SF per manhour	Material coverage pounds/SF	Material cost per pound	Labor cost per 100 SF	Labor burden 100 SF	Material cost per 100 SF	Overhead per 100 SF	Profit per 100 SF	Total price per 100 SF
Sandblast, white blast (100% uniform white stage)									
Surface condition basis - large projects & surface areas (material #46)									
Loose mill scale & fine powder rust									
Slow	50	7.0	.66	46.00	11.04	462.00	98.62	98.83	716.49
Medium	75	8.5	.58	39.33	11.36	493.00	135.93	81.56	761.18
Fast	100	10.0	.50	36.00	12.71	500.00	170.10	50.32	769.13
Tight mill scale & little or no rust									
Slow	40	8.0	.66	57.50	13.80	528.00	113.87	114.11	827.28
Medium	60	9.5	.58	49.17	14.23	551.00	153.60	92.16	860.16
Fast	80	11.0	.50	45.00	15.89	550.00	189.38	56.02	856.29
Hard scale, blistered, rusty surface									
Slow	30	10.0	.66	76.67	18.39	660.00	143.46	143.76	1042.28
Medium	45	12.5	.58	65.56	18.94	725.00	202.38	121.43	1133.31
Fast	60	15.0	.50	60.00	21.20	750.00	257.67	76.22	1165.09
Rust nodules and pitted surface									
Slow	25	15.0	.66	92.00	22.08	990.00	209.78	210.22	1524.08
Medium	35	17.5	.58	84.29	24.36	1015.00	280.91	168.55	1573.11
Fast	45	20.0	.50	80.00	28.23	1000.00	343.55	101.63	1553.41
Surface area basis - large projects & surface areas (material #46)									
Pipe up to 12" O/D									
Slow	30	10.0	.66	76.67	18.39	660.00	143.46	143.76	1042.28
Medium	40	11.5	.58	73.75	21.33	667.00	190.52	114.31	1066.91
Fast	50	13.0	.50	72.00	25.42	650.00	231.70	68.54	1047.66
Structural steel									
Sizes up to 2 SF/LF									
Slow	40	9.0	.66	57.50	13.80	594.00	126.41	126.67	918.38
Medium	50	10.5	.58	59.00	17.06	609.00	171.26	102.76	959.08
Fast	60	12.0	.50	60.00	21.20	600.00	211.17	62.46	954.83
Sizes from 2 to 5 SF/LF									
Slow	45	8.0	.66	51.11	12.26	528.00	112.36	112.60	816.33
Medium	55	9.5	.58	53.64	15.50	551.00	155.04	93.02	868.20
Fast	65	11.0	.50	55.38	19.54	550.00	193.73	57.31	875.96
Sizes over 5 SF/LF									
Slow	50	8.0	.66	46.00	11.04	528.00	111.16	111.39	807.59
Medium	60	9.5	.58	49.17	14.23	551.00	153.60	92.16	860.16
Fast	70	11.0	.50	51.43	18.18	550.00	192.07	56.82	868.50

	Labor SF per manhour	Material coverage pounds/SF	Material cost per pound	Labor cost per 100 SF	Labor burden 100 SF	Material cost per 100 SF	Overhead per 100 SF	Profit per 100 SF	Total price per 100 SF
Tanks and vessels									
Sizes up to 12'0" O/D									
Slow	60	8.0	.66	38.33	9.21	528.00	109.35	109.58	794.47
Medium	70	9.5	.58	42.14	12.20	551.00	151.33	90.80	847.47
Fast	80	11.0	.50	45.00	15.89	550.00	189.38	56.02	856.29
Sizes over 12'0" O/D									
Slow	70	8.0	.66	32.86	7.90	528.00	108.06	108.29	785.11
Medium	80	9.5	.58	36.88	10.66	551.00	149.64	89.78	837.96
Fast	90	11.0	.50	40.00	14.12	550.00	187.28	55.40	846.80

For heights above 8 feet, use the High Time Difficulty Factors on page 139. "Slow" work is based on an hourly wage of $23.00, "Medium" work on an hourly wage of $29.50, and "Fast" work on an hourly wage of $36.00. Other qualifications that apply to this table are on page 9.

	Labor SF per manhour	Material coverage pounds/SF	Material cost per pound	Labor cost per 100 SF	Labor burden 100 SF	Material cost per 100 SF	Overhead per 100 SF	Profit per 100 SF	Total price per 100 SF

Scribing (edge scraping) and back-painting, horizontal, interior or exterior

	Labor SF per manhour	Material coverage pounds/SF	Material cost per pound	Labor cost per 100 SF	Labor burden 100 SF	Material cost per 100 SF	Overhead per 100 SF	Profit per 100 SF	Total price per 100 SF
Scribing, horizontal, heights up to 6'8"									
Five point tool scribing (edge scraping)									
Slow	25	--	--	92.00	22.08	--	21.68	21.72	157.48
Medium	34	--	--	85.82	24.81	--	27.66	16.59	154.88
Fast	44	--	--	82.29	29.06	--	34.52	10.21	156.08
Scribing, horizontal, heights from 6'8" to 9'0" (1.3 High Time Difficulty Factor included)									
Five point tool scribing (edge scraping)									
Slow	19	--	--	121.05	29.05	--	28.52	28.58	207.20
Medium	26	--	--	113.46	32.80	--	36.56	21.94	204.76
Fast	34	--	--	105.88	37.38	--	44.41	13.14	200.81
Scribing, horizontal, heights from 9'0" to 13'0" (1.6 High Time Difficulty Factor included)									
Five point tool scribing (edge scraping)									
Slow	16	--	--	147.20	35.33	--	34.68	34.75	251.96
Medium	21	--	--	137.31	39.72	--	44.25	26.55	247.83
Fast	27	--	--	131.66	46.47	--	55.22	16.34	249.69
Scribing, horizontal, heights from 13'0" to 17'0" (1.9 High Time Difficulty Factor included)									
Five point tool scribing (edge scraping)									
Slow	13	--	--	176.92	42.46	--	41.68	41.77	302.83
Medium	18	--	--	163.89	47.40	--	52.81	31.69	295.79
Fast	23	--	--	156.52	55.27	--	65.65	19.42	296.86
Scribing, horizontal, heights from 17'0" to 19'0" (2.2 High Time Difficulty Factor included)									
Five point tool scribing (edge scraping)									
Slow	11	--	--	209.09	50.19	--	49.26	49.36	357.90
Medium	16	--	--	188.80	54.59	--	60.84	36.50	340.73
Fast	20	--	--	180.00	63.55	--	75.50	22.33	341.38

	Labor SF per manhour	Material coverage pounds/SF	Material cost per pound	Labor cost per 100 SF	Labor burden 100 SF	Material cost per 100 SF	Overhead per 100 SF	Profit per 100 SF	Total price per 100 SF
Scribing, horizontal, heights from 19'0" to 21'0" (2.5 High Time Difficulty Factor included)									
Five point tool scribing (edge scraping)									
Slow	10	--	--	230.00	55.20	--	54.19	54.30	393.69
Medium	14	--	--	214.55	62.04	--	69.14	41.48	387.21
Fast	18	--	--	205.71	72.62	--	86.28	25.52	390.13

	Labor SF per manhour	Material coverage pounds/SF	Material cost per pound	Labor cost per 100 SF	Labor burden 100 SF	Material cost per 100 SF	Overhead per 100 SF	Profit per 100 SF	Total price per 100 SF

Scribing (edge scraping) and back-painting, vertical, interior or exterior

	Labor SF per manhour	Material coverage pounds/SF	Material cost per pound	Labor cost per 100 SF	Labor burden 100 SF	Material cost per 100 SF	Overhead per 100 SF	Profit per 100 SF	Total price per 100 SF
Scribing and back-painting, vertical, heights up to 6'8"									
Five point tool scribing (edge scraping)									
Slow	31	--	--	74.19	17.82	--	17.48	17.52	127.01
Medium	40	--	--	73.20	21.15	--	23.59	14.15	132.09
Fast	50	--	--	72.58	25.62	--	30.44	9.00	137.64
Scribing and back-painting, vertical, heights from 6'8" to 9'0" (1.3 High Time Difficulty Factor included)									
Five point tool scribing (edge scraping)									
Slow	24	--	--	95.83	23.01	--	22.58	22.63	164.05
Medium	31	--	--	95.16	27.52	--	30.67	18.40	171.75
Fast	38	--	--	94.74	33.46	--	39.74	11.75	179.69
Scribing and back-painting, vertical, heights from 9'0" to 13'0" (1.6 High Time Difficulty Factor included)									
Five point tool scribing (edge scraping)									
Slow	19	--	--	118.71	28.48	--	27.97	28.03	203.19
Medium	25	--	--	117.12	33.86	--	37.74	22.65	211.37
Fast	31	--	--	116.13	41.01	--	48.71	14.41	220.26
Scribing and back-painting, vertical, heights from 13'0" to 17'0" (1.9 High Time Difficulty Factor included)									
Five point tool scribing (edge scraping)									
Slow	16	--	--	143.75	34.50	--	33.87	33.94	246.06
Medium	21	--	--	140.48	40.62	--	45.27	27.16	253.53
Fast	26	--	--	138.46	48.88	--	58.08	17.18	262.60
Scribing and back-painting, vertical, heights from 17'0" to 19'0" (2.2 High Time Difficulty Factor included)									
Five point tool scribing (edge scraping)									
Slow	14	--	--	164.29	39.43	--	38.71	38.79	281.22
Medium	18	--	--	163.89	47.40	--	52.81	31.69	295.79
Fast	23	--	--	156.52	55.27	--	65.65	19.42	296.86
Scribing and back-painting, vertical, heights from 19'0" to 21'0" (2.5 High Time Difficulty Factor included)									
Five point tool scribing (edge scraping)									
Slow	12	--	--	185.48	44.53	--	43.70	43.79	317.50
Medium	16	--	--	183.00	52.90	--	58.97	35.38	330.25
Fast	20	--	--	181.45	64.05	--	76.11	22.51	344.12

Use these figures in combination with the cutting-in figures to achieve a clean edge on textured surfaces prior to the cutting-in operation. Scribing or edge scraping by hand with a five point tool, then back-painting, is common and necessary where medium to heavy texture has been applied to vertical walls, horizontal ceilings, etc., in preparation for cutting-in at walls and ceilings where different colors or different sheens (i.e. flat vs. semi-gloss) are used on the adjacent surfaces. For example, assume a medium or heavy texture has been applied to the walls and ceiling - the ceiling is painted white and the walls painted an earth tone color. The first step is to scrape or scribe the texture off the wall, down to the drywall tape (approximately 5/16" to 1/2" from the ceiling), to create a smooth surface for cutting-in. Then, the white ceiling color would be painted back (back-painted) on the wall and ceiling where the texture was removed. Now, with this surface smooth and repainted, cut-in the earth tone wall color to the ceiling. "Slow" work is based on an hourly wage of $23.00, "Medium" work on an hourly wage of $29.50, and "Fast" work on an hourly wage of $36.00. "Slow" applies to residential repaints with heavy texture. "Medium" applies to residential or commercial repaints with light-to-medium texture. "Fast" applies to new construction with a light textured surface.

Notes:

1 - Material consumption for back-painting is minimal or zero (0) since the material cost is actually calculated in the wall painting or ceiling painting line item.

2 - High Time Difficulty Factors are built into these figures to allow for up and down time and moving ladders or scaffolding.

3 - Horizontal scribing is typically more difficult and consumes more time than vertical scribing, as the figures indicate.

4 - In new construction, it's best to have the drywall texture applicators scribe the edges where different colors occur while the texture is still wet, thus eliminating this painting operation.

	Labor SF per manhour	Material coverage SF/gallon	Material cost per gallon	Labor cost per 100 SF	Labor burden 100 SF	Material cost per 100 SF	Overhead per 100 SF	Profit per 100 SF	Total price per 100 SF
Strip, remove, or bleach									
Remove wallcover by hand									
Slow	50	--	--	46.00	11.04	--	10.84	10.86	78.74
Medium	70	--	--	42.14	12.20	--	13.58	8.15	76.07
Fast	90	--	--	40.00	14.12	--	16.78	4.96	75.86
Stripping flat, vertical, varnished surfaces									
Light duty liquid remover (material #43)									
Slow	25	175	44.20	92.00	22.08	25.26	34.84	27.87	202.05
Medium	35	158	38.70	84.29	24.36	24.49	33.29	19.97	186.40
Fast	45	140	33.20	80.00	28.23	23.71	40.90	12.10	184.94
Heavy duty liquid remover (material #44)									
Slow	20	150	55.40	115.00	27.60	36.93	44.88	35.91	260.32
Medium	30	138	48.40	98.33	28.42	35.07	40.46	24.27	226.55
Fast	40	125	41.50	90.00	31.78	33.20	48.04	14.21	217.23
Stripping flat, horizontal floor surfaces									
Paint removal with light duty liquid remover (material #43)									
Slow	20	180	44.20	115.00	27.60	24.56	41.79	33.43	242.38
Medium	30	175	38.70	98.33	28.42	22.11	37.22	22.33	208.41
Fast	40	170	33.20	90.00	31.78	19.53	43.80	12.96	198.07

	Labor SF per manhour	Material coverage SF/gallon	Material cost per gallon	Labor cost per 100 SF	Labor burden 100 SF	Material cost per 100 SF	Overhead per 100 SF	Profit per 100 SF	Total price per 100 SF
Paint removal with heavy duty liquid remover (material #44)									
Slow	15	170	55.40	153.33	36.81	32.59	55.68	44.54	322.95
Medium	25	160	48.40	118.00	34.12	30.25	45.59	27.35	255.31
Fast	35	150	41.50	102.86	36.30	27.67	51.72	15.30	233.85
Varnish removal with light duty liquid remover (material #43)									
Slow	30	185	44.20	76.67	18.39	23.89	29.74	23.79	172.48
Medium	40	180	38.70	73.75	21.33	21.50	29.14	17.48	163.20
Fast	50	175	33.20	72.00	25.42	18.97	36.08	10.67	163.14
Varnish removal with heavy duty liquid remover (material #44)									
Slow	25	180	55.40	92.00	22.08	30.78	36.22	28.97	210.05
Medium	35	170	48.40	84.29	24.36	28.47	34.28	20.57	191.97
Fast	45	160	41.50	80.00	28.23	25.94	41.60	12.30	188.07

All widths less than 12", consider as 1 square foot per linear foot. For heights above 8 feet, use the High Time Difficulty Factors on page 139. "Slow" work is based on an hourly wage of $23.00, "Medium" work on an hourly wage of $29.50, and "Fast" work on an hourly wage of $36.00. Other qualifications that apply to this table are on page 9.

	Labor LF per manhour	Material coverage LF/gallon	Material cost per gallon	Labor cost per 100 LF	Labor burden 100 LF	Material cost per 100 LF	Overhead per 100 LF	Profit per 100 LF	Total price per 100 LF
Tape gypsum wallboard									
Preparation									
Bead, spot nail heads & sand									
Slow	85	--	--	27.06	6.48	--	6.37	6.39	46.30
Medium	100	--	--	29.50	8.53	--	9.51	5.70	53.24
Fast	115	--	--	31.30	11.08	--	13.13	3.88	59.39
Taping									
Hand operation									
Slow	125	--	--	18.40	4.42	--	4.34	4.35	31.51
Medium	150	--	--	19.67	5.70	--	6.34	3.80	35.51
Fast	175	--	--	20.57	7.24	--	8.63	2.55	38.99
Mechanical tools									
Slow	200	--	--	11.50	2.76	--	2.71	2.72	19.69
Medium	225	--	--	13.11	3.78	--	4.23	2.54	23.66
Fast	250	--	--	14.40	5.08	--	6.04	1.79	27.31
Joint cement (premixed) per 75 lb bag									
Slow	--	650	10.00	--	--	1.54	--	--	1.54
Medium	--	600	10.00	--	--	1.67	--	--	1.67
Fast	--	550	10.00	--	--	1.82	--	--	1.82

For heights above 8 feet, use the High Time Difficulty Factors on page 139.

Unstick windows

On repaint jobs, test all windows during your estimating walk-through and allow approximately 15 minutes (.25 hours) for each stuck window. But don't price yourself out of the job with this extra time.

	Labor SF per manhour	Material coverage SF/gallon	Material cost per gallon	Labor cost per 100 SF	Labor burden 100 SF	Material cost per 100 SF	Overhead per 100 SF	Profit per 100 SF	Total price per 100 SF
Wash									
Interior flatwall (smooth surfaces)									
Wash only									
Slow	175	--	--	13.14	3.14	--	3.10	3.10	22.48
Medium	200	--	--	14.75	4.27	--	4.75	2.85	26.62
Fast	225	--	--	16.00	5.63	--	6.71	1.99	30.33
Wash & touchup									
Slow	135	--	--	17.04	4.09	--	4.01	4.02	29.16
Medium	160	--	--	18.44	5.33	--	5.94	3.57	33.28
Fast	185	--	--	19.46	6.89	--	8.16	2.41	36.92
Interior flatwall (rough surfaces)									
Wash only									
Slow	125	--	--	18.40	4.42	--	4.34	4.35	31.51
Medium	150	--	--	19.67	5.70	--	6.34	3.80	35.51
Fast	175	--	--	20.57	7.24	--	8.63	2.55	38.99
Wash & touchup									
Slow	55	--	--	41.82	10.03	--	9.85	9.87	71.57
Medium	95	--	--	31.05	9.00	--	10.01	6.00	56.06
Fast	135	--	--	26.67	9.42	--	11.18	3.31	50.58
Interior enamel (wall surfaces)									
Wash only									
Slow	190	--	--	12.11	2.89	--	2.85	2.86	20.71
Medium	215	--	--	13.72	3.96	--	4.42	2.65	24.75
Fast	240	--	--	15.00	5.31	--	6.29	1.86	28.46
Wash & touchup									
Slow	85	--	--	27.06	6.48	--	6.37	6.39	46.30
Medium	113	--	--	26.11	7.55	--	8.42	5.05	47.13
Fast	140	--	--	25.71	9.07	--	10.78	3.19	48.75
Interior enamel trim									
Wash only									
Slow	100	--	--	23.00	5.52	--	5.42	5.43	39.37
Medium	150	--	--	19.67	5.70	--	6.34	3.80	35.51
Fast	200	--	--	18.00	6.36	--	7.55	2.23	34.14
Wash & touchup									
Slow	90	--	--	25.56	6.13	--	6.02	6.03	43.74
Medium	120	--	--	24.58	7.10	--	7.92	4.75	44.35
Fast	150	--	--	24.00	8.49	--	10.07	2.98	45.54

	Labor SF per manhour	Material coverage SF/gallon	Material cost per gallon	Labor cost per 100 SF	Labor burden 100 SF	Material cost per 100 SF	Overhead per 100 SF	Profit per 100 SF	Total price per 100 SF
Interior varnish trim									
Wash only									
Slow	150	--	--	15.33	3.69	--	3.61	3.62	26.25
Medium	195	--	--	15.13	4.38	--	4.88	2.93	27.32
Fast	240	--	--	15.00	5.31	--	6.29	1.86	28.46
Wash & touchup									
Slow	120	--	--	19.17	4.59	--	4.52	4.53	32.81
Medium	140	--	--	21.07	6.08	--	6.79	4.07	38.01
Fast	160	--	--	22.50	7.94	--	9.44	2.79	42.67
Interior varnish floors									
Wash only									
Slow	160	--	--	14.38	3.45	--	3.39	3.40	24.62
Medium	210	--	--	14.05	4.05	--	4.53	2.72	25.35
Fast	260	--	--	13.85	4.90	--	5.81	1.72	26.28
Wash & touchup									
Slow	130	--	--	17.69	4.24	--	4.17	4.18	30.28
Medium	153	--	--	19.28	5.59	--	6.21	3.73	34.81
Fast	175	--	--	20.57	7.24	--	8.63	2.55	38.99
Interior plaster (smooth)									
Wash only									
Slow	150	--	--	15.33	3.69	--	3.61	3.62	26.25
Medium	175	--	--	16.86	4.86	--	5.43	3.26	30.41
Fast	200	--	--	18.00	6.36	--	7.55	2.23	34.14
Wash & touchup									
Slow	125	--	--	18.40	4.42	--	4.34	4.35	31.51
Medium	140	--	--	21.07	6.08	--	6.79	4.07	38.01
Fast	155	--	--	23.23	8.19	--	9.74	2.88	44.04
Interior plaster (sand finish)									
Wash only									
Slow	110	--	--	20.91	5.01	--	4.93	4.94	35.79
Medium	135	--	--	21.85	6.33	--	7.04	4.22	39.44
Fast	160	--	--	22.50	7.94	--	9.44	2.79	42.67
Wash & touchup									
Slow	85	--	--	27.06	6.48	--	6.37	6.39	46.30
Medium	110	--	--	26.82	7.75	--	8.64	5.19	48.40
Fast	140	--	--	25.71	9.07	--	10.78	3.19	48.75

Because the type of surface and type of material being removed will alter rates, it is best to wash surfaces with calcium deposits, excess debris or other unusual surface markings on a Time and Material or a Cost Plus Fee basis at a pre-set hourly rate. Consider all trim which is less than 12" wide to be 12" wide and figured as 1 square foot per linear foot. For heights above 8 feet, use the High Time Difficulty Factors on page 139. "Slow" work is based on an hourly wage of $23.00, "Medium" work on an hourly wage of $29.50, and "Fast" work on an hourly wage of $36.00. Other qualifications that apply to this table are on page 9.

	Labor SF per manhour	Material coverage SF/gallon	Material cost per gallon	Labor cost per 100 SF	Labor burden 100 SF	Material cost per 100 SF	Overhead per 100 SF	Profit per 100 SF	Total price per 100 SF
Waterblast (Power wash)									
Power wash									
Slow	700	--	--	3.29	.79	--	.78	.78	5.64
Medium	1000	--	--	2.95	.85	--	.95	.57	5.32
Fast	1500	--	--	2.40	.86	--	1.01	.30	4.57

Power wash to clean surfaces prior to painting and to remove deteriorated, cracked, flaking paint from accessible wood, concrete, brick, block, plaster or stucco surfaces. Use the above rates for larger jobs, such as apartments or commercial buildings and large homes. For heights above 8 feet, use the High Time Difficulty Factors on page 139. For average-sized single-family homes, a half-day minimum might apply for move on, set up, power wash, clean up and move off. Rates assume a 1/4" diameter nozzle with 2500 lbs. pressure. "Slow" work is based on an hourly wage of $23.00, "Medium" work on an hourly wage of $29.50, and "Fast" work on an hourly wage of $36.00. Other qualifications that apply to this table are on page 9.

	Labor SF per manhour	Material coverage SF/roll	Material cost per roll	Labor cost per 100 SF	Labor burden 100 SF	Material cost per 100 SF	Overhead per 100 SF	Profit per 100 SF	Total price per 100 SF
Window protection (visqueen, 1.5 mil)									
Window protection - visqueen (material #47)									
Hand application									
Slow	90	2400	53.60	25.56	6.13	2.23	6.44	6.46	46.82
Medium	100	2400	46.90	29.50	8.53	1.95	10.00	6.00	55.98
Fast	110	2400	40.20	32.73	11.55	1.68	14.25	4.21	64.42

Masking tape is included in the sundry (15%) and escalation (10%) allowances within the material pricing columns in Figure 9. For heights above 8 feet, use the High Time Difficulty Factors on page 139. "Slow" work is based on an hourly wage of $23.00, "Medium" work on an hourly wage of $29.50, and "Fast" work on an hourly wage of $36.00. Other qualifications that apply to this table are on page 9.

	Labor SF per manhour	Material coverage SF/gallon	Material cost per gallon	Labor cost per 100 SF	Labor burden 100 SF	Material cost per 100 SF	Overhead per 100 SF	Profit per 100 SF	Total price per 100 SF
Wire brush									
Surface area basis - large projects & surface areas									
Pipe up to 12" O/D									
Slow	50	--	--	46.00	11.04	--	10.84	10.86	78.74
Medium	75	--	--	39.33	11.36	--	12.68	7.61	70.98
Fast	100	--	--	36.00	12.71	--	15.10	4.47	68.28
Structural steel									
Sizes up to 2 SF/LF									
Slow	90	--	--	25.56	6.13	--	6.02	6.03	43.74
Medium	110	--	--	26.82	7.75	--	8.64	5.19	48.40
Fast	125	--	--	28.80	10.17	--	12.08	3.57	54.62
Sizes from 2 to 5 SF/LF									
Slow	100	--	--	23.00	5.52	--	5.42	5.43	39.37
Medium	120	--	--	24.58	7.10	--	7.92	4.75	44.35
Fast	140	--	--	25.71	9.07	--	10.78	3.19	48.75
Sizes over 5 SF/LF									
Slow	110	--	--	20.91	5.01	--	4.93	4.94	35.79
Medium	130	--	--	22.69	6.56	--	7.31	4.39	40.95
Fast	150	--	--	24.00	8.49	--	10.07	2.98	45.54
Tanks and vessels									
Sizes up to 12'0" O/D									
Slow	110	--	--	20.91	5.01	--	4.93	4.94	35.79
Medium	130	--	--	22.69	6.56	--	7.31	4.39	40.95
Fast	150	--	--	24.00	8.49	--	10.07	2.98	45.54
Sizes over 12'0" O/D									
Slow	120	--	--	19.17	4.59	--	4.52	4.53	32.81
Medium	140	--	--	21.07	6.08	--	6.79	4.07	38.01
Fast	160	--	--	22.50	7.94	--	9.44	2.79	42.67

For heights above 8 feet, use the High Time Difficulty Factors on page 139. "Slow" work is based on an hourly wage of $23.00, "Medium" work on an hourly wage of $29.50, and "Fast" work on an hourly wage of $36.00. Other qualifications that apply to this table are on page 9.

Part III

- INDUSTRIAL
- INSTITUTIONAL
- HEAVY COMMERCIAL

Painting Costs

	Labor SF per manhour	Material coverage SF/gallon	Material cost per gallon	Labor cost per 100 SF	Labor burden 100 SF	Material cost per 100 SF	Overhead per 100 SF	Profit per 100 SF	Total price per 100 SF
Conduit, electric, brush application									
Acid wash coat, muriatic acid (material #49)									
Brush each coat									
Slow	60	700	21.20	38.33	9.21	3.03	9.61	9.63	69.81
Medium	80	650	18.50	36.88	10.66	2.85	12.60	7.56	70.55
Fast	100	600	15.90	36.00	12.71	2.65	15.92	4.71	71.99
Metal primer, rust inhibitor - clean metal (material #35)									
Brush prime coat									
Slow	60	450	59.70	38.33	9.21	13.27	11.55	11.58	83.94
Medium	80	425	52.30	36.88	10.66	12.31	14.96	8.98	83.79
Fast	100	400	44.80	36.00	12.71	11.20	18.57	5.49	83.97
Metal primer, rust inhibitor - rusty metal (material #36)									
Brush prime coat									
Slow	60	400	80.10	38.33	9.21	20.03	12.84	12.86	93.27
Medium	80	375	70.10	36.88	10.66	18.69	16.56	9.93	92.72
Fast	100	350	60.10	36.00	12.71	17.17	20.42	6.04	92.34
Industrial enamel, oil base, high gloss - light colors (material #56)									
Brush 1st or additional finish coats									
Slow	100	450	65.40	23.00	5.52	14.53	8.18	8.20	59.43
Medium	125	425	57.20	23.60	6.82	13.46	10.97	6.58	61.43
Fast	150	400	49.00	24.00	8.49	12.25	13.86	4.10	62.70
Industrial enamel, oil base, high gloss - dark (OSHA) colors (material #57)									
Brush 1st or additional finish coats									
Slow	100	500	78.30	23.00	5.52	15.66	8.39	8.41	60.98
Medium	125	475	68.50	23.60	6.82	14.42	11.21	6.73	62.78
Fast	150	450	58.70	24.00	8.49	13.04	14.11	4.17	63.81
Epoxy coating, 2 part system - clear (material #51)									
Brush 1st coat									
Slow	60	425	160.60	38.33	9.21	37.79	16.21	16.24	117.78
Medium	80	400	140.50	36.88	10.66	35.13	20.67	12.40	115.74
Fast	100	375	120.50	36.00	12.71	32.13	25.06	7.41	113.31
Brush 2nd or additional finish coats									
Slow	100	475	160.60	23.00	5.52	33.81	11.84	11.87	86.04
Medium	125	450	140.50	23.60	6.82	31.22	15.41	9.25	86.30
Fast	150	425	120.50	24.00	8.49	28.35	18.85	5.58	85.27

	Labor SF per manhour	Material coverage SF/gallon	Material cost per gallon	Labor cost per 100 SF	Labor burden 100 SF	Material cost per 100 SF	Overhead per 100 SF	Profit per 100 SF	Total price per 100 SF
Epoxy coating, 2 part system - white (material #52)									
Brush 1st coat									
Slow	60	425	157.00	38.33	9.21	36.94	16.05	16.08	116.61
Medium	80	400	137.40	36.88	10.66	34.35	20.47	12.28	114.64
Fast	100	375	117.70	36.00	12.71	31.39	24.83	7.35	112.28
Brush 2nd or additional finish coats									
Slow	100	475	157.00	23.00	5.52	33.05	11.70	11.72	84.99
Medium	125	450	137.40	23.60	6.82	30.53	15.24	9.14	85.33
Fast	150	425	117.70	24.00	8.49	27.69	18.65	5.52	84.35

This table is based on square feet of conduit surface area. See Figure 21 on page 325 to convert from linear feet of various conduit or pipe sizes to square feet of surface area. For heights above 8 feet, use the High Time Difficulty Factors on page 139. Note: A two coat system, prime and finish, using oil base material is recommended for any metal surface. Using water base material may cause oxidation, corrosion and rust. Using one coat of oil base paint on metal surfaces may result in cracking, peeling or chipping without the proper prime coat application. If off white or another light colored finish paint is specified, make sure the prime coat is also a light color, or more than one finish coat will be necessary. "Slow" work is based on an hourly wage of $23.00, "Medium" work on an hourly wage of $29.50, and "Fast" work on an hourly wage of $36.00. Other qualifications that apply to this table are on page 9.

	Labor SF per manhour	Material coverage SF/gallon	Material cost per gallon	Labor cost per 100 SF	Labor burden 100 SF	Material cost per 100 SF	Overhead per 100 SF	Profit per 100 SF	Total price per 100 SF
Conduit, electric, roll application									
Acid wash coat, muriatic acid (material #49)									
Roll each coat									
Slow	175	700	21.20	13.14	3.14	3.03	3.67	3.68	26.66
Medium	200	650	18.50	14.75	4.27	2.85	5.47	3.28	30.62
Fast	225	600	15.90	16.00	5.63	2.65	7.53	2.23	34.04
Metal primer, rust inhibitor - clean metal (material #35)									
Roll prime coat									
Slow	175	425	59.70	13.14	3.14	14.05	5.76	5.78	41.87
Medium	200	400	52.30	14.75	4.27	13.08	8.02	4.81	44.93
Fast	225	375	44.80	16.00	5.63	11.95	10.42	3.08	47.08
Metal primer, rust inhibitor - rusty metal (material #36)									
Roll prime coat									
Slow	175	375	80.10	13.14	3.14	21.36	7.15	7.17	51.96
Medium	200	350	70.10	14.75	4.27	20.03	9.76	5.86	54.67
Fast	225	325	60.10	16.00	5.63	18.49	12.44	3.68	56.24
Industrial enamel, oil base, high gloss - light colors (material #56)									
Roll 1st or additional finish coats									
Slow	225	425	65.40	10.22	2.44	15.39	5.33	5.34	38.72
Medium	250	400	57.20	11.80	3.41	14.30	7.38	4.43	41.32
Fast	275	375	49.00	13.09	4.64	13.07	9.54	2.82	43.16
Industrial enamel, oil base, high gloss - dark (OSHA) colors (material #57)									
Roll 1st or additional finish coats									
Slow	225	475	78.30	10.22	2.44	16.48	5.54	5.55	40.23
Medium	250	450	68.50	11.80	3.41	15.22	7.61	4.56	42.60
Fast	275	425	58.70	13.09	4.64	13.81	9.77	2.89	44.20
Epoxy coating, 2 part system - clear (material #51)									
Roll 1st coat									
Slow	175	400	160.60	13.14	3.14	40.15	10.72	10.75	77.90
Medium	200	375	140.50	14.75	4.27	37.47	14.12	8.47	79.08
Fast	225	350	120.50	16.00	5.63	34.43	17.38	5.14	78.58
Roll 2nd or additional finish coats									
Slow	225	450	160.60	10.22	2.44	35.69	9.19	9.21	66.75
Medium	250	425	140.50	11.80	3.41	33.06	12.07	7.24	67.58
Fast	275	400	120.50	13.09	4.64	30.13	14.83	4.39	67.08

	Labor SF per manhour	Material coverage SF/gallon	Material cost per gallon	Labor cost per 100 SF	Labor burden 100 SF	Material cost per 100 SF	Overhead per 100 SF	Profit per 100 SF	Total price per 100 SF
Epoxy coating, 2 part system - white (material #52)									
Roll 1st coat									
Slow	175	400	157.00	13.14	3.14	39.25	10.55	10.57	76.65
Medium	200	375	137.40	14.75	4.27	36.64	13.91	8.35	77.92
Fast	225	350	117.70	16.00	5.63	33.63	17.14	5.07	77.47
Roll 2nd or additional finish coats									
Slow	225	450	157.00	10.22	2.44	34.89	9.04	9.06	65.65
Medium	250	425	137.40	11.80	3.41	32.33	11.89	7.13	66.56
Fast	275	400	117.70	13.09	4.64	29.43	14.61	4.32	66.09

This table is based on square feet of conduit surface area. See Figure 21 on page 325 to convert from linear feet of various conduit or pipe sizes to square feet of surface area. For heights above 8 feet, use the High Time Difficulty Factors on page 139. Note: A two coat system, prime and finish, using oil base material is recommended for any metal surface. Using water base material may cause oxidation, corrosion and rust. Using one coat of oil base paint on metal surfaces may result in cracking, peeling or chipping without the proper prime coat application. If off white or another light colored finish paint is specified, make sure the prime coat is also a light color, or more than one finish coat will be necessary. "Slow" work is based on an hourly wage of $23.00, "Medium" work on an hourly wage of $29.50, and "Fast" work on an hourly wage of $36.00. Other qualifications that apply to this table are on page 9.

	Labor SF per manhour	Material coverage SF/gallon	Material cost per gallon	Labor cost per 100 SF	Labor burden 100 SF	Material cost per 100 SF	Overhead per 100 SF	Profit per 100 SF	Total price per 100 SF
Conduit, electric, spray application									
Acid wash coat, muriatic acid (material #49)									
Spray each coat									
Slow	350	350	21.20	6.57	1.59	6.06	2.70	2.71	19.63
Medium	400	325	18.50	7.38	2.13	5.69	3.80	2.28	21.28
Fast	450	300	15.90	8.00	2.81	5.30	5.00	1.48	22.59
Metal primer, rust inhibitor - clean metal (material #35)									
Spray prime coat									
Slow	350	275	59.70	6.57	1.59	21.71	5.67	5.68	41.22
Medium	400	263	52.30	7.38	2.13	19.89	7.35	4.41	41.16
Fast	450	250	44.80	8.00	2.81	17.92	8.91	2.64	40.28
Metal primer, rust inhibitor - rusty metal (material #36)									
Spray prime coat									
Slow	350	225	80.10	6.57	1.59	35.60	8.31	8.33	60.40
Medium	400	213	70.10	7.38	2.13	32.91	10.61	6.36	59.39
Fast	450	200	60.10	8.00	2.81	30.05	12.67	3.75	57.28
Industrial enamel, oil base, high gloss - light colors (material #56)									
Spray 1st or additional finish coats									
Slow	450	275	65.40	5.11	1.22	23.78	5.72	5.73	41.56
Medium	500	263	57.20	5.90	1.71	21.75	7.34	4.40	41.10
Fast	550	250	49.00	6.55	2.32	19.60	8.82	2.61	39.90
Industrial enamel, oil base, high gloss - dark (OSHA) colors (material #57)									
Spray 1st or additional finish coats									
Slow	450	300	78.30	5.11	1.22	26.10	6.16	6.18	44.77
Medium	500	288	68.50	5.90	1.71	23.78	7.85	4.71	43.95
Fast	550	275	58.70	6.55	2.32	21.35	9.37	2.77	42.36
Epoxy coating, 2 part system - clear (material #51)									
Spray 1st coat									
Slow	350	250	160.60	6.57	1.59	64.24	13.75	13.78	99.93
Medium	400	238	140.50	7.38	2.13	59.03	17.14	10.28	95.96
Fast	450	225	120.50	8.00	2.81	53.56	19.96	5.90	90.23
Spray 2nd or additional finish coats									
Slow	450	285	160.60	5.11	1.22	56.35	11.91	11.94	86.53
Medium	500	273	140.50	5.90	1.71	51.47	14.77	8.86	82.71
Fast	550	260	120.50	6.55	2.32	46.35	17.12	5.06	77.40

	Labor SF per manhour	Material coverage SF/gallon	Material cost per gallon	Labor cost per 100 SF	Labor burden 100 SF	Material cost per 100 SF	Overhead per 100 SF	Profit per 100 SF	Total price per 100 SF
Epoxy coating, 2 part system - white (material #52)									
Spray 1st coat									
Slow	350	250	157.00	6.57	1.59	62.80	13.48	13.51	97.95
Medium	400	238	137.40	7.38	2.13	57.73	16.81	10.09	94.14
Fast	450	225	117.70	8.00	2.81	52.31	19.57	5.79	88.48
Spray 2nd or additional finish coats									
Slow	450	285	157.00	5.11	1.22	55.09	11.67	11.70	84.79
Medium	500	273	137.40	5.90	1.71	50.33	14.49	8.69	81.12
Fast	550	260	117.70	6.55	2.32	45.27	16.78	4.96	75.88

This table is based on square feet of conduit surface area. See Figure 21 on page 325 to convert from linear feet of various conduit or pipe sizes to square feet of surface area. For heights above 8 feet, use the High Time Difficulty Factors on page 139. Note: A two coat system, prime and finish, using oil base material is recommended for any metal surface. Using water base material may cause oxidation, corrosion and rust. Using one coat of oil base paint on metal surfaces may result in cracking, peeling or chipping without the proper prime coat application. If off white or another light colored finish paint is specified, make sure the prime coat is also a light color, or more than one finish coat will be necessary. "Slow" work is based on an hourly wage of $23.00, "Medium" work on an hourly wage of $29.50, and "Fast" work on an hourly wage of $36.00. Other qualifications that apply to this table are on page 9.

	Labor SF per manhour	Material coverage SF/gallon	Material cost per gallon	Labor cost per 100 SF	Labor burden 100 SF	Material cost per 100 SF	Overhead per 100 SF	Profit per 100 SF	Total price per 100 SF
Conduit, electric, mitt or glove application									
Acid wash coat, muriatic acid (material #49)									
Mitt or glove each coat									
Slow	175	700	21.20	13.14	3.14	3.03	3.67	3.68	26.66
Medium	200	650	18.50	14.75	4.27	2.85	5.47	3.28	30.62
Fast	225	600	15.90	16.00	5.63	2.65	7.53	2.23	34.04
Metal primer, rust inhibitor - clean metal (material #35)									
Mitt or glove prime coat									
Slow	175	450	59.70	13.14	3.14	13.27	5.62	5.63	40.80
Medium	200	438	52.30	14.75	4.27	11.94	7.74	4.64	43.34
Fast	225	425	44.80	16.00	5.63	10.54	9.98	2.95	45.10
Metal primer, rust inhibitor - rusty metal (material #36)									
Mitt or glove prime coat									
Slow	175	400	80.10	13.14	3.14	20.03	6.90	6.92	50.13
Medium	200	388	70.10	14.75	4.27	18.07	9.27	5.56	51.92
Fast	225	375	60.10	16.00	5.63	16.03	11.68	3.46	52.80
Industrial enamel, oil base, high gloss - light colors (material #56)									
Mitt or glove 1st or additional finish coats									
Slow	225	450	65.40	10.22	2.44	14.53	5.17	5.18	37.54
Medium	250	438	57.20	11.80	3.41	13.06	7.07	4.24	39.58
Fast	275	425	49.00	13.09	4.64	11.53	9.06	2.68	41.00
Industrial enamel, oil base, high gloss - dark (OSHA) colors (material #57)									
Mitt or glove 1st or additional finish coats									
Slow	225	500	78.30	10.22	2.44	15.66	5.38	5.39	39.09
Medium	250	488	68.50	11.80	3.41	14.04	7.31	4.39	40.95
Fast	275	475	58.70	13.09	4.64	12.36	9.32	2.76	42.17
Epoxy coating, 2 part system - clear (material #51)									
Mitt or glove 1st coat									
Slow	175	425	160.60	13.14	3.14	37.79	10.28	10.30	74.65
Medium	200	413	140.50	14.75	4.27	34.02	13.26	7.95	74.25
Fast	225	400	120.50	16.00	5.63	30.13	16.05	4.75	72.56
Mitt or glove 2nd or additional finish coats									
Slow	225	475	160.60	10.22	2.44	33.81	8.83	8.85	64.15
Medium	250	463	140.50	11.80	3.41	30.35	11.39	6.83	63.78
Fast	275	450	120.50	13.09	4.64	26.78	13.79	4.08	62.38

	Labor SF per manhour	Material coverage SF/gallon	Material cost per gallon	Labor cost per 100 SF	Labor burden 100 SF	Material cost per 100 SF	Overhead per 100 SF	Profit per 100 SF	Total price per 100 SF
Epoxy coating, 2 part system - white (material #52)									
Mitt or glove 1st coat									
Slow	175	425	157.00	13.14	3.14	36.94	10.11	10.13	73.46
Medium	200	413	137.40	14.75	4.27	33.27	13.07	7.84	73.20
Fast	225	400	117.70	16.00	5.63	29.43	15.83	4.68	71.57
Mitt or glove 2nd or additional finish coats									
Slow	225	475	157.00	10.22	2.44	33.05	8.69	8.71	63.11
Medium	250	463	137.40	11.80	3.41	29.68	11.22	6.73	62.84
Fast	275	450	117.70	13.09	4.64	26.16	13.60	4.02	61.51

This table is based on square feet of conduit surface area. See Figure 21 below to convert from linear feet of various conduit or pipe sizes to square feet of surface area. For heights above 8 feet, use the High Time Difficulty Factors on page 139. Note: A two coat system, prime and finish, using oil base material is recommended for any metal surface. Using water base material may cause oxidation, corrosion and rust. Using one coat of oil base paint on metal surfaces may result in cracking, peeling or chipping without the proper prime coat application. If off white or another light colored finish paint is specified, make sure the prime coat is also a light color, or more than one finish coat will be necessary. "Slow" work is based on an hourly wage of $23.00, "Medium" work on an hourly wage of $29.50, and "Fast" work on an hourly wage of $36.00. Other qualifications that apply to this table are on page 9.

Pipe O/D (inches)	Conversion Factor (SF per measured LF)
1" to 3"	1 SF for each 1 LF
4" to 7"	2 SF for each 1 LF
8" to 11"	3 SF for each 1 LF
12" to 15"	4 SF for each 1 LF
16" to 19"	5 SF for each 1 LF
20" to 22"	6 SF for each 1 LF
23" to 26"	7 SF for each 1 LF
27" to 30"	8 SF for each 1 LF

Figure 21
Conduit/pipe area conversions

	Labor SF per manhour	Material coverage SF/gallon	Material cost per gallon	Labor cost per 100 SF	Labor burden 100 SF	Material cost per 100 SF	Overhead per 100 SF	Profit per 100 SF	Total price per 100 SF
Decking and siding, metal, corrugated metal									
Acid wash coat, muriatic acid (material #49)									
Spray each coat									
Slow	700	500	21.20	3.29	.79	4.24	1.58	1.58	11.48
Medium	750	450	18.50	3.93	1.13	4.11	2.30	1.38	12.85
Fast	800	400	15.90	4.50	1.59	3.98	3.12	.92	14.11
Metal primer, rust inhibitor - clean metal (material #35)									
Spray prime coat									
Slow	700	325	59.70	3.29	.79	18.37	4.27	4.28	31.00
Medium	750	300	52.30	3.93	1.13	17.43	5.63	3.38	31.50
Fast	800	275	44.80	4.50	1.59	16.29	6.94	2.05	31.37
Metal primer, rust inhibitor - rusty metal (material #36)									
Spray prime coat									
Slow	700	275	80.10	3.29	.79	29.13	6.31	6.32	45.84
Medium	750	250	70.10	3.93	1.13	28.04	8.28	4.97	46.35
Fast	800	225	60.10	4.50	1.59	26.71	10.17	3.01	45.98
Industrial enamel, oil base, high gloss - light colors (material #56)									
Spray 1st or additional finish coats									
Slow	850	325	65.40	2.71	.66	20.12	4.46	4.47	32.42
Medium	900	300	57.20	3.28	.94	19.07	5.83	3.50	32.62
Fast	950	275	49.00	3.79	1.32	17.82	7.11	2.10	32.14
Industrial enamel, oil base, high gloss - dark (OSHA) colors (material #57)									
Spray 1st or additional finish coats									
Slow	850	425	78.30	2.71	.66	18.42	4.14	4.15	30.08
Medium	900	400	68.50	3.28	.94	17.13	5.34	3.20	29.89
Fast	950	375	58.70	3.79	1.32	15.65	6.44	1.91	29.11
Epoxy coating, 2 part system - clear (material #51)									
Spray 1st coat									
Slow	700	300	160.60	3.29	.79	53.53	10.95	10.97	79.53
Medium	750	275	140.50	3.93	1.13	51.09	14.04	8.42	78.61
Fast	800	250	120.50	4.50	1.59	48.20	16.83	4.98	76.10
Spray 2nd or additional finish coats									
Slow	850	400	160.60	2.71	.66	40.15	8.27	8.28	60.07
Medium	900	375	140.50	3.28	.94	37.47	10.43	6.26	58.38
Fast	950	350	120.50	3.79	1.32	34.43	12.26	3.63	55.43

	Labor SF per manhour	Material coverage SF/gallon	Material cost per gallon	Labor cost per 100 SF	Labor burden 100 SF	Material cost per 100 SF	Overhead per 100 SF	Profit per 100 SF	Total price per 100 SF
Epoxy coating, 2 part system - white (material #52)									
Spray 1st coat									
Slow	700	300	157.00	3.29	.79	52.33	10.72	10.74	77.87
Medium	750	275	137.40	3.93	1.13	49.96	13.76	8.25	77.03
Fast	800	250	117.70	4.50	1.59	47.08	16.48	4.88	74.53
Spray 2nd or additional finish coats									
Slow	850	400	157.00	2.71	.66	39.25	8.10	8.11	58.83
Medium	900	375	137.40	3.28	.94	36.64	10.22	6.13	57.21
Fast	950	350	117.70	3.79	1.32	33.63	12.02	3.55	54.31

The figures in the table above are based on overall dimensions (length times width). But all decking and siding has corrugations, peaks and valleys that increase the surface that has to be painted. For example, corrugated siding with 2-1/2" center to center corrugations has a surface area 10% greater than the width times length dimension. For corrugated siding with 1-1/4" center to center corrugations, increase the surface area by 15%. For square corner decking, Figure 22 below shows how much area must be added to allow for peaks and valleys. For heights above 8 feet, use the High Time Difficulty Factors on page 139. Note: A two coat system, prime and finish, using oil base material is recommended for any metal surface. Using water base material may cause oxidation, corrosion and rust. Using one coat of oil base paint on metal surfaces may result in cracking, peeling or chipping without the proper prime coat application. If off white or another light colored finish paint is specified, make sure the prime coat is also a light color, or more than one finish coat will be necessary. "Slow" work is based on an hourly wage of $23.00, "Medium" work on an hourly wage of $29.50, and "Fast" work on an hourly wage of $36.00. Other qualifications that apply to this table are on page 9.

C	2C	D	PW	VW	Factor
--	12"	4-1/2"	3"	2"	2.50
--	12"	1-1/2"	3-1/8"	2"	1.50
--	12"	1-1/2"	5-1/16"	1"	1.45
12"	--	3"	9-5/8"	1"	1.50
12"	--	4-1/2"	9-5/8"	1"	1.75
24"	--	4-1/2"	12"	12"	1.60
24"	--	6"	12"	12"	1.75
24"	--	8"	12"	12"	1.95

For square corner decking, calculate the overall (length times width) deck area. Then measure the peaks and valleys on the deck. Select the row in the table above that most nearly matches the deck you're painting. Multiply the overall deck area by the number listed in the column headed "factor" to find the actual area you're painting. Use this actual area when calculating labor and material requirements. In the table above, figures in the column headed C show the distance between the center of the peaks. Column 2C shows the distance between every second center of peak (2 centers). Column D shows the depth of corrugation. Column PW shows the peak width. Column VW shows the valley width. If the deck you're painting doesn't match any deck listed in this table, use the factor for the most similar deck in the table.

Figure 22
Square corner decking factors

	Labor SF per manhour	Material coverage SF/gallon	Material cost per gallon	Labor cost per 100 SF	Labor burden 100 SF	Material cost per 100 SF	Overhead per 100 SF	Profit per 100 SF	Total price per 100 SF
Decking and siding, metal, flat pan metal									
Acid wash coat, muriatic acid (material #49)									
Spray each coat									
Slow	800	600	21.20	2.88	.69	3.53	1.35	1.35	9.80
Medium	850	550	18.50	3.47	1.02	3.36	1.96	1.17	10.98
Fast	900	500	15.90	4.00	1.41	3.18	2.66	.79	12.04
Metal primer, rust inhibitor - clean metal (material #35)									
Spray prime coat									
Slow	800	375	59.70	2.88	.69	15.92	3.70	3.71	26.90
Medium	850	350	52.30	3.47	1.02	14.94	4.85	2.91	27.19
Fast	900	325	44.80	4.00	1.41	13.78	5.95	1.76	26.90
Metal primer, rust inhibitor - rusty metal (material #36)									
Spray prime coat									
Slow	800	300	80.10	2.88	.69	26.70	5.75	5.76	41.78
Medium	850	275	70.10	3.47	1.02	25.49	7.49	4.49	41.96
Fast	900	250	60.10	4.00	1.41	24.04	9.13	2.70	41.28
Industrial enamel, oil base, high gloss - light colors (material #56)									
Spray 1st or additional finish coats									
Slow	1000	375	65.40	2.30	.55	17.44	3.86	3.86	28.01
Medium	1050	350	57.20	2.81	.80	16.34	4.99	2.99	27.93
Fast	1100	325	49.00	3.27	1.16	15.08	6.05	1.79	27.35
Industrial enamel, oil base, high gloss - dark (OSHA) colors (material #57)									
Spray 1st or additional finish coats									
Slow	1000	425	78.30	2.30	.55	18.42	4.04	4.05	29.36
Medium	1050	400	68.50	2.81	.80	17.13	5.19	3.11	29.04
Fast	1100	375	58.70	3.27	1.16	15.65	6.22	1.84	28.14
Epoxy coating, 2 part system - clear (material #51)									
Spray 1st coat									
Slow	800	350	160.60	2.88	.69	45.89	9.40	9.42	68.28
Medium	850	325	140.50	3.47	1.02	43.23	11.93	7.16	66.81
Fast	900	300	120.50	4.00	1.41	40.17	14.13	4.18	63.89
Spray 2nd or additional finish coats									
Slow	1000	400	160.60	2.30	.55	40.15	8.17	8.19	59.36
Medium	1050	375	140.50	2.81	.80	37.47	10.27	6.16	57.51
Fast	1100	350	120.50	3.27	1.16	34.43	12.04	3.56	54.46

	Labor SF per manhour	Material coverage SF/gallon	Material cost per gallon	Labor cost per 100 SF	Labor burden 100 SF	Material cost per 100 SF	Overhead per 100 SF	Profit per 100 SF	Total price per 100 SF
Epoxy coating, 2 part system - white (material #52)									
Spray 1st coat									
Slow	800	350	157.00	2.88	.69	44.86	9.20	9.22	66.85
Medium	850	325	137.40	3.47	1.02	42.28	11.69	7.01	65.47
Fast	900	300	117.70	4.00	1.41	39.23	13.84	4.09	62.57
Spray 2nd or additional coats									
Slow	1000	400	157.00	2.30	.55	39.25	8.00	8.02	58.12
Medium	1050	375	137.40	2.81	.80	36.64	10.07	6.04	56.36
Fast	1100	350	117.70	3.27	1.16	33.63	11.80	3.49	53.35

The figures in the table above are based on overall dimensions (length times width). But all decking and siding has corrugations, peaks and valleys that increase the surface that has to be painted. For example, corrugated siding with 2-1/2" center to center corrugations has a surface area 10% greater than the width times length dimension. For corrugated siding with 1-1/4" center to center corrugations, increase the surface area by 15%. For square corner decking, Figure 22 on page 327 shows how much area must be added to allow for peaks and valleys. For heights above 8 feet, use the High Time Difficulty Factors on page 139. Note: A two coat system, prime and finish, using oil base material is recommended for any metal surface. Using water base material may cause oxidation, corrosion and rust. Using one coat of oil base paint on metal surfaces may result in cracking, peeling or chipping without the proper prime coat application. If off white or another light colored finish paint is specified, make sure the prime coat is also a light color, or more than one finish coat will be necessary. "Slow" work is based on an hourly wage of $23.00, "Medium" work on an hourly wage of $29.50, and "Fast" work on an hourly wage of $36.00. Other qualifications that apply to this table are on page 9.

	Labor SF per manhour	Material coverage SF/gallon	Material cost per gallon	Labor cost per 100 SF	Labor burden 100 SF	Material cost per 100 SF	Overhead per 100 SF	Profit per 100 SF	Total price per 100 SF

Doors, hollow metal, brush application, square foot basis

	Labor SF per manhour	Material coverage SF/gallon	Material cost per gallon	Labor cost per 100 SF	Labor burden 100 SF	Material cost per 100 SF	Overhead per 100 SF	Profit per 100 SF	Total price per 100 SF
Metal primer, rust inhibitor - clean metal (material #35)									
Roll and brush prime coat									
Slow	160	450	59.70	14.38	3.45	13.27	5.91	5.92	42.93
Medium	180	438	52.30	16.39	4.75	11.94	8.27	4.96	46.31
Fast	200	425	44.80	18.00	6.36	10.54	10.82	3.20	48.92
Metal primer, rust inhibitor - rusty metal (material #36)									
Roll and brush prime coat									
Slow	160	425	80.10	14.38	3.45	18.85	6.97	6.98	50.63
Medium	180	413	70.10	16.39	4.75	16.97	9.53	5.72	53.36
Fast	200	400	60.10	18.00	6.36	15.03	12.21	3.61	55.21
Metal finish - synthetic enamel, gloss, interior or exterior - off white (material #37)									
Roll and brush 1st or additional finish coats									
Slow	175	450	63.60	13.14	3.14	14.13	5.78	5.79	41.98
Medium	195	438	55.70	15.13	4.38	12.72	8.06	4.83	45.12
Fast	215	425	47.70	16.74	5.91	11.22	10.50	3.11	47.48
Metal finish - synthetic enamel, gloss, interior or exterior - colors, except orange & red (material #38)									
Roll and brush 1st or additional finish coats									
Slow	175	475	66.30	13.14	3.14	13.96	5.75	5.76	41.75
Medium	195	463	58.00	15.13	4.38	12.53	8.01	4.80	44.85
Fast	215	450	49.70	16.74	5.91	11.04	10.44	3.09	47.22

To calculate the cost per door, figure the square feet for each side based on a 3 x 7 door. Add 3 feet to the width and 1 foot to the top of the door to allow for the time and material necessary to finish the door edges, frame and jamb. The result is a 6 x 8 door or 48 square feet per side, times 2 is 96 square feet per door. The cost per door is simply the 96 square feet times the cost per square foot indicated in the table above. Typically, hollow metal doors are pre-primed by the manufacturer and shipped ready to install and paint. These doors usually require only one coat of finish paint to cover. The metal finish figures include minor touchup of the prime coat. Note: A two coat system, prime and finish, using oil base material is recommended for any metal surface. Using water base material may cause oxidation, corrosion and rust. Using one coat of oil base paint on exterior metal may result in cracking, peeling or chipping without the proper prime coat application. If off white or other light colored finish paint is specified, make sure the prime coat is a light color also, or more than one finish coat will be necessary. "Slow" work is based on an hourly wage of $23.00, "Medium" work on an hourly wage of $29.50, and "Fast" work on an hourly wage of $36.00. Other qualifications that apply to this table are on page 9.

	Labor SF per manhour	Material coverage SF/gallon	Material cost per gallon	Labor cost per 100 SF	Labor burden 100 SF	Material cost per 100 SF	Overhead per 100 SF	Profit per 100 SF	Total price per 100 SF
Ductwork, bare duct, brush application									
Acid wash coat, muriatic acid (material #49)									
Brush each coat									
Slow	80	750	21.20	28.75	6.90	2.83	7.31	7.33	53.12
Medium	100	700	18.50	29.50	8.53	2.64	10.17	6.10	56.94
Fast	120	650	15.90	30.00	10.58	2.45	13.34	3.95	60.32
Metal primer, rust inhibitor - clean metal (material #35)									
Brush prime coat									
Slow	80	400	59.70	28.75	6.90	14.93	9.61	9.63	69.82
Medium	100	375	52.30	29.50	8.53	13.95	13.00	7.80	72.78
Fast	120	350	44.80	30.00	10.58	12.80	16.55	4.90	74.83
Metal primer, rust inhibitor - rusty metal (material #36)									
Brush prime coat									
Slow	80	350	80.10	28.75	6.90	22.89	11.12	11.15	80.81
Medium	100	325	70.10	29.50	8.53	21.57	14.90	8.94	83.44
Fast	120	300	60.10	30.00	10.58	20.03	18.79	5.56	84.96
Industrial enamel, oil base, high gloss - light colors (material #56)									
Brush 1st or additional finish coats									
Slow	90	400	65.40	25.56	6.13	16.35	9.13	9.15	66.32
Medium	115	375	57.20	25.65	7.44	15.25	12.08	7.25	67.67
Fast	140	350	49.00	25.71	9.07	14.00	15.12	4.47	68.37
Industrial enamel, oil base, high gloss - dark (OSHA) colors (material #57)									
Brush 1st or additional finish coats									
Slow	90	450	78.30	25.56	6.13	17.40	9.33	9.35	67.77
Medium	115	425	68.50	25.65	7.44	16.12	12.30	7.38	68.89
Fast	140	400	58.70	25.71	9.07	14.68	15.34	4.54	69.34
Epoxy coating, 2 part system - clear (material #51)									
Brush 1st coat									
Slow	80	350	160.60	28.75	6.90	45.89	15.49	15.52	112.55
Medium	100	325	140.50	29.50	8.53	43.23	20.32	12.19	113.77
Fast	120	300	120.50	30.00	10.58	40.17	25.04	7.41	113.20
Brush 2nd or additional finish coats									
Slow	90	425	160.60	25.56	6.13	37.79	13.20	13.23	95.91
Medium	115	400	140.50	25.65	7.44	35.13	17.05	10.23	95.50
Fast	140	375	120.50	25.71	9.07	32.13	20.75	6.14	93.80

	Labor SF per manhour	Material coverage SF/gallon	Material cost per gallon	Labor cost per 100 SF	Labor burden 100 SF	Material cost per 100 SF	Overhead per 100 SF	Profit per 100 SF	Total price per 100 SF
Epoxy coating, 2 part system - white (material #52)									
Brush 1st coat									
Slow	80	350	157.00	28.75	6.90	44.86	15.30	15.33	111.14
Medium	100	325	137.40	29.50	8.53	42.28	20.08	12.05	112.44
Fast	120	300	117.70	30.00	10.58	39.23	24.74	7.32	111.87
Brush 2nd or additional finish coats									
Slow	90	425	157.00	25.56	6.13	36.94	13.04	13.07	94.74
Medium	115	400	137.40	25.65	7.44	34.35	16.85	10.11	94.40
Fast	140	375	117.70	25.71	9.07	31.39	20.52	6.07	92.76

For heights over 8 feet, use the High Time Difficulty Factors on page 139. Note: A two coat system, prime and finish, using oil base material is recommended for any metal surface. Using water base material may cause oxidation, corrosion and rust. Using one coat of oil base paint on metal surfaces may result in cracking, peeling or chipping without the proper prime coat application. If off white or another light colored finish paint is specified, make sure the prime coat is also a light color, or more than one finish coat will be necessary. "Slow" work is based on an hourly wage of $23.00, "Medium" work on an hourly wage of $29.50, and "Fast" work on an hourly wage of $36.00. Other qualifications that apply to this table are on page 9.

	Labor SF per manhour	Material coverage SF/gallon	Material cost per gallon	Labor cost per 100 SF	Labor burden 100 SF	Material cost per 100 SF	Overhead per 100 SF	Profit per 100 SF	Total price per 100 SF
Ductwork, bare duct, roll application									
Acid wash coat, muriatic acid (material #49)									
Roll each coat									
Slow	225	700	21.20	10.22	2.44	3.03	2.98	2.99	21.66
Medium	250	650	18.50	11.80	3.41	2.85	4.52	2.71	25.29
Fast	275	600	15.90	13.09	4.64	2.65	6.31	1.87	28.56
Metal primer, rust inhibitor - clean metal (material #35)									
Roll prime coat									
Slow	225	375	59.70	10.22	2.44	15.92	5.43	5.44	39.45
Medium	250	350	52.30	11.80	3.41	14.94	7.54	4.52	42.21
Fast	275	325	44.80	13.09	4.64	13.78	9.76	2.89	44.16
Metal primer, rust inhibitor - rusty metal (material #36)									
Roll prime coat									
Slow	225	325	80.10	10.22	2.44	24.65	7.09	7.11	51.51
Medium	250	300	70.10	11.80	3.41	23.37	9.65	5.79	54.02
Fast	275	275	60.10	13.09	4.64	21.85	12.26	3.63	55.47
Industrial enamel, oil base, high gloss - light colors (material #56)									
Roll 1st or additional finish coats									
Slow	275	425	65.40	8.36	2.02	15.39	4.89	4.90	35.56
Medium	300	400	57.20	9.83	2.83	14.30	6.74	4.05	37.75
Fast	325	375	49.00	11.08	3.92	13.07	8.70	2.57	39.34
Industrial enamel, oil base, high gloss - dark (OSHA) colors (material #57)									
Roll 1st or additional finish coats									
Slow	275	475	78.30	8.36	2.02	16.48	5.10	5.11	37.07
Medium	300	450	68.50	9.83	2.83	15.22	6.97	4.18	39.03
Fast	325	425	58.70	11.08	3.92	13.81	8.93	2.64	40.38
Epoxy coating, 2 part system - clear (material #51)									
Roll 1st coat									
Slow	225	350	160.60	10.22	2.44	45.89	11.13	11.15	80.83
Medium	250	325	140.50	11.80	3.41	43.23	14.61	8.77	81.82
Fast	275	300	120.50	13.09	4.64	40.17	17.94	5.31	81.15
Roll 2nd or additional finish coats									
Slow	275	450	160.60	8.36	2.02	35.69	8.75	8.77	63.59
Medium	300	425	140.50	9.83	2.83	33.06	11.43	6.86	64.01
Fast	325	400	120.50	11.08	3.92	30.13	13.99	4.14	63.26

	Labor SF per manhour	Material coverage SF/gallon	Material cost per gallon	Labor cost per 100 SF	Labor burden 100 SF	Material cost per 100 SF	Overhead per 100 SF	Profit per 100 SF	Total price per 100 SF
Epoxy coating, 2 part system - white (material #52)									
Roll 1st coat									
Slow	225	350	157.00	10.22	2.44	44.86	10.93	10.95	79.40
Medium	250	325	137.40	11.80	3.41	42.28	14.37	8.62	80.48
Fast	275	300	117.70	13.09	4.64	39.23	17.65	5.22	79.83
Roll 2nd or additional finish coats									
Slow	275	450	157.00	8.36	2.02	34.89	8.60	8.62	62.49
Medium	300	425	137.40	9.83	2.83	32.33	11.25	6.75	62.99
Fast	325	400	117.70	11.08	3.92	29.43	13.77	4.07	62.27

For heights over 8 feet, use the High Time Difficulty Factors on page 139. Note: A two coat system, prime and finish, using oil base material is recommended for any metal surface. Using water base material may cause oxidation, corrosion and rust. Using one coat of oil base paint on metal surfaces may result in cracking, peeling or chipping without the proper prime coat application. If off white or another light colored finish paint is specified, make sure the prime coat is also a light color, or more than one finish coat will be necessary. "Slow" work is based on an hourly wage of $23.00, "Medium" work on an hourly wage of $29.50, and "Fast" work on an hourly wage of $36.00. Other qualifications that apply to this table are on page 9.

	Labor SF per manhour	Material coverage SF/gallon	Material cost per gallon	Labor cost per 100 SF	Labor burden 100 SF	Material cost per 100 SF	Overhead per 100 SF	Profit per 100 SF	Total price per 100 SF
Ductwork, bare duct, spray application									
Acid wash coat, muriatic acid (material #49)									
Spray each coat									
Slow	550	450	21.20	4.18	1.01	4.71	1.88	1.88	13.66
Medium	600	400	18.50	4.92	1.43	4.63	2.74	1.65	15.37
Fast	650	350	15.90	5.54	1.96	4.54	3.73	1.10	16.87
Metal primer, rust inhibitor - clean metal (material #35)									
Spray prime coat									
Slow	550	250	59.70	4.18	1.01	23.88	5.52	5.53	40.12
Medium	600	225	52.30	4.92	1.43	23.24	7.40	4.44	41.43
Fast	650	200	44.80	5.54	1.96	22.40	9.27	2.74	41.91
Metal primer, rust inhibitor - rusty metal (material #36)									
Spray prime coat									
Slow	550	200	80.10	4.18	1.01	40.05	8.59	8.61	62.44
Medium	600	188	70.10	4.92	1.43	37.29	10.91	6.54	61.09
Fast	650	175	60.10	5.54	1.96	34.34	12.97	3.84	58.65
Industrial enamel, oil base, high gloss - light colors (material #56)									
Spray 1st or additional finish coats									
Slow	700	225	65.40	3.29	.79	29.07	6.30	6.31	45.76
Medium	750	213	57.20	3.93	1.13	26.85	7.98	4.79	44.68
Fast	800	200	49.00	4.50	1.59	24.50	9.48	2.80	42.87
Industrial enamel, oil base, high gloss - dark (OSHA) colors (material #57)									
Spray 1st or additional finish coats									
Slow	700	275	78.30	3.29	.79	28.47	6.18	6.20	44.93
Medium	750	250	68.50	3.93	1.13	27.40	8.12	4.87	45.45
Fast	800	225	58.70	4.50	1.59	26.09	9.98	2.95	45.11
Epoxy coating, 2 part system - clear (material #51)									
Spray 1st coat									
Slow	550	225	160.60	4.18	1.01	71.38	14.55	14.58	105.70
Medium	600	213	140.50	4.92	1.43	65.96	18.08	10.85	101.24
Fast	650	200	120.50	5.54	1.96	60.25	21.00	6.21	94.96
Spray 2nd or additional finish coats									
Slow	700	250	160.60	3.29	.79	64.24	12.98	13.01	94.31
Medium	750	238	140.50	3.93	1.13	59.03	16.03	9.62	89.74
Fast	800	225	120.50	4.50	1.59	53.56	18.49	5.47	83.61

	Labor SF per manhour	Material coverage SF/gallon	Material cost per gallon	Labor cost per 100 SF	Labor burden 100 SF	Material cost per 100 SF	Overhead per 100 SF	Profit per 100 SF	Total price per 100 SF
Epoxy coating, 2 part system - white (material #52)									
Spray 1st coat									
Slow	550	225	157.00	4.18	1.01	69.78	14.24	14.27	103.48
Medium	600	213	137.40	4.92	1.43	64.51	17.71	10.63	99.20
Fast	650	200	117.70	5.54	1.96	58.85	20.57	6.08	93.00
Spray 2nd or additional finish coats									
Slow	700	250	157.00	3.29	.79	62.80	12.71	12.73	92.32
Medium	750	238	137.40	3.93	1.13	57.73	15.70	9.42	87.91
Fast	800	225	117.70	4.50	1.59	52.31	18.10	5.36	81.86

For heights over 8 feet, use the High Time Difficulty Factors on page 139. Note: A two coat system, prime and finish, using oil base material is recommended for any metal surface. Using water base material may cause oxidation, corrosion and rust. Using one coat of oil base paint on metal surfaces may result in cracking, peeling or chipping without the proper prime coat application. If off white or another light colored finish paint is specified, make sure the prime coat is also a light color, or more than one finish coat will be necessary. "Slow" work is based on an hourly wage of $23.00, "Medium" work on an hourly wage of $29.50, and "Fast" work on an hourly wage of $36.00. Other qualifications that apply to this table are on page 9.

	Labor SF per manhour	Material coverage SF/gallon	Material cost per gallon	Labor cost per 100 SF	Labor burden 100 SF	Material cost per 100 SF	Overhead per 100 SF	Profit per 100 SF	Total price per 100 SF
Ductwork, bare duct, mitt or glove application									
Acid wash coat, muriatic acid (material #49)									
Mitt or glove each coat									
Slow	175	700	21.20	13.14	3.14	3.03	3.67	3.68	26.66
Medium	200	650	18.50	14.75	4.27	2.85	5.47	3.28	30.62
Fast	225	600	15.90	16.00	5.63	2.65	7.53	2.23	34.04
Metal primer, rust inhibitor - clean metal (material #35)									
Mitt or glove prime coat									
Slow	175	400	59.70	13.14	3.14	14.93	5.93	5.94	43.08
Medium	200	388	52.30	14.75	4.27	13.48	8.12	4.87	45.49
Fast	225	375	44.80	16.00	5.63	11.95	10.42	3.08	47.08
Metal primer, rust inhibitor - rusty metal (material #36)									
Mitt or glove prime coat									
Slow	175	400	80.10	13.14	3.14	20.03	6.90	6.92	50.13
Medium	200	350	70.10	14.75	4.27	20.03	9.76	5.86	54.67
Fast	225	325	60.10	16.00	5.63	18.49	12.44	3.68	56.24
Industrial enamel, oil base, high gloss - light colors (material #56)									
Mitt or glove 1st or additional finish coats									
Slow	225	425	65.40	10.22	2.44	15.39	5.33	5.34	38.72
Medium	250	400	57.20	11.80	3.41	14.30	7.38	4.43	41.32
Fast	275	375	49.00	13.09	4.64	13.07	9.54	2.82	43.16
Industrial enamel, oil base, high gloss - dark (OSHA) colors (material #57)									
Mitt or glove 1st or additional finish coats									
Slow	225	475	78.30	10.22	2.44	16.48	5.54	5.55	40.23
Medium	250	450	68.50	11.80	3.41	15.22	7.61	4.56	42.60
Fast	275	425	58.70	13.09	4.64	13.81	9.77	2.89	44.20
Epoxy coating, 2 part system - clear (material #51)									
Mitt or glove 1st coat									
Slow	175	400	160.60	13.14	3.14	40.15	10.72	10.75	77.90
Medium	200	375	140.50	14.75	4.27	37.47	14.12	8.47	79.08
Fast	225	350	120.50	16.00	5.63	34.43	17.38	5.14	78.58
Mitt or glove 2nd or additional finish coats									
Slow	225	425	160.60	10.22	2.44	37.79	9.59	9.61	69.65
Medium	250	400	140.50	11.80	3.41	35.13	12.59	7.55	70.48
Fast	275	375	120.50	13.09	4.64	32.13	15.45	4.57	69.88

	Labor SF per manhour	Material coverage SF/gallon	Material cost per gallon	Labor cost per 100 SF	Labor burden 100 SF	Material cost per 100 SF	Overhead per 100 SF	Profit per 100 SF	Total price per 100 SF
Epoxy coating, 2 part system - white (material #52)									
Mitt or glove 1st coat									
Slow	175	400	157.00	13.14	3.14	39.25	10.55	10.57	76.65
Medium	200	375	137.40	14.75	4.27	36.64	13.91	8.35	77.92
Fast	225	350	117.70	16.00	5.63	33.63	17.14	5.07	77.47
Mitt or glove 2nd or additional finish coats									
Slow	225	425	157.00	10.22	2.44	36.94	9.43	9.45	68.48
Medium	250	400	137.40	11.80	3.41	34.35	12.39	7.43	69.38
Fast	275	375	117.70	13.09	4.64	31.39	15.22	4.50	68.84

For heights over 8 feet, use the High Time Difficulty Factors on page 139. Note: A two coat system, prime and finish, using oil base material is recommended for any metal surface. Using water base material may cause oxidation, corrosion and rust. Using one coat of oil base paint on metal surfaces may result in cracking, peeling or chipping without the proper prime coat application. If off white or another light colored finish paint is specified, make sure the prime coat is also a light color, or more than one finish coat will be necessary. "Slow" work is based on an hourly wage of $23.00, "Medium" work on an hourly wage of $29.50, and "Fast" work on an hourly wage of $36.00. Other qualifications that apply to this table are on page 9.

	Labor SF per manhour	Material coverage SF/gallon	Material cost per gallon	Labor cost per 100 SF	Labor burden 100 SF	Material cost per 100 SF	Overhead per 100 SF	Profit per 100 SF	Total price per 100 SF
Ductwork, canvas insulated, brush application									
Flat latex, water base (material #5)									
Brush 1st coat									
Slow	60	250	36.00	38.33	9.21	14.40	11.77	11.79	85.50
Medium	75	238	31.50	39.33	11.36	13.24	15.99	9.59	89.51
Fast	90	225	27.00	40.00	14.12	12.00	20.50	6.06	92.68
Brush 2nd coat									
Slow	85	275	36.00	27.06	6.48	13.09	8.86	8.88	64.37
Medium	105	238	31.50	28.10	8.10	13.24	12.37	7.42	69.23
Fast	125	250	27.00	28.80	10.17	10.80	15.43	4.56	69.76
Brush 3rd or additional coats									
Slow	100	325	36.00	23.00	5.52	11.08	7.52	7.54	54.66
Medium	125	313	31.50	23.60	6.82	10.06	10.12	6.07	56.67
Fast	150	300	27.00	24.00	8.49	9.00	12.86	3.80	58.15
Sealer, off white, water base (material #1)									
Brush 1 coat									
Slow	60	250	36.00	38.33	9.21	14.40	11.77	11.79	85.50
Medium	75	238	31.50	39.33	11.36	13.24	15.99	9.59	89.51
Fast	90	225	27.00	40.00	14.12	12.00	20.50	6.06	92.68
Sealer, off white, oil base (material #2)									
Brush 1 coat									
Slow	60	275	46.20	38.33	9.21	16.80	12.22	12.25	88.81
Medium	75	263	40.50	39.33	11.36	15.40	16.53	9.92	92.54
Fast	90	250	34.70	40.00	14.12	13.88	21.08	6.24	95.32
Enamel, water base (material #9)									
Brush 1st finish coat									
Slow	85	275	53.30	27.06	6.48	19.38	10.06	10.08	73.06
Medium	105	238	46.70	28.10	8.10	19.62	13.96	8.38	78.16
Fast	125	250	40.00	28.80	10.17	16.00	17.04	5.04	77.05
Brush 2nd or additional finish coats									
Slow	100	325	53.30	23.00	5.52	16.40	8.53	8.55	62.00
Medium	125	313	46.70	23.60	6.82	14.92	11.34	6.80	63.48
Fast	150	300	40.00	24.00	8.49	13.33	14.20	4.20	64.22
Industrial enamel, oil base, high gloss - light colors (material #56)									
Brush 1st finish coat									
Slow	85	300	65.40	27.06	6.48	21.80	10.52	10.54	76.40
Medium	105	288	57.20	28.10	8.10	19.86	14.02	8.41	78.49
Fast	125	275	49.00	28.80	10.17	17.82	17.60	5.21	79.60

	Labor SF per manhour	Material coverage SF/gallon	Material cost per gallon	Labor cost per 100 SF	Labor burden 100 SF	Material cost per 100 SF	Overhead per 100 SF	Profit per 100 SF	Total price per 100 SF
Brush 2nd or additional finish coats									
Slow	100	350	65.40	23.00	5.52	18.69	8.97	8.99	65.17
Medium	125	338	57.20	23.60	6.82	16.92	11.84	7.10	66.28
Fast	150	325	49.00	24.00	8.49	15.08	14.74	4.36	66.67
Industrial enamel, oil base, high gloss - dark (OSHA) colors (material #57)									
Brush 1st finish coat									
Slow	85	300	78.30	27.06	6.48	26.10	11.33	11.36	82.33
Medium	105	288	68.50	28.10	8.10	23.78	15.00	9.00	83.98
Fast	125	275	58.70	28.80	10.17	21.35	18.70	5.53	84.55
Brush 2nd or additional finish coats									
Slow	100	350	78.30	23.00	5.52	22.37	9.67	9.69	70.25
Medium	125	338	68.50	23.60	6.82	20.27	12.67	7.60	70.96
Fast	150	325	58.70	24.00	8.49	18.06	15.66	4.63	70.84
Epoxy coating, 2 part system - clear (material #51)									
Brush 1st coat									
Slow	60	260	160.60	38.33	9.21	61.77	20.77	20.81	150.89
Medium	75	243	140.50	39.33	11.36	57.82	27.13	16.28	151.92
Fast	90	235	120.50	40.00	14.12	51.28	32.67	9.66	147.73
Brush 2nd coat									
Slow	85	285	160.60	27.06	6.48	56.35	17.08	17.12	124.09
Medium	105	273	140.50	28.10	8.10	51.47	21.92	13.15	122.74
Fast	125	260	120.50	28.80	10.17	46.35	26.45	7.82	119.59
Brush 3rd or additional coats									
Slow	100	335	160.60	23.00	5.52	47.94	14.53	14.56	105.55
Medium	125	323	140.50	23.60	6.82	43.50	18.48	11.09	103.49
Fast	150	310	120.50	24.00	8.49	38.87	22.12	6.54	100.02

	Labor SF per manhour	Material coverage SF/gallon	Material cost per gallon	Labor cost per 100 SF	Labor burden 100 SF	Material cost per 100 SF	Overhead per 100 SF	Profit per 100 SF	Total price per 100 SF
Epoxy coating, 2 part system - white (material #52)									
Brush 1st coat									
Slow	60	260	157.00	38.33	9.21	60.38	20.50	20.55	148.97
Medium	75	243	137.40	39.33	11.36	56.54	26.81	16.09	150.13
Fast	90	235	117.70	40.00	14.12	50.09	32.31	9.56	146.08
Brush 2nd coat									
Slow	85	285	157.00	27.06	6.48	55.09	16.84	16.88	122.35
Medium	100	273	137.40	29.50	8.53	50.33	22.09	13.25	123.70
Fast	125	260	117.70	28.80	10.17	45.27	26.11	7.72	118.07
Brush 3rd or additional coats									
Slow	100	335	157.00	23.00	5.52	46.87	14.32	14.35	104.06
Medium	125	323	137.40	23.60	6.82	42.54	18.24	10.94	102.14
Fast	150	310	117.70	24.00	8.49	37.97	21.84	6.46	98.76

For heights over 8 feet, use the High Time Difficulty Factors on page 139. "Slow" work is based on an hourly wage of $23.00, "Medium" work on an hourly wage of $29.50, and "Fast" work on an hourly wage of $36.00. Other qualifications that apply to this table are on page 9.

	Labor SF per manhour	Material coverage SF/gallon	Material cost per gallon	Labor cost per 100 SF	Labor burden 100 SF	Material cost per 100 SF	Overhead per 100 SF	Profit per 100 SF	Total price per 100 SF
Ductwork, canvas insulated, roll application									
Flat latex, water base (material #5)									
Roll 1st coat									
Slow	125	250	36.00	18.40	4.42	14.40	7.07	7.09	51.38
Medium	150	238	31.50	19.67	5.70	13.24	9.65	5.79	54.05
Fast	175	225	27.00	20.57	7.24	12.00	12.35	3.65	55.81
Roll 2nd coat									
Slow	175	350	36.00	13.14	3.14	10.29	5.05	5.06	36.68
Medium	200	325	31.50	14.75	4.27	9.69	7.18	4.31	40.20
Fast	225	300	27.00	16.00	5.63	9.00	9.50	2.81	42.94
Roll 3rd or additional coats									
Slow	225	400	36.00	10.22	2.44	9.00	4.12	4.13	29.91
Medium	250	388	31.50	11.80	3.41	8.12	5.83	3.50	32.66
Fast	275	375	27.00	13.09	4.64	7.20	7.72	2.28	34.93
Sealer, off white, water base (material #1)									
Roll 1 coat									
Slow	125	250	36.00	18.40	4.42	14.40	7.07	7.09	51.38
Medium	150	238	31.50	19.67	5.70	13.24	9.65	5.79	54.05
Fast	175	225	27.00	20.57	7.24	12.00	12.35	3.65	55.81
Sealer, off white, oil base (material #2)									
Roll 1 coat									
Slow	125	275	46.20	18.40	4.42	16.80	7.53	7.54	54.69
Medium	150	263	40.50	19.67	5.70	15.40	10.19	6.11	57.07
Fast	175	250	34.70	20.57	7.24	13.88	12.93	3.82	58.44
Enamel, water base (material #9)									
Roll 1st finish coat									
Slow	175	350	53.30	13.14	3.14	15.23	5.99	6.00	43.50
Medium	200	325	46.70	14.75	4.27	14.37	8.35	5.01	46.75
Fast	225	300	40.00	16.00	5.63	13.33	10.84	3.21	49.01
Roll 2nd or additional finish coats									
Slow	225	400	53.30	10.22	2.44	13.33	4.94	4.95	35.88
Medium	250	388	46.70	11.80	3.41	12.04	6.81	4.09	38.15
Fast	275	375	40.00	13.09	4.64	10.67	8.80	2.60	39.80

	Labor SF per manhour	Material coverage SF/gallon	Material cost per gallon	Labor cost per 100 SF	Labor burden 100 SF	Material cost per 100 SF	Overhead per 100 SF	Profit per 100 SF	Total price per 100 SF
Industrial enamel, oil base, high gloss - light colors (material #56)									
Roll 1st finish coat									
Slow	175	375	65.40	13.14	3.14	17.44	6.41	6.42	46.55
Medium	200	350	57.20	14.75	4.27	16.34	8.84	5.30	49.50
Fast	225	325	49.00	16.00	5.63	15.08	11.39	3.37	51.47
Roll 2nd or additional finish coats									
Slow	225	450	65.40	10.22	2.44	14.53	5.17	5.18	37.54
Medium	250	438	57.20	11.80	3.41	13.06	7.07	4.24	39.58
Fast	275	425	49.00	13.09	4.64	11.53	9.06	2.68	41.00
Industrial enamel, oil base, high gloss - dark (OSHA) colors (material #57)									
Roll 1st finish coat									
Slow	175	300	78.30	13.14	3.14	26.10	8.05	8.07	58.50
Medium	200	288	68.50	14.75	4.27	23.78	10.70	6.42	59.92
Fast	225	275	58.70	16.00	5.63	21.35	13.33	3.94	60.25
Roll 2nd or additional finish coats									
Slow	225	450	78.30	10.22	2.44	17.40	5.71	5.72	41.49
Medium	250	438	68.50	11.80	3.41	15.64	7.71	4.63	43.19
Fast	275	425	58.70	13.09	4.64	13.81	9.77	2.89	44.20
Epoxy coating, 2 part system - clear (material #51)									
Roll 1st coat									
Slow	125	250	160.60	18.40	4.42	64.24	16.54	16.58	120.18
Medium	150	238	140.50	19.67	5.70	59.03	21.10	12.66	118.16
Fast	175	225	120.50	20.57	7.24	53.56	25.23	7.46	114.06
Roll 2nd coat									
Slow	175	350	160.60	13.14	3.14	45.89	11.81	11.84	85.82
Medium	200	325	140.50	14.75	4.27	43.23	15.56	9.34	87.15
Fast	225	300	120.50	16.00	5.63	40.17	19.16	5.67	86.63
Roll 3rd or additional coats									
Slow	225	425	160.60	10.22	2.44	37.79	9.59	9.61	69.65
Medium	250	413	140.50	11.80	3.41	34.02	12.31	7.38	68.92
Fast	275	400	120.50	13.09	4.64	30.13	14.83	4.39	67.08

	Labor SF per manhour	Material coverage SF/gallon	Material cost per gallon	Labor cost per 100 SF	Labor burden 100 SF	Material cost per 100 SF	Overhead per 100 SF	Profit per 100 SF	Total price per 100 SF
Epoxy coating, 2 part system - white (material #52)									
Roll 1st coat									
Slow	125	250	157.00	18.40	4.42	62.80	16.27	16.30	118.19
Medium	150	238	137.40	19.67	5.70	57.73	20.77	12.46	116.33
Fast	175	225	117.70	20.57	7.24	52.31	24.84	7.35	112.31
Roll 2nd coat									
Slow	175	375	157.00	13.14	3.14	41.87	11.05	11.07	80.27
Medium	200	338	137.40	14.75	4.27	40.65	14.92	8.95	83.54
Fast	225	300	117.70	16.00	5.63	39.23	18.87	5.58	85.31
Roll 3rd or additional coats									
Slow	225	425	157.00	10.22	2.44	36.94	9.43	9.45	68.48
Medium	250	413	137.40	11.80	3.41	33.27	12.12	7.27	67.87
Fast	275	400	117.70	13.09	4.64	29.43	14.61	4.32	66.09

For heights over 8 feet, use the High Time Difficulty Factors on page 139. "Slow" work is based on an hourly wage of $23.00, "Medium" work on an hourly wage of $29.50, and "Fast" work on an hourly wage of $36.00. Other qualifications that apply to this table are on page 9.

	Labor SF per manhour	Material coverage SF/gallon	Material cost per gallon	Labor cost per 100 SF	Labor burden 100 SF	Material cost per 100 SF	Overhead per 100 SF	Profit per 100 SF	Total price per 100 SF

Ductwork, canvas insulated, spray application

Flat latex, water base (material #5)									
Spray prime coat									
Slow	450	200	36.00	5.11	1.22	18.00	4.62	4.63	33.58
Medium	500	188	31.50	5.90	1.71	16.76	6.09	3.66	34.12
Fast	550	175	27.00	6.55	2.32	15.43	7.53	2.23	34.06
Spray 1st finish coat									
Slow	550	225	36.00	4.18	1.01	16.00	4.02	4.03	29.24
Medium	625	213	31.50	4.72	1.36	14.79	5.22	3.13	29.22
Fast	700	200	27.00	5.14	1.83	13.50	6.34	1.88	28.69
Spray 2nd or additional finish coats									
Slow	700	250	36.00	3.29	.79	14.40	3.51	3.52	25.51
Medium	750	238	31.50	3.93	1.13	13.24	4.58	2.75	25.63
Fast	800	225	27.00	4.50	1.59	12.00	5.61	1.66	25.36
Sealer, off white, water base (material #1)									
Spray 1 coat									
Slow	450	200	36.00	5.11	1.22	18.00	4.62	4.63	33.58
Medium	500	188	31.50	5.90	1.71	16.76	6.09	3.66	34.12
Fast	550	175	27.00	6.55	2.32	15.43	7.53	2.23	34.06
Sealer, off white, oil base (material #2)									
Spray 1 coat									
Slow	450	225	46.20	5.11	1.22	20.53	5.11	5.12	37.09
Medium	500	213	40.50	5.90	1.71	19.01	6.66	3.99	37.27
Fast	550	200	34.70	6.55	2.32	17.35	8.13	2.40	36.75
Enamel, water base (material #9)									
Spray 1st finish coat									
Slow	550	225	53.30	4.18	1.01	23.69	5.49	5.50	39.87
Medium	625	213	46.70	4.72	1.36	21.92	7.00	4.20	39.20
Fast	700	200	40.00	5.14	1.83	20.00	8.35	2.47	37.79
Spray 2nd or additional finish coats									
Slow	700	250	53.30	3.29	.79	21.32	4.83	4.84	35.07
Medium	750	238	46.70	3.93	1.13	19.62	6.17	3.70	34.55
Fast	800	225	40.00	4.50	1.59	17.78	7.40	2.19	33.46

	Labor SF per manhour	Material coverage SF/gallon	Material cost per gallon	Labor cost per 100 SF	Labor burden 100 SF	Material cost per 100 SF	Overhead per 100 SF	Profit per 100 SF	Total price per 100 SF
Industrial enamel, oil base, high gloss - light colors (material #56)									
Spray 1st finish coat									
Slow	700	250	65.40	3.29	.79	26.16	5.75	5.76	41.75
Medium	625	238	57.20	4.72	1.36	24.03	7.53	4.52	42.16
Fast	550	225	49.00	6.55	2.32	21.78	9.50	2.81	42.96
Spray 2nd or additional finish coats									
Slow	800	275	65.40	2.88	.69	23.78	5.20	5.21	37.76
Medium	750	263	57.20	3.93	1.13	21.75	6.71	4.02	37.54
Fast	700	250	49.00	5.14	1.83	19.60	8.23	2.43	37.23
Industrial enamel, oil base, high gloss - dark (OSHA) colors (material #57)									
Spray 1st finish coat									
Slow	700	250	78.30	3.29	.79	31.32	6.73	6.74	48.87
Medium	625	238	68.50	4.72	1.36	28.78	8.72	5.23	48.81
Fast	550	225	58.70	6.55	2.32	26.09	10.83	3.20	48.99
Spray 2nd or additional finish coats									
Slow	800	275	78.30	2.88	.69	28.47	6.09	6.10	44.23
Medium	750	263	68.50	3.93	1.13	26.05	7.78	4.67	43.56
Fast	700	250	58.70	5.14	1.83	23.48	9.43	2.79	42.67
Epoxy coating, 2 part system - clear (material #51)									
Spray prime coat									
Slow	450	200	160.60	5.11	1.22	80.30	16.46	16.50	119.59
Medium	500	188	140.50	5.90	1.71	74.73	20.59	12.35	115.28
Fast	550	175	120.50	6.55	2.32	68.86	24.09	7.13	108.95
Spray 1st finish coat									
Slow	550	235	160.60	4.18	1.01	68.34	13.97	14.00	101.50
Medium	625	225	140.50	4.72	1.36	62.44	17.13	10.28	95.93
Fast	700	215	120.50	5.14	1.83	56.05	19.53	5.78	88.33
Spray 2nd or additional finish coats									
Slow	700	265	160.60	3.29	.79	60.60	12.29	12.32	89.29
Medium	750	253	140.50	3.93	1.13	55.53	15.15	9.09	84.83
Fast	800	240	120.50	4.50	1.59	50.21	17.45	5.16	78.91

	Labor SF per manhour	Material coverage SF/gallon	Material cost per gallon	Labor cost per 100 SF	Labor burden 100 SF	Material cost per 100 SF	Overhead per 100 SF	Profit per 100 SF	Total price per 100 SF
Epoxy coating, 2 part system - white (material #52)									
Spray prime coat									
Slow	450	200	157.00	5.11	1.22	78.50	16.12	16.15	117.10
Medium	500	188	137.40	5.90	1.71	73.09	20.18	12.11	112.99
Fast	550	175	117.70	6.55	2.32	67.26	23.60	6.98	106.71
Spray 1st finish coat									
Slow	550	235	157.00	4.18	1.01	66.81	13.68	13.71	99.39
Medium	625	225	137.40	4.72	1.36	61.07	16.79	10.07	94.01
Fast	700	215	117.70	5.14	1.83	54.74	19.12	5.66	86.49
Spray 2nd or additional finish coats									
Slow	700	265	157.00	3.29	.79	59.25	12.03	12.06	87.42
Medium	750	253	137.40	3.93	1.13	54.31	14.85	8.91	83.13
Fast	800	240	117.70	4.50	1.59	49.04	17.09	5.06	77.28

For heights over 8 feet, use the High Time Difficulty Factors on page 139. "Slow" work is based on an hourly wage of $23.00, "Medium" work on an hourly wage of $29.50, and "Fast" work on an hourly wage of $36.00. Other qualifications that apply to this table are on page 9.

	Manhours per flight	Flights per gallon	Material cost per gallon	Labor cost per flight	Labor burden flight	Material cost per flight	Overhead per flight	Profit per flight	Total price per flight
Fire escapes									
Solid (plain) deck									
Spray each coat									
Metal primer, rust inhibitor - clean metal (material #35)									
Slow	2.0	1.25	59.70	46.00	11.04	47.76	19.91	19.95	144.66
Medium	1.5	1.00	52.30	44.25	12.80	52.30	27.34	16.40	153.09
Fast	1.0	0.75	44.80	36.00	12.71	59.73	33.62	9.94	152.00
Metal primer, rust inhibitor - rusty metal (material #36)									
Slow	2.0	1.25	80.10	46.00	11.04	64.08	23.01	23.06	167.19
Medium	1.5	1.00	70.10	44.25	12.80	70.10	31.79	19.07	178.01
Fast	1.0	0.75	60.10	36.00	12.71	80.13	39.94	11.81	180.59
Industrial enamel, oil base, high gloss - light colors (material #56)									
Slow	2.25	1.50	65.40	51.75	12.42	43.60	20.48	20.52	148.77
Medium	1.75	1.25	57.20	51.63	14.92	45.76	28.08	16.85	157.24
Fast	1.25	1.00	49.00	45.00	15.89	49.00	34.07	10.08	154.04
Industrial enamel, oil base, high gloss - dark (OSHA) colors (material #57)									
Slow	2.25	1.50	78.30	51.75	12.42	52.20	22.11	22.16	160.64
Medium	1.75	1.25	68.50	51.63	14.92	54.80	30.34	18.20	169.89
Fast	1.25	1.00	58.70	45.00	15.89	58.70	37.07	10.97	167.63
Grating deck									
Spray each coat									
Metal primer, rust inhibitor - clean metal (material #35)									
Slow	3.0	1.75	59.70	69.00	16.56	34.11	22.74	22.79	165.20
Medium	2.5	1.50	52.30	73.75	21.33	34.87	32.48	19.49	181.92
Fast	2.0	1.25	44.80	72.00	25.42	35.84	41.31	12.22	186.79
Metal primer, rust inhibitor - rusty metal (material #36)									
Slow	3.0	1.75	80.10	69.00	16.56	45.77	24.95	25.00	181.28
Medium	2.5	1.50	70.10	73.75	21.33	46.73	35.45	21.27	198.53
Fast	2.0	1.25	60.10	72.00	25.42	48.08	45.11	13.34	203.95

	Manhours per flight	Flights per gallon	Material cost per gallon	Labor cost per flight	Labor burden flight	Material cost per flight	Overhead per flight	Profit per flight	Total price per flight
Industrial enamel, oil base, high gloss - light colors (material #56)									
Slow	3.25	2.00	65.40	74.75	17.94	32.70	23.82	23.87	173.08
Medium	2.75	1.75	57.20	81.13	23.45	32.69	34.32	20.59	192.18
Fast	2.25	1.50	49.00	81.00	28.60	32.67	44.10	13.05	199.42
Industrial enamel, oil base, high gloss - dark (OSHA) colors (material #57)									
Slow	3.25	2.00	78.30	74.75	17.94	39.15	25.05	25.10	181.99
Medium	2.75	1.75	68.50	81.13	23.45	39.14	35.93	21.56	201.21
Fast	2.25	1.50	58.70	81.00	28.60	39.13	46.10	13.64	208.47

Fire escapes can also be estimated by the square foot. Calculate the actual area to be coated. For continuous solid (plain) deck, use the rates listed under Decking and siding. For continuous grating deck, use the rates listed under Grates, steel. Note: A two coat system, prime and finish, using oil base material is recommended for any metal surface. Using water base material may cause oxidation, corrosion and rust. Using one coat of oil base paint on metal surfaces may result cracking, peeling or chipping without the proper prime coat application. If off white or another light colored finish paint is specified, make sure the prime coat is also a light color, or more than one finish coat will be necessary. "Slow" work is based on an hourly wage of $23.00, "Medium" work on an hourly wage of $29.50, and "Fast" work on an hourly wage of $36.00. Other qualifications that apply to this table are on page 9.

Fire sprinkler systems

Use the costs listed for 1" to 4" pipe. For painting sprinkler heads at 12 feet on center at a ceiling height of 12 feet, figure 3 minutes per head (20 per hour). Very little paint is needed. Your material estimate for the sprinkler pipe will include enough to cover the heads. Note: A two coat system, prime and finish, using oil base material is recommended for any metal surface. Using water base material may cause oxidation, corrosion and rust. One coat of oil base paint on metal surfaces may result in cracking, peeling or chipping without the proper prime coat application. If off white or another light colored finish paint is specified, make sure the prime coat is also a light color, or more than one finish coat will be necessary.

	Labor SF per manhour	Material coverage SF/gallon	Material cost per gallon	Labor cost per 100 SF	Labor burden 100 SF	Material cost per 100 SF	Overhead per 100 SF	Profit per 100 SF	Total price per 100 SF
Grates, steel, over 1" thick									
Without supports									
Brush each coat									
Metal primer, rust inhibitor - clean metal (material #35)									
Slow	60	175	59.70	38.33	9.21	.34	9.10	9.12	66.10
Medium	85	150	52.30	34.71	10.01	.35	11.27	6.76	63.10
Fast	110	125	44.80	32.73	11.55	.36	13.84	4.09	62.57
Metal primer, rust inhibitor - rusty metal (material #36)									
Slow	60	175	80.10	38.33	9.21	.46	9.12	9.14	66.26
Medium	85	150	70.10	34.71	10.01	.47	11.30	6.78	63.27
Fast	110	125	60.10	32.73	11.55	.48	13.88	4.10	62.74
Industrial enamel, oil base, high gloss - light colors (material #56)									
Slow	75	225	65.40	30.67	7.35	.29	7.28	7.30	52.89
Medium	100	200	57.20	29.50	8.53	.29	9.58	5.75	53.65
Fast	125	175	49.00	28.80	10.17	.28	12.17	3.60	55.02
Industrial enamel, oil base, high gloss - dark (OSHA) colors (material #57)									
Slow	75	225	78.30	30.67	7.35	.35	7.29	7.31	52.97
Medium	100	200	68.50	29.50	8.53	.34	9.59	5.76	53.72
Fast	125	175	58.70	28.80	10.17	.34	12.19	3.61	55.11
Spray each coat									
Metal primer, rust inhibitor - clean metal (material #35)									
Slow	190	125	59.70	12.11	2.89	.48	2.95	2.95	21.38
Medium	208	113	52.30	14.18	4.11	.46	4.69	2.81	26.25
Fast	225	100	44.80	16.00	5.63	.45	6.85	2.03	30.96
Metal primer, rust inhibitor - rusty metal (material #36)									
Slow	190	125	80.10	12.11	2.89	.64	2.98	2.98	21.60
Medium	208	113	70.10	14.18	4.11	.62	4.73	2.84	26.48
Fast	225	100	60.10	16.00	5.63	.60	6.90	2.04	31.17
Industrial enamel, oil base, high gloss - light colors (material #56)									
Slow	215	160	65.40	10.70	2.56	.41	2.60	2.60	18.87
Medium	233	148	57.20	12.66	3.65	.39	4.18	2.51	23.39
Fast	250	135	49.00	14.40	5.08	.36	6.15	1.82	27.81
Industrial enamel, oil base, high gloss - dark (OSHA) colors (material #57)									
Slow	215	160	78.30	10.70	2.56	.49	2.61	2.62	18.98
Medium	233	148	68.50	12.66	3.65	.46	4.20	2.52	23.49
Fast	250	135	58.70	14.40	5.08	.43	6.17	1.83	27.91

	Labor SF per manhour	Material coverage SF/gallon	Material cost per gallon	Labor cost per 100 SF	Labor burden 100 SF	Material cost per 100 SF	Overhead per 100 SF	Profit per 100 SF	Total price per 100 SF
Including typical supports									
Brush each coat									
Metal primer, rust inhibitor - clean metal (material #35)									
Slow	40	125	59.70	57.50	13.80	.48	13.64	13.67	99.09
Medium	55	113	52.30	53.64	15.50	.46	17.40	10.44	97.44
Fast	70	100	44.80	51.43	18.18	.45	21.71	6.42	98.19
Metal primer, rust inhibitor - rusty metal (material #36)									
Slow	40	125	80.10	57.50	13.80	.64	13.67	13.70	99.31
Medium	55	113	70.10	53.64	15.50	.62	17.44	10.46	97.66
Fast	70	100	60.10	51.43	18.18	.60	21.76	6.44	98.41
Industrial enamel, oil base, high gloss - light colors (material #56)									
Slow	50	160	65.40	46.00	11.04	.41	10.92	10.94	79.31
Medium	75	148	57.20	39.33	11.36	.39	12.77	7.66	71.51
Fast	100	135	49.00	36.00	12.71	.36	15.21	4.50	68.78
Industrial enamel, oil base, high gloss - dark (OSHA) colors (material #57)									
Slow	50	160	78.30	46.00	11.04	.49	10.93	10.95	79.41
Medium	75	148	68.50	39.33	11.36	.46	12.79	7.67	71.61
Fast	100	135	58.70	36.00	12.71	.43	15.23	4.51	68.88
Spray each coat									
Metal primer, rust inhibitor - clean metal (material #35)									
Slow	120	100	59.70	19.17	4.59	.60	4.63	4.64	33.63
Medium	135	88	52.30	21.85	6.33	.59	7.19	4.31	40.27
Fast	150	75	44.80	24.00	8.49	.60	10.25	3.03	46.37
Metal primer, rust inhibitor - rusty metal (material #36)									
Slow	120	100	80.10	19.17	4.59	.80	4.67	4.68	33.91
Medium	135	88	70.10	21.85	6.33	.80	7.24	4.34	40.56
Fast	150	75	60.10	24.00	8.49	.80	10.31	3.05	46.65
Industrial enamel, oil base, high gloss - light colors (material #56)									
Slow	150	125	65.40	15.33	3.69	.52	3.71	3.72	26.97
Medium	163	113	57.20	18.10	5.21	.51	5.96	3.58	33.36
Fast	175	100	49.00	20.57	7.24	.49	8.78	2.60	39.68
Industrial enamel, oil base, high gloss - dark (OSHA) colors (material #57)									
Slow	150	125	78.30	15.33	3.69	.63	3.73	3.74	27.12
Medium	163	113	68.50	18.10	5.21	.61	5.99	3.59	33.50
Fast	175	100	58.70	20.57	7.24	.59	8.81	2.61	39.82

Use these figures when estimating steel grates over 1" thick. The figures will apply when both sides are painted with oil or water base paint. Square feet calculations for grates are based on overall (length times width) dimensions. For grilles under 1" thick, see the following table. Note: A two coat system, prime and finish, using oil base material is recommended for any metal surface. Water base material may cause oxidation, corrosion and rust. Using one coat of oil base paint on metal surfaces may result in cracking, peeling, or chipping without the proper prime coat application. If off white or another light colored finish paint is specified, make sure the prime coat is also a light color, or more than one finish coat will be necessary. "Slow" work is based on an hourly wage of $23.00, "Medium" work on an hourly wage of $29.50, and "Fast" work on an hourly wage of $36.00. Other qualifications that apply to this table are on page 9.

	Labor SF per manhour	Material coverage SF/gallon	Material cost per gallon	Labor cost per 100 SF	Labor burden 100 SF	Material cost per 100 SF	Overhead per 100 SF	Profit per 100 SF	Total price per 100 SF
Grilles, steel, under 1" thick									
Without supports									
Brush each coat									
Metal primer, rust inhibitor, clean metal (material #35)									
Slow	175	200	59.70	13.14	3.14	29.85	8.77	8.79	63.69
Medium	200	175	52.30	14.75	4.27	29.89	12.23	7.34	68.48
Fast	225	150	44.80	16.00	5.63	29.87	15.97	4.72	72.19
Metal primer, rust inhibitor, rusty metal (material #36)									
Slow	175	200	80.10	13.14	3.14	40.05	10.70	10.73	77.76
Medium	200	175	70.10	14.75	4.27	40.06	14.77	8.86	82.71
Fast	225	150	60.10	16.00	5.63	40.07	19.13	5.66	86.49
Industrial enamel, oil base, high gloss, light colors (material #56)									
Slow	200	250	65.40	11.50	2.76	26.16	7.68	7.70	55.80
Medium	225	225	57.20	13.11	3.78	25.42	10.58	6.35	59.24
Fast	250	200	49.00	14.40	5.08	24.50	13.63	4.03	61.64
Industrial enamel, oil base, high gloss, dark (OSHA) colors (material #57)									
Slow	200	250	78.30	11.50	2.76	31.32	8.66	8.68	62.92
Medium	225	225	68.50	13.11	3.78	30.44	11.84	7.10	66.27
Fast	250	200	58.70	14.40	5.08	29.35	15.14	4.48	68.45
Spray each coat									
Metal primer, rust inhibitor, clean metal (material #35)									
Slow	400	150	59.70	5.75	1.38	39.80	8.92	8.94	64.79
Medium	450	138	52.30	6.56	1.88	37.90	11.59	6.95	64.88
Fast	500	125	44.80	7.20	2.54	35.84	14.13	4.18	63.89
Metal primer, rust inhibitor, rusty metal (material #36)									
Slow	400	150	80.10	5.75	1.38	53.40	11.50	11.52	83.55
Medium	450	138	70.10	6.56	1.88	50.80	14.82	8.89	82.95
Fast	500	125	60.10	7.20	2.54	48.08	17.92	5.30	81.04
Industrial enamel, oil base, high gloss, light colors (material #56)									
Slow	425	175	65.40	5.41	1.29	37.37	8.38	8.39	60.84
Medium	475	163	57.20	6.21	1.81	35.09	10.77	6.46	60.34
Fast	525	150	49.00	6.86	2.39	32.67	13.00	3.85	58.77
Industrial enamel, oil base, high gloss, dark (OSHA) colors (material #57)									
Slow	425	175	78.30	5.41	1.29	44.74	9.78	9.80	71.02
Medium	475	163	68.50	6.21	1.81	42.02	12.51	7.50	70.05
Fast	525	150	58.70	6.86	2.39	39.13	15.01	4.44	67.83

	Labor SF per manhour	Material coverage SF/gallon	Material cost per gallon	Labor cost per 100 SF	Labor burden 100 SF	Material cost per 100 SF	Overhead per 100 SF	Profit per 100 SF	Total price per 100 SF
Including typical supports									
Brush each coat									
Metal primer, rust inhibitor, clean metal (material #35)									
Slow	125	150	59.70	18.40	4.42	39.80	11.90	11.92	86.44
Medium	150	135	52.30	19.67	5.70	38.74	16.02	9.61	89.74
Fast	175	120	44.80	20.57	7.24	37.33	20.20	5.98	91.32
Metal primer, rust inhibitor, rusty metal (material #36)									
Slow	125	150	80.10	18.40	4.42	53.40	14.48	14.51	105.21
Medium	150	135	70.10	19.67	5.70	51.93	19.32	11.59	108.21
Fast	175	120	60.10	20.57	7.24	50.08	24.15	7.14	109.18
Industrial enamel, oil base, high gloss, light colors (material #56)									
Slow	150	200	65.40	15.33	3.69	32.70	9.82	9.84	71.38
Medium	175	175	57.20	16.86	4.86	32.69	13.61	8.16	76.18
Fast	200	150	49.00	18.00	6.36	32.67	17.68	5.23	79.94
Industrial enamel, oil base, high gloss, dark (OSHA) colors (material #57)									
Slow	150	200	78.30	15.33	3.69	39.15	11.05	11.07	80.29
Medium	175	175	68.50	16.86	4.86	39.14	15.22	9.13	85.21
Fast	200	150	58.70	18.00	6.36	39.13	19.68	5.82	88.99
Spray each coat									
Metal primer, rust inhibitor, clean metal (material #35)									
Slow	325	125	59.70	7.08	1.70	47.76	10.74	10.76	78.04
Medium	375	113	52.30	7.87	2.28	46.28	14.11	8.46	79.00
Fast	425	100	44.80	8.47	2.98	44.80	17.44	5.16	78.85
Metal primer, rust inhibitor, rusty metal (material #36)									
Slow	325	125	80.10	7.08	1.70	64.08	13.84	13.87	100.57
Medium	375	113	70.10	7.87	2.28	62.04	18.05	10.83	101.07
Fast	425	100	60.10	8.47	2.98	60.10	22.18	6.56	100.29
Industrial enamel, oil base, high gloss, light colors (material #56)									
Slow	350	150	65.40	6.57	1.59	43.60	9.83	9.85	71.44
Medium	400	138	57.20	7.38	2.13	41.45	12.74	7.64	71.34
Fast	450	125	49.00	8.00	2.81	39.20	15.51	4.59	70.11
Industrial enamel, oil base, high gloss, dark (OSHA) colors (material #57)									
Slow	350	150	78.30	6.57	1.59	52.20	11.47	11.49	83.32
Medium	400	138	68.50	7.38	2.13	49.64	14.79	8.87	82.81
Fast	450	125	58.70	8.00	2.81	46.96	17.91	5.30	80.98

Use these figures to estimate steel grilles under 1" thick. The figures will apply when both sides are painted with oil or water base paint. Square foot calculations for grates are based on overall (length times width) dimensions. For grates over 1" thick, see the previous table. Note: A two coat system, prime and finish, using oil base material is recommended for any metal surface. Water base material may cause oxidation, corrosion and rust. Using one coat of oil base paint on metal surfaces may result in cracking, peeling, or chipping without the proper prime coat application. If off white or another light colored finish paint is specified, make sure the prime coat is also a light color, or more than one finish coat will be necessary. "Slow" work is based on an hourly wage of $23.00, "Medium" work on an hourly wage of $29.50, and "Fast" work on an hourly wage of $36.00. Other qualifications that apply to this table are on page 9.

Ladders

Measure the length of the ladder rungs and vertical members. Then multiply by a difficulty factor of 1.5 (Length x 1.5) to allow for limited access to the back of the ladder. Then use the rates in the Bare Pipe tables to figure the labor and material costs.

	Labor SF per manhour	Material coverage SF/gallon	Material cost per gallon	Labor cost per 100 SF	Labor burden 100 SF	Material cost per 100 SF	Overhead per 100 SF	Profit per 100 SF	Total price per 100 SF
Masonry, Concrete Masonry Units (CMU), rough, porous surface									
Industrial bonding & penetrating oil paint (material #55)									
Brush 1st coat									
Slow	225	200	66.30	10.22	2.44	33.15	8.71	8.72	63.24
Medium	250	188	58.00	11.80	3.41	30.85	11.52	6.91	64.49
Fast	275	175	49.70	13.09	4.64	28.40	14.29	4.23	64.65
Brush 2nd coat									
Slow	230	275	66.30	10.00	2.41	24.11	6.94	6.95	50.41
Medium	260	250	58.00	11.35	3.29	23.20	9.46	5.67	52.97
Fast	290	225	49.70	12.41	4.39	22.09	12.05	3.57	54.51
Industrial waterproofing (material #58)									
Brush 1st coat									
Slow	65	65	53.70	35.38	8.48	82.62	24.03	24.08	174.59
Medium	90	55	47.00	32.78	9.47	85.45	31.93	19.16	178.79
Fast	115	45	40.30	31.30	11.08	89.56	40.89	12.10	184.93
Brush 2nd or additional coats									
Slow	90	150	53.70	25.56	6.13	35.80	12.82	12.85	93.16
Medium	115	125	47.00	25.65	7.44	37.60	17.67	10.60	98.96
Fast	145	100	40.30	24.83	8.78	40.30	22.91	6.78	103.60
Roll 1st coat									
Slow	100	125	53.70	23.00	5.52	42.96	13.58	13.61	98.67
Medium	125	108	47.00	23.60	6.82	43.52	18.49	11.09	103.52
Fast	150	90	40.30	24.00	8.49	44.78	23.95	7.08	108.30
Roll 2nd or additional coats									
Slow	150	175	53.70	15.33	3.69	30.69	9.44	9.46	68.61
Medium	180	150	47.00	16.39	4.75	31.33	13.12	7.87	73.46
Fast	210	125	40.30	17.14	6.05	32.24	17.18	5.08	77.69

Use these figures for Concrete Masonry Units (CMU) where the block surfaces are rough, porous or unfilled, with joints struck to average depth. The more porous the surface, the rougher the texture, the more time and material will be required. For heights above 8 feet, use the High Time Difficulty Factors on page 139. Also refer to other masonry applications under Masonry in the "General Painting Operations" section of this book. "Slow" work is based on an hourly wage of $23.00, "Medium" work on an hourly wage of $29.50, "Fast" work on an hourly wage of $36.00. Other qualifications that apply to this table are on page 9.

	Labor SF per manhour	Material coverage SF/gallon	Material cost per gallon	Labor cost per 100 SF	Labor burden 100 SF	Material cost per 100 SF	Overhead per 100 SF	Profit per 100 SF	Total price per 100 SF
Masonry, Concrete Masonry Units (CMU), smooth surface									
Industrial bonding & penetrating oil paint (material #55)									
Brush 1st coat									
Slow	325	240	66.30	7.08	1.70	27.63	6.92	6.93	50.26
Medium	350	230	58.00	8.43	2.45	25.22	9.02	5.41	50.53
Fast	375	220	49.70	9.60	3.41	22.59	11.03	3.26	49.89
Brush 2nd coat									
Slow	340	300	66.30	6.76	1.62	22.10	5.79	5.80	42.07
Medium	370	275	58.00	7.97	2.30	21.09	7.84	4.70	43.90
Fast	400	250	49.70	9.00	3.18	19.88	9.94	2.94	44.94
Industrial waterproofing (material #58)									
Brush 1st coat									
Slow	75	100	53.70	30.67	7.35	53.70	17.43	17.47	126.62
Medium	100	95	47.00	29.50	8.53	49.47	21.88	13.13	122.51
Fast	125	90	40.30	28.80	10.17	44.78	25.96	7.68	117.39
Brush 2nd or additional coats									
Slow	100	150	53.70	23.00	5.52	35.80	12.22	12.25	88.79
Medium	125	138	47.00	23.60	6.82	34.06	16.12	9.67	90.27
Fast	150	125	40.30	24.00	8.49	32.24	20.06	5.93	90.72
Roll 1st coat									
Slow	125	150	53.70	18.40	4.42	35.80	11.14	11.16	80.92
Medium	150	138	47.00	19.67	5.70	34.06	14.85	8.91	83.19
Fast	175	125	40.30	20.57	7.24	32.24	18.62	5.51	84.18
Roll 2nd or additional coats									
Slow	175	200	53.70	13.14	3.14	26.85	8.20	8.21	59.54
Medium	200	175	47.00	14.75	4.27	26.86	11.47	6.88	64.23
Fast	225	150	40.30	16.00	5.63	26.87	15.04	4.45	67.99

Use these figures for Concrete Masonry Units (CMU), where block surfaces are smooth, as in precision block, filled block or slump stone with joints struck to average depth. The more smooth the surface texture, the less time and material will be required. For heights above 8 feet, use the High Time Difficulty Factors on page 139. Also refer to other masonry applications under Masonry in the "General Painting Operations" section of this book. "Slow" work is based on an hourly wage of $23.00, "Medium" work on an hourly wage of $29.50, "Fast" work on an hourly wage of $36.00. Other qualifications that apply to this table are on page 9.

	Labor SF per manhour	Material coverage SF/gallon	Material cost per gallon	Labor cost per 100 SF	Labor burden 100 SF	Material cost per 100 SF	Overhead per 100 SF	Profit per 100 SF	Total price per 100 SF
Mechanical equipment									
Brush each coat									
Metal primer, rust inhibitor, clean metal (material #35)									
Slow	175	275	59.70	13.14	3.14	21.71	7.22	7.24	52.45
Medium	200	263	52.30	14.75	4.27	19.89	9.73	5.84	54.48
Fast	225	250	44.80	16.00	5.63	17.92	12.27	3.63	55.45
Metal primer, rust inhibitor, rusty metal (material #36)									
Slow	175	275	80.10	13.14	3.14	29.13	8.63	8.65	62.69
Medium	200	263	70.10	14.75	4.27	26.65	11.42	6.85	63.94
Fast	225	250	60.10	16.00	5.63	24.04	14.16	4.19	64.02
Industrial enamel, oil base, high gloss, light colors (material #56)									
Slow	200	375	65.40	11.50	2.76	17.44	6.02	6.04	43.76
Medium	225	363	57.20	13.11	3.78	15.76	8.17	4.90	45.72
Fast	250	350	49.00	14.40	5.08	14.00	10.38	3.07	46.93
Industrial enamel, oil base, high gloss, dark (OSHA) colors (material #57)									
Slow	200	375	78.30	11.50	2.76	20.88	6.68	6.69	48.51
Medium	225	363	68.50	13.11	3.78	18.87	8.94	5.37	50.07
Fast	250	350	58.70	14.40	5.08	16.77	11.24	3.32	50.81
Spray each coat									
Metal primer, rust inhibitor, clean metal (material #35)									
Slow	350	200	59.70	6.57	1.59	29.85	7.22	7.24	52.47
Medium	375	175	52.30	7.87	2.28	29.89	10.01	6.00	56.05
Fast	400	150	44.80	9.00	3.18	29.87	13.04	3.86	58.95
Metal primer, rust inhibitor, rusty metal (material #36)									
Slow	350	200	80.10	6.57	1.59	40.05	9.16	9.18	66.55
Medium	375	175	70.10	7.87	2.28	40.06	12.55	7.53	70.29
Fast	400	150	60.10	9.00	3.18	40.07	16.20	4.79	73.24
Industrial enamel, oil base, high gloss, light colors (material #56)									
Slow	375	275	65.40	6.13	1.48	23.78	5.96	5.97	43.32
Medium	400	250	57.20	7.38	2.13	22.88	8.10	4.86	45.35
Fast	425	225	49.00	8.47	2.98	21.78	10.30	3.05	46.58
Industrial enamel, oil base, high gloss, dark (OSHA) colors (material #57)									
Slow	375	275	78.30	6.13	1.48	28.47	6.85	6.87	49.80
Medium	400	250	68.50	7.38	2.13	27.40	9.23	5.54	51.68
Fast	425	225	58.70	8.47	2.98	26.09	11.64	3.44	52.62

Use these figures to estimate the cost of painting mechanical equipment (such as compressors and mixing boxes). Measurements are based on the overall dimension (length times width) of the surface area to be covered in square feet. For heights above 8 feet, use the High Time Difficulty Factors on page 139. Note: A two coat system, prime and finish, using oil base material is recommended for any metal surface. Water base material may cause oxidation, corrosion and rust. Using one coat of oil base paint on metal surfaces may result in cracking, peeling or chipping without the proper prime coat application. If off white or another light colored finish paint is specified, make sure the prime coat is also a light color, or more than one finish coat will be necessary. "Slow" work is based on an hourly wage of $23.00, "Medium" work on an hourly wage of $29.50, "Fast" work on an hourly wage of $36.00. Other qualifications that apply to this table are on page 9.

Mechanical equipment, boiler room

Don't bother figuring the exact area of boiler room equipment that has to be coated. Instead, take the area as equal to 1/2 the wall height times the ceiling area in the room. Figure a painter will coat 125 square feet per hour and a gallon of paint will cover 300 square feet. This rate does not include time needed to paint walls, ceiling or floor around mechanical equipment. In any case, you'll need to rely on judgment when painting boiler room equipment.

	Labor SF per manhour	Material coverage SF/gallon	Material cost per gallon	Labor cost per 100 SF	Labor burden 100 SF	Material cost per 100 SF	Overhead per 100 SF	Profit per 100 SF	Total price per 100 SF
Piping, bare pipe, brush application									
Metal primer, rust inhibitor, clean metal (material #35)									
Brush prime coat									
Slow	75	360	59.70	30.67	7.35	16.58	10.38	10.40	75.38
Medium	100	335	52.30	29.50	8.53	15.61	13.41	8.05	75.10
Fast	125	310	44.80	28.80	10.17	14.45	16.56	4.90	74.88
Metal primer, rust inhibitor, rusty metal (material #36)									
Brush prime coat									
Slow	75	360	80.10	30.67	7.35	22.25	11.45	11.48	83.20
Medium	100	335	70.10	29.50	8.53	20.93	14.74	8.84	82.54
Fast	125	310	60.10	28.80	10.17	19.39	18.09	5.35	81.80
Industrial enamel, oil base, high gloss, light colors (material #56)									
Brush 1st finish coat									
Slow	90	450	65.40	25.56	6.13	14.53	8.78	8.80	63.80
Medium	115	425	57.20	25.65	7.44	13.46	11.63	6.98	65.16
Fast	140	400	49.00	25.71	9.07	12.25	14.58	4.31	65.92
Brush 2nd or additional finish coats									
Slow	125	500	65.40	18.40	4.42	13.08	6.82	6.84	49.56
Medium	150	475	57.20	19.67	5.70	12.04	9.35	5.61	52.37
Fast	175	450	49.00	20.57	7.24	10.89	12.00	3.55	54.25
Industrial enamel, oil base, high gloss, dark (OSHA) colors (material #57)									
Brush 1st finish coat									
Slow	90	450	78.30	25.56	6.13	17.40	9.33	9.35	67.77
Medium	115	425	68.50	25.65	7.44	16.12	12.30	7.38	68.89
Fast	140	400	58.70	25.71	9.07	14.68	15.34	4.54	69.34
Brush 2nd or additional finish coats									
Slow	125	500	78.30	18.40	4.42	15.66	7.31	7.33	53.12
Medium	150	475	68.50	19.67	5.70	14.42	9.94	5.97	55.70
Fast	175	450	58.70	20.57	7.24	13.04	12.67	3.75	57.27

	Labor SF per manhour	Material coverage SF/gallon	Material cost per gallon	Labor cost per 100 SF	Labor burden 100 SF	Material cost per 100 SF	Overhead per 100 SF	Profit per 100 SF	Total price per 100 SF
Epoxy coating, 2 part system, clear (material #51)									
Brush 1st coat									
Slow	75	375	160.60	30.67	7.35	42.83	15.36	15.40	111.61
Medium	100	350	140.50	29.50	8.53	40.14	19.54	11.73	109.44
Fast	125	325	120.50	28.80	10.17	37.08	23.58	6.97	106.60
Brush 2nd coat									
Slow	90	425	160.60	25.56	6.13	37.79	13.20	13.23	95.91
Medium	115	400	140.50	25.65	7.44	35.13	17.05	10.23	95.50
Fast	140	375	120.50	25.71	9.07	32.13	20.75	6.14	93.80
Brush 3rd or additional coats									
Slow	125	475	160.60	18.40	4.42	33.81	10.76	10.78	78.17
Medium	150	450	140.50	19.67	5.70	31.22	14.14	8.49	79.22
Fast	175	425	120.50	20.57	7.24	28.35	17.42	5.15	78.73
Epoxy coating, 2 part system, white (material #52)									
Brush 1st coat									
Slow	75	375	157.00	30.67	7.35	41.87	15.18	15.21	110.28
Medium	100	350	137.40	29.50	8.53	39.26	19.32	11.59	108.20
Fast	125	325	117.70	28.80	10.17	36.22	23.31	6.90	105.40
Brush 2nd coat									
Slow	90	425	157.00	25.56	6.13	36.94	13.04	13.07	94.74
Medium	115	400	137.40	25.65	7.44	34.35	16.85	10.11	94.40
Fast	140	375	117.70	25.71	9.07	31.39	20.52	6.07	92.76
Brush 3rd or additional coats									
Slow	125	475	157.00	18.40	4.42	33.05	10.62	10.64	77.13
Medium	150	450	137.40	19.67	5.70	30.53	13.97	8.38	78.25
Fast	175	425	117.70	20.57	7.24	27.69	17.21	5.09	77.80
Aluminum base paint (material #50)									
Brush each coat									
Slow	50	600	109.70	46.00	11.04	18.28	14.31	14.34	103.97
Medium	75	575	96.00	39.33	11.36	16.70	16.85	10.11	94.35
Fast	100	550	82.30	36.00	12.71	14.96	19.74	5.84	89.25

	Labor SF per manhour	Material coverage SF/gallon	Material cost per gallon	Labor cost per 100 SF	Labor burden 100 SF	Material cost per 100 SF	Overhead per 100 SF	Profit per 100 SF	Total price per 100 SF
Heat resistant enamel, 800 to 1200 degree range (material #53)									
Brush each coat									
Slow	50	600	149.40	46.00	11.04	24.90	15.57	15.60	113.11
Medium	75	575	130.70	39.33	11.36	22.73	18.36	11.01	102.79
Fast	100	550	112.00	36.00	12.71	20.36	21.41	6.33	96.81
Heat resistant enamel, 300 to 800 degree range (material #54)									
Brush each coat									
Slow	50	600	136.30	46.00	11.04	22.72	15.15	15.19	110.10
Medium	75	575	119.30	39.33	11.36	20.75	17.86	10.72	100.02
Fast	100	550	102.20	36.00	12.71	18.58	20.86	6.17	94.32

Use the pipe conversion factors in Figure 21 on page 325 to convert linear feet of pipe to square feet of surface. Vertical pipe runs require 2 to 3 times the manhours plus 10% more material. Solid color coded piping requires 15% to 25% more labor and material. For color bands on piping at 10' to 15' intervals, add the cost of an additional 1st coat. For heights above 8 feet, use the High Time Difficulty Factors on page 139. Note: A two coat system, prime and finish, using oil base material is recommended for any metal surface. Water base material may cause oxidation, corrosion and rust. One coat of oil base paint on metal surfaces may result in cracking, peeling or chipping without the proper prime coat application. If off white or another light colored finish paint is specified, make sure the prime coat is also a light color, or more than one finish coat will be necessary. "Slow" work is based on an hourly wage of $23.00, "Medium" work on an hourly wage of $29.50, "Fast" work on an hourly wage of $36.00. Other qualifications that apply to this table are on page 9.

	Labor SF per manhour	Material coverage SF/gallon	Material cost per gallon	Labor cost per 100 SF	Labor burden 100 SF	Material cost per 100 SF	Overhead per 100 SF	Profit per 100 SF	Total price per 100 SF
Piping, bare pipe, roll application									
Metal primer, rust inhibitor, clean metal (material #35)									
Roll prime coat									
Slow	175	400	59.70	13.14	3.14	14.93	5.93	5.94	43.08
Medium	200	375	52.30	14.75	4.27	13.95	8.24	4.94	46.15
Fast	225	350	44.80	16.00	5.63	12.80	10.68	3.16	48.27
Metal primer, rust inhibitor, rusty metal (material #36)									
Roll prime coat									
Slow	175	400	80.10	13.14	3.14	20.03	6.90	6.92	50.13
Medium	200	375	70.10	14.75	4.27	18.69	9.43	5.66	52.80
Fast	225	350	60.10	16.00	5.63	17.17	12.03	3.56	54.39
Industrial enamel, oil base, high gloss, light colors (material #56)									
Roll 1st finish coat									
Slow	200	450	65.40	11.50	2.76	14.53	5.47	5.48	39.74
Medium	225	425	57.20	13.11	3.78	13.46	7.59	4.55	42.49
Fast	250	400	49.00	14.40	5.08	12.25	9.84	2.91	44.48
Roll 2nd or additional finish coats									
Slow	275	500	65.40	8.36	2.02	13.08	4.46	4.47	32.39
Medium	300	475	57.20	9.83	2.83	12.04	6.18	3.71	34.59
Fast	325	450	49.00	11.08	3.92	10.89	8.02	2.37	36.28
Industrial enamel, oil base, high gloss, dark (OSHA) colors (material #57)									
Roll 1st finish coat									
Slow	200	450	78.30	11.50	2.76	17.40	6.02	6.03	43.71
Medium	225	425	68.50	13.11	3.78	16.12	8.26	4.95	46.22
Fast	250	400	58.70	14.40	5.08	14.68	10.59	3.13	47.88
Roll 2nd or additional finish coats									
Slow	275	500	78.30	8.36	2.02	15.66	4.95	4.96	35.95
Medium	300	475	68.50	9.83	2.83	14.42	6.77	4.06	37.91
Fast	325	450	58.70	11.08	3.92	13.04	8.69	2.57	39.30

Use the pipe conversion factors in Figure 21 on page 325 to convert linear feet of pipe to square feet of surface. Vertical pipe runs require 2 to 3 times the manhours plus 10% more material. Solid color coded piping requires 15% to 25% more labor and material. For color bands on piping at 10' to 15' intervals, add the cost of an additional 1st coat. For heights above 8 feet, use the High Time Difficulty Factors on page 139. Note: A two coat system, prime and finish, using oil base material is recommended for any metal surface. Water base material may cause oxidation, corrosion and rust. One coat of oil base paint on metal surfaces may result in cracking, peeling or chipping without the proper prime coat application. If off white or another light colored finish paint is specified, make sure the prime coat is also a light color, or more than one finish coat will be necessary. "Slow" work is based on an hourly wage of $23.00, "Medium" work on an hourly wage of $29.50, "Fast" work on an hourly wage of $36.00. Other qualifications that apply to this table are on page 9.

	Labor SF per manhour	Material coverage SF/gallon	Material cost per gallon	Labor cost per 100 SF	Labor burden 100 SF	Material cost per 100 SF	Overhead per 100 SF	Profit per 100 SF	Total price per 100 SF
Piping, bare pipe, spray application									
Metal primer, rust inhibitor, clean metal (material #35)									
Spray prime coat									
Slow	300	175	59.70	7.67	1.83	34.11	8.29	8.31	60.21
Medium	350	163	52.30	8.43	2.45	32.09	10.74	6.44	60.15
Fast	400	150	44.80	9.00	3.18	29.87	13.04	3.86	58.95
Metal primer, rust inhibitor, rusty metal (material #36)									
Spray prime coat									
Slow	300	175	80.10	7.67	1.83	45.77	10.50	10.52	76.29
Medium	350	163	70.10	8.43	2.45	43.01	13.47	8.08	75.44
Fast	400	150	60.10	9.00	3.18	40.07	16.20	4.79	73.24
Industrial enamel, oil base, high gloss, light colors (material #56)									
Spray 1st finish coat									
Slow	375	210	65.40	6.13	1.48	31.14	7.36	7.38	53.49
Medium	425	198	57.20	6.94	2.00	28.89	9.46	5.68	52.97
Fast	475	185	49.00	7.58	2.70	26.49	11.39	3.37	51.53
Spray 2nd or additional finish coats									
Slow	425	300	65.40	5.41	1.29	21.80	5.42	5.43	39.35
Medium	475	288	57.20	6.21	1.81	19.86	6.97	4.18	39.03
Fast	525	275	49.00	6.86	2.39	17.82	8.40	2.49	37.96
Industrial enamel, oil base, high gloss, dark (OSHA) colors (material #57)									
Spray 1st finish coat									
Slow	375	210	78.30	6.13	1.48	37.29	8.53	8.55	61.98
Medium	425	198	68.50	6.94	2.00	34.60	10.89	6.53	60.96
Fast	475	185	58.70	7.58	2.70	31.73	13.02	3.85	58.88
Spray 2nd or additional finish coats									
Slow	425	300	78.30	5.41	1.29	26.10	6.23	6.25	45.28
Medium	475	288	68.50	6.21	1.81	23.78	7.95	4.77	44.52
Fast	525	275	58.70	6.86	2.39	21.35	9.50	2.81	42.91
Epoxy coating, 2 part system, clear (material #51)									
Spray 1st coat									
Slow	300	185	160.60	7.67	1.83	86.81	18.30	18.34	132.95
Medium	350	173	140.50	8.43	2.45	81.21	23.02	13.81	128.92
Fast	400	160	120.50	9.00	3.18	75.31	27.12	8.02	122.63
Spray 2nd coat									
Slow	375	200	160.60	6.13	1.48	80.30	16.70	16.74	121.35
Medium	425	188	140.50	6.94	2.00	74.73	20.92	12.55	117.14
Fast	475	175	120.50	7.58	2.70	68.86	24.53	7.26	110.93

	Labor SF per manhour	Material coverage SF/gallon	Material cost per gallon	Labor cost per 100 SF	Labor burden 100 SF	Material cost per 100 SF	Overhead per 100 SF	Profit per 100 SF	Total price per 100 SF
Spray 3rd or additional coats									
Slow	425	275	160.60	5.41	1.29	58.40	12.37	12.40	89.87
Medium	475	263	140.50	6.21	1.81	53.42	15.36	9.21	86.01
Fast	525	250	120.50	6.86	2.39	48.20	17.82	5.27	80.54
Epoxy coating, 2 part system, white (material #52)									
Spray 1st coat									
Slow	300	185	157.00	7.67	1.83	84.86	17.93	17.97	130.26
Medium	350	173	137.40	8.43	2.45	79.42	22.57	13.54	126.41
Fast	400	160	117.70	9.00	3.18	73.56	26.58	7.86	120.18
Spray 2nd coat									
Slow	375	200	157.00	6.13	1.48	78.50	16.36	16.39	118.86
Medium	425	188	137.40	6.94	2.00	73.09	20.51	12.31	114.85
Fast	475	175	117.70	7.58	2.70	67.26	24.03	7.11	108.68
Spray 3rd or additional coats									
Slow	425	275	157.00	5.41	1.29	57.09	12.12	12.15	88.06
Medium	475	263	137.40	6.21	1.81	52.24	15.06	9.04	84.36
Fast	525	250	117.70	6.86	2.39	47.08	17.47	5.17	78.97

Use the pipe conversion factors in Figure 21 on page 325 to convert linear feet of pipe to square feet of surface. Vertical pipe runs require 2 to 3 times the manhours plus 10% more material. Solid color coded piping requires 15% to 25% more labor and material. For color bands on piping at 10' to 15' intervals, add the cost of an additional 1st coat. For heights above 8 feet, use the High Time Difficulty Factors on page 139. Note: A two coat system, prime and finish, using oil base material is recommended for any metal surface. Water base material may cause oxidation, corrosion and rust. One coat of oil base paint on metal surfaces may result in cracking, peeling or chipping without the proper prime coat application. If off white or another light colored finish paint is specified, make sure the prime coat is also a light color, or more than one finish coat will be necessary. "Slow" work is based on an hourly wage of $23.00, "Medium" work on an hourly wage of $29.50, "Fast" work on an hourly wage of $36.00. Other qualifications that apply to this table are on page 9.

	Labor SF per manhour	Material coverage SF/gallon	Material cost per gallon	Labor cost per 100 SF	Labor burden 100 SF	Material cost per 100 SF	Overhead per 100 SF	Profit per 100 SF	Total price per 100 SF
Piping, bare pipe, mitt or glove application									
Metal primer, rust inhibitor, clean metal (material #35)									
Mitt or glove prime coat									
Slow	175	325	59.70	13.14	3.14	18.37	6.59	6.60	47.84
Medium	200	313	52.30	14.75	4.27	16.71	8.93	5.36	50.02
Fast	225	300	44.80	16.00	5.63	14.93	11.34	3.35	51.25
Metal primer, rust inhibitor, rusty metal (material #36)									
Mitt or glove prime coat									
Slow	175	325	80.10	13.14	3.14	24.65	7.78	7.80	56.51
Medium	200	313	70.10	14.75	4.27	22.40	10.35	6.21	57.98
Fast	225	300	60.10	16.00	5.63	20.03	12.92	3.82	58.40
Industrial enamel, oil base, high gloss, light colors (material #56)									
Mitt or glove 1st finish coat									
Slow	200	375	65.40	11.50	2.76	17.44	6.02	6.04	43.76
Medium	225	363	57.20	13.11	3.78	15.76	8.17	4.90	45.72
Fast	250	350	49.00	14.40	5.08	14.00	10.38	3.07	46.93
Mitt or glove 2nd or additional finish coats									
Slow	275	400	65.40	8.36	2.02	16.35	5.08	5.09	36.90
Medium	300	388	57.20	9.83	2.83	14.74	6.85	4.11	38.36
Fast	325	375	49.00	11.08	3.92	13.07	8.70	2.57	39.34
Industrial enamel, oil base, high gloss, dark (OSHA) colors (material #57)									
Mitt or glove 1st finish coat									
Slow	200	375	78.30	11.50	2.76	20.88	6.68	6.69	48.51
Medium	225	363	68.50	13.11	3.78	18.87	8.94	5.37	50.07
Fast	250	350	58.70	14.40	5.08	16.77	11.24	3.32	50.81
Mitt or glove 2nd or additional finish coats									
Slow	275	400	78.30	8.36	2.02	19.58	5.69	5.70	41.35
Medium	300	388	68.50	9.83	2.83	17.65	7.58	4.55	42.44
Fast	325	375	58.70	11.08	3.92	15.65	9.50	2.81	42.96
Epoxy coating, 2 part system, clear (material #51)									
Mitt or glove 1st coat									
Slow	175	360	160.60	13.14	3.14	44.61	11.57	11.60	84.06
Medium	200	348	140.50	14.75	4.27	40.37	14.85	8.91	83.15
Fast	225	335	120.50	16.00	5.63	35.97	17.86	5.28	80.74
Mitt or glove 2nd coat									
Slow	200	375	160.60	11.50	2.76	42.83	10.85	10.87	78.81
Medium	225	363	140.50	13.11	3.78	38.71	13.90	8.34	77.84
Fast	250	350	120.50	14.40	5.08	34.43	16.71	4.94	75.56

	Labor SF per manhour	Material coverage SF/gallon	Material cost per gallon	Labor cost per 100 SF	Labor burden 100 SF	Material cost per 100 SF	Overhead per 100 SF	Profit per 100 SF	Total price per 100 SF
Mitt or glove 3rd or additional coats									
Slow	275	390	160.60	8.36	2.02	41.18	9.79	9.81	71.16
Medium	300	378	140.50	9.83	2.83	37.17	12.46	7.48	69.77
Fast	325	365	120.50	11.08	3.92	33.01	14.88	4.40	67.29
Epoxy coating, 2 part system, white (material #52)									
Mitt or glove 1st coat									
Slow	175	360	157.00	13.14	3.14	43.61	11.38	11.40	82.67
Medium	200	348	137.40	14.75	4.27	39.48	14.62	8.77	81.89
Fast	225	335	117.70	16.00	5.63	35.13	17.60	5.21	79.57
Mitt or glove 2nd coat									
Slow	200	375	157.00	11.50	2.76	41.87	10.66	10.69	77.48
Medium	225	363	137.40	13.11	3.78	37.85	13.69	8.21	76.64
Fast	250	350	117.70	14.40	5.08	33.63	16.46	4.87	74.44
Mitt or glove 3rd or additional coats									
Slow	275	390	157.00	8.36	2.02	40.26	9.62	9.64	69.90
Medium	300	378	137.40	9.83	2.83	36.35	12.26	7.35	68.62
Fast	325	365	117.70	11.08	3.92	32.25	14.64	4.33	66.22

Use the pipe conversion factors in Figure 21 on page 325 to convert linear feet of pipe to square feet of surface. Vertical pipe runs require 2 to 3 times the manhours plus 10% more material. Solid color coded piping requires 15% to 25% more labor and material. For color bands on piping at 10' to 15' intervals, add the cost of an additional 1st coat. For heights above 8 feet, use the High Time Difficulty Factors on page 139. Note: A two coat system, prime and finish, using oil base material is recommended for any metal surface. Water base material may cause oxidation, corrosion and rust. One coat of oil base paint on metal surfaces may result in cracking, peeling or chipping without the proper prime coat application. If off white or another light colored finish paint is specified, make sure the prime coat is also a light color, or more than one finish coat will be necessary. "Slow" work is based on an hourly wage of $23.00, "Medium" work on an hourly wage of $29.50, "Fast" work on an hourly wage of $36.00. Other qualifications that apply to this table are on page 9.

	Labor SF per manhour	Material coverage SF/gallon	Material cost per gallon	Labor cost per 100 SF	Labor burden 100 SF	Material cost per 100 SF	Overhead per 100 SF	Profit per 100 SF	Total price per 100 SF
Piping, insulated, canvas jacket, brush									
Flat latex, water base (material #5)									
Brush 1st coat									
Slow	60	150	36.00	38.33	9.21	24.00	13.59	13.62	98.75
Medium	80	138	31.50	36.88	10.66	22.83	17.59	10.56	98.52
Fast	100	125	27.00	36.00	12.71	21.60	21.80	6.45	98.56
Brush 2nd coat									
Slow	75	300	36.00	30.67	7.35	12.00	9.51	9.53	69.06
Medium	100	288	31.50	29.50	8.53	10.94	12.24	7.35	68.56
Fast	125	275	27.00	28.80	10.17	9.82	15.12	4.47	68.38
Brush 3rd or additional coats									
Slow	100	400	36.00	23.00	5.52	9.00	7.13	7.14	51.79
Medium	138	375	31.50	21.38	6.19	8.40	8.99	5.39	50.35
Fast	175	350	27.00	20.57	7.24	7.71	11.02	3.26	49.80
Sealer, off white, water base (material #1)									
Brush 1 coat									
Slow	60	150	36.00	38.33	9.21	24.00	13.59	13.62	98.75
Medium	80	138	31.50	36.88	10.66	22.83	17.59	10.56	98.52
Fast	100	125	27.00	36.00	12.71	21.60	21.80	6.45	98.56
Sealer, off white, oil base (material #2)									
Brush 1 coat									
Slow	60	200	46.20	38.33	9.21	23.10	13.42	13.45	97.51
Medium	80	188	40.50	36.88	10.66	21.54	17.27	10.36	96.71
Fast	100	175	34.70	36.00	12.71	19.83	21.25	6.29	96.08
Enamel, water base latex (material #9)									
Brush 1st finish coat									
Slow	75	300	53.30	30.67	7.35	17.77	10.60	10.62	77.01
Medium	100	288	46.70	29.50	8.53	16.22	13.56	8.14	75.95
Fast	125	275	40.00	28.80	10.17	14.55	16.59	4.91	75.02
Brush 2nd or additional finish coats									
Slow	100	400	53.30	23.00	5.52	13.33	7.95	7.97	57.77
Medium	138	375	46.70	21.38	6.19	12.45	10.00	6.00	56.02
Fast	175	350	40.00	20.57	7.24	11.43	12.17	3.60	55.01
Industrial enamel, oil base, high gloss, light colors (material #56)									
Brush 1st finish coat									
Slow	75	335	65.40	30.67	7.35	19.52	10.93	10.96	79.43
Medium	100	323	57.20	29.50	8.53	17.71	13.94	8.36	78.04
Fast	125	310	49.00	28.80	10.17	15.81	16.98	5.02	76.78

	Labor SF per manhour	Material coverage SF/gallon	Material cost per gallon	Labor cost per 100 SF	Labor burden 100 SF	Material cost per 100 SF	Overhead per 100 SF	Profit per 100 SF	Total price per 100 SF
Brush 2nd or additional finish coats									
Slow	100	450	65.40	23.00	5.52	14.53	8.18	8.20	59.43
Medium	150	438	57.20	19.67	5.70	13.06	9.60	5.76	53.79
Fast	175	425	49.00	20.57	7.24	11.53	12.20	3.61	55.15
Industrial enamel, oil base, high gloss, dark (OSHA) colors (material #57)									
Brush 1st finish coat									
Slow	75	335	78.30	30.67	7.35	23.37	11.67	11.69	84.75
Medium	100	323	68.50	29.50	8.53	21.21	14.81	8.89	82.94
Fast	125	310	58.70	28.80	10.17	18.94	17.95	5.31	81.17
Brush 2nd or additional finish coats									
Slow	100	450	78.30	23.00	5.52	17.40	8.72	8.74	63.38
Medium	150	438	68.50	19.67	5.70	15.64	10.25	6.15	57.41
Fast	175	425	58.70	20.57	7.24	13.81	12.91	3.82	58.35
Epoxy coating, 2 part system, clear (material #51)									
Brush 1st coat									
Slow	75	325	160.60	30.67	7.35	49.42	16.62	16.65	120.71
Medium	100	313	140.50	29.50	8.53	44.89	20.73	12.44	116.09
Fast	125	300	120.50	28.80	10.17	40.17	24.53	7.26	110.93
Brush 2nd or additional coats									
Slow	100	450	160.60	23.00	5.52	35.69	12.20	12.23	88.64
Medium	150	438	140.50	19.67	5.70	32.08	14.36	8.61	80.42
Fast	175	425	120.50	20.57	7.24	28.35	17.42	5.15	78.73
Epoxy coating, 2 part system, white (material #52)									
Brush 1st coat									
Slow	75	325	157.00	30.67	7.35	48.31	16.40	16.44	119.17
Medium	100	313	137.40	29.50	8.53	43.90	20.48	12.29	114.70
Fast	125	300	117.70	28.80	10.17	39.23	24.24	7.17	109.61
Brush 2nd or additional coats									
Slow	100	450	157.00	23.00	5.52	34.89	12.05	12.07	87.53
Medium	150	438	137.40	19.67	5.70	31.37	14.18	8.51	79.43
Fast	175	425	117.70	20.57	7.24	27.69	17.21	5.09	77.80

Use the pipe conversion factors in Figure 21 on page 325 to convert linear feet of pipe to square feet of surface. Vertical pipe runs require 2 to 3 times the manhours plus 10% more material. Solid color coded piping requires 15% to 25% more labor and material. For color bands on piping at 10' to 15' intervals, add the cost of an additional 1st coat. For heights above 8 feet, use the High Time Difficulty Factors on page 139. "Slow" work is based on an hourly wage of $23.00, "Medium" work on an hourly wage of $29.50, "Fast" work on an hourly wage of $36.00. Other qualifications that apply to this table are on page 9.

	Labor SF per manhour	Material coverage SF/gallon	Material cost per gallon	Labor cost per 100 SF	Labor burden 100 SF	Material cost per 100 SF	Overhead per 100 SF	Profit per 100 SF	Total price per 100 SF
Piping, insulated, canvas jacket, roll									
Flat latex, water base (material #5)									
Roll 1st coat									
Slow	135	150	36.00	17.04	4.09	24.00	8.57	8.59	62.29
Medium	160	138	31.50	18.44	5.33	22.83	11.65	6.99	65.24
Fast	185	125	27.00	19.46	6.89	21.60	14.86	4.40	67.21
Roll 2nd coat									
Slow	175	300	36.00	13.14	3.14	12.00	5.38	5.39	39.05
Medium	200	288	31.50	14.75	4.27	10.94	7.49	4.49	41.94
Fast	225	275	27.00	16.00	5.63	9.82	9.76	2.89	44.10
Roll 3rd or additional coats									
Slow	275	400	36.00	8.36	2.02	9.00	3.68	3.69	26.75
Medium	300	388	31.50	9.83	2.83	8.12	5.20	3.12	29.10
Fast	325	375	27.00	11.08	3.92	7.20	6.88	2.03	31.11
Sealer, off white, water base (material #1)									
Roll 1 coat									
Slow	135	150	36.00	17.04	4.09	24.00	8.57	8.59	62.29
Medium	160	138	31.50	18.44	5.33	22.83	11.65	6.99	65.24
Fast	185	125	27.00	19.46	6.89	21.60	14.86	4.40	67.21
Sealer, off white, oil base (material #2)									
Roll 1 coat									
Slow	135	225	46.20	17.04	4.09	20.53	7.92	7.93	57.51
Medium	160	213	40.50	18.44	5.33	19.01	10.70	6.42	59.90
Fast	185	200	34.70	19.46	6.89	17.35	13.54	4.01	61.25
Enamel, water base latex (material #9)									
Roll 1st finish coat									
Slow	175	300	53.30	13.14	3.14	17.77	6.47	6.48	47.00
Medium	200	288	46.70	14.75	4.27	16.22	8.81	5.28	49.33
Fast	225	275	40.00	16.00	5.63	14.55	11.22	3.32	50.72
Roll 2nd or additional finish coats									
Slow	275	400	53.30	8.36	2.02	13.33	4.50	4.51	32.72
Medium	300	388	46.70	9.83	2.83	12.04	6.18	3.71	34.59
Fast	325	375	40.00	11.08	3.92	10.67	7.95	2.35	35.97
Industrial enamel, oil base, high gloss, light colors (material #56)									
Roll 1st finish coat									
Slow	175	335	65.40	13.14	3.14	19.52	6.80	6.82	49.42
Medium	200	323	57.20	14.75	4.27	17.71	9.18	5.51	51.42
Fast	225	310	49.00	16.00	5.63	15.81	11.61	3.43	52.48

	Labor SF per manhour	Material coverage SF/gallon	Material cost per gallon	Labor cost per 100 SF	Labor burden 100 SF	Material cost per 100 SF	Overhead per 100 SF	Profit per 100 SF	Total price per 100 SF
Roll 2nd or additional finish coats									
Slow	275	450	65.40	8.36	2.02	14.53	4.73	4.74	34.38
Medium	300	438	57.20	9.83	2.83	13.06	6.43	3.86	36.01
Fast	325	425	49.00	11.08	3.92	11.53	8.22	2.43	37.18
Industrial enamel, oil base, high gloss, dark (OSHA) colors (material #57)									
Roll 1st finish coat									
Slow	175	335	78.30	13.14	3.14	23.37	7.54	7.55	54.74
Medium	200	323	68.50	14.75	4.27	21.21	10.06	6.03	56.32
Fast	225	310	58.70	16.00	5.63	18.94	12.58	3.72	56.87
Roll 2nd or additional finish coats									
Slow	275	450	78.30	8.36	2.02	17.40	5.28	5.29	38.35
Medium	300	438	68.50	9.83	2.83	15.64	7.08	4.25	39.63
Fast	325	425	58.70	11.08	3.92	13.81	8.93	2.64	40.38
Epoxy coating, 2 part system, clear (material #51)									
Roll 1st coat									
Slow	175	325	160.60	13.14	3.14	49.42	12.48	12.51	90.69
Medium	200	313	140.50	14.75	4.27	44.89	15.98	9.59	89.48
Fast	225	300	120.50	16.00	5.63	40.17	19.16	5.67	86.63
Roll 2nd or additional coats									
Slow	275	425	160.60	8.36	2.02	37.79	9.15	9.17	66.49
Medium	300	413	140.50	9.83	2.83	34.02	11.67	7.00	65.35
Fast	325	400	120.50	11.08	3.92	30.13	13.99	4.14	63.26
Epoxy coating, 2 part system, white (material #52)									
Roll 1st coat									
Slow	175	325	157.00	13.14	3.14	48.31	12.27	12.30	89.16
Medium	200	313	137.40	14.75	4.27	43.90	15.73	9.44	88.09
Fast	225	300	117.70	16.00	5.63	39.23	18.87	5.58	85.31
Roll 2nd or additional coats									
Slow	275	425	157.00	8.36	2.02	36.94	8.99	9.01	65.32
Medium	300	413	137.40	9.83	2.83	33.27	11.49	6.89	64.31
Fast	325	400	117.70	11.08	3.92	29.43	13.77	4.07	62.27

Use the pipe conversion factors in Figure 21 on page 325 to convert linear feet of pipe to square feet of surface. Vertical pipe runs require 2 to 3 times the manhours plus 10% more material. Solid color coded piping requires 15% to 25% more labor and material. For color bands on piping at 10' to 15' intervals, add the cost of an additional 1st coat. For heights above 8 feet, use the High Time Difficulty Factors on page 139. "Slow" work is based on an hourly wage of $23.00, "Medium" work on an hourly wage of $29.50, "Fast" work on an hourly wage of $36.00. Other qualifications that apply to this table are on page 9.

	Labor SF per manhour	Material coverage SF/gallon	Material cost per gallon	Labor cost per 100 SF	Labor burden 100 SF	Material cost per 100 SF	Overhead per 100 SF	Profit per 100 SF	Total price per 100 SF
Piping, insulated, canvas jacket, spray									
Flat latex, water base (material #5)									
Spray 1st coat									
Slow	225	100	36.00	10.22	2.44	36.00	9.25	9.27	67.18
Medium	250	88	31.50	11.80	3.41	35.80	12.75	7.65	71.41
Fast	275	75	27.00	13.09	4.64	36.00	16.65	4.93	75.31
Spray 2nd coat									
Slow	275	200	36.00	8.36	2.02	18.00	5.39	5.40	39.17
Medium	300	188	31.50	9.83	2.83	16.76	7.36	4.41	41.19
Fast	325	175	27.00	11.08	3.92	15.43	9.43	2.79	42.65
Spray 3rd or additional coats									
Slow	375	300	36.00	6.13	1.48	12.00	3.72	3.73	27.06
Medium	400	288	31.50	7.38	2.13	10.94	5.11	3.07	28.63
Fast	425	275	27.00	8.47	2.98	9.82	6.60	1.95	29.82
Sealer, off white, water base (material #1)									
Spray 1 coat									
Slow	225	100	36.00	10.22	2.44	36.00	9.25	9.27	67.18
Medium	250	88	31.50	11.80	3.41	35.80	12.75	7.65	71.41
Fast	275	75	27.00	13.09	4.64	36.00	16.65	4.93	75.31
Sealer, off white, oil base (material #2)									
Spray 1 coat									
Slow	225	150	46.20	10.22	2.44	30.80	8.26	8.28	60.00
Medium	250	138	40.50	11.80	3.41	29.35	11.14	6.68	62.38
Fast	275	125	34.70	13.09	4.64	27.76	14.10	4.17	63.76
Enamel, water base latex (material #9)									
Spray 1st finish coat									
Slow	275	200	53.30	8.36	2.02	26.65	7.03	7.05	51.11
Medium	300	188	46.70	9.83	2.83	24.84	9.38	5.63	52.51
Fast	325	175	40.00	11.08	3.92	22.86	11.73	3.47	53.06
Spray 2nd or additional finish coats									
Slow	375	300	53.30	6.13	1.48	17.77	4.82	4.83	35.03
Medium	400	288	46.70	7.38	2.13	16.22	6.43	3.86	36.02
Fast	425	275	40.00	8.47	2.98	14.55	8.06	2.38	36.44
Industrial enamel, oil base, high gloss, light colors (material #56)									
Spray 1st finish coat									
Slow	275	235	65.40	8.36	2.02	27.83	7.26	7.27	52.74
Medium	300	223	57.20	9.83	2.83	25.65	9.58	5.75	53.64
Fast	325	210	49.00	11.08	3.92	23.33	11.88	3.51	53.72

	Labor SF per manhour	Material coverage SF/gallon	Material cost per gallon	Labor cost per 100 SF	Labor burden 100 SF	Material cost per 100 SF	Overhead per 100 SF	Profit per 100 SF	Total price per 100 SF
Spray 2nd or additional finish coats									
Slow	375	335	65.40	6.13	1.48	19.52	5.15	5.16	37.44
Medium	400	323	57.20	7.38	2.13	17.71	6.81	4.08	38.11
Fast	425	310	49.00	8.47	2.98	15.81	8.45	2.50	38.21
Industrial enamel, oil base, high gloss, dark (OSHA) colors (material #57)									
Spray 1st finish coat									
Slow	275	235	78.30	8.36	2.02	33.32	8.30	8.32	60.32
Medium	300	223	68.50	9.83	2.83	30.72	10.85	6.51	60.74
Fast	325	210	58.70	11.08	3.92	27.95	13.31	3.94	60.20
Spray 2nd or additional finish coats									
Slow	375	335	78.30	6.13	1.48	23.37	5.88	5.90	42.76
Medium	400	323	68.50	7.38	2.13	21.21	7.68	4.61	43.01
Fast	425	310	58.70	8.47	2.98	18.94	9.42	2.79	42.60
Epoxy coating, 2 part system, clear (material #51)									
Spray 1st coat									
Slow	275	225	160.60	8.36	2.02	71.38	15.53	15.56	112.85
Medium	300	213	140.50	9.83	2.83	65.96	19.66	11.79	110.07
Fast	325	200	120.50	11.08	3.92	60.25	23.32	6.90	105.47
Spray 2nd or additional coats									
Slow	375	325	160.60	6.13	1.48	49.42	10.83	10.86	78.72
Medium	400	313	140.50	7.38	2.13	44.89	13.60	8.16	76.16
Fast	425	300	120.50	8.47	2.98	40.17	16.01	4.73	72.36
Epoxy coating, 2 part system, white (material #52)									
Spray 1st coat									
Slow	275	225	157.00	8.36	2.02	69.78	15.23	15.26	110.65
Medium	300	213	137.40	9.83	2.83	64.51	19.30	11.58	108.05
Fast	325	200	117.70	11.08	3.92	58.85	22.89	6.77	103.51
Spray 2nd or additional coats									
Slow	375	325	157.00	6.13	1.48	48.31	10.62	10.64	77.18
Medium	400	313	137.40	7.38	2.13	43.90	13.35	8.01	74.77
Fast	425	300	117.70	8.47	2.98	39.23	15.71	4.65	71.04

Use the pipe conversion factors in Figure 21 on page 325 to convert linear feet of pipe to square feet of surface. Vertical pipe runs require 2 to 3 times the manhours plus 10% more material. Solid color coded piping requires 15% to 25% more labor and material. For color bands on piping at 10' to 15' intervals, add the cost of an additional 1st coat. For heights above 8 feet, use the High Time Difficulty Factors on page 139. "Slow" work is based on an hourly wage of $23.00, "Medium" work on an hourly wage of $29.50, "Fast" work on an hourly wage of $36.00. Other qualifications that apply to this table are on page 9.

	Labor SF per manhour	Material coverage SF/gallon	Material cost per gallon	Labor cost per 100 SF	Labor burden 100 SF	Material cost per 100 SF	Overhead per 100 SF	Profit per 100 SF	Total price per 100 SF
Radiators									
Brush each coat									
Metal primer, rust inhibitor, clean metal (material #35)									
Slow	50	100	59.70	46.00	11.04	59.70	22.18	22.23	161.15
Medium	70	95	52.30	42.14	12.20	55.05	27.34	16.41	153.14
Fast	90	90	44.80	40.00	14.12	49.78	32.21	9.53	145.64
Metal primer, rust inhibitor, rusty metal (material #36)									
Slow	50	100	80.10	46.00	11.04	80.10	26.06	26.11	189.31
Medium	70	95	70.10	42.14	12.20	73.79	32.03	19.22	179.38
Fast	90	90	60.10	40.00	14.12	66.78	37.48	11.09	169.47
Industrial enamel, oil base, high gloss, light colors (material #56)									
Slow	60	150	65.40	38.33	9.21	43.60	17.31	17.35	125.80
Medium	80	138	57.20	36.88	10.66	41.45	22.25	13.35	124.59
Fast	100	125	49.00	36.00	12.71	39.20	27.25	8.06	123.22
Industrial enamel, oil base, high gloss, dark (OSHA) colors (material #57)									
Slow	60	150	78.30	38.33	9.21	52.20	18.95	18.99	137.68
Medium	80	138	68.50	36.88	10.66	49.64	24.30	14.58	136.06
Fast	100	125	58.70	36.00	12.71	46.96	29.66	8.77	134.10
Spray each coat									
Metal primer, rust inhibitor, clean metal (material #35)									
Slow	225	90	59.70	10.22	2.44	66.33	15.01	15.04	109.04
Medium	250	83	52.30	11.80	3.41	63.01	19.56	11.73	109.51
Fast	275	75	44.80	13.09	4.64	59.73	24.01	7.10	108.57
Metal primer, rust inhibitor, rusty metal (material #36)									
Slow	225	90	80.10	10.22	2.44	89.00	19.32	19.36	140.34
Medium	250	83	70.10	11.80	3.41	84.46	24.92	14.95	139.54
Fast	275	75	60.10	13.09	4.64	80.13	30.33	8.97	137.16
Industrial enamel, oil base, high gloss, light colors (material #56)									
Slow	250	110	65.40	9.20	2.21	59.45	13.46	13.49	97.81
Medium	275	100	57.20	10.73	3.11	57.20	17.76	10.65	99.45
Fast	300	90	49.00	12.00	4.22	54.44	21.91	6.48	99.05
Industrial enamel, oil base, high gloss, dark (OSHA) colors (material #57)									
Slow	250	110	78.30	9.20	2.21	71.18	15.69	15.72	114.00
Medium	275	100	68.50	10.73	3.11	68.50	20.58	12.35	115.27
Fast	300	90	58.70	12.00	4.22	65.22	25.25	7.47	114.16

Use these figures to estimate the cost of painting both sides of 6" to 18" deep hot water or steam radiators with oil or water base paint. Measurements are per square foot of area measured, one side (length times width). For heights above 8 feet, use the High Time Difficulty Factors on page 139. Note: A two coat system, prime and finish, using oil base material is recommended for any metal surface. Using water base material may cause oxidation, corrosion and rust. Using one coat of oil base paint on metal surfaces may result in cracking, peeling or chipping and rust without the proper prime coat application. If off white or another light colored finish paint is specified, make sure the prime coat is also a light color, or more than one finish coat will be necessary. "Slow" work is based on an hourly wage of $23.00, "Medium" work on an hourly wage of $29.50, "Fast" work on an hourly wage of $36.00. Other qualifications that apply to this table are on page 9.

Structural steel

Fabrication and erection estimates are usually based on weight of the steel in tons. As a paint estimator you need to convert tons of steel to square feet of surface. Of course, the conversion factor depends on the size of the steel members. On larger jobs and where accuracy is essential, use the Structural Steel conversion table in Figure 23 on pages 391 through 399 at the end of this Structural steel section to make exact conversions. On smaller jobs, a *rule of thumb* is that there are 225 square feet of paintable surface per ton of steel.

	Labor SF per manhour	Material coverage SF/gallon	Material cost per gallon	Labor cost per 100 SF	Labor burden 100 SF	Material cost per 100 SF	Overhead per 100 SF	Profit per 100 SF	Total price per 100 SF
Structural steel, heavy, brush application									
Metal primer, rust inhibitor - clean metal (material #35)									
Brush prime coat									
Slow	130	400	59.70	17.69	4.24	14.93	7.01	7.02	50.89
Medium	150	388	52.30	19.67	5.70	13.48	9.71	5.82	54.38
Fast	170	375	44.80	21.18	7.46	11.95	12.59	3.72	56.90
Metal primer, rust inhibitor - rusty metal (material #36)									
Brush prime coat									
Slow	130	400	80.10	17.69	4.24	20.03	7.97	7.99	57.92
Medium	150	388	70.10	19.67	5.70	18.07	10.86	6.51	60.81
Fast	170	375	60.10	21.18	7.46	16.03	13.85	4.10	62.62
Industrial enamel, oil base, high gloss - light colors (material #56)									
Brush 1st or additional finish coats									
Slow	175	425	65.40	13.14	3.14	15.39	6.02	6.03	43.72
Medium	200	413	57.20	14.75	4.27	13.85	8.22	4.93	46.02
Fast	225	400	49.00	16.00	5.63	12.25	10.51	3.11	47.50
Industrial enamel, oil base, high gloss - dark (OSHA) colors (material #57)									
Brush 1st or additional finish coats									
Slow	175	475	78.30	13.14	3.14	16.48	6.23	6.24	45.23
Medium	200	463	68.50	14.75	4.27	14.79	8.45	5.07	47.33
Fast	225	450	58.70	16.00	5.63	13.04	10.75	3.18	48.60
Epoxy coating, 2 part system, clear (material #51)									
Brush 1st coat									
Slow	130	425	160.60	17.69	4.24	37.79	11.35	11.37	82.44
Medium	150	413	140.50	19.67	5.70	34.02	14.84	8.91	83.14
Fast	170	400	120.50	21.18	7.46	30.13	18.22	5.39	82.38
Brush 2nd or additional coats									
Slow	175	450	160.60	13.14	3.14	35.69	9.88	9.90	71.75
Medium	200	438	140.50	14.75	4.27	32.08	12.77	7.66	71.53
Fast	225	425	120.50	16.00	5.63	28.35	15.50	4.59	70.07

	Labor SF per manhour	Material coverage SF/gallon	Material cost per gallon	Labor cost per 100 SF	Labor burden 100 SF	Material cost per 100 SF	Overhead per 100 SF	Profit per 100 SF	Total price per 100 SF
Epoxy coating, 2 part system, white (material #52)									
Brush 1st coat									
Slow	130	425	157.00	17.69	4.24	36.94	11.19	11.21	81.27
Medium	150	413	137.40	19.67	5.70	33.27	14.66	8.79	82.09
Fast	170	400	117.70	21.18	7.46	29.43	18.01	5.33	81.41
Brush 2nd or additional coats									
Slow	175	450	157.00	13.14	3.14	34.89	9.72	9.74	70.63
Medium	200	438	137.40	14.75	4.27	31.37	12.60	7.56	70.55
Fast	225	425	117.70	16.00	5.63	27.69	15.30	4.52	69.14

For field painting at heights above 8 feet, use the High Time Difficulty Factors on page 139. Heavy structural steel has from 100 to 150 square feet of surface area per ton. Extra heavy structural steel has from 50 to 100 square feet of surface area per ton. *Rule of thumb*: When coatings are applied by brush, a journeyman painter will apply a first coat on 6 to 7 tons per 8 hour day. When coatings are applied by spray, figure output at 0.2 hours per ton and material use at about 0.2 gallons per ton. Use Figure 23 on pages 391 to 399 to convert structural steel linear feet or tonnage to surface area. Note: A two coat system, prime and finish, using oil base material is recommended for any metal surface. Using water base material may cause oxidation, corrosion and rust. One coat of oil base paint on metal surfaces may result in cracking, peeling or chipping without the proper prime coat application. If off white or another light-color finish paint is specified, make sure the prime coat is also a light color, or more than one finish coat will be necessary. "Slow" work is based on an hourly wage of $23.00, "Medium" work on an hourly wage of $29.50, and "Fast" work on an hourly wage of $36.00. Other qualifications that apply to this table are on page 9.

	Labor SF per manhour	Material coverage SF/gallon	Material cost per gallon	Labor cost per 100 SF	Labor burden 100 SF	Material cost per 100 SF	Overhead per 100 SF	Profit per 100 SF	Total price per 100 SF
Structural steel, heavy, roll application									
Metal primer, rust inhibitor - clean metal (material #35)									
Roll prime coat									
Slow	250	390	59.70	9.20	2.21	15.31	5.08	5.09	36.89
Medium	275	378	52.30	10.73	3.11	13.84	6.92	4.15	38.75
Fast	300	365	44.80	12.00	4.22	12.27	8.84	2.61	39.94
Metal primer, rust inhibitor - rusty metal (material #36)									
Roll prime coat									
Slow	250	380	80.10	9.20	2.21	21.08	6.17	6.19	44.85
Medium	275	363	70.10	10.73	3.11	19.31	8.29	4.97	46.41
Fast	300	355	60.10	12.00	4.22	16.93	10.28	3.04	46.47
Industrial enamel, oil base, high gloss - light colors (material #56)									
Roll 1st or additional finish coats									
Slow	275	425	65.40	8.36	2.02	15.39	4.89	4.90	35.56
Medium	300	413	57.20	9.83	2.83	13.85	6.63	3.98	37.12
Fast	325	400	49.00	11.08	3.92	12.25	8.44	2.50	38.19
Industrial enamel, oil base, high gloss - dark (OSHA) colors (material #57)									
Roll 1st or additional finish coats									
Slow	275	440	78.30	8.36	2.02	17.80	5.35	5.36	38.89
Medium	300	428	68.50	9.83	2.83	16.00	7.17	4.30	40.13
Fast	325	415	58.70	11.08	3.92	14.14	9.03	2.67	40.84
Epoxy coating, 2 part system, clear (material #51)									
Roll 1st coat									
Slow	250	400	160.60	9.20	2.21	40.15	9.80	9.82	71.18
Medium	275	388	140.50	10.73	3.11	36.21	12.51	7.51	70.07
Fast	300	375	120.50	12.00	4.22	32.13	14.99	4.44	67.78
Roll 2nd or additional coats									
Slow	275	425	160.60	8.36	2.02	37.79	9.15	9.17	66.49
Medium	300	413	140.50	9.83	2.83	34.02	11.67	7.00	65.35
Fast	325	400	120.50	11.08	3.92	30.13	13.99	4.14	63.26

	Labor SF per manhour	Material coverage SF/gallon	Material cost per gallon	Labor cost per 100 SF	Labor burden 100 SF	Material cost per 100 SF	Overhead per 100 SF	Profit per 100 SF	Total price per 100 SF
Epoxy coating, 2 part system, white (material #52)									
Roll 1st coat									
Slow	250	400	157.00	9.20	2.21	39.25	9.63	9.65	69.94
Medium	275	388	137.40	10.73	3.11	35.41	12.31	7.39	68.95
Fast	300	375	117.70	12.00	4.22	31.39	14.77	4.37	66.75
Roll 2nd or additional coats									
Slow	275	425	157.00	8.36	2.02	36.94	8.99	9.01	65.32
Medium	300	413	137.40	9.83	2.83	33.27	11.49	6.89	64.31
Fast	325	400	117.70	11.08	3.92	29.43	13.77	4.07	62.27

For field painting at heights above 8 feet, use the High Time Difficulty Factors on page 139. Heavy structural steel has from 100 to 150 square feet of surface area per ton. Extra heavy structural steel has from 50 to 100 square feet of surface area per ton. *Rule of thumb*: When coatings are applied by brush, a journeyman painter will apply a first coat on 6 to 7 tons per 8 hour day. When coatings are applied by spray, figure output at 0.2 hours per ton and material use at about 0.2 gallons per ton. Use Figure 23 on pages 391 to 399 to convert structural steel linear feet or tonnage to surface area. Note: A two coat system, prime and finish, using oil base material is recommended for any metal surface. Using water base material may cause oxidation, corrosion and rust. One coat of oil base paint on metal surfaces may result in cracking, peeling or chipping without the proper prime coat application. If off white or another light-color finish paint is specified, make sure the prime coat is also a light color, or more than one finish coat will be necessary. "Slow" work is based on an hourly wage of $23.00, "Medium" work on an hourly wage of $29.50, and "Fast" work on an hourly wage of $36.00. Other qualifications that apply to this table are on page 9.

	Labor SF per manhour	Material coverage SF/gallon	Material cost per gallon	Labor cost per 100 SF	Labor burden 100 SF	Material cost per 100 SF	Overhead per 100 SF	Profit per 100 SF	Total price per 100 SF
Structural steel, heavy, spray application									
Metal primer, rust inhibitor - clean metal (material #35)									
Spray prime coat									
Slow	650	325	59.70	3.54	.85	18.37	4.32	4.33	31.41
Medium	750	313	52.30	3.93	1.13	16.71	5.45	3.27	30.49
Fast	850	300	44.80	4.24	1.51	14.93	6.41	1.90	28.99
Metal primer, rust inhibitor - rusty metal (material #36)									
Spray prime coat									
Slow	650	300	80.10	3.54	.85	26.70	5.91	5.92	42.92
Medium	750	288	70.10	3.93	1.13	24.34	7.35	4.41	41.16
Fast	850	275	60.10	4.24	1.51	21.85	8.55	2.53	38.68
Industrial enamel, oil base, high gloss - light colors (material #56)									
Spray 1st or additional finish coats									
Slow	750	340	65.40	3.07	.72	19.24	4.38	4.39	31.80
Medium	850	325	57.20	3.47	1.02	17.60	5.52	3.31	30.92
Fast	950	310	49.00	3.79	1.32	15.81	6.49	1.92	29.33
Industrial enamel, oil base, high gloss - dark (OSHA) colors (material #57)									
Spray 1st or additional finish coats									
Slow	750	365	78.30	3.07	.72	21.45	4.80	4.81	34.85
Medium	850	350	68.50	3.47	1.02	19.57	6.01	3.61	33.68
Fast	950	335	58.70	3.79	1.32	17.52	7.02	2.08	31.73
Epoxy coating, 2 part system, clear (material #51)									
Spray 1st coat									
Slow	650	325	160.60	3.54	.85	49.42	10.22	10.24	74.27
Medium	750	313	140.50	3.93	1.13	44.89	12.49	7.49	69.93
Fast	850	300	120.50	4.24	1.51	40.17	14.23	4.21	64.36
Spray 2nd or additional coats									
Slow	750	350	160.60	3.07	.72	45.89	9.44	9.46	68.58
Medium	850	338	140.50	3.47	1.02	41.57	11.51	6.91	64.48
Fast	950	325	120.50	3.79	1.32	37.08	13.09	3.87	59.15

	Labor SF per manhour	Material coverage SF/gallon	Material cost per gallon	Labor cost per 100 SF	Labor burden 100 SF	Material cost per 100 SF	Overhead per 100 SF	Profit per 100 SF	Total price per 100 SF
Epoxy coating, 2 part system, white (material #52)									
Spray 1st coat									
Slow	650	325	157.00	3.54	.85	48.31	10.01	10.03	72.74
Medium	750	313	137.40	3.93	1.13	43.90	12.24	7.35	68.55
Fast	850	300	117.70	4.24	1.51	39.23	13.94	4.12	63.04
Spray 2nd or additional coats									
Slow	750	350	157.00	3.07	.72	44.86	9.25	9.27	67.17
Medium	850	338	137.40	3.47	1.02	40.65	11.28	6.77	63.19
Fast	950	325	117.70	3.79	1.32	36.22	12.82	3.79	57.94

For field painting at heights above 8 feet, use the High Time Difficulty Factors on page 139. Heavy structural steel has from 100 to 150 square feet of surface area per ton. Extra heavy structural steel has from 50 to 100 square feet of surface area per ton. *Rule of thumb*: When coatings are applied by brush, a journeyman painter will apply a first coat on 6 to 7 tons per 8 hour day. When coatings are applied by spray, figure output at 0.2 hours per ton and material use at about 0.2 gallons per ton. Use Figure 23 on pages 391 to 399 to convert structural steel linear feet or tonnage to surface area. Note: A two coat system, prime and finish, using oil base material is recommended for any metal surface. Using water base material may cause oxidation, corrosion and rust. One coat of oil base paint on metal surfaces may result in cracking, peeling or chipping without the proper prime coat application. If off white or another light-color finish paint is specified, make sure the prime coat is also a light color, or more than one finish coat will be necessary. "Slow" work is based on an hourly wage of $23.00, "Medium" work on an hourly wage of $29.50, and "Fast" work on an hourly wage of $36.00. Other qualifications that apply to this table are on page 9.

	Labor SF per manhour	Material coverage SF/gallon	Material cost per gallon	Labor cost per 100 SF	Labor burden 100 SF	Material cost per 100 SF	Overhead per 100 SF	Profit per 100 SF	Total price per 100 SF
Structural steel, light, brush application									
Metal primer, rust inhibitor, clean metal (material #35)									
Brush prime coat									
Slow	60	425	59.70	38.33	9.21	14.05	11.70	11.72	85.01
Medium	80	413	52.30	36.88	10.66	12.66	15.05	9.03	84.28
Fast	100	400	44.80	36.00	12.71	11.20	18.57	5.49	83.97
Metal primer, rust inhibitor, rusty metal (material #36)									
Brush prime coat									
Slow	60	400	80.10	38.33	9.21	20.03	12.84	12.86	93.27
Medium	80	388	70.10	36.88	10.66	18.07	16.40	9.84	91.85
Fast	100	375	60.10	36.00	12.71	16.03	20.07	5.94	90.75
Industrial enamel, oil base, high gloss, light colors (material #56)									
Brush 1st or additional finish coats									
Slow	80	425	65.40	28.75	6.90	15.39	9.70	9.72	70.46
Medium	100	413	57.20	29.50	8.53	13.85	12.97	7.78	72.63
Fast	120	400	49.00	30.00	10.58	12.25	16.38	4.85	74.06
Industrial enamel, oil base, high gloss, dark (OSHA) colors (material #57)									
Brush 1st or additional finish coats									
Slow	80	475	78.30	28.75	6.90	16.48	9.90	9.92	71.95
Medium	100	463	68.50	29.50	8.53	14.79	13.21	7.92	73.95
Fast	120	450	58.70	30.00	10.58	13.04	16.63	4.92	75.17
Epoxy coating, 2 part system, clear (material #51)									
Brush 1st coat									
Slow	60	425	160.60	38.33	9.21	37.79	16.21	16.24	117.78
Medium	80	413	140.50	36.88	10.66	34.02	20.39	12.23	114.18
Fast	100	400	120.50	36.00	12.71	30.13	24.44	7.23	110.51
Brush 2nd or additional coats									
Slow	80	450	160.60	28.75	6.90	35.69	13.55	13.58	98.47
Medium	100	438	140.50	29.50	8.53	32.08	17.53	10.52	98.16
Fast	120	425	120.50	30.00	10.58	28.35	21.37	6.32	96.62

	Labor SF per manhour	Material coverage SF/gallon	Material cost per gallon	Labor cost per 100 SF	Labor burden 100 SF	Material cost per 100 SF	Overhead per 100 SF	Profit per 100 SF	Total price per 100 SF
Epoxy coating, 2 part system, white (material #52)									
Brush 1st coat									
Slow	60	425	157.00	38.33	9.21	36.94	16.05	16.08	116.61
Medium	80	413	137.40	36.88	10.66	33.27	20.20	12.12	113.13
Fast	100	400	117.70	36.00	12.71	29.43	24.22	7.17	109.53
Brush 2nd or additional coats									
Slow	80	450	157.00	28.75	6.90	34.89	13.40	13.43	97.37
Medium	100	438	137.40	29.50	8.53	31.37	17.35	10.41	97.16
Fast	120	425	117.70	30.00	10.58	27.69	21.17	6.26	95.70

For field painting at heights above 8 feet, use the High Time Difficulty Factors on page 139. Light structural steel has from 300 to 500 square feet of surface per ton. As a comparison, when coatings are applied by brush, a journeyman painter will apply a first coat on from 2 to 3 tons per 8 hour day. A second and subsequent coats can be applied on 3 to 4 tons per day. Use Figure 23 on pages 391 to 399 to convert structural steel linear feet or tonnage to surface area. Note: A two coat system, prime and finish, using oil base material is recommended for any metal surface. Using water base material may cause oxidation, corrosion and rust. One coat of oil base paint on metal surfaces may result in cracking, peeling or chipping without the proper prime coat application. If off white or another light colored finish paint is specified, make sure the prime coat is also a light color, or more than one finish coat will be necessary. "Slow" work is based on an hourly wage of $23.00, "Medium" work on an hourly wage of $29.50, and "Fast" work on an hourly wage of $36.00. Other qualifications that apply to this table are on page 9.

	Labor SF per manhour	Material coverage SF/gallon	Material cost per gallon	Labor cost per 100 SF	Labor burden 100 SF	Material cost per 100 SF	Overhead per 100 SF	Profit per 100 SF	Total price per 100 SF
Structural steel, light, roll application									
Metal primer, rust inhibitor, clean metal (material #35)									
Roll and brush prime coat									
Slow	125	390	59.70	18.40	4.42	15.31	7.24	7.26	52.63
Medium	150	378	52.30	19.67	5.70	13.84	9.80	5.88	54.89
Fast	175	365	44.80	20.57	7.24	12.27	12.43	3.68	56.19
Metal primer, rust inhibitor, rusty metal (material #36)									
Roll and brush prime coat									
Slow	125	380	80.10	18.40	4.42	21.08	8.34	8.36	60.60
Medium	150	368	70.10	19.67	5.70	19.05	11.10	6.66	62.18
Fast	175	355	60.10	20.57	7.24	16.93	13.88	4.10	62.72
Industrial enamel, oil base, high gloss, light colors (material #56)									
Roll and brush 1st or additional finish coats									
Slow	175	390	65.40	13.14	3.14	16.77	6.28	6.29	45.62
Medium	200	378	57.20	14.75	4.27	15.13	8.54	5.12	47.81
Fast	225	365	49.00	16.00	5.63	13.42	10.87	3.22	49.14
Industrial enamel, oil base, high gloss, dark (OSHA) colors (material #57)									
Roll and brush 1st or additional finish coats									
Slow	175	440	78.30	13.14	3.14	17.80	6.48	6.49	47.05
Medium	200	428	68.50	14.75	4.27	16.00	8.75	5.25	49.02
Fast	225	415	58.70	16.00	5.63	14.14	11.09	3.28	50.14
Epoxy coating, 2 part system, clear (material #51)									
Roll and brush 1st coat									
Slow	125	400	160.60	18.40	4.42	40.15	11.96	11.99	86.92
Medium	150	388	140.50	19.67	5.70	36.21	15.39	9.23	86.20
Fast	175	375	120.50	20.57	7.24	32.13	18.59	5.50	84.03
Roll and brush 2nd or additional coats									
Slow	175	425	160.60	13.14	3.14	37.79	10.28	10.30	74.65
Medium	200	413	140.50	14.75	4.27	34.02	13.26	7.95	74.25
Fast	225	400	120.50	16.00	5.63	30.13	16.05	4.75	72.56

	Labor SF per manhour	Material coverage SF/gallon	Material cost per gallon	Labor cost per 100 SF	Labor burden 100 SF	Material cost per 100 SF	Overhead per 100 SF	Profit per 100 SF	Total price per 100 SF
Epoxy coating, 2 part system, white (material #52)									
Roll and brush 1st coat									
Slow	125	400	157.00	18.40	4.42	39.25	11.79	11.82	85.68
Medium	150	388	137.40	19.67	5.70	35.41	15.19	9.11	85.08
Fast	175	375	117.70	20.57	7.24	31.39	18.36	5.43	82.99
Roll and brush 2nd or additional coats									
Slow	175	425	157.00	13.14	3.14	36.94	10.11	10.13	73.46
Medium	200	413	137.40	14.75	4.27	33.27	13.07	7.84	73.20
Fast	225	400	117.70	16.00	5.63	29.43	15.83	4.68	71.57

For field painting at heights above 8 feet, use the High Time Difficulty Factors on page 139. Light structural steel has from 300 to 500 square feet of surface per ton. As a comparison, when coatings are applied by brush, a journeyman painter will apply a first coat on from 2 to 3 tons per 8 hour day. A second and subsequent coats can be applied on 3 to 4 tons per day. Use Figure 23 on pages 391 to 399 to convert structural steel linear feet or tonnage to surface area. Note: A two coat system, prime and finish, using oil base material is recommended for any metal surface. Using water base material may cause oxidation, corrosion and rust. One coat of oil base paint on metal surfaces may result in cracking, peeling or chipping without the proper prime coat application. If off white or another light colored finish paint is specified, make sure the prime coat is also a light color, or more than one finish coat will be necessary. "Slow" work is based on an hourly wage of $23.00, "Medium" work on an hourly wage of $29.50, and "Fast" work on an hourly wage of $36.00. Other qualifications that apply to this table are on page 9.

	Labor SF per manhour	Material coverage SF/gallon	Material cost per gallon	Labor cost per 100 SF	Labor burden 100 SF	Material cost per 100 SF	Overhead per 100 SF	Profit per 100 SF	Total price per 100 SF
Structural steel, light, spray application									
Metal primer, rust inhibitor, clean metal (material #35)									
Spray prime coat									
Slow	400	325	59.70	5.75	1.38	18.37	4.85	4.86	35.21
Medium	500	313	52.30	5.90	1.71	16.71	6.08	3.65	34.05
Fast	600	300	44.80	6.00	2.13	14.93	7.15	2.11	32.32
Metal primer, rust inhibitor, rusty metal (material #36)									
Spray prime coat									
Slow	400	300	80.10	5.75	1.38	26.70	6.43	6.44	46.70
Medium	500	288	70.10	5.90	1.71	24.34	7.99	4.79	44.73
Fast	600	275	60.10	6.00	2.13	21.85	9.29	2.75	42.02
Industrial enamel, oil base, high gloss, light colors (material #56)									
Spray 1st or additional finish coats									
Slow	500	325	65.40	4.60	1.10	20.12	4.91	4.92	35.65
Medium	600	313	57.20	4.92	1.43	18.27	6.15	3.69	34.46
Fast	700	300	49.00	5.14	1.83	16.33	7.22	2.14	32.66
Industrial enamel, oil base, high gloss, dark (OSHA) colors (material #57)									
Spray 1st or additional finish coats									
Slow	500	365	78.30	4.60	1.10	21.45	5.16	5.17	37.48
Medium	600	350	68.50	4.92	1.43	19.57	6.48	3.89	36.29
Fast	700	335	58.70	5.14	1.83	17.52	7.59	2.24	34.32
Epoxy coating, 2 part system, clear (material #51)									
Spray 1st coat									
Slow	400	325	160.60	5.75	1.38	49.42	10.74	10.77	78.06
Medium	500	313	140.50	5.90	1.71	44.89	13.13	7.88	73.51
Fast	600	300	120.50	6.00	2.13	40.17	14.97	4.43	67.70
Spray 2nd or additional coats									
Slow	500	350	160.60	4.60	1.10	45.89	9.80	9.82	71.21
Medium	600	338	140.50	4.92	1.43	41.57	11.98	7.19	67.09
Fast	700	325	120.50	5.14	1.83	37.08	13.65	4.04	61.74

	Labor SF per manhour	Material coverage SF/gallon	Material cost per gallon	Labor cost per 100 SF	Labor burden 100 SF	Material cost per 100 SF	Overhead per 100 SF	Profit per 100 SF	Total price per 100 SF
Epoxy coating, 2 part system, white (material #52)									
Spray 1st coat									
Slow	400	325	157.00	5.75	1.38	48.31	10.53	10.56	76.53
Medium	500	313	137.40	5.90	1.71	43.90	12.88	7.73	72.12
Fast	600	300	117.70	6.00	2.13	39.23	14.68	4.34	66.38
Spray 2nd or additional coats									
Slow	500	350	157.00	4.60	1.10	44.86	9.61	9.63	69.80
Medium	600	338	137.40	4.92	1.43	40.65	11.75	7.05	65.80
Fast	700	325	117.70	5.14	1.83	36.22	13.38	3.96	60.53

For field painting at heights above 8 feet, use the High Time Difficulty Factors on page 139. Light structural steel has from 300 to 500 square feet of surface per ton. The *rule of thumb* for labor output and material usage for spray application on light structural steel is shown in the table on the next page. Use Figure 23 on pages 391 to 399 to convert structural steel linear feet or tonnage to surface area. Note: A two coat system, prime and finish, using oil base material is recommended for any metal surface. Using water base material may cause oxidation, corrosion and rust. One coat of oil base paint on metal surfaces may result in cracking, peeling or chipping without the proper prime coat application. If off white or another light-color finish paint is specified, make sure the prime coat is also a light color, or more than one finish coat will be necessary. "Slow" work is based on an hourly wage of $23.00, "Medium" work on an hourly wage of $29.50, and "Fast" work on an hourly wage of $36.00. Other qualifications that apply to this table are on page 9.

	Labor manhours per ton	Material gallons per ton	Material cost per gallon	Labor cost per ton	Labor burden per ton	Material cost per ton	Overhead per ton	Profit per ton	Total price per ton

Structural steel, light, coating *rule of thumb*, spray, per ton

	Labor manhours per ton	Material gallons per ton	Material cost per gallon	Labor cost per ton	Labor burden per ton	Material cost per ton	Overhead per ton	Profit per ton	Total price per ton
Metal primer, rust inhibitor, clean metal (material #35)									
Spray prime coat									
Slow	1.8	1.0	59.70	41.40	9.94	59.70	21.10	21.14	153.28
Medium	1.6	1.1	52.30	47.20	13.65	57.53	29.59	17.76	165.73
Fast	1.4	1.2	44.80	50.40	17.79	53.76	37.80	11.18	170.93
Metal primer, rust inhibitor, rusty metal (material #36)									
Spray prime coat									
Slow	1.8	1.0	80.10	41.40	9.94	80.10	24.97	25.03	181.44
Medium	1.6	1.1	70.10	47.20	13.65	77.11	34.49	20.69	193.14
Fast	1.4	1.2	60.10	50.40	17.79	72.12	43.50	12.87	196.68
Industrial enamel, oil base, high gloss, light colors (material #56)									
Spray 1st finish coat									
Slow	1.5	0.9	65.40	34.50	8.28	58.86	19.31	19.35	140.30
Medium	1.3	1.0	57.20	38.35	11.09	57.20	26.66	15.99	149.29
Fast	1.1	1.1	49.00	39.60	13.98	53.90	33.32	9.86	150.66
Spray 2nd or additional finish coats									
Slow	1.1	0.8	65.40	25.30	6.07	52.32	15.90	15.93	115.52
Medium	1.0	0.9	57.20	29.50	8.53	51.48	22.38	13.43	125.32
Fast	0.9	1.0	49.00	32.40	11.44	49.00	28.78	8.51	130.13
Industrial enamel, oil base, high gloss, dark (OSHA) colors (material #57)									
Spray 1st finish coat									
Slow	1.5	0.9	78.30	34.50	8.28	70.47	21.52	21.56	156.33
Medium	1.3	1.0	68.50	38.35	11.09	68.50	29.48	17.69	165.11
Fast	1.1	1.1	58.70	39.60	13.98	64.57	36.63	10.83	165.61
Spray 2nd or additional finish coats									
Slow	1.1	0.8	78.30	25.30	6.07	62.64	17.86	17.90	129.77
Medium	1.0	0.9	68.50	29.50	8.53	61.65	24.92	14.95	139.55
Fast	0.9	1.0	58.70	32.40	11.44	58.70	31.79	9.40	143.73

For field painting at heights above 8 feet, use the High Time Difficulty Factors on page 139. Note: A two coat system, prime and finish, using oil base material is recommended for any metal surface. Using water base material may cause oxidation, corrosion and rust. One coat of oil base paint on metal surfaces may result in cracking, peeling or chipping without the proper prime coat application. If off white or another light colored finish paint is specified, make sure the prime coat is also a light color, or more than one finish coat will be necessary. "Slow" work is based on an hourly wage of $23.00, "Medium" work on an hourly wage of $29.50, and "Fast" work on an hourly wage of $36.00. Other qualifications that apply to this table are on page 9.

	Labor SF per manhour	Material coverage SF/gallon	Material cost per gallon	Labor cost per 100 SF	Labor burden 100 SF	Material cost per 100 SF	Overhead per 100 SF	Profit per 100 SF	Total price per 100 SF
Structural steel, medium, brush application									
Metal primer, rust inhibitor, clean metal (material #35)									
Brush prime coat									
Slow	80	425	59.70	28.75	6.90	14.05	9.44	9.46	68.60
Medium	100	413	52.30	29.50	8.53	12.66	12.67	7.60	70.96
Fast	120	400	44.80	30.00	10.58	11.20	16.05	4.75	72.58
Metal primer, rust inhibitor, rusty metal (material #36)									
Brush prime coat									
Slow	80	400	80.10	28.75	6.90	20.03	10.58	10.60	76.86
Medium	100	388	70.10	29.50	8.53	18.07	14.03	8.42	78.55
Fast	120	375	60.10	30.00	10.58	16.03	17.55	5.19	79.35
Industrial enamel, oil base, high gloss, light colors (material #56)									
Brush 1st or additional finish coats									
Slow	100	425	65.40	23.00	5.52	15.39	8.34	8.36	60.61
Medium	125	413	57.20	23.60	6.82	13.85	11.07	6.64	61.98
Fast	150	400	49.00	24.00	8.49	12.25	13.86	4.10	62.70
Industrial enamel, oil base, high gloss, dark (OSHA) colors (material #57)									
Brush 1st or additional finish coats									
Slow	100	475	78.30	23.00	5.52	16.48	8.55	8.57	62.12
Medium	125	463	68.50	23.60	6.82	14.79	11.30	6.78	63.29
Fast	150	450	58.70	24.00	8.49	13.04	14.11	4.17	63.81
Epoxy coating, 2 part system, clear (material #51)									
Brush 1st coat									
Slow	80	425	160.60	28.75	6.90	37.79	13.95	13.98	101.37
Medium	100	413	140.50	29.50	8.53	34.02	18.01	10.81	100.87
Fast	120	400	120.50	30.00	10.58	30.13	21.92	6.48	99.11
Brush 2nd or additional coats									
Slow	100	450	160.60	23.00	5.52	35.69	12.20	12.23	88.64
Medium	125	438	140.50	23.60	6.82	32.08	15.63	9.38	87.51
Fast	150	425	120.50	24.00	8.49	28.35	18.85	5.58	85.27

	Labor SF per manhour	Material coverage SF/gallon	Material cost per gallon	Labor cost per 100 SF	Labor burden 100 SF	Material cost per 100 SF	Overhead per 100 SF	Profit per 100 SF	Total price per 100 SF
Epoxy coating, 2 part system, white (material #52)									
Brush 1st coat									
Slow	80	425	157.00	28.75	6.90	36.94	13.79	13.82	100.20
Medium	100	413	137.40	29.50	8.53	33.27	17.83	10.70	99.83
Fast	120	400	117.70	30.00	10.58	29.43	21.71	6.42	98.14
Brush 2nd or additional coats									
Slow	100	450	157.00	23.00	5.52	34.89	12.05	12.07	87.53
Medium	125	438	137.40	23.60	6.82	31.37	15.45	9.27	86.51
Fast	150	425	117.70	24.00	8.49	27.69	18.65	5.52	84.35

For field painting at heights above 8 feet, use the High Time Difficulty Factors on page 139. Medium structural steel has from 150 to 300 square feet of surface per ton. As a comparison, when coatings are applied by brush, a journeyman painter will apply a first coat on 4 to 5 tons per 8 hour day. A second and subsequent coat can be applied on 5 to 6 tons per day. When coatings are applied by spray, figure output at 0.6 hours per ton and material use at about 0.6 gallons per ton. Use Figure 23 on pages 391 to 399 to convert structural steel linear feet or tonnage to surface area. Note: A two coat system, prime and finish, using oil base material is recommended for any metal surface. Using water base material may cause oxidation, corrosion and rust. One coat of oil base paint on metal surfaces may result in cracking, peeling or chipping without the proper prime coat application. If off white or another light colored finish paint is specified, make sure the prime coat is also a light color, or more than one finish coat will be necessary. "Slow" work is based on an hourly wage of $23.00, "Medium" work on an hourly wage of $29.50, and "Fast" work on an hourly wage of $36.00. Other qualifications that apply to this table are on page 9.

	Labor SF per manhour	Material coverage SF/gallon	Material cost per gallon	Labor cost per 100 SF	Labor burden 100 SF	Material cost per 100 SF	Overhead per 100 SF	Profit per 100 SF	Total price per 100 SF
Structural steel, medium, roll and brush application									
Metal primer, rust inhibitor, clean metal (material #35)									
Roll and brush prime coat									
Slow	200	390	59.70	11.50	2.76	15.31	5.62	5.63	40.82
Medium	225	378	52.30	13.11	3.78	13.84	7.69	4.61	43.03
Fast	250	365	44.80	14.40	5.08	12.27	9.84	2.91	44.50
Metal primer, rust inhibitor, rusty metal (material #36)									
Roll and brush prime coat									
Slow	200	380	80.10	11.50	2.76	21.08	6.71	6.73	48.78
Medium	225	368	70.10	13.11	3.78	19.05	8.99	5.39	50.32
Fast	250	355	60.10	14.40	5.08	16.93	11.29	3.34	51.04
Industrial enamel, oil base, high gloss, light colors (material #56)									
Roll and brush 1st or additional finish coats									
Slow	225	390	65.40	10.22	2.44	16.77	5.59	5.60	40.62
Medium	250	378	57.20	11.80	3.41	15.13	7.59	4.55	42.48
Fast	275	365	49.00	13.09	4.64	13.42	9.65	2.85	43.65
Industrial enamel, oil base, high gloss, dark (OSHA) colors (material #57)									
Roll and brush 1st or additional finish coats									
Slow	225	440	78.30	10.22	2.44	17.80	5.79	5.80	42.05
Medium	250	428	68.50	11.80	3.41	16.00	7.80	4.68	43.69
Fast	275	415	58.70	13.09	4.64	14.14	9.87	2.92	44.66
Epoxy coating, 2 part system, clear (material #51)									
Roll and brush 1st coat									
Slow	200	400	160.60	11.50	2.76	40.15	10.34	10.36	75.11
Medium	225	388	140.50	13.11	3.78	36.21	13.28	7.97	74.35
Fast	250	375	120.50	14.40	5.08	32.13	16.00	4.73	72.34
Roll and brush 2nd or additional coats									
Slow	225	425	160.60	10.22	2.44	37.79	9.59	9.61	69.65
Medium	250	413	140.50	11.80	3.41	34.02	12.31	7.38	68.92
Fast	275	400	120.50	13.09	4.64	30.13	14.83	4.39	67.08

	Labor SF per manhour	Material coverage SF/gallon	Material cost per gallon	Labor cost per 100 SF	Labor burden 100 SF	Material cost per 100 SF	Overhead per 100 SF	Profit per 100 SF	Total price per 100 SF
Epoxy coating, 2 part system, white (material #52)									
Roll and brush 1st coat									
Slow	200	400	157.00	11.50	2.76	39.25	10.17	10.19	73.87
Medium	225	388	137.40	13.11	3.78	35.41	13.08	7.85	73.23
Fast	250	375	117.70	14.40	5.08	31.39	15.77	4.66	71.30
Roll and brush 2nd or additional coats									
Slow	225	425	157.00	10.22	2.44	36.94	9.43	9.45	68.48
Medium	250	413	137.40	11.80	3.41	33.27	12.12	7.27	67.87
Fast	275	400	117.70	13.09	4.64	29.43	14.61	4.32	66.09

For field painting at heights above 8 feet, use the High Time Difficulty Factors on page 139. Medium structural steel has from 150 to 300 square feet of surface per ton. As a comparison, when coatings are applied by brush, a journeyman painter will apply a first coat on 4 to 5 tons per 8 hour day. A second and subsequent coat can be applied on 5 to 6 tons per day. When coatings are applied by spray, figure output at 0.6 hours per ton and material use at about 0.6 gallons per ton. Use Figure 23 on pages 391 to 399 to convert structural steel linear feet or tonnage to surface area. Note: A two coat system, prime and finish, using oil base material is recommended for any metal surface. Using water base material may cause oxidation, corrosion and rust. One coat of oil base paint on metal surfaces may result in cracking, peeling or chipping without the proper prime coat application. If off white or another light colored finish paint is specified, make sure the prime coat is also a light color, or more than one finish coat will be necessary. "Slow" work is based on an hourly wage of $23.00, "Medium" work on an hourly wage of $29.50, and "Fast" work on an hourly wage of $36.00. Other qualifications that apply to this table are on page 9.

	Labor SF per manhour	Material coverage SF/gallon	Material cost per gallon	Labor cost per 100 SF	Labor burden 100 SF	Material cost per 100 SF	Overhead per 100 SF	Profit per 100 SF	Total price per 100 SF
Structural steel, medium, spray application									
Metal primer, rust inhibitor, clean metal (material #35)									
Spray prime coat									
Slow	500	325	59.70	4.60	1.10	18.37	4.57	4.58	33.22
Medium	600	313	52.30	4.92	1.43	16.71	5.76	3.46	32.28
Fast	700	300	44.80	5.14	1.83	14.93	6.78	2.01	30.69
Metal primer, rust inhibitor, rusty metal (material #36)									
Spray prime coat									
Slow	500	300	80.10	4.60	1.10	26.70	6.16	6.17	44.73
Medium	600	288	70.10	4.92	1.43	24.34	7.67	4.60	42.96
Fast	700	275	60.10	5.14	1.83	21.85	8.93	2.64	40.39
Industrial enamel, oil base, high gloss, light colors (material #56)									
Spray 1st or additional finish coats									
Slow	600	325	65.40	3.83	.93	20.12	4.73	4.74	34.35
Medium	700	313	57.20	4.21	1.23	18.27	5.93	3.56	33.20
Fast	800	300	49.00	4.50	1.59	16.33	6.95	2.06	31.43
Industrial enamel, oil base, high gloss, dark (OSHA) colors (material #57)									
Spray 1st or additional finish coats									
Slow	600	365	78.30	3.83	.93	21.45	4.98	4.99	36.18
Medium	700	350	68.50	4.21	1.23	19.57	6.25	3.75	35.01
Fast	800	335	58.70	4.50	1.59	17.52	7.32	2.17	33.10
Epoxy coating, 2 part system, clear (material #51)									
Spray 1st coat									
Slow	500	325	160.60	4.60	1.10	49.42	10.47	10.49	76.08
Medium	600	313	140.50	4.92	1.43	44.89	12.81	7.68	71.73
Fast	700	300	120.50	5.14	1.83	40.17	14.61	4.32	66.07
Spray 2nd or additional coats									
Slow	600	350	160.60	3.83	.93	45.89	9.62	9.64	69.91
Medium	700	338	140.50	4.21	1.23	41.57	11.75	7.05	65.81
Fast	800	325	120.50	4.50	1.59	37.08	13.38	3.96	60.51

	Labor SF per manhour	Material coverage SF/gallon	Material cost per gallon	Labor cost per 100 SF	Labor burden 100 SF	Material cost per 100 SF	Overhead per 100 SF	Profit per 100 SF	Total price per 100 SF
Epoxy coating, 2 part system, white (material #52)									
Spray 1st coat									
Slow	500	325	157.00	4.60	1.10	48.31	10.26	10.28	74.55
Medium	600	313	137.40	4.92	1.43	43.90	12.56	7.54	70.35
Fast	700	300	117.70	5.14	1.83	39.23	14.32	4.24	64.76
Spray 2nd or additional coats									
Slow	600	350	157.00	3.83	.93	44.86	9.43	9.45	68.50
Medium	700	338	137.40	4.21	1.23	40.65	11.52	6.91	64.52
Fast	800	325	117.70	4.50	1.59	36.22	13.12	3.88	59.31

For field painting at heights above 8 feet, use the High Time Difficulty Factors on page 139. Medium structural steel has from 150 to 300 square feet of surface per ton. As a comparison, when coatings are applied by brush, a journeyman painter will apply a first coat on 4 to 5 tons per 8 hour day. A second and subsequent coat can be applied on 5 to 6 tons per day. When coatings are applied by spray, figure output at 0.6 hours per ton and material use at about 0.6 gallons per ton. Use Figure 23 on pages 391 to 399 to convert structural steel linear feet or tonnage to surface area. Note: A two coat system, prime and finish, using oil base material is recommended for any metal surface. Using water base material may cause oxidation, corrosion and rust. One coat of oil base paint on metal surfaces may result in cracking, peeling or chipping without the proper prime coat application. If off white or another light colored finish paint is specified, make sure the prime coat is also a light color, or more than one fin coat will be coat will be necessary. "Slow" work is based on an hourly wage of $23.00, "Medium" work on an hourly wage of $29.50, and "Fast" work on an hourly wage of $36.00. Other qualifications that apply to this table are on page 9.

Section designation		Square feet of surface area per foot of length: Minus one flange side	Square feet of surface area per foot of length: All around	Square feet of surface area per ton: Minus one flange side	Square feet of surface area per ton: All around
W 30	x 99	7.56	8.43	152.7	170.3
W 27	x 94	6.99	7.81	148.7	166.2
	x 84	6.99	7.79	166.4	185.5
W 24	x 100	7.00	8.00	140.0	160.0
	x 94	6.29	7.04	133.8	149.8
	x 84	6.27	7.02	149.3	167.1
	x 76	6.23	6.98	163.9	183.7
	x 68	6.21	6.96	182.6	204.7
	x 61	5.71	6.29	187.2	206.2
	x 55	5.67	6.25	206.2	227.2
W 21	x 96	5.77	6.52	120.2	135.8
	x 82	5.73	6.48	139.8	158.0
	x 73	5.60	6.29	153.4	172.3
	x 68	5.58	6.57	164.1	184.4
	x 62	5.56	6.25	179.4	201.6
	x 55	5.52	6.21	200.7	225.8
	x 49	5.10	5.65	208.2	230.6
	x 44	5.08	5.63	230.9	255.9
W 18	x 96	5.96	6.94	124.2	144.6
	x 85	5.28	6.02	124.2	141.6
	x 77	5.22	5.95	135.6	154.5
	x 70	5.19	5.92	148.3	169.1
	x 64	5.17	5.90	161.6	184.4
	x 60	4.92	5.54	164.0	184.7
	x 55	4.90	5.52	178.1	200.7
	x 50	4.88	5.50	195.2	220.0
	x 45	4.85	5.48	215.6	243.6
	x 40	4.48	4.98	224.0	249.0
	x 35	4.46	4.96	254.9	283.4
W 16	x 96	5.60	6.56	116.7	136.7
	x 88	5.57	6.53	126.6	148.4
	x 78	4.88	5.60	125.1	143.6
	x 71	4.82	5.53	135.8	155.8
	x 64	4.79	5.50	149.7	171.9
	x 58	4.77	5.48	164.5	189.0
	x 50	4.51	5.10	180.4	204.0
	x 45	4.45	5.03	197.8	223.6
	x 40	4.42	5.00	221.0	250.0
	x 36	4.40	4.98	244.4	276.7
	x 31	4.02	4.48	259.4	289.0
	x 26	3.98	4.44	306.2	341.5
W 14	x 95	5.99	7.20	126.1	151.6
	x 87	5.96	7.17	137.0	164.8
	x 84	5.36	6.36	127.6	151.4
	x 78	5.33	6.33	136.7	162.3
	x 74	4.92	5.77	133.0	155.9
	x 68	4.83	5.67	142.1	166.8
	x 61	4.81	5.65	157.7	185.2
	x 53	4.33	5.00	163.4	188.7
	x 48	4.29	4.96	178.8	206.7
	x 43	4.27	4.94	198.6	229.8
	x 38	4.04	4.60	212.6	242.1
	x 34	4.02	4.58	236.5	269.4
	x 30	4.00	4.56	266.7	304.0
	x 26	3.56	3.98	273.8	306.2
	x 22	3.54	3.96	321.8	360.0
W 12	x 99	5.19	6.21	104.8	125.4
	x 92	5.14	6.15	111.7	133.7
	x 85	5.12	6.13	120.5	144.2
	x 79	5.09	6.10	128.9	154.4
	x 72	5.04	6.04	140.0	167.8
	x 65	5.02	6.02	154.5	185.2
	x 58	4.54	5.38	156.6	185.5
	x 53	4.50	5.33	169.8	201.1
	x 50	4.07	4.75	162.8	190.0
	x 45	4.00	4.67	177.8	207.6
	x 40	4.00	4.67	200.0	233.5
	x 36	3.70	4.25	205.6	236.1
	x 31	3.65	4.19	235.5	270.3
	x 27	3.63	4.17	268.9	308.9
	x 22	3.04	3.38	276.4	307.3
	x 19	3.02	3.35	317.9	352.6
	x 16.5	3.00	3.33	363.6	403.6
	x 14	2.98	3.31	425.7	472.9
W 10	x 100	4.45	5.31	89.0	106.2
	x 89	4.38	5.23	98.4	117.5
	x 77	4.33	5.19	112.5	134.8
	x 72	4.28	5.13	118.9	142.5
	x 66	4.26	5.10	129.1	154.5
	x 60	4.24	5.08	141.3	169.3

Figure 23
Structural steel conversion table

Section designation		Square feet of surface area per foot of length: Minus one flange side	Square feet of surface area per foot of length: All around	Square feet of surface area per ton: Minus one flange side	Square feet of surface area per ton: All around
W 10	x 54	4.19	5.02	155.2	185.9
	x 49	4.17	5.00	170.2	204.0
	x 45	3.69	4.35	164.0	193.3
	x 39	3.67	4.33	188.2	222.1
	x 33	3.63	4.29	220.0	260.0
	x 29	3.15	3.63	217.2	250.3
	x 25	3.13	3.60	250.4	288.0
	x 21	3.08	3.56	293.3	339.0
	x 19	2.71	3.04	285.3	320.0
	x 17	2.69	3.02	316.5	355.3
	x 15	2.67	3.00	356.0	400.0
	x 11.5	2.65	2.98	460.9	518.3
W 8	x 67	3.56	4.25	106.3	126.9
	x 58	3.52	4.21	121.4	145.2
	x 48	3.45	4.13	143.8	172.1
	x 40	3.41	4.08	170.5	204.0
	x 35	3.35	4.02	191.4	229.7
	x 31	3.33	4.00	214.8	258.1
	x 28	2.96	3.50	211.4	250.0
	x 24	2.94	3.48	245.0	290.0
	x 20	2.67	3.10	267.0	310.0
	x 17	2.65	3.08	311.8	362.4
	x 15	2.35	2.69	313.3	358.7
	x 13	2.33	2.67	358.5	410.8
	x 10	2.31	2.65	462.0	530.0
W 6	x 25	2.59	3.10	207.2	248.0
	x 20	2.54	3.04	254.0	304.0
	x 15.5	2.50	3.00	322.6	387.1
	x 16	2.04	2.38	255.0	297.5
	x 12	2.00	2.33	333.3	388.3
	x 8.5	1.98	2.31	465.9	543.5
W 5	x 18.5	2.10	2.52	227.0	272.4
	x 16	2.08	2.50	260.0	312.5
W 4	x 13	1.69	2.02	260.0	310.8
S-24	x 90	5.78	6.38	128.4	141.8
	x 79.9	5.75	6.33	143.9	158.4
S-20	x 95	5.15	5.75	108.4	121.1
	x 85	5.08	5.67	119.5	133.4
	x 75	4.93	5.46	131.5	145.6
	x 65.4	4.90	5.42	149.8	165.7
S-18	x 70	4.56	5.08	130.3	145.1
	x 54.7	4.50	5.00	164.5	182.8
S-15	x 50	3.91	4.38	156.4	175.2
	x 42.9	3.88	4.33	180.9	201.9
S-12	x 50	3.38	3.83	135.2	153.2
	x 40.8	3.31	3.75	162.3	183.8
	x 35	3.28	3.71	187.4	212.0
	x 31.8	3.25	3.67	204.4	230.8
S-10	x 35	2.92	3.33	166.9	190.2
	x 25.4	2.82	3.21	222.0	252.8
S-8	x 23	2.36	2.71	205.2	235.7
	x 18.4	2.33	2.67	253.3	290.2
S-7	x 20	2.14	2.46	214.0	246.0
	x 15.3	2.07	2.38	270.6	311.1
S-6	x 17.25	1.91	2.21	221.4	256.2
	x 12.5	1.84	2.13	294.4	340.8
S-5	x 14.75	1.65	1.92	223.7	260.3
	x 10	1.58	1.83	316.0	366.0
S-4	x 9.5	1.35	1.58	284.2	332.6
	x 7.7	1.32	1.54	342.9	400.0
S-3	x 7.5	1.13	1.33	301.3	354.7
	x 5.7	1.09	1.29	382.5	452.6
Miscellaneous shape					
M-5	x 18.9	2.08	2.50	220.1	264.6

Figure 23 (cont'd)
Structural steel conversion table

Section designation		Square feet of surface area per foot of length: Minus one flange side	Square feet of surface area per foot of length: All around	Square feet of surface area per ton: Minus one flange side	Square feet of surface area per ton: All around
C-15	x 50	3.44	3.75	137.6	150.0
	x 40	3.38	3.67	169.0	183.5
	x 33.9	3.34	3.63	197.1	214.2
C-12	x 30	2.78	3.04	185.3	202.7
	x 25	2.75	3.00	220.0	240.0
	x 20.7	2.75	3.00	265.7	289.9
C-10	x 30	2.42	2.67	161.3	178.0
	x 25	2.39	2.63	191.2	210.4
	x 20	2.35	2.58	235.0	258.0
	x 15.3	2.32	2.54	305.3	334.2
C-9	x 20	2.16	2.38	216.0	238.0
	x 15	2.13	2.33	284.0	310.7
	x 13.4	2.09	2.29	311.9	341.8
C-8	x 18.75	1.96	2.17	209.1	231.5
	x 13.75	1.93	2.13	280.7	309.8
	x 11.5	1.90	2.08	330.4	361.7
C-7	x 14.75	1.73	1.92	234.6	260.3
	x 12.25	1.73	1.92	282.4	313.5
	x 9.8	1.70	1.88	346.9	383.7
C-6	x 13	1.53	1.71	235.4	263.1
	x 10.5	1.50	1.67	285.7	318.1
	x 8.2	1.47	1.63	358.5	397.6
C-5	x 9	1.30	1.46	288.9	324.4
	x 6.7	1.27	1.42	379.1	423.9
C-4	x 7.25	1.10	1.25	295.2	344.8
	x 5.4	1.07	1.21	396.3	448.1
C-3	x 6	.91	1.04	303.3	346.7
	x 5	.88	1.00	352.0	400.0
	x 4.1	.84	.96	409.8	468.3
MC-18	x 58	4.06	4.42	140.0	152.4
	x 51.9	4.03	4.38	155.3	168.8
	x 45.8	4.00	4.33	176.7	189.1
	x 42.7	4.00	4.33	187.4	202.8
MC-13	x 50	3.26	3.63	130.4	145.2
	x 40	3.20	3.54	160.0	177.0
	x 35	3.20	3.54	182.9	202.3
	x 31.8	3.17	3.50	199.4	220.1

Section designation		Square feet of surface area per foot of length: Minus one flange side	Square feet of surface area per foot of length: All around	Square feet of surface area per ton: Minus one flange side	Square feet of surface area per ton: All around
MC-12	x 50	3.03	3.38	121.2	135.2
	x 45	3.00	3.33	133.3	148.0
	x 40	2.97	3.29	148.5	164.5
	x 35	2.94	3.25	168.0	185.7
	x 37	2.91	3.21	157.3	173.5
	x 32.9	2.88	3.17	175.1	192.7
	x 30.9	2.88	3.17	186.4	205.2
MC-10	x 41.1	2.76	3.13	134.3	152.3
	x 33.6	2.70	3.04	160.7	181.0
	x 28.5	2.67	3.00	187.4	210.5
	x 28.3	2.54	2.83	179.5	200.0
	x 25.3	2.54	2.83	200.8	223.7
	x 24.9	2.51	2.79	201.6	224.1
	x 21.9	2.54	2.83	232.0	258.4
MC-9	x 25.4	2.38	2.67	187.4	210.2
	x 23.9	2.38	2.67	199.2	223.4
MC-8	x 22.8	2.21	2.50	193.9	219.3
	x 21.4	2.21	2.50	206.5	233.6
	x 20	2.08	2.33	208.0	233.0
	x 18.7	2.08	2.33	222.5	249.2
MC-7	x 22.7	2.07	2.38	182.4	209.7
	x 19.1	2.04	2.33	213.6	244.0
	x 17.6	1.92	2.17	218.2	246.6
MC-6	x 18	1.88	2.17	208.9	241.1
	x 15.3	1.88	2.17	245.8	283.7
	x 16.3	1.75	2.00	214.7	245.4
	x 15.1	1.75	2.00	231.8	264.9
	x 12	1.63	1.83	271.7	305.0
MC-3	x 9	1.03	1.21	228.9	268.9
	x 7.1	1.00	1.17	281.7	329.6

Figure 23 (cont'd)
Structural steel conversion table

Section designation		Surface area per foot of length	Surface area per ton
ST 18	x 97	5.06	104.3
	x 91	5.04	110.8
	x 85	5.02	118.1
	x 80	5.00	125.0
	x 75	4.98	132.8
	x 67.5	4.95	147.0
ST 16.5	x 76	4.72	124.2
	x 70.5	4.70	133.3
	x 65	4.38	144.0
	x 59	4.65	157.6
ST 15	x 95	5.02	105.7
	x 86	4.99	116.0
	x 66	4.28	129.7
	x 62	4.27	137.7
	x 58	4.25	146.6
	x 54	4.23	156.7
	x 49.5	4.21	170.1
ST 13.5	x 88.5	4.63	104.6
	x 80	4.59	114.8
	x 72.5	4.57	126.1
	x 57	3.95	138.6
	x 51	3.93	154.1
	x 47	3.91	166.4
	x 42	3.89	185.2
ST 12	x 80	4.41	110.3
	x 72.5	4.38	120.8
	x 65	4.36	134.1
	x 60	4.04	134.7
	x 55	4.02	146.2
	x 50	4.00	160.0
	x 47	3.54	150.6
	x 42	3.51	167.1
	x 38	3.49	183.7
	x 34	3.47	204.1
	x 30.5	3.15	206.6
	x 27.5	3.13	227.6

Section designation		Surface area per foot of length	Surface area per ton
ST 10.5	x 71	3.97	111.8
	x 63.5	3.95	124.4
	x 56	3.92	140.0
	x 48	3.27	136.3
	x 41	3.23	157.6
	x 36.5	3.15	172.6
	x 34	3.14	184.7
	x 31	3.12	201.3
	x 27.5	3.10	225.5
	x 24.5	2.82	230.2
	x 22	2.81	255.5
ST 9	x 57	3.51	123.2
	x 52.5	3.49	133.0
	x 48	3.47	144.6
	x 42.5	3.00	141.2
	x 38.5	2.98	154.8
	x 35	2.96	169.1
	x 32	2.94	183.8
	x 30	2.78	185.3
	x 27.5	2.77	201.5
	x 25	2.75	220.0
	x 22.5	2.73	242.7
	x 20	2.49	249.0
	x 17.5	2.48	283.4
ST 8	x 48	3.28	136.7
	x 44	3.26	148.2
	x 39	2.79	143.1
	x 35.5	2.77	156.1
	x 32	2.76	172.5
	x 29	2.73	188.3
	x 25	2.53	202.4
	x 22.5	2.52	224.0
	x 20	2.50	250.0
	x 18	2.49	276.7
	x 15.5	2.24	289.0
	x 13	2.22	341.5

Figure 23 (cont'd)
Structural steel conversion table

Section designation		Surface area per foot of length	Surface area per ton
ST 7	x 88	3.88	88.2
	x 83.5	3.86	92.5
	x 79	3.84	97.2
	x 75	3.83	102.1
	x 71	3.81	107.3
	x 68	3.69	108.5
	x 63.5	3.67	115.6
	x 59.5	3.65	122.7
	x 55.5	3.64	131.2
	x 51.5	3.62	140.6
	x 47.5	3.60	151.6
	x 43.5	3.58	164.6
	x 42	3.19	151.9
	x 39	3.17	162.6
	x 37	2.86	154.6
	x 34	2.85	167.6
	x 30.5	2.83	185.6
	x 26.5	2.51	189.4
	x 24	2.49	207.5
	x 21.5	2.47	229.8
	x 19	2.31	243.2
	x 17	2.29	269.4
	x 15	2.28	304.0
	x 13	2.00	307.7
	x 11	1.98	360.0
ST 6	x 95	3.31	69.7
	x 80.5	3.24	80.5
	x 66.5	3.18	97.8
	x 60	3.15	105.6
	x 53	3.11	117.4
	x 49.5	3.10	125.3
	x 46	3.08	133.9
	x 42.5	3.06	144.0
	x 39.5	3.05	154.4
	x 36	3.03	168.3
	x 32.5	3.01	185.2
	x 29	2.69	185.5
	x 26.5	2.67	201.5
	x 25	2.36	188.8
	x 22.5	2.35	208.9

Section designation		Surface area per foot of length	Surface area per ton
ST 6	x 20	2.33	233.0
	x 18	2.11	234.4
	x 15.5	2.10	271.0
	x 13.5	2.08	308.1
	x 11	1.70	309.1
	x 9.5	1.68	353.7
	x 8.25	1.67	404.8
	x 7	1.65	471.4
ST 5	x 56	2.68	95.7
	x 50	2.65	106.0
	x 44.5	2.62	117.8
	x 38.5	2.58	134.0
	x 36	2.57	142.8
	x 33	2.55	154.5
	x 30	2.53	168.7
	x 27	2.51	185.9
	x 24.5	2.50	204.1
	x 22.5	2.18	193.8
	x 19.5	2.16	221.5
	x 16.5	2.14	259.4
	x 14.5	1.82	251.0
	x 12.5	1.80	288.0
	x 10.5	1.78	339.0
	x 9.5	1.53	322.1
	x 8.5	1.51	355.3
	x 7.5	1.50	400.0
	x 5.75	1.48	514.8
ST 4	x 33.5	2.13	127.7
	x 29	2.10	144.8
	x 24	2.06	171.7
	x 20	2.03	203.0
	x 17.5	2.01	229.7
	x 15.5	2.00	258.1
	x 14	1.76	251.4
	x 12	1.75	291.7
	x 10	1.56	312.0
	x 8.5	1.54	362.4
	x 7.5	1.35	360.0
	x 6.5	1.33	409.2
	x 5	1.31	524.0

Figure 23 (cont'd)
Structural steel conversion table

Section designation		Surface area per foot of length	Surface area per ton
ST 3	x 12.5	1.55	248.0
	x 10	1.52	304.0
	x 7.75	1.50	387.1
	x 8	1.19	297.5
	x 6	1.17	390.0
	x 4.25	1.14	536.5
ST 2.5	x 9.25	1.26	272.4
	x 8	1.25	312.5
ST 2	x 6.5	1.02	313.8

Tees cut from American standard shapes

Section designation		Surface area per foot of length	Surface area per ton
ST 12	x 60	3.34	111.3
	x 52.95	3.31	125.0
	x 50	3.21	128.4
	x 45	3.19	141.8
	x 39.95	3.17	158.7
ST 10	x47.5	2.87	120.8
	x 42.5	2.84	133.6
	x 37.5	2.73	145.6
	x 32.7	2.70	165.1
ST 9	x 35	2.54	145.1
	x 27.35	2.50	182.8
ST 7.5	x 25	2.19	175.2
	x 21.45	2.17	202.3

Section designation		Surface area per foot of length	Surface area per ton
ST 6	x 25	1.91	152.8
	x 20.4	1.88	184.3
	x 17.5	1.85	211.4
	x 15.9	1.83	230.2
ST 5	x 17.5	1.66	189.7
	x 12.7	1.61	253.5
ST 4	x 11.5	1.36	236.5
	x 9.2	1.33	289.1
ST 3.5	x 10	1.23	246.0
	x 7.65	1.19	311.0
ST 3	x 8.625	1.09	252.8
	x 6.25	1.06	339.2
ST 2.5	x 7.375	.96	260.3
	x 5	.92	368.0
ST 2	x 4.75	.80	336.8
	x 3.85	.78	405.2

Miscellaneous tee (cut from M5 x 18.9)

MT 2.5	x 9.45	.83	175.7

Figure 23 (cont'd)
Structural steel conversion table

Section designation			Surface area per foot of length	Surface area per ton
L 8	x 8	x 1-1/8	2.67	93.8
	x 8	x 1	2.67	104.7
	x 8	x 1	2.67	118.7
	x 8	x 7/8	2.67	118.7
	x 8	x 3/4	2.67	137.3
	x 8	x 5/8	2.67	163.3
	x 8	x 9/16	2.67	180.4
	x 8	x 1/2	2.67	202.3
L 6	x 6	x 1	2.00	107.0
	x 6	x 7/8	2.00	120.8
	x 6	x 3/4	2.00	139.4
	x 6	x 5/8	2.00	165.3
	x 6	x 9/16	2.00	182.6
	x 6	x 1/2	2.00	204.1
	x 6	x 7/16	2.00	232.6
	x 6	x 3/8	2.00	268.5
	x 6	x 5/16	2.00	322.6
L5	x 5	x 7/8	1.67	122.8
	x 5	x 3/4	1.67	141.5
	x 5	x 5/8	1.67	167.0
	x 5	x 1/2	1.67	206.2
	x 5	x 7/16	1.67	233.6
	x 5	x 3/8	1.67	271.5
	x 5	x 5/16	1.67	324.3
L 4	x 4	x 3/4	1.33	143.8
	x 4	x 5/8	1.33	169.4
	x 4	x 1/2	1.33	207.8
	x 4	x 7/16	1.33	235.4
	x 4	x 3/8	1.33	271.4
	x 4	x 5/16	1.33	324.4
	x 4	x 1/4	1.33	403.0
L 3-1/2	x 3/1/2	x 1/2	1.17	210.8
	x 3-1/2	x 7/16	1.17	238.8
	x 3-1/2	x 3/8	1.17	275.3
	x 3-1/2	x 5/16	1.17	325.0
	x 3-1/2	x 1/4	1.17	403.4
L 3	x 3	x 1/2	1.00	212.8
	x 3	x 7/16	1.00	241.0
	x 3	x 3/8	1.00	277.8
	x 3	x 5/16	1.00	327.9
	x 3	x 1/4	1.00	408.2
	x 3	x 3/16	1.00	539.1
L 2-1/2	x 2-1/2	x 1/2	.83	215.6
	x 2-1/2	x 3/8	.83	281.4
	x 2-1/2	x 5/16	.83	332.0
	x 2-1/2	x 1/4	.83	404.9
	x 2-1/2	x 3/16	.83	540.7
L 2	x 2	x 3/8	.67	285.1
	x 2	x 5/16	.67	341.8
	x 2	x 1/4	.67	420.1
	x 2	x 3/16	.67	549.2
	x 2	x 1/8	.67	812.1
L 1-3/4	x 1-3/4	x 1/4	.58	418.8
	x 1-3/4	x 3/16	.58	547.2
	x 1-3/4	x 1/8	.58	805.6
L 1-1/2	x 1-1/2	x 1/4	.50	427.4
	x 1-1/2	x 3/16	.50	555.6
	x 1-1/2	x 5/32	.50	657.9
	x 1-1/2	x 1/8	.50	813.0
L 1-1/4	x 1-1/4	x 1/4	.42	437.5
	x 1-1/4	x 3/16	.42	567.6
	x 1-1/4	x 1/8	.42	831.7
L 1	x 1	x 1/4	.33	443.0
	x 1	x 3/16	.33	569.0
	x 1	x 1/8	.33	825.0

Figure 23 (cont'd)
Structural steel conversion table

Section designation			Surface area per foot of length	Surface area per ton
L 9	x 4	x 1	2.17	106.4
	x 4	x 7/8	2.17	120.2
	x 4	x 3/4	2.17	138.7
	x 4	x 5/8	2.17	165.0
	x 4	x 9/16	2.17	182.4
	x 4	x 1/2	2.17	203.8
L 8	x 6	x 1	2.33	105.4
	x 6	x 7/8	2.33	119.2
	x 6	x 3/4	2.33	137.9
	x 6	x 5/8	2.33	163.5
	x 6	x 9/16	2.33	181.3
	x 6	x 1/2	2.33	202.6
	x 6	x 7/16	2.33	230.7
L 8	x 4	x 1	2.00	107.0
	x 4	x 7/8	2.00	120.8
	x 4	x 3/4	2.00	139.4
	x 4	x 5/8	2.00	165.3
	x 4	x 9/16	2.00	182.6
	x 4	x 1/2	2.00	204.1
	x 4	x 7/16	2.00	232.6
L 7	x 4	x 7/8	1.83	121.2
	x 4	x 3/4	1.83	139.7
	x 4	x 5/8	1.83	165.6
	x 4	x 9/16	1.83	183.0
	x 4	x 1/2	1.83	204.5
	x 4	x 7/16	1.83	231.6
	x 4	x 3/8	1.83	269.1
L 6	x 4	x 7/8	1.67	122.8
	x 4	x 3/4	1.67	141.5
	x 4	x 5/8	1.67	167.0
	x 4	x 9/16	1.67	184.5
	x 4	x 1/2	1.67	206.2
	x 4	x 7/16	1.67	233.6
	x 4	x 3/8	1.67	271.5
	x 4	x 5/16	1.67	324.3
	x 4	x 1/4	1.67	402.4
L 6	x 3-1/2	x 1/2	1.58	206.5
	x 3-1/2	x 3/8	1.58	270.1
	x 3-1/2	x 5/16	1.58	322.4
	x 3-1/2	x 1/4	1.58	400.0

Section designation			Surface area per foot of length	Surface area per ton
L 5	x 3-1/2	x 3/4	1.42	143.4
	x 3-1/2	x 5/8	1.42	169.0
	x 3-1/2	x 1/2	1.42	208.8
	x 3-1/2	x 7/16	1.42	236.7
	x 3-1/2	x 3/8	1.42	273.1
	x 3-1/2	x 5/16	1.42	326.4
	x 3-1/2	x 1/4	1.42	405.7
L 5	x 3	x 1/2	1.33	207.8
	x 3	x 7/16	1.33	235.4
	x 3	x 3/8	1.33	271.4
	x 3	x 5/16	1.33	324.4
	x 3	x 1/4	1.33	403.0
L 4	x 3-1/2	x 5/8	1.25	170.1
	x 3-1/2	x 1/2	1.25	210.1
	x 3-1/2	x 7/16	1.25	235.8
	x 3-1/2	x 3/8	1.25	274.7
	x 3-1/2	x 5/16	1.25	324.7
	x 3-1/2	x 1/4	1.25	403.2
L 4	x 3	x 5/8	1.17	172.1
	x 3	x 1/2	1.17	210.8
	x 3	x 7/16	1.17	238.8
	x 3	x 3/8	1.17	275.3
	x 3	x 5/16	1.17	325.0
	x 3	x 1/4	1.17	403.4
L 3-1/2	x 3	x 1/2	1.08	211.8
	x 3	x 7/16	1.08	237.4
	x 3	x 3/8	1.08	273.4
	x 3	x 5/16	1.08	327.3
	x 3	x 1/4	1.08	400.0
L 3-1/2	x 2-1/2	x 1/2	1.00	212.8
	x 2-1/2	x 7/16	1.00	241.0
	x 2-1/2	x 3/8	1.00	277.8
	x 2-1/2	x 5/16	1.00	327.9
	x 2-1/2	x 1/4	1.00	408.2
L 3	x 2-1/2	x 1/2	.92	216.5
	x 2-1/2	x 7/16	.92	242.1
	x 2-1/2	x 3/8	.92	278.8
	x 2-1/2	x 5/16	.92	328.6
	x 2-1/2	x 1/4	.92	408.9
	x 2-1/2	x 3/16	.92	542.8

Figure 23 (cont'd)
Structural steel conversion table

Section designation			Surface area per foot of length	Surface area per ton
L 3	x 2	x 1/2	.83	215.6
	x 2	x 7/16	.83	244.1
	x 2	x 3/8	.83	281.4
	x 2	x 5/16	.83	332.0
	x 2	x 1/4	.83	404.9
	x 2	x 3/16	.83	540.7
L 2-1/2	x 2	x 3/8	.75	283.0
	x 2	x 5/16	.75	333.3
	x 2	x 1/4	.75	414.4
	x 2	x 3/16	.75	545.5
L 2-1/2	x 1-1/2	x 5/16	.67	341.8
	x 1-1/2	x 1/4	.67	420.1
	x 1-1/2	x 3/16	.67	549.2

Courtesy: Richardson Engineering Services, Inc.

Section designation			Surface area per foot of length	Surface area per ton
L 2	x 1-1/2	x 1/4	.58	418.8
	x 1-1/2	x 3/16	.58	547.2
	x 1-1/2	x 1/8	.58	805.6
L 2	x 1-1/4	x 1/4	.54	423.5
	x 1-1/4	x 3/16	.54	551.0
L 1-3/4	x 1-1/4	x 1/4	.50	427.4
	x 1-1/4	x 3/16	.50	555.6
	x 1-1/4	x 1/8	.50	813.0

Figure 23 (cont'd)
Structural steel conversion table

Diameter (in feet)	Area (SF)
10	314
15	707
20	1,257
25	1,963
30	2,827
35	3,848
40	5,027
45	6,362
50	7,854
55	9,503
60	11,310
65	13,273
70	15,394

Figure 24
Surface area of spheres

	Labor SF per manhour	Material coverage SF/gallon	Material cost per gallon	Labor cost per 100 SF	Labor burden 100 SF	Material cost per 100 SF	Overhead per 100 SF	Profit per 100 SF	Total price per 100 SF
Tank, silo, vessel, or hopper, brush, exterior walls only									
Metal primer, rust inhibitor - clean metal (material #35)									
Brush prime coat									
Slow	150	425	59.70	15.33	3.69	14.05	6.28	6.29	45.64
Medium	175	400	52.30	16.86	4.86	13.08	8.70	5.22	48.72
Fast	200	375	44.80	18.00	6.36	11.95	11.25	3.33	50.89
Metal primer, rust inhibitor - rusty metal (material #36)									
Brush prime coat									
Slow	150	400	80.10	15.33	3.69	20.03	7.42	7.43	53.90
Medium	175	375	70.10	16.86	4.86	18.69	10.11	6.06	56.58
Fast	200	350	60.10	18.00	6.36	17.17	12.87	3.81	58.21
Industrial enamel, oil base, high gloss - light colors (material #56)									
Brush 1st or additional finish coats									
Slow	200	450	65.40	11.50	2.76	14.53	5.47	5.48	39.74
Medium	225	425	57.20	13.11	3.78	13.46	7.59	4.55	42.49
Fast	250	400	49.00	14.40	5.08	12.25	9.84	2.91	44.48
Industrial enamel, oil base, high gloss - dark (OSHA) colors (material #57)									
Brush 1st or additional finish coats									
Slow	200	475	78.30	11.50	2.76	16.48	5.84	5.85	42.43
Medium	225	450	68.50	13.11	3.78	15.22	8.03	4.82	44.96
Fast	250	425	58.70	14.40	5.08	13.81	10.32	3.05	46.66
Epoxy coating, 2 part system, clear (material #51)									
Brush 1st coat									
Slow	150	425	160.60	15.33	3.69	37.79	10.79	10.81	78.41
Medium	175	400	140.50	16.86	4.86	35.13	14.22	8.53	79.60
Fast	200	375	120.50	18.00	6.36	32.13	17.51	5.18	79.18
Brush 2nd or additional coats									
Slow	200	450	160.60	11.50	2.76	35.69	9.49	9.51	68.95
Medium	225	425	140.50	13.11	3.78	33.06	12.49	7.49	69.93
Fast	250	400	120.50	14.40	5.08	30.13	15.38	4.55	69.54
Epoxy coating, 2 part system, white (material #52)									
Brush 1st coat									
Slow	150	425	157.00	15.33	3.69	36.94	10.63	10.65	77.24
Medium	175	400	137.40	16.86	4.86	34.35	14.02	8.41	78.50
Fast	200	375	117.70	18.00	6.36	31.39	17.28	5.11	78.14
Brush 2nd or additional coats									
Slow	200	450	157.00	11.50	2.76	34.89	9.34	9.36	67.85
Medium	225	425	137.40	13.11	3.78	32.33	12.31	7.38	68.91
Fast	250	400	117.70	14.40	5.08	29.43	15.16	4.48	68.55

	Labor SF per manhour	Material coverage SF/gallon	Material cost per gallon	Labor cost per 100 SF	Labor burden 100 SF	Material cost per 100 SF	Overhead per 100 SF	Profit per 100 SF	Total price per 100 SF
Vinyl coating (material #59)									
Brush 1st coat									
Slow	125	250	151.20	18.40	4.42	60.48	15.83	15.86	114.99
Medium	150	238	132.30	19.67	5.70	55.59	20.24	12.14	113.34
Fast	175	225	113.40	20.57	7.24	50.40	24.25	7.17	109.63
Brush 2nd or additional coats									
Slow	165	150	151.20	13.94	3.34	100.80	22.44	22.48	163.00
Medium	190	125	132.30	15.53	4.47	105.84	31.47	18.88	176.19
Fast	215	100	113.40	16.74	5.91	113.40	42.18	12.48	190.71

See Figure 24 on page 399 to find the surface area of a spherical vessel. Use this table when estimating walls only. The cost tables for painting steel tank, silo, vessel or hopper roofs follow those for painting walls. For heights above 8 feet, use the High Time Difficulty Factors on page 139. Note: A two coat system, prime and finish, using oil base material is recommended for any metal surface. Using water base material may cause oxidation, corrosion and rust. One coat of oil base paint on metal surfaces may result in cracking, peeling or chipping without the proper prime coat application. If off-white or another light-colored finish paint is specified, make sure the prime coat is also a light color, or more than one finish coat will be necessary. "Slow" work is based on an hourly wage of $23.00, "Medium" work on an hourly wage of $29.50, and "Fast" work on an hourly wage of $36.00. Other qualifications that apply to this table are on page 9.

	Labor SF per manhour	Material coverage SF/gallon	Material cost per gallon	Labor cost per 100 SF	Labor burden 100 SF	Material cost per 100 SF	Overhead per 100 SF	Profit per 100 SF	Total price per 100 SF
Tank, silo, vessel, or hopper, roll, exterior walls only									
Metal primer, rust inhibitor - clean metal (material #35)									
Roll prime coat									
Slow	275	400	59.70	8.36	2.02	14.93	4.81	4.82	34.94
Medium	300	375	52.30	9.83	2.83	13.95	6.66	3.99	37.26
Fast	325	350	44.80	11.08	3.92	12.80	8.61	2.55	38.96
Metal primer, rust inhibitor - rusty metal (material #36)									
Roll prime coat									
Slow	275	380	80.10	8.36	2.02	21.08	5.98	5.99	43.43
Medium	300	355	70.10	9.83	2.83	19.75	8.11	4.86	45.38
Fast	325	330	60.10	11.08	3.92	18.21	10.29	3.04	46.54
Industrial enamel, oil base, high gloss - light colors (material #56)									
Roll 1st or additional finish coats									
Slow	375	425	65.40	6.13	1.48	15.39	4.37	4.38	31.75
Medium	400	400	57.20	7.38	2.13	14.30	5.95	3.57	33.33
Fast	425	375	49.00	8.47	2.98	13.07	7.60	2.25	34.37
Industrial enamel, oil base, high gloss - dark (OSHA) colors (material #57)									
Roll 1st or additional finish coats									
Slow	375	450	78.30	6.13	1.48	17.40	4.75	4.76	34.52
Medium	400	425	68.50	7.38	2.13	16.12	6.41	3.84	35.88
Fast	425	400	58.70	8.47	2.98	14.68	8.10	2.40	36.63
Epoxy coating, 2 part system, clear (material #51)									
Roll 1st coat									
Slow	275	425	160.60	8.36	2.02	37.79	9.15	9.17	66.49
Medium	300	400	140.50	9.83	2.83	35.13	11.95	7.17	66.91
Fast	325	375	120.50	11.08	3.92	32.13	14.61	4.32	66.06
Roll 2nd or additional coats									
Slow	375	450	160.60	6.13	1.48	35.69	8.23	8.24	59.77
Medium	400	425	140.50	7.38	2.13	33.06	10.64	6.39	59.60
Fast	425	400	120.50	8.47	2.98	30.13	12.89	3.81	58.28

	Labor SF per manhour	Material coverage SF/gallon	Material cost per gallon	Labor cost per 100 SF	Labor burden 100 SF	Material cost per 100 SF	Overhead per 100 SF	Profit per 100 SF	Total price per 100 SF
Epoxy coating, 2 part system, white (material #52)									
Roll 1st coat									
Slow	275	425	157.00	8.36	2.02	36.94	8.99	9.01	65.32
Medium	300	400	137.40	9.83	2.83	34.35	11.76	7.05	65.82
Fast	325	375	117.70	11.08	3.92	31.39	14.38	4.25	65.02
Roll 2nd or additional coats									
Slow	375	450	157.00	6.13	1.48	34.89	8.07	8.09	58.66
Medium	400	425	137.40	7.38	2.13	32.33	10.46	6.28	58.58
Fast	425	400	117.70	8.47	2.98	29.43	12.68	3.75	57.31

See Figure 24 on page 399 to find the surface area of a spherical vessel. Use this table when estimating walls only. The cost tables for painting steel tank, silo, vessel or hopper roofs follow those for painting walls. For heights above 8 feet, use the High Time Difficulty Factors on page 139. Note: A two coat system, prime and finish, using oil base material is recommended for any metal surface. Using water base material may cause oxidation, corrosion and rust. One coat of oil base paint on metal surfaces may result in cracking, peeling or chipping without the proper prime coat application. If off-white or another light-colored finish paint is specified, make sure the prime coat is also a light color, or more than one finish coat will be necessary. "Slow" work is based on an hourly wage of $23.00, "Medium" work on an hourly wage of $29.50, and "Fast" work on an hourly wage of $36.00. Other qualifications that apply to this table are on page 9.

	Labor SF per manhour	Material coverage SF/gallon	Material cost per gallon	Labor cost per 100 SF	Labor burden 100 SF	Material cost per 100 SF	Overhead per 100 SF	Profit per 100 SF	Total price per 100 SF
Tank, silo, vessel, or hopper, spray, exterior walls only									
Metal primer, rust inhibitor - clean metal (material #35)									
Spray prime coat									
Slow	700	325	59.70	3.29	.79	18.37	4.27	4.28	31.00
Medium	750	300	52.30	3.93	1.13	17.43	5.63	3.38	31.50
Fast	800	275	44.80	4.50	1.59	16.29	6.94	2.05	31.37
Metal primer, rust inhibitor - rusty metal (material #36)									
Spray prime coat									
Slow	700	300	80.10	3.29	.79	26.70	5.85	5.86	42.49
Medium	750	275	70.10	3.93	1.13	25.49	7.64	4.58	42.77
Fast	800	250	60.10	4.50	1.59	24.04	9.34	2.76	42.23
Industrial enamel, oil base, high gloss - light colors (material #56)									
Spray 1st or additional finish coats									
Slow	850	350	65.40	2.71	.66	18.69	4.19	4.20	30.45
Medium	900	325	57.20	3.28	.94	17.60	5.46	3.27	30.55
Fast	950	300	49.00	3.79	1.32	16.33	6.65	1.97	30.06
Industrial enamel, oil base, high gloss - dark (OSHA) colors (material #57)									
Spray 1st or additional finish coats									
Slow	850	375	78.30	2.71	.66	20.88	4.61	4.62	33.48
Medium	900	350	68.50	3.28	.94	19.57	5.95	3.57	33.31
Fast	950	325	58.70	3.79	1.32	18.06	7.19	2.13	32.49
Epoxy coating, 2 part system, clear (material #51)									
Spray 1st coat									
Slow	700	325	160.60	3.29	.79	49.42	10.17	10.19	73.86
Medium	750	313	140.50	3.93	1.13	44.89	12.49	7.49	69.93
Fast	800	300	120.50	4.50	1.59	40.17	14.34	4.24	64.84
Spray 2nd or additional coats									
Slow	850	350	160.60	2.71	.66	45.89	9.36	9.38	68.00
Medium	900	338	140.50	3.28	.94	41.57	11.45	6.87	64.11
Fast	950	325	120.50	3.79	1.32	37.08	13.09	3.87	59.15
Epoxy coating, 2 part system, white (material #52)									
Spray 1st coat									
Slow	700	325	157.00	3.29	.79	48.31	9.95	9.97	72.31
Medium	750	313	137.40	3.93	1.13	43.90	12.24	7.35	68.55
Fast	800	300	117.70	4.50	1.59	39.23	14.05	4.16	63.53
Spray 2nd or additional coats									
Slow	850	350	157.00	2.71	.66	44.86	9.16	9.18	66.57
Medium	900	338	137.40	3.28	.94	40.65	11.22	6.73	62.82
Fast	950	325	117.70	3.79	1.32	36.22	12.82	3.79	57.94

	Labor SF per manhour	Material coverage SF/gallon	Material cost per gallon	Labor cost per 100 SF	Labor burden 100 SF	Material cost per 100 SF	Overhead per 100 SF	Profit per 100 SF	Total price per 100 SF
Vinyl coating (material #59)									
Spray 1st coat									
Slow	600	225	151.20	3.83	.93	67.20	13.67	13.70	99.33
Medium	625	213	132.30	4.72	1.36	62.11	17.05	10.23	95.47
Fast	650	200	113.40	5.54	1.96	56.70	19.90	5.89	89.99
Spray 2nd or additional coats									
Slow	725	130	151.20	3.17	.77	116.31	22.85	22.89	165.99
Medium	750	105	132.30	3.93	1.13	126.00	32.77	19.66	183.49
Fast	775	80	113.40	4.65	1.63	141.75	45.89	13.58	207.50

See Figure 24 on page 399 to find the surface area of a spherical vessel. Use this table when estimating walls only. The cost tables for painting steel tank, silo, vessel or hopper roofs follow those for painting walls. For heights above 8 feet, use the High Time Difficulty Factors on page 139. Note: A two coat system, prime and finish, using oil base material is recommended for any metal surface. Using water base material may cause oxidation, corrosion and rust. One coat of oil base paint on metal surfaces may result in cracking, peeling or chipping without the proper prime coat application. If off-white or another light-colored finish paint is specified, make sure the prime coat is also a light color, or more than one finish coat will be necessary. "Slow" work is based on an hourly wage of $23.00, "Medium" work on an hourly wage of $29.50, and "Fast" work on an hourly wage of $36.00. Other qualifications that apply to this table are on page 9.

	Labor SF per manhour	Material coverage SF/gallon	Material cost per gallon	Labor cost per 100 SF	Labor burden 100 SF	Material cost per 100 SF	Overhead per 100 SF	Profit per 100 SF	Total price per 100 SF
Tank, silo, vessel, or hopper, brush, exterior roof only									
Metal primer, rust inhibitor - clean metal (material #35)									
Brush prime coat									
Slow	175	425	59.70	13.14	3.14	14.05	5.76	5.78	41.87
Medium	200	400	52.30	14.75	4.27	13.08	8.02	4.81	44.93
Fast	225	375	44.80	16.00	5.63	11.95	10.42	3.08	47.08
Metal primer, rust inhibitor - rusty metal (material #36)									
Brush prime coat									
Slow	175	400	80.10	13.14	3.14	20.03	6.90	6.92	50.13
Medium	200	375	70.10	14.75	4.27	18.69	9.43	5.66	52.80
Fast	225	350	60.10	16.00	5.63	17.17	12.03	3.56	54.39
Industrial enamel, oil base, high gloss - light colors (material #56)									
Brush 1st or additional finish coats									
Slow	225	450	65.40	10.22	2.44	14.53	5.17	5.18	37.54
Medium	250	425	57.20	11.80	3.41	13.46	7.17	4.30	40.14
Fast	275	400	49.00	13.09	4.64	12.25	9.29	2.75	42.02
Industrial enamel, oil base, high gloss - dark (OSHA) colors (material #57)									
Brush 1st or additional finish coats									
Slow	225	475	78.30	10.22	2.44	16.48	5.54	5.55	40.23
Medium	250	450	68.50	11.80	3.41	15.22	7.61	4.56	42.60
Fast	275	425	58.70	13.09	4.64	13.81	9.77	2.89	44.20
Epoxy coating, 2 part system, clear (material #51)									
Brush 1st coat									
Slow	175	425	160.60	13.14	3.14	37.79	10.28	10.30	74.65
Medium	200	400	140.50	14.75	4.27	35.13	13.54	8.12	75.81
Fast	225	375	120.50	16.00	5.63	32.13	16.67	4.93	75.36
Brush 2nd or additional coats									
Slow	225	450	160.60	10.22	2.44	35.69	9.19	9.21	66.75
Medium	250	425	140.50	11.80	3.41	33.06	12.07	7.24	67.58
Fast	275	400	120.50	13.09	4.64	30.13	14.83	4.39	67.08
Epoxy coating, 2 part system, white (material #52)									
Brush 1st coat									
Slow	175	425	157.00	13.14	3.14	36.94	10.11	10.13	73.46
Medium	200	400	137.40	14.75	4.27	34.35	13.34	8.00	74.71
Fast	225	375	117.70	16.00	5.63	31.39	16.44	4.86	74.32
Brush 2nd or additional coats									
Slow	225	450	157.00	10.22	2.44	34.89	9.04	9.06	65.65
Medium	250	425	137.40	11.80	3.41	32.33	11.89	7.13	66.56
Fast	275	400	117.70	13.09	4.64	29.43	14.61	4.32	66.09

	Labor SF per manhour	Material coverage SF/gallon	Material cost per gallon	Labor cost per 100 SF	Labor burden 100 SF	Material cost per 100 SF	Overhead per 100 SF	Profit per 100 SF	Total price per 100 SF
Vinyl coating (material #59)									
Brush 1st coat									
Slow	125	250	151.20	18.40	4.42	60.48	15.83	15.86	114.99
Medium	150	238	132.30	19.67	5.70	55.59	20.24	12.14	113.34
Fast	175	225	113.40	20.57	7.24	50.40	24.25	7.17	109.63
Brush 2nd or additional coats									
Slow	200	150	151.20	11.50	2.76	100.80	21.86	21.91	158.83
Medium	225	125	132.30	13.11	3.78	105.84	30.69	18.41	171.83
Fast	250	100	113.40	14.40	5.08	113.40	41.19	12.18	186.25

Use these figures to estimate labor and material costs for painting the exterior surface of a flat roof on a steel tank, silo, vessel or hopper. *Rule of thumb*: For a vaulted, peaked or sloping roof, figure the roof area as though it were flat and add 5%. Note: A two coat system, prime and finish, using oil base material is recommended for any metal surface. Using water base material may cause oxidation, corrosion and rust. One coat of oil base paint on metal surfaces may result in cracking, peeling or chipping without the proper prime coat application. If off-white or another light-colored finish paint is specified, make sure the prime coat is also a light color, or more than one finish coat will be necessary. "Slow" work is based on an hourly wage of $23.00, "Medium" work on an hourly wage of $29.50, and "Fast" work on an hourly wage of $36.00. Other qualifications that apply to this table are on page 9.

	Labor SF per manhour	Material coverage SF/gallon	Material cost per gallon	Labor cost per 100 SF	Labor burden 100 SF	Material cost per 100 SF	Overhead per 100 SF	Profit per 100 SF	Total price per 100 SF
Tank, silo, vessel, or hopper, roll, exterior roof only									
Metal primer, rust inhibitor - clean metal (material #35)									
Roll prime coat									
Slow	325	400	59.70	7.08	1.70	14.93	4.50	4.51	32.72
Medium	350	375	52.30	8.43	2.45	13.95	6.21	3.72	34.76
Fast	375	350	44.80	9.60	3.41	12.80	7.99	2.36	36.16
Metal primer, rust inhibitor - rusty metal (material #36)									
Roll prime coat									
Slow	325	380	80.10	7.08	1.70	21.08	5.67	5.68	41.21
Medium	350	355	70.10	8.43	2.45	19.75	7.66	4.59	42.88
Fast	375	330	60.10	9.60	3.41	18.21	9.67	2.86	43.75
Industrial enamel, oil base, high gloss - light colors (material #56)									
Roll 1st or additional finish coats									
Slow	400	425	65.40	5.75	1.38	15.39	4.28	4.29	31.09
Medium	425	400	57.20	6.94	2.00	14.30	5.81	3.49	32.54
Fast	450	375	49.00	8.00	2.81	13.07	7.41	2.19	33.48
Industrial enamel, oil base, high gloss - dark (OSHA) colors (material #57)									
Roll 1st or additional finish coats									
Slow	400	450	78.30	5.75	1.38	17.40	4.66	4.67	33.86
Medium	425	425	68.50	6.94	2.00	16.12	6.27	3.76	35.09
Fast	450	400	58.70	8.00	2.81	14.68	7.91	2.34	35.74
Epoxy coating, 2 part system, clear (material #51)									
Roll 1st coat									
Slow	325	425	160.60	7.08	1.70	37.79	8.85	8.87	64.29
Medium	350	400	140.50	8.43	2.45	35.13	11.50	6.90	64.41
Fast	375	375	120.50	9.60	3.41	32.13	13.99	4.14	63.27
Roll 2nd or additional coats									
Slow	400	450	160.60	5.75	1.38	35.69	8.14	8.15	59.11
Medium	425	425	140.50	6.94	2.00	33.06	10.50	6.30	58.80
Fast	450	400	120.50	8.00	2.81	30.13	12.69	3.75	57.38

	Labor SF per manhour	Material coverage SF/gallon	Material cost per gallon	Labor cost per 100 SF	Labor burden 100 SF	Material cost per 100 SF	Overhead per 100 SF	Profit per 100 SF	Total price per 100 SF
Epoxy coating, 2 part system, white (material #52)									
Roll 1st coat									
Slow	325	425	157.00	7.08	1.70	36.94	8.69	8.71	63.12
Medium	350	400	137.40	8.43	2.45	34.35	11.31	6.78	63.32
Fast	375	375	117.70	9.60	3.41	31.39	13.76	4.07	62.23
Roll 2nd or additional coats									
Slow	400	450	157.00	5.75	1.38	34.89	7.98	8.00	58.00
Medium	425	425	137.40	6.94	2.00	32.33	10.32	6.19	57.78
Fast	450	400	117.70	8.00	2.81	29.43	12.48	3.69	56.41

Use these figures to estimate labor and material costs for painting the exterior surface of a flat roof on a steel tank, silo, vessel or hopper. *Rule of thumb*: For a vaulted, peaked or sloping roof, figure the roof area as though it were flat and add 5%. Note: A two coat system, prime and finish, using oil base material is recommended for any metal surface. Using water base material may cause oxidation, corrosion and rust. One coat of oil base paint on metal surfaces may result in cracking, peeling or chipping without the proper prime coat application. If off-white or another light-colored finish paint is specified, make sure the prime coat is also a light color, or more than one finish coat will be necessary. "Slow" work is based on an hourly wage of $23.00, "Medium" work on an hourly wage of $29.50, and "Fast" work on an hourly wage of $36.00. Other qualifications that apply to this table are on page 9.

	Labor SF per manhour	Material coverage SF/gallon	Material cost per gallon	Labor cost per 100 SF	Labor burden 100 SF	Material cost per 100 SF	Overhead per 100 SF	Profit per 100 SF	Total price per 100 SF
Tank, silo, vessel, or hopper, spray, exterior roof only									
Metal primer, rust inhibitor - clean metal (material #35)									
Spray prime coat									
Slow	850	325	59.70	2.71	.66	18.37	4.13	4.14	30.01
Medium	900	300	52.30	3.28	.94	17.43	5.42	3.25	30.32
Fast	950	275	44.80	3.79	1.32	16.29	6.64	1.96	30.00
Metal primer, rust inhibitor - rusty metal (material #36)									
Spray prime coat									
Slow	850	300	80.10	2.71	.66	26.70	5.71	5.72	41.50
Medium	900	275	70.10	3.28	.94	25.49	7.43	4.46	41.60
Fast	950	250	60.10	3.79	1.32	24.04	9.04	2.67	40.86
Industrial enamel, oil base, high gloss - light colors (material #56)									
Spray 1st or additional finish coats									
Slow	950	300	65.40	2.42	.57	21.80	4.71	4.72	34.22
Medium	1025	275	57.20	2.88	.85	20.80	6.13	3.68	34.34
Fast	1100	250	49.00	3.27	1.16	19.60	7.45	2.20	33.68
Industrial enamel, oil base, high gloss - dark (OSHA) colors (material #57)									
Spray 1st or additional finish coats									
Slow	950	325	78.30	2.42	.57	24.09	5.15	5.16	37.39
Medium	1025	300	68.50	2.88	.85	22.83	6.64	3.98	37.18
Fast	1100	275	58.70	3.27	1.16	21.35	7.99	2.36	36.13
Epoxy coating, 2 part system, clear (material #51)									
Spray 1st coat									
Slow	850	325	160.60	2.71	.66	49.42	10.03	10.05	72.87
Medium	900	313	140.50	3.28	.94	44.89	12.28	7.37	68.76
Fast	950	300	120.50	3.79	1.32	40.17	14.04	4.15	63.47
Spray 2nd or additional coats									
Slow	950	350	160.60	2.42	.57	45.89	9.29	9.31	67.48
Medium	1025	338	140.50	2.88	.85	41.57	11.32	6.79	63.41
Fast	1100	325	120.50	3.27	1.16	37.08	12.87	3.81	58.19
Epoxy coating, 2 part system, white (material #52)									
Spray 1st coat									
Slow	850	325	157.00	2.71	.66	48.31	9.82	9.84	71.34
Medium	900	313	137.40	3.28	.94	43.90	12.03	7.22	67.37
Fast	950	300	117.70	3.79	1.32	39.23	13.75	4.07	62.16
Spray 2nd or additional coats									
Slow	950	350	157.00	2.42	.57	44.86	9.09	9.11	66.05
Medium	1025	338	137.40	2.88	.85	40.65	11.09	6.65	62.12
Fast	1100	325	117.70	3.27	1.16	36.22	12.60	3.73	56.98

	Labor SF per manhour	Material coverage SF/gallon	Material cost per gallon	Labor cost per 100 SF	Labor burden 100 SF	Material cost per 100 SF	Overhead per 100 SF	Profit per 100 SF	Total price per 100 SF
Vinyl coating (material #59)									
Spray 1st coat									
Slow	750	225	151.20	3.07	.72	67.20	13.49	13.52	98.00
Medium	775	213	132.30	3.81	1.10	62.11	16.76	10.05	93.83
Fast	800	200	113.40	4.50	1.59	56.70	19.46	5.76	88.01
Spray 2nd or additional coats									
Slow	900	130	151.20	2.56	.61	116.31	22.70	22.75	164.93
Medium	950	105	132.30	3.11	.88	126.00	32.50	19.50	181.99
Fast	1000	80	113.40	3.60	1.27	141.75	45.45	13.44	205.51

Use these figures to estimate labor and material costs for painting the exterior surface of a flat roof on a steel tank, silo, vessel or hopper. *Rule of thumb*: For a vaulted, peaked or sloping roof, figure the roof area as though it were flat and add 5%. Note: A two coat system, prime and finish, using oil base material is recommended for any metal surface. Using water base material may cause oxidation, corrosion and rust. One coat of oil base paint on metal surfaces may result in cracking, peeling or chipping without the proper prime coat application. If off-white or another light-colored finish paint is specified, make sure the prime coat is also a light color, or more than one finish coat will be necessary. "Slow" work is based on an hourly wage of $23.00, "Medium" work on an hourly wage of $29.50, and "Fast" work on an hourly wage of $36.00. Other qualifications that apply to this table are on page 9.

	Labor SF per manhour	Material coverage SF/gallon	Material cost per gallon	Labor cost per 100 SF	Labor burden 100 SF	Material cost per 100 SF	Overhead per 100 SF	Profit per 100 SF	Total price per 100 SF
Walls, concrete tilt-up, brush application									
Flat latex, water base (material #5)									
Brush 1st coat									
Slow	150	300	36.00	15.33	3.69	12.00	5.89	5.90	42.81
Medium	188	263	31.50	15.69	4.54	11.98	8.05	4.83	45.09
Fast	225	225	27.00	16.00	5.63	12.00	10.43	3.09	47.15
Brush 2nd or additional coats									
Slow	200	360	36.00	11.50	2.76	10.00	4.61	4.62	33.49
Medium	225	305	31.50	13.11	3.78	10.33	6.81	4.08	38.11
Fast	250	250	27.00	14.40	5.08	10.80	9.39	2.78	42.45
Enamel, water base (material #9)									
Brush 1st coat									
Slow	150	275	53.30	15.33	3.69	19.38	7.29	7.31	53.00
Medium	188	238	46.70	15.69	4.54	19.62	9.96	5.98	55.79
Fast	225	200	40.00	16.00	5.63	20.00	12.91	3.82	58.36
Brush 2nd or additional coats									
Slow	200	360	53.30	11.50	2.76	14.81	5.52	5.53	40.12
Medium	225	243	46.70	13.11	3.78	19.22	9.03	5.42	50.56
Fast	250	225	40.00	14.40	5.08	17.78	11.55	3.42	52.23
Enamel, oil base (material #10)									
Brush 1st coat									
Slow	150	300	61.50	15.33	3.69	20.50	7.51	7.52	54.55
Medium	188	250	53.80	15.69	4.54	21.52	10.44	6.26	58.45
Fast	225	200	46.10	16.00	5.63	23.05	13.86	4.10	62.64
Brush 2nd or additional coats									
Slow	200	400	61.50	11.50	2.76	15.38	5.63	5.64	40.91
Medium	225	325	53.80	13.11	3.78	16.55	8.36	5.02	46.82
Fast	250	250	46.10	14.40	5.08	18.44	11.76	3.48	53.16
Epoxy coating, 2 part system, clear (material #51)									
Brush 1st coat									
Slow	150	330	160.60	15.33	3.69	48.67	12.86	12.89	93.44
Medium	188	290	140.50	15.69	4.54	48.45	17.17	10.30	96.15
Fast	225	250	120.50	16.00	5.63	48.20	21.65	6.41	97.89
Brush 2nd or additional coats									
Slow	200	380	160.60	11.50	2.76	42.26	10.74	10.76	78.02
Medium	225	340	140.50	13.11	3.78	41.32	14.56	8.73	81.50
Fast	250	300	120.50	14.40	5.08	40.17	18.49	5.47	83.61

	Labor SF per manhour	Material coverage SF/gallon	Material cost per gallon	Labor cost per 100 SF	Labor burden 100 SF	Material cost per 100 SF	Overhead per 100 SF	Profit per 100 SF	Total price per 100 SF
Epoxy coating, 2 part system, white (material #52)									
Brush 1st coat									
Slow	150	330	157.00	15.33	3.69	47.58	12.65	12.68	91.93
Medium	188	290	137.40	15.69	4.54	47.38	16.90	10.14	94.65
Fast	225	250	117.70	16.00	5.63	47.08	21.31	6.30	96.32
Brush 2nd or additional coats									
Slow	200	380	157.00	11.50	2.76	41.32	10.56	10.58	76.72
Medium	225	340	137.40	13.11	3.78	40.41	14.33	8.60	80.23
Fast	250	300	117.70	14.40	5.08	39.23	18.20	5.38	82.29
Waterproofing, clear hydro sealer (material #34)									
Brush 1st coat									
Slow	150	160	40.60	15.33	3.69	25.38	8.43	8.45	61.28
Medium	175	140	35.50	16.86	4.86	25.36	11.77	7.06	65.91
Fast	200	120	30.40	18.00	6.36	25.33	15.40	4.56	69.65
Brush 2nd or additional coats									
Slow	230	200	40.60	10.00	2.41	20.30	6.21	6.23	45.15
Medium	275	188	35.50	10.73	3.11	18.88	8.18	4.91	45.81
Fast	295	175	30.40	12.20	4.31	17.37	10.50	3.11	47.49
Industrial waterproofing (material #58)									
Brush 1st coat									
Slow	90	100	53.70	25.56	6.13	53.70	16.22	16.26	117.87
Medium	100	95	47.00	29.50	8.53	49.47	21.88	13.13	122.51
Fast	110	90	40.30	32.73	11.55	44.78	27.61	8.17	124.84
Brush 2nd or additional coats									
Slow	150	200	53.70	15.33	3.69	26.85	8.71	8.73	63.31
Medium	180	188	47.00	16.39	4.75	25.00	11.53	6.92	64.59
Fast	195	175	40.30	18.46	6.53	23.03	14.88	4.40	67.30

Use these figures to estimate the costs for finishing concrete walls which have a smooth surface (trowel), rough texture, or exposed aggregate finish. For wall heights above 10', increase the computed area by 50%. "Slow" work is based on an hourly wage of $23.00, "Medium" work on an hourly wage of $29.50, and "Fast" work on an hourly wage of $36.00. Other qualifications that apply to this table are on page 9.

	Labor SF per manhour	Material coverage SF/gallon	Material cost per gallon	Labor cost per 100 SF	Labor burden 100 SF	Material cost per 100 SF	Overhead per 100 SF	Profit per 100 SF	Total price per 100 SF
Walls, concrete tilt-up, roll application									
Flat latex, water base (material #5)									
Roll 1st coat									
Slow	275	275	36.00	8.36	2.02	13.09	4.46	4.47	32.40
Medium	300	238	31.50	9.83	2.83	13.24	6.48	3.89	36.27
Fast	325	225	27.00	11.08	3.92	12.00	8.37	2.48	37.85
Roll 2nd or additional coats									
Slow	300	375	36.00	7.67	1.83	9.60	3.63	3.64	26.37
Medium	338	325	31.50	8.73	2.53	9.69	5.24	3.14	29.33
Fast	375	275	27.00	9.60	3.41	9.82	7.07	2.09	31.99
Enamel, water base (material #9)									
Roll 1st coat									
Slow	275	275	53.30	8.36	2.02	19.38	5.65	5.66	41.07
Medium	300	238	46.70	9.83	2.83	19.62	8.07	4.84	45.19
Fast	325	200	40.00	11.08	3.92	20.00	10.85	3.21	49.06
Roll 2nd or additional coats									
Slow	300	325	53.30	7.67	1.83	16.40	4.92	4.93	35.75
Medium	338	275	46.70	8.73	2.53	16.98	7.06	4.23	39.53
Fast	375	225	40.00	9.60	3.41	17.78	9.54	2.82	43.15
Enamel, oil base (material #10)									
Roll 1st coat									
Slow	275	260	61.50	8.36	2.02	23.65	6.46	6.48	46.97
Medium	300	210	53.80	9.83	2.83	25.62	9.57	5.74	53.59
Fast	325	160	46.10	11.08	3.92	28.81	13.58	4.02	61.41
Roll 2nd or additional coats									
Slow	300	350	61.50	7.67	1.83	17.57	5.15	5.16	37.38
Medium	338	300	53.80	8.73	2.53	17.93	7.30	4.38	40.87
Fast	375	250	46.10	9.60	3.41	18.44	9.74	2.88	44.07
Epoxy coating, 2 part system, clear (material #51)									
Roll 1st coat									
Slow	275	290	160.60	8.36	2.02	55.38	12.49	12.52	90.77
Medium	300	250	140.50	9.83	2.83	56.20	17.22	10.33	96.41
Fast	325	210	120.50	11.08	3.92	57.38	22.43	6.64	101.45
Roll 2nd or additional coats									
Slow	300	380	160.60	7.67	1.83	42.26	9.84	9.86	71.46
Medium	338	340	140.50	8.73	2.53	41.32	13.14	7.89	73.61
Fast	375	300	120.50	9.60	3.41	40.17	16.48	4.87	74.53

	Labor SF per manhour	Material coverage SF/gallon	Material cost per gallon	Labor cost per 100 SF	Labor burden 100 SF	Material cost per 100 SF	Overhead per 100 SF	Profit per 100 SF	Total price per 100 SF
Epoxy coating, 2 part system, white (material #52)									
Roll 1st coat									
Slow	275	290	157.00	8.36	2.02	54.14	12.26	12.28	89.06
Medium	300	250	137.40	9.83	2.83	54.96	16.91	10.14	94.67
Fast	325	210	117.70	11.08	3.92	56.05	22.02	6.51	99.58
Roll 2nd or additional coats									
Slow	300	380	157.00	7.67	1.83	41.32	9.66	9.68	70.16
Medium	338	340	137.40	8.73	2.53	40.41	12.92	7.75	72.34
Fast	375	300	117.70	9.60	3.41	39.23	16.19	4.79	73.22
Waterproofing, clear hydro sealer (material #34)									
Roll 1st coat									
Slow	170	200	40.60	13.53	3.24	20.30	7.05	7.06	51.18
Medium	200	165	35.50	14.75	4.27	21.52	10.13	6.08	56.75
Fast	245	130	30.40	14.69	5.18	23.38	13.41	3.97	60.63
Roll 2nd or additional coats									
Slow	275	325	40.60	8.36	2.02	12.49	4.34	4.35	31.56
Medium	300	275	35.50	9.83	2.83	12.91	6.40	3.84	35.81
Fast	325	225	30.40	11.08	3.92	13.51	8.84	2.61	39.96
Industrial waterproofing (material #58)									
Roll 1st coat									
Slow	100	125	53.70	23.00	5.52	42.96	13.58	13.61	98.67
Medium	113	113	47.00	26.11	7.55	41.59	18.81	11.29	105.35
Fast	125	100	40.30	28.80	10.17	40.30	24.57	7.27	111.11
Roll 2nd or additional coats									
Slow	180	200	53.70	12.78	3.08	26.85	8.11	8.13	58.95
Medium	198	188	47.00	14.90	4.31	25.00	11.05	6.63	61.89
Fast	215	175	40.30	16.74	5.91	23.03	14.16	4.19	64.03

Use these figures to estimate the costs for finishing concrete walls which have a smooth surface (trowel), rough texture, or exposed aggregate finish. For wall heights above 10', increase the computed area by 50%. "Slow" work is based on an hourly wage of $23.00, "Medium" work on an hourly wage of $29.50, and "Fast" work on an hourly wage of $36.00. Other qualifications that apply to this table are on page 9.

	Labor SF per manhour	Material coverage SF/gallon	Material cost per gallon	Labor cost per 100 SF	Labor burden 100 SF	Material cost per 100 SF	Overhead per 100 SF	Profit per 100 SF	Total price per 100 SF
Walls, concrete tilt-up, spray application									
Flat latex, water base (material #5)									
Spray 1st coat									
Slow	500	225	36.00	4.60	1.10	16.00	4.12	4.13	29.95
Medium	600	188	31.50	4.92	1.43	16.76	5.78	3.47	32.36
Fast	700	150	27.00	5.14	1.83	18.00	7.73	2.29	34.99
Spray 2nd or additional coats									
Slow	600	275	36.00	3.83	.93	13.09	3.39	3.40	24.64
Medium	700	238	31.50	4.21	1.23	13.24	4.67	2.80	26.15
Fast	800	200	27.00	4.50	1.59	13.50	6.07	1.80	27.46
Enamel, water base (material #9)									
Spray 1st coat									
Slow	500	225	53.30	4.60	1.10	23.69	5.58	5.60	40.57
Medium	550	188	46.70	5.36	1.56	24.84	7.94	4.76	44.46
Fast	600	150	40.00	6.00	2.13	26.67	10.78	3.19	48.77
Spray 2nd or additional coats									
Slow	600	275	53.30	3.83	.93	19.38	4.58	4.59	33.31
Medium	700	238	46.70	4.21	1.23	19.62	6.26	3.76	35.08
Fast	800	200	40.00	4.50	1.59	20.00	8.09	2.39	36.57
Enamel, oil base (material #10)									
Spray 1st coat									
Slow	500	200	61.50	4.60	1.10	30.75	6.93	6.94	50.32
Medium	550	163	53.80	5.36	1.56	33.01	9.98	5.99	55.90
Fast	600	125	46.10	6.00	2.13	36.88	13.95	4.13	63.09
Spray 2nd or additional coats									
Slow	600	300	61.50	3.83	.93	20.50	4.80	4.81	34.87
Medium	700	243	53.80	4.21	1.23	22.14	6.89	4.14	38.61
Fast	800	175	46.10	4.50	1.59	26.34	10.05	2.97	45.45
Epoxy coating, 2 part system, clear (material #51)									
Spray 1st coat									
Slow	500	270	160.60	4.60	1.10	59.48	12.38	12.41	89.97
Medium	700	253	140.50	4.21	1.23	55.53	15.24	9.14	85.35
Fast	900	235	120.50	4.00	1.41	51.28	17.57	5.20	79.46
Spray 2nd or additional coats									
Slow	600	375	160.60	3.83	.93	42.83	9.04	9.06	65.69
Medium	800	338	140.50	3.69	1.06	41.57	11.58	6.95	64.85
Fast	1000	300	120.50	3.60	1.27	40.17	13.96	4.13	63.13

	Labor SF per manhour	Material coverage SF/gallon	Material cost per gallon	Labor cost per 100 SF	Labor burden 100 SF	Material cost per 100 SF	Overhead per 100 SF	Profit per 100 SF	Total price per 100 SF
Epoxy coating, 2 part system, white (material #52)									
Spray 1st coat									
Slow	500	270	157.00	4.60	1.10	58.15	12.13	12.16	88.14
Medium	700	253	137.40	4.21	1.23	54.31	14.94	8.96	83.65
Fast	900	235	117.70	4.00	1.41	50.09	17.21	5.09	77.80
Spray 2nd or additional coats									
Slow	600	375	157.00	3.83	.93	41.87	8.86	8.88	64.37
Medium	800	338	137.40	3.69	1.06	40.65	11.35	6.81	63.56
Fast	1000	300	117.70	3.60	1.27	39.23	13.67	4.04	61.81
Waterproofing, clear hydro sealer (material #34)									
Spray 1st coat									
Slow	600	125	40.60	3.83	.93	32.48	7.07	7.09	51.40
Medium	650	113	35.50	4.54	1.32	31.42	9.32	5.59	52.19
Fast	700	100	30.40	5.14	1.83	30.40	11.58	3.43	52.38
Spray 2nd or additional coats									
Slow	700	200	40.60	3.29	.79	20.30	4.63	4.64	33.65
Medium	800	170	35.50	3.69	1.06	20.88	6.41	3.85	35.89
Fast	900	140	30.40	4.00	1.41	21.71	8.41	2.49	38.02

Use these figures to estimate the costs for finishing concrete walls which have a smooth surface (trowel), rough texture, or exposed aggregate finish. For wall heights above 10', increase the computed area by 50%. "Slow" work is based on an hourly wage of $23.00, "Medium" work on an hourly wage of $29.50, and "Fast" work on an hourly wage of $36.00. Other qualifications that apply to this table are on page 9.

Walls, gypsum drywall

Smooth-wall finish, no texture: See General Painting Costs, page 240

	Labor LF per manhour	Material coverage LF/gallon	Material cost per gallon	Labor cost per 100 LF	Labor burden 100 LF	Material cost per 100 LF	Overhead per 100 LF	Profit per 100 LF	Total price per 100 LF
Windows, steel factory sash, brush application									
Metal primer, rust inhibitor - clean metal (material #35)									
Brush prime coat									
Slow	100	850	59.70	23.00	5.52	7.02	6.75	6.77	49.06
Medium	125	800	52.30	23.60	6.82	6.54	9.24	5.54	51.74
Fast	150	750	44.80	24.00	8.49	5.97	11.92	3.53	53.91
Metal primer, rust inhibitor - rusty metal (material #36)									
Brush prime coat									
Slow	100	800	80.10	23.00	5.52	10.01	7.32	7.34	53.19
Medium	125	750	70.10	23.60	6.82	9.35	9.94	5.97	55.68
Fast	150	700	60.10	24.00	8.49	8.59	12.73	3.77	57.58
Metal finish - synthetic enamel, gloss, interior or exterior, off white (material #37)									
Brush 1st coat									
Slow	125	900	63.60	18.40	4.42	7.07	5.68	5.69	41.26
Medium	150	850	55.70	19.67	5.70	6.55	7.98	4.79	44.69
Fast	175	800	47.70	20.57	7.24	5.96	10.47	3.10	47.34
Brush 2nd or additional coats									
Slow	150	1000	63.60	15.33	3.69	6.36	4.82	4.83	35.03
Medium	175	950	55.70	16.86	4.86	5.86	6.90	4.14	38.62
Fast	200	900	47.70	18.00	6.36	5.30	9.19	2.72	41.57
Metal finish - synthetic enamel, gloss, interior or exterior, colors, except orange & red (material #38)									
Brush 1st coat									
Slow	125	950	66.30	18.40	4.42	6.98	5.66	5.67	41.13
Medium	150	900	58.00	19.67	5.70	6.44	7.95	4.77	44.53
Fast	175	850	49.70	20.57	7.24	5.85	10.44	3.09	47.19
Brush 2nd or additional coats									
Slow	150	1050	66.30	15.33	3.69	6.31	4.81	4.82	34.96
Medium	175	1000	58.00	16.86	4.86	5.80	6.88	4.13	38.53
Fast	200	950	49.70	18.00	6.36	5.23	9.17	2.71	41.47

These figures will apply when painting steel factory sash but do not include work on glazing or frame. For heights above 8 feet, use the High Time Difficulty Factors on page 139. Note: A two coat system, prime and finish, using oil base material is recommended for any metal surface. Using water base material may cause oxidation, corrosion and rust. Using one coat of oil base paint on metal surfaces may result in cracking, peeling or chipping without the proper prime coat application. If off-white or another light-colored finish paint is specified, make sure the prime coat is also a light color, or more than one finish coat will be necessary. "Slow" work is based on an hourly wage of $23.00, "Medium" work on an hourly wage of $29.50, and "Fast" work on an hourly wage of $36.00. Other qualifications that apply to this table are on page 9.

Field Production Times and Rates

Date	>	10-15-20XX	Start	Finish	Total	LF or SF	LF or SF	Book	Your
Painter		Operation	times	times	hours	completed	per hour	rate	rate
David H.	1	Cutting-in ceiling	7:05	8:35	1.5	28 LF	18.67		
	2	8' 0" height	8:40	9:40	1	20 LF	20		
	3		9:40	11:55	2.25	38 LF	16.89		
	4		12:45	2:00	1.25	20 LF	16		
	5								
	6								
	7								
	8								
	9								
	10								
	11								
	12								
	13								
	14								
	15								
	16								
	17								
	18								
	19								
	20								
	21								
	22								
	23								
	24								
	25								
	26								
	27								
	28								
	29								
	30								
	31								
	32								
	33								
	34								
	35								
	36								
	37								

Figure 25
Sample field production times and rates form

Field Production Times and Rates

Date Painter	>	Operation	Start times	Finish times	Total hours	LF or SF completed	LF or SF per hour	Average rate	Your rate
	1								
	2								
	3								
	4								
	5								
	6								
	7								
	8								
	9								
	10								
	11								
	12								
	13								
	14								
	15								
	16								
	17								
	18								
	19								
	20								
	21								
	22								
	23								
	24								
	25								
	26								
	27								
	28								
	29								
	30								
	31								
	32								
	33								
	34								
	35								
	36								
	37								

Figure 26
Blank field production times and rates form

Part IV

Wallcovering Costs

	American rolls per gallon	Linear yards per gallon	Adhesive cost per gallon
Adhesive coverage, rolls to yards conversion (2.5 yards per roll)			
Ready-mix			
Light weight vinyl (material #60)			
Slow	8.0	20.0	14.40
Medium	7.5	18.8	12.60
Fast	7.0	17.5	10.80
Heavy weight vinyl (material #61)			
Slow	6.0	15.0	16.30
Medium	5.5	13.8	14.30
Fast	5.0	12.5	12.20
Cellulose (material #62)			
Slow	8.0	20.0	17.30
Medium	7.0	17.5	15.20
Fast	6.0	15.0	13.00
Vinyl to vinyl (material #63)			
Slow	7.0	17.5	25.20
Medium	6.5	16.3	22.00
Fast	6.0	15.0	18.90
Powdered cellulose (material #64)			
Slow	12.0	30.0	8.40
Medium	11.0	27.5	7.40
Fast	10.0	25.0	6.30
Powdered vinyl (material #65)			
Slow	12.0	30.0	10.10
Medium	11.0	27.5	8.90
Fast	10.0	25.0	7.60
Powdered wheat paste (material #66)			
Slow	13.0	32.5	7.70
Medium	12.0	30.0	6.80
Fast	11.0	27.5	5.80

These figures are based on rolls or yards per gallon of liquid paste. Vinyl to vinyl ready-mix is usually distributed in pint containers which have been converted to gallons. One pint makes 6 gallons. Powdered adhesive coverage is based on water added to powder to prepare gallon quantities. One pound of powdered vinyl or wheat paste makes 1.75 gallons of liquid paste when mixed with about 1-1/2 gallons of cold water. Two ounces of powdered cellulose makes 1.75 gallons of liquid paste.

	Labor SF per manhour	Material coverage SF/gallon	Material cost per gallon	Labor cost per 100 SF	Labor burden 100 SF	Material cost per 100 SF	Overhead per 100 SF	Profit per 100 SF	Total cost per 100 SF
Adhesive coverage and application rates, square foot basis									
Ready-mix									
Light weight vinyl (material #60)									
Slow	350	275	14.40	6.43	1.55	5.24	2.51	2.52	18.25
Medium	375	250	12.60	7.73	2.25	5.04	3.75	2.25	21.02
Fast	400	225	10.80	8.88	3.13	4.80	5.21	1.54	23.56
Heavy weight vinyl (material #61)									
Slow	275	200	16.30	8.18	1.98	8.15	3.48	3.48	25.27
Medium	300	175	14.30	9.67	2.78	8.17	5.16	3.09	28.87
Fast	325	150	12.20	10.92	3.87	8.13	7.10	2.10	32.12
Cellulose (material #62)									
Slow	325	250	17.30	6.92	1.67	6.92	2.95	2.95	21.41
Medium	350	225	15.20	8.29	2.40	6.76	4.36	2.62	24.43
Fast	375	200	13.00	9.47	3.35	6.50	5.99	1.77	27.08
Vinyl to vinyl (material #63)									
Slow	300	225	25.20	7.50	1.79	11.20	3.90	3.90	28.29
Medium	325	200	22.00	8.92	2.59	11.00	5.63	3.38	31.52
Fast	350	175	18.90	10.14	3.60	10.80	7.60	2.25	34.39
Powdered cellulose (material #64)									
Slow	425	400	8.40	5.29	1.27	2.10	1.65	1.65	11.96
Medium	450	370	7.40	6.44	1.86	2.00	2.58	1.55	14.43
Fast	475	335	6.30	7.47	2.66	1.88	3.72	1.10	16.83
Powdered vinyl (material #65)									
Slow	425	400	10.10	5.29	1.27	2.53	1.73	1.73	12.55
Medium	450	370	8.90	6.44	1.86	2.41	2.68	1.61	15.00
Fast	475	335	7.60	7.47	2.66	2.27	3.84	1.14	17.38
Powdered wheat paste (material #66)									
Slow	450	425	7.70	4.99	1.20	1.81	15.70	1.52	11.05
Medium	475	400	6.80	6.12	1.77	1.70	16.39	1.44	13.42
Fast	500	375	5.80	7.10	2.51	1.55	16.96	1.02	15.64

These figures are based on gallon quantities of liquid paste. Vinyl to vinyl ready-mix is usually distributed in pint containers which have been converted to gallons. Powdered adhesive coverage is based on water added to powder to prepare gallon quantities. Typically, powdered ready-mix material is in 2 to 4 ounce packages which will adhere 6 to 12 rolls of wallcovering. See the Adhesive coverage table on the previous page for conversion to rolls and yards. "Slow" work is based on an hourly rate of $22.50, "Medium" work on an hourly rate of $29.00, and "Fast" work on an hourly rate of $35.50.

Wallcovering application, average labor production, medium rooms

	Labor rolls per day	Material by others	Labor cost per 10 rolls	Labor burden 10 rolls	Overhead per 10 rolls	Profit per 10 rolls	Total cost per 10 rolls	Total price per roll
Walls								
Slow	17	--	105.88	25.42	24.95	25.00	181.25	18.12
Medium	19	--	122.11	35.30	39.35	23.61	220.37	22.04
Fast	21	--	135.24	47.75	56.72	16.78	256.49	25.65
Ceilings								
Slow	15	--	120.00	28.79	28.27	28.33	205.39	20.54
Medium	17	--	136.47	39.44	43.98	26.39	246.28	24.63
Fast	19	--	149.47	52.78	62.69	18.54	283.48	28.35

The table above assumes that residential rolls are hand pasted. Add surface preparation time on page 425 as needed. For heights above 8 feet, use the High Time Difficulty Factors on page 139. "Slow" work is based on an hourly wage of $22.50, "Medium" work on an hourly wage of $29.00, and "Fast" work on an hourly wage of $35.50.

Wallcovering application, average labor production, small rooms

	Labor rolls per day	Material by others	Labor cost per 10 rolls	Labor burden 10 rolls	Overhead per 10 rolls	Profit per 10 rolls	Total cost per 10 rolls	Total price per roll
Walls								
Slow	7	--	257.14	61.73	60.58	60.71	440.16	44.01
Medium	9	--	257.78	74.49	83.07	49.84	465.18	46.52
Fast	11	--	258.18	91.14	108.29	32.03	489.64	48.96
Ceilings								
Slow	6	--	300.00	71.99	70.68	70.83	513.50	51.35
Medium	8	--	290.00	83.80	93.45	56.07	523.32	52.33
Fast	10	--	284.00	100.24	119.12	35.24	538.60	53.86

The table above assumes that residential rolls are hand pasted. Add surface preparation time on page 425 as needed. For heights above 8 feet, use the High Time Difficulty Factors on page 139. "Slow" work is based on an hourly wage of $22.50, "Medium" work on an hourly wage of $29.00, and "Fast" work on an hourly wage of $35.50.

Borders 3" to 8" wide, commercial, machine pasted

	Labor LF per manhour	Material by others	Labor cost per 100 LF	Labor burden 100 LF	Overhead per 100 LF	Profit per 100 LF	Total cost per 100 LF	Total price per LF
Medium size rooms (10 x 10 range)								
Slow	140	--	15.36	3.68	3.62	3.63	26.29	.26
Medium	158	--	17.41	5.03	4.26	4.27	30.97	.31
Fast	175	--	19.14	6.74	4.92	4.93	35.73	.36

The table above assumes that commercial rolls are machine pasted. ADD for surface preparation time from page 425 or from the tables in Part II, Preparation costs beginning on page 295, as needed. For heights above 8 feet, use the High Time Difficulty Factors on page 139. "Slow" work is based on an hourly wage of $21.50, "Medium" work on an hourly wage of $27.50, and "Fast" work on an hourly wage of $33.50.

	Labor LF per manhour	Material by others	Labor cost per 100 LF	Labor burden 100 LF	Overhead per 100 LF	Profit per 100 LF	Total cost per 100 LF	Total price per LF

Borders 3" to 8" wide, residential, hand pasted

Medium size rooms (bedrooms, dining rooms)

Slow	100	--	22.50	5.40	5.30	5.31	38.51	.39
Medium	113	--	25.66	7.42	8.27	4.96	46.31	.46
Fast	125	--	28.40	10.02	11.91	3.52	53.85	.54

The table above assumes that residential rolls are hand pasted. ADD for surface preparation time from table at bottom or from the tables in Part II, Preparation costs beginning on page 295, as needed. For heights above 8 feet, use the High Time Difficulty Factors on page 139. "Slow" work is based on an hourly wage of $22.50, "Medium" work on an hourly wage of $29.00, and "Fast" work on an hourly wage of $35.50.

	Labor SF per manhour	Material by others	Labor cost per 100 SF	Labor burden 100 SF	Overhead per 100 SF	Profit per 100 SF	Total cost per 100 SF	Total price per SF

Flexible wood sheet and veneer

Wood veneer flexwood (residential or commercial)

Medium size rooms (bedrooms, dining rooms, offices, reception areas)

Slow	14	--	157.14	37.72	37.02	37.10	268.98	2.69
Medium	20	--	141.25	40.80	45.52	27.31	254.88	2.55
Fast	26	--	132.69	46.84	55.65	16.46	251.64	2.52

Flexi-wall systems (residential or commercial)

Medium size rooms (bedrooms, dining rooms, offices, reception areas)

Slow	12	--	183.33	43.99	43.19	43.28	313.79	3.14
Medium	18	--	156.94	45.35	50.58	30.35	283.22	2.83
Fast	24	--	143.75	50.77	60.29	17.83	272.64	2.73

Flexible wood sheet and veneer appears under section 097416 in the Construction Specifications Institute (CSI) indexing system. For heights above 8 feet, use the High Time Difficulty Factors on page 139. The labor rates in the table above are an average of the residential and commercial rates. Thus, "Slow" work is based on an hourly rate of $22.00, "Medium" work on an hourly rate of $28.25, and "Fast" work on an hourly rate of $34.50.

	Labor SF per manhour	Material by others	Labor cost per 100 SF	Labor burden 100 SF	Overhead per 100 SF	Profit per 100 SF	Total cost per 100 SF	Total price per SF

Surface preparation, wallcovering

Rule of thumb, typical preparation

Slow	100	--	22.50	5.40	8.65	2.56	39.11	.39
Medium	125	--	23.20	6.70	9.27	2.74	41.91	.42
Fast	150	--	23.67	8.37	9.93	2.94	44.91	.45

Putty cracks, sand and wash

Slow	120	--	18.75	4.49	7.21	2.13	32.58	.33
Medium	135	--	21.48	6.22	8.58	2.54	38.82	.39
Fast	150	--	23.67	8.37	9.93	2.94	44.91	.45

For additional preparation tasks see the Preparation operation tables beginning on page 295. For heights above 8 feet, use the High Time Difficulty Factors on page 139. "Slow" work is based on an hourly rate of $22.50, "Medium" work on an hourly rate of $29.00, and "Fast" work on an hourly rate of $35.50.

	Labor yards per day	Material by others	Labor cost per 10 yards	Labor burden 10 yards	Overhead per 10 yards	Profit per 10 yards	Total cost per 10 yards	Total price per yard

Vinyl wallcover, commercial, machine pasted, yards per day, 48" to 54" width

	Labor yards per day	Material by others	Labor cost per 10 yards	Labor burden 10 yards	Overhead per 10 yards	Profit per 10 yards	Total cost per 10 yards	Total price per yard
Cut-up areas (stair halls, landing areas)								
Walls								
Slow	40	--	43.00	10.32	10.13	10.15	73.60	7.36
Medium	45	--	48.89	14.14	15.76	9.45	88.24	8.82
Fast	50	--	53.60	18.93	22.48	6.65	101.66	10.17
Ceilings								
Slow	33	--	52.12	12.50	12.28	12.31	89.21	8.92
Medium	38	--	57.89	16.73	18.66	11.19	104.47	10.45
Fast	43	--	62.33	21.98	26.14	7.73	118.18	11.82
Small rooms (restrooms, utility rooms)								
Walls								
Slow	42	--	40.95	9.84	9.65	9.67	70.11	7.01
Medium	50	--	44.00	12.72	14.18	8.51	79.41	7.94
Fast	55	--	48.73	17.23	20.44	6.05	92.45	9.24
Ceilings								
Slow	38	--	45.26	10.86	10.66	10.68	77.46	7.75
Medium	46	--	47.83	13.82	15.41	9.25	86.31	8.63
Fast	51	--	52.55	18.57	22.04	6.52	99.68	9.97
Medium rooms (offices)								
Walls								
Slow	60	--	28.67	6.87	6.75	6.77	49.06	4.91
Medium	75	--	29.33	8.50	9.45	5.67	52.95	5.29
Fast	90	--	29.78	10.52	12.49	3.69	56.48	5.65
Ceilings								
Slow	56	--	30.71	7.39	7.24	7.25	52.59	5.26
Medium	69	--	31.88	9.21	10.27	6.16	57.52	5.75
Fast	81	--	33.09	11.70	13.88	4.11	62.78	6.28
Large rooms (conference rooms)								
Walls								
Slow	76	--	22.63	5.44	5.33	5.34	38.74	3.87
Medium	89	--	24.72	7.15	7.97	4.78	44.62	4.46
Fast	102	--	26.27	9.27	11.02	3.26	49.82	4.98
Ceilings								
Slow	66	--	26.06	6.25	6.14	6.15	44.60	4.46
Medium	79	--	27.85	8.06	8.98	5.39	50.28	5.03
Fast	91	--	29.45	10.40	12.35	3.65	55.85	5.59

	Labor yards per day	Material by others	Labor cost per 10 yards	Labor burden 10 yards	Overhead per 10 yards	Profit per 10 yards	Total cost per 10 yards	Total price per yard
Large wall areas (corridors, long hallways)								
Walls								
Slow	89	--	19.33	4.64	4.55	4.56	33.08	3.31
Medium	102	--	21.57	6.22	6.95	4.17	38.91	3.89
Fast	114	--	23.51	8.31	9.86	2.92	44.60	4.46
Ceilings								
Slow	76	--	22.63	5.44	5.33	5.34	38.74	3.87
Medium	89	--	24.72	7.15	7.97	4.78	44.62	4.46
Fast	102	--	26.27	9.27	11.02	3.26	49.82	4.98
Paper-backed vinyl on medium room walls								
Bedrooms, dining rooms								
Slow	43	--	40.00	9.59	9.42	9.44	68.45	6.85
Medium	53	--	41.51	11.98	13.38	8.03	74.90	7.49
Fast	64	--	41.88	14.78	17.56	5.20	79.42	7.94
Cork wallcovering on medium room walls								
Bedrooms, dining rooms								
Slow	43	--	40.00	9.59	9.42	9.44	68.45	6.85
Medium	53	--	41.51	11.98	13.38	8.03	74.90	7.49
Fast	64	--	41.88	14.78	17.56	5.20	79.42	7.94

Vinyl and vinyl-coated wallcovering appear under section 097216 in the Construction Specifications Institute (CSI) indexing system. Cork wallcovering appears in section 097213. The table above assumes that commercial rolls are machine pasted. ADD for surface preparation time from page 425 or from the tables in Part II, Preparation costs beginning on page 295, as needed. For heights above 8 feet, use the High Time Difficulty Factors on page 139. "Slow" work is based on an hourly wage of $21.50, "Medium" work on an hourly wage of $27.50, and "Fast" work on an hourly wage of $33.50.

	Labor rolls per day	Material by others	Labor cost per 10 rolls	Labor burden 10 rolls	Overhead per 10 rolls	Profit per 10 rolls	Total cost per 10 rolls	Total price per roll

Vinyl wallcovering, residential, hand pasted, single rolls per day 18" to 27" wide

	Labor rolls per day	Material by others	Labor cost per 10 rolls	Labor burden 10 rolls	Overhead per 10 rolls	Profit per 10 rolls	Total cost per 10 rolls	Total price per roll
Cut-up areas (stair halls, landing areas)								
Walls								
Slow	10	--	180.00	43.20	42.41	42.50	308.11	30.81
Medium	11	--	210.91	60.95	67.97	40.78	380.61	38.06
Fast	12	--	236.67	83.55	99.27	29.36	448.85	44.88
Ceilings								
Slow	7	--	257.14	61.73	60.58	60.71	440.16	44.01
Medium	8	--	290.00	83.80	93.45	56.07	523.32	52.33
Fast	9	--	315.56	111.38	132.35	39.15	598.44	59.85
Small rooms (baths, utility rooms)								
Walls								
Slow	11	--	163.64	39.28	38.55	38.63	280.10	28.01
Medium	12	--	193.33	55.88	62.30	37.38	348.89	34.89
Fast	13	--	218.46	77.12	91.63	27.10	414.31	41.43
Ceilings								
Slow	9	--	200.00	48.00	47.12	47.22	342.34	34.23
Medium	10	--	232.00	67.04	74.76	44.86	418.66	41.87
Fast	11	--	258.18	91.14	108.29	32.03	489.64	48.96
Medium rooms (bedrooms, dining rooms)								
Walls								
Slow	17	--	105.88	25.42	24.95	25.00	181.25	18.12
Medium	19	--	122.11	35.30	39.35	23.61	220.37	22.04
Fast	21	--	135.24	47.75	56.72	16.78	256.49	25.65
Ceilings								
Slow	15	--	120.00	28.79	28.27	28.33	205.39	20.54
Medium	17	--	136.47	39.44	43.98	26.39	246.28	24.63
Fast	19	--	149.47	52.78	62.69	18.54	283.48	28.35
Large rooms (living rooms)								
Walls								
Slow	20	--	90.00	21.60	21.20	21.25	154.05	15.41
Medium	23	--	100.87	29.14	32.51	19.50	182.02	18.20
Fast	26	--	109.23	38.56	45.81	13.55	207.15	20.72
Ceilings								
Slow	18	--	100.00	23.99	23.56	23.61	171.16	17.12
Medium	20	--	116.00	33.52	37.38	22.43	209.33	20.93
Fast	22	--	129.09	45.55	54.14	16.02	244.80	24.48

	Labor rolls per day	Material by others	Labor cost per 10 rolls	Labor burden 10 rolls	Overhead per 10 rolls	Profit per 10 rolls	Total cost per 10 rolls	Total price per roll
Large wall areas (corridors, long hallways)								
Walls								
Slow	24	--	75.00	17.99	17.67	17.71	128.37	12.84
Medium	27	--	85.93	24.83	27.69	16.61	155.06	15.51
Fast	30	--	94.67	33.43	39.71	11.75	179.56	17.96
Ceilings								
Slow	21	--	85.71	20.59	20.19	20.24	146.73	14.67
Medium	23	--	100.87	29.14	32.51	19.50	182.02	18.20
Fast	25	--	113.60	40.10	47.65	14.09	215.44	21.54
Paper-backed vinyl on medium room walls								
Bedrooms, dining rooms								
Slow	8	--	225.00	54.00	53.01	53.12	385.13	38.51
Medium	10	--	232.00	67.04	74.76	44.86	418.66	41.87
Fast	12	--	236.67	83.55	99.27	29.36	448.85	44.88
Cork wallcovering on medium room walls								
Bedrooms, dining rooms								
Slow	8	--	225.00	54.00	53.01	53.12	385.13	38.51
Medium	10	--	232.00	67.04	74.76	44.86	418.66	41.87
Fast	12	--	236.67	83.55	99.27	29.36	448.85	44.88

The table above assumes that residential rolls are hand pasted. ADD for surface preparation time from page 425 or from the tables in Part II, Preparation costs beginning on page 295, as needed. For heights above 8 feet, use the High Time Difficulty Factors on page 139. "Slow" work is based on an hourly wage of $22.50, "Medium" work on an hourly wage of $29.00, and "Fast" work on an hourly wage of $35.50. Add for ready-mix paste.

	Labor yards per day	Material by others	Labor cost per 10 yards	Labor burden 10 yards	Overhead per 10 yards	Profit per 10 yards	Total cost per 10 yards	Total price per yard

Wall fabric, commercial, machine pasted, yards per day, 48" to 54" width

	Labor yards per day	Material by others	Labor cost per 10 yards	Labor burden 10 yards	Overhead per 10 yards	Profit per 10 yards	Total cost per 10 yards	Total price per yard
Coated fabric								
Cut-up areas (stair halls, landing areas)								
Walls								
Slow	58	--	29.66	7.10	6.99	7.00	50.75	5.08
Medium	63	--	34.92	10.10	11.25	6.75	63.02	6.30
Fast	68	--	39.41	13.90	16.53	4.89	74.73	7.47
Ceilings								
Slow	35	--	49.14	11.80	11.58	11.60	84.12	8.41
Medium	40	--	55.00	15.90	17.73	10.64	99.27	9.93
Fast	45	--	59.56	21.04	24.98	7.39	112.97	11.30
Small rooms (restrooms, utility rooms)								
Walls								
Slow	60	--	28.67	6.87	6.75	6.77	49.06	4.91
Medium	65	--	33.85	9.79	10.91	6.54	61.09	6.11
Fast	70	--	38.29	13.52	16.06	4.75	72.62	7.26
Ceilings								
Slow	40	--	43.00	10.32	10.13	10.15	73.60	7.36
Medium	48	--	45.83	13.27	14.77	8.86	82.73	8.27
Fast	55	--	48.73	17.23	20.44	6.05	92.45	9.24
Medium rooms (offices)								
Walls								
Slow	68	--	25.29	6.06	5.96	5.97	43.28	4.33
Medium	80	--	27.50	7.95	8.86	5.32	49.63	4.96
Fast	93	--	28.82	10.16	12.09	3.58	54.65	5.47
Ceilings								
Slow	60	--	28.67	6.87	6.75	6.77	49.06	4.91
Medium	73	--	30.14	8.71	9.71	5.83	54.39	5.44
Fast	85	--	31.53	11.13	13.22	3.91	59.79	5.98
Large rooms (conference rooms)								
Walls								
Slow	85	--	20.24	4.85	4.77	4.78	34.64	3.47
Medium	98	--	22.45	6.48	7.24	4.34	40.51	4.05
Fast	110	--	24.36	8.59	10.22	3.02	46.19	4.62
Ceilings								
Slow	68	--	25.29	6.06	5.96	5.97	43.28	4.33
Medium	80	--	27.50	7.95	8.86	5.32	49.63	4.96
Fast	93	--	28.82	10.16	12.09	3.58	54.65	5.47

	Labor yards per day	Material by others	Labor cost per 10 yards	Labor burden 10 yards	Overhead per 10 yards	Profit per 10 yards	Total cost per 10 yards	Total price per yard
Large wall areas (corridors, long hallways)								
Walls								
Slow	80	--	21.50	5.16	5.07	5.08	36.81	3.68
Medium	110	--	20.00	5.77	6.45	3.87	36.09	3.61
Fast	125	--	21.44	7.57	8.99	2.66	40.66	4.07
Ceilings								
Slow	78	--	22.05	5.30	5.19	5.20	37.74	3.77
Medium	90	--	24.44	7.08	7.88	4.73	44.13	4.41
Fast	103	--	26.02	9.20	10.92	3.23	49.37	4.94

Canvas sheeting: see Wall fabric, residential, hand pasted

Grasscloth: see Wall fabric, residential, hand pasted

Burlap: see Wall fabric, residential, hand pasted

Natural fabric, silk: see Wall fabric, residential, hand pasted

Natural fabric, felt, linen, cotton: see Wall fabric, residential, hand pasted

The table above assumes that commercial rolls are machine pasted. ADD for surface preparation time from page 425 or from the tables in Part II, Preparation costs beginning on page 295 as needed. For heights above 8 feet, use the High Time Difficulty Factors on page 139. "Slow" work is based on an hourly wage of $21.50, "Medium" work on an hourly wage of $27.50, and "Fast" work on an hourly wage of $33.50. Add for ready-mix paste.

	Labor rolls per day	Material by others	Labor cost per 10 rolls	Labor burden 10 rolls	Overhead per 10 rolls	Profit per 10 rolls	Total cost per 10 rolls	Total price per roll
Wall fabric, residential, hand pasted, single rolls per day 18" to 27" wide								
Coated fabrics								
Cut-up areas (stair halls, landing areas)								
Walls								
Slow	12	--	150.00	36.01	35.34	35.41	256.76	25.68
Medium	13	--	178.46	51.58	57.51	34.50	322.05	32.20
Fast	14	--	202.86	71.58	85.09	25.17	384.70	38.47
Ceilings								
Slow	8	--	225.00	54.00	53.01	53.12	385.13	38.51
Medium	9	--	257.78	74.49	83.07	49.84	465.18	46.52
Fast	10	--	284.00	100.24	119.12	35.24	538.60	53.86
Small rooms (baths, utility rooms)								
Walls								
Slow	13	--	138.46	33.24	32.62	32.69	237.01	23.70
Medium	14	--	165.71	47.88	53.40	32.04	299.03	29.90
Fast	15	--	189.33	66.81	79.41	23.49	359.04	35.91
Ceilings								
Slow	10	--	180.00	43.20	42.41	42.50	308.11	30.81
Medium	11	--	210.91	60.95	67.97	40.78	380.61	38.06
Fast	12	--	236.67	83.55	99.27	29.36	448.85	44.88
Medium rooms (bedrooms, dining rooms)								
Walls								
Slow	18	--	100.00	23.99	23.56	23.61	171.16	17.12
Medium	20	--	116.00	33.52	37.38	22.43	209.33	20.93
Fast	22	--	129.09	45.55	54.14	16.02	244.80	24.48
Ceilings								
Slow	16	--	112.50	27.00	26.51	26.56	192.57	19.26
Medium	18	--	128.89	37.23	41.54	24.92	232.58	23.26
Fast	20	--	142.00	50.12	59.56	17.62	269.30	26.93
Large rooms (living rooms)								
Walls								
Slow	22	--	81.82	19.62	19.28	19.32	140.04	14.01
Medium	25	--	92.80	26.82	29.91	17.94	167.47	16.75
Fast	28	--	101.43	35.79	42.54	12.58	192.34	19.24
Ceilings								
Slow	19	--	94.74	22.75	22.32	22.37	162.18	16.22
Medium	21	--	110.48	31.94	35.60	21.36	199.38	19.94
Fast	23	--	123.48	43.57	51.79	15.32	234.16	23.42

	Labor rolls per day	Material by others	Labor cost per 10 rolls	Labor burden 10 rolls	Overhead per 10 rolls	Profit per 10 rolls	Total cost per 10 rolls	Total price per roll
Large wall areas (corridors, long hallways)								
Walls								
Slow	26	--	69.23	16.62	16.31	16.35	118.51	11.85
Medium	29	--	80.00	23.13	25.78	15.47	144.38	14.44
Fast	32	--	88.75	31.33	37.22	11.01	168.31	16.83
Ceilings								
Slow	22	--	81.82	19.62	19.28	19.32	140.04	14.01
Medium	24	--	96.67	27.92	31.15	18.69	174.43	17.45
Fast	26	--	109.23	38.56	45.81	13.55	207.15	20.72
Canvas sheeting								
Medium room walls (bedrooms, dining rooms)								
Walls								
Slow	14	--	128.57	30.85	30.29	30.36	220.07	22.01
Medium	16	--	145.00	41.90	46.73	28.04	261.67	26.17
Fast	17	--	167.06	58.97	70.07	20.73	316.83	31.68
Grasscloth								
Medium room walls (bedrooms, dining rooms)								
Walls								
Slow	15	--	120.00	28.79	28.27	28.33	205.39	20.54
Medium	20	--	116.00	33.52	37.38	22.43	209.33	20.93
Fast	24	--	118.33	41.75	49.63	14.68	224.39	22.44
Burlap								
Medium rooms (bedrooms, dining rooms)								
Walls								
Slow	10	--	180.00	43.20	42.41	42.50	308.11	30.81
Medium	14	--	165.71	47.88	53.40	32.04	299.03	29.90
Fast	18	--	157.78	55.67	66.18	19.58	299.21	29.92
Natural fabric, silk								
Medium rooms (bedrooms, dining rooms)								
Walls								
Slow	8	--	225.00	54.00	53.01	53.12	385.13	38.51
Medium	12	--	193.33	55.88	62.30	37.38	348.89	34.89
Fast	15	--	189.33	66.81	79.41	23.49	359.04	35.91
Natural fabric, felt, linen, cotton								
Medium rooms (bedrooms, dining rooms)								
Walls								
Slow	7	--	257.14	61.73	60.58	60.71	440.16	44.01
Medium	9	--	257.78	74.49	83.07	49.84	465.18	46.52
Fast	11	--	258.18	91.14	108.29	32.03	489.64	48.96

The table above assumes that residential rolls are hand pasted. ADD for surface preparation time from page 425 or from the tables in Part II, Preparation costs beginning on page 295 as needed. For heights above 8 feet, use the High Time Difficulty Factors on page 139. "Slow" work is based on an hourly wage of $22.50, "Medium" work on an hourly wage of $29.00, and "Fast" work on an hourly wage of $35.50. Add for ready-mix paste.

	Labor rolls per day	Material by others	Labor cost per 10 rolls	Labor burden 10 rolls	Overhead per 10 rolls	Profit per 10 rolls	Total cost per 10 rolls	Total price per roll
Wallpaper, commercial, machine pasted, single rolls per day, 27" width								
Blind stock (lining)								
Cut-up areas (stair halls, landing areas)								
Walls								
Slow	19	--	90.53	21.74	21.33	21.37	154.97	15.50
Medium	21	--	104.76	30.30	33.76	20.26	189.08	18.91
Fast	23	--	116.52	41.14	48.87	14.46	220.99	22.10
Ceilings								
Slow	16	--	107.50	25.80	25.33	25.38	184.01	18.40
Medium	18	--	122.22	35.32	39.39	23.63	220.56	22.06
Fast	20	--	134.00	47.32	56.20	16.63	254.15	25.41
Small rooms (restrooms, utility rooms)								
Walls								
Slow	21	--	81.90	19.67	19.30	19.34	140.21	14.02
Medium	25	--	88.00	25.44	28.36	17.01	158.81	15.88
Fast	29	--	92.41	32.66	38.76	11.47	175.30	17.53
Ceilings								
Slow	17	--	101.18	24.28	23.84	23.89	173.19	17.32
Medium	20	--	110.00	31.80	35.45	21.27	198.52	19.85
Fast	22	--	121.82	43.00	51.09	15.11	231.02	23.10
Medium rooms (offices)								
Walls								
Slow	29	--	59.31	14.24	13.97	14.00	101.52	10.15
Medium	35	--	62.86	18.18	20.26	12.15	113.45	11.34
Fast	41	--	65.37	23.07	27.42	8.11	123.97	12.40
Ceilings								
Slow	25	--	68.80	16.51	16.21	16.24	117.76	11.78
Medium	30	--	73.33	21.22	23.63	14.18	132.36	13.23
Fast	34	--	78.82	27.84	33.06	9.78	149.50	14.95
Large rooms (conference rooms)								
Walls								
Slow	35	--	49.14	11.80	11.58	11.60	84.12	8.41
Medium	41	--	53.66	15.50	17.29	10.38	96.83	9.68
Fast	47	--	57.02	20.13	23.92	7.07	108.14	10.81
Ceilings								
Slow	29	--	59.31	14.24	13.97	14.00	101.52	10.15
Medium	34	--	64.71	18.70	20.85	12.51	116.77	11.68
Fast	38	--	70.53	24.89	29.58	8.75	133.75	13.38

	Labor rolls per day	Material by others	Labor cost per 10 rolls	Labor burden 10 rolls	Overhead per 10 rolls	Profit per 10 rolls	Total cost per 10 rolls	Total price per roll
Large wall areas (corridors, long hallways)								
Walls								
Slow	40	--	43.00	10.32	10.13	10.15	73.60	7.36
Medium	45	--	48.89	14.14	15.76	9.45	88.24	8.82
Fast	50	--	53.60	18.93	22.48	6.65	101.66	10.17
Ceilings								
Slow	33	--	52.12	12.50	12.28	12.31	89.21	8.92
Medium	38	--	57.89	16.73	18.66	11.19	104.47	10.45
Fast	42	--	63.81	22.54	26.76	7.92	121.03	12.10
Ordinary pre-trimmed wallpaper or butt joint work								
Medium room walls (offices)								
Walls								
Slow	26	--	66.15	15.88	15.59	15.62	113.24	11.32
Medium	31	--	70.97	20.53	22.87	13.72	128.09	12.81
Fast	36	--	74.44	26.28	31.22	9.24	141.18	14.12
Ceilings								
Slow	23	--	74.78	17.94	17.62	17.66	128.00	12.80
Medium	28	--	78.57	22.71	25.32	15.19	141.79	14.18
Fast	32	--	83.75	29.58	35.13	10.39	158.85	15.88
Hand-crafted wallpaper								
Medium room walls (offices)								
Walls								
Slow	20	--	86.00	20.64	20.26	20.30	147.20	14.72
Medium	24	--	91.67	26.48	29.54	17.72	165.41	16.54
Fast	28	--	95.71	33.80	40.15	11.88	181.54	18.15
Ceilings								
Slow	18	--	95.56	22.92	22.51	22.56	163.55	16.36
Medium	21	--	104.76	30.30	33.76	20.26	189.08	18.91
Fast	23	--	116.52	41.14	48.87	14.46	220.99	22.10
Flock wallpaper, medium rooms (offices)								
Walls								
Slow	14	--	122.86	29.48	28.95	29.01	210.30	21.03
Medium	17	--	129.41	37.42	41.70	25.02	233.55	23.35
Fast	20	--	134.00	47.32	56.20	16.63	254.15	25.41
Foil wallpaper, medium rooms (offices)								
Walls								
Slow	13	--	132.31	31.76	31.17	31.24	226.48	22.65
Medium	16	--	137.50	39.75	44.31	26.59	248.15	24.81
Fast	19	--	141.05	49.83	59.16	17.50	267.54	26.75

	Labor rolls per day	Material by others	Labor cost per 10 rolls	Labor burden 10 rolls	Overhead per 10 rolls	Profit per 10 rolls	Total cost per 10 rolls	Total price per roll
Canvas wallpaper, medium rooms (offices)								
Walls								
Slow	12	--	143.33	34.41	33.77	33.84	245.35	24.53
Medium	14	--	157.14	45.42	50.64	30.38	283.58	28.36
Fast	16	--	167.50	59.15	70.26	20.78	317.69	31.77
Scenic wallpaper, medium rooms (offices)								
Walls								
Slow	16	--	107.50	25.80	25.33	25.38	184.01	18.40
Medium	18	--	122.22	35.32	39.39	23.63	220.56	22.06
Fast	20	--	134.00	47.32	56.20	16.63	254.15	25.41

The table above assumes that commercial rolls are machine pasted. ADD for surface preparation time from page 425 or from the tables in Part II, Preparation costs beginning on page 295, as needed. For heights above 8 feet, use the High Time Difficulty Factors on page 139. "Slow" work is based on an hourly wage of $21.50, "Medium" work on an hourly wage of $27.50, and "Fast" work on an hourly wage of $33.50. Add for ready-mix paste.

	Labor rolls per day	Material by others	Labor cost per 10 rolls	Labor burden 10 rolls	Overhead per 10 rolls	Profit per 10 rolls	Total cost per 10 rolls	Total price per roll

Wallpaper, residential, hand pasted, single rolls per day 18" to 27" wide

	Labor rolls per day	Material by others	Labor cost per 10 rolls	Labor burden 10 rolls	Overhead per 10 rolls	Profit per 10 rolls	Total cost per 10 rolls	Total price per roll
Blind stock (lining)								
Cut-up areas (stair halls, landing areas)								
Walls								
Slow	11	--	163.64	39.28	38.55	38.63	280.10	28.01
Medium	13	--	178.46	51.58	57.51	34.50	322.05	32.20
Fast	14	--	202.86	71.58	85.09	25.17	384.70	38.47
Ceilings								
Slow	10	--	180.00	43.20	42.41	42.50	308.11	30.81
Medium	11	--	210.91	60.95	67.97	40.78	380.61	38.06
Fast	12	--	236.67	83.55	99.27	29.36	448.85	44.88
Small rooms (baths, utility rooms)								
Walls								
Slow	12	--	150.00	36.01	35.34	35.41	256.76	25.68
Medium	14	--	165.71	47.88	53.40	32.04	299.03	29.90
Fast	16	--	177.50	62.65	74.45	22.02	336.62	33.66
Ceilings								
Slow	11	--	163.64	39.28	38.55	38.63	280.10	28.01
Medium	13	--	178.46	51.58	57.51	34.50	322.05	32.20
Fast	14	--	202.86	71.58	85.09	25.17	384.70	38.47
Medium rooms (bedrooms, dining rooms)								
Walls								
Slow	20	--	90.00	21.60	21.20	21.25	154.05	15.41
Medium	23	--	100.87	29.14	32.51	19.50	182.02	18.20
Fast	26	--	109.23	38.56	45.81	13.55	207.15	20.72
Ceilings								
Slow	18	--	100.00	23.99	23.56	23.61	171.16	17.12
Medium	20	--	116.00	33.52	37.38	22.43	209.33	20.93
Fast	22	--	129.09	45.55	54.14	16.02	244.80	24.48
Large rooms (living rooms)								
Walls								
Slow	23	--	78.26	18.78	18.44	18.48	133.96	13.40
Medium	27	--	85.93	24.83	27.69	16.61	155.06	15.51
Fast	31	--	91.61	32.36	38.42	11.37	173.76	17.37
Ceilings								
Slow	20	--	90.00	21.60	21.20	21.25	154.05	15.41
Medium	22	--	105.45	30.46	33.98	20.39	190.28	19.03
Fast	24	--	118.33	41.75	49.63	14.68	224.39	22.44

	Labor rolls per day	Material by others	Labor cost per 10 rolls	Labor burden 10 rolls	Overhead per 10 rolls	Profit per 10 rolls	Total cost per 10 rolls	Total price per roll
Large wall areas (corridors, long hallways)								
Walls								
Slow	29	--	62.07	14.91	14.62	14.65	106.25	10.62
Medium	33	--	70.30	20.31	22.66	13.59	126.86	12.69
Fast	36	--	78.89	27.83	33.09	9.79	149.60	14.96
Ceilings								
Slow	23	--	78.26	18.78	18.44	18.48	133.96	13.40
Medium	25	--	92.80	26.82	29.91	17.94	167.47	16.75
Fast	27	--	105.19	37.12	44.12	13.05	199.48	19.95
Ordinary pre-trimmed wallpaper or butt joint work								
Medium room walls (bedrooms, dining rooms)								
Walls								
Slow	18	--	100.00	23.99	23.56	23.61	171.16	17.12
Medium	21	--	110.48	31.94	35.60	21.36	199.38	19.94
Fast	24	--	118.33	41.75	49.63	14.68	224.39	22.44
Ceilings								
Slow	16	--	112.50	27.00	26.51	26.56	192.57	19.26
Medium	18	--	128.89	37.23	41.54	24.92	232.58	23.26
Fast	20	--	142.00	50.12	59.56	17.62	269.30	26.93
Hand-crafted wallpaper								
Medium room walls (bedrooms, dining rooms)								
Walls								
Slow	12	--	150.00	36.01	35.34	35.41	256.76	25.68
Medium	15	--	154.67	44.68	49.84	29.91	279.10	27.91
Fast	18	--	157.78	55.67	66.18	19.58	299.21	29.92
Ceilings								
Slow	10	--	180.00	43.20	42.41	42.50	308.11	30.81
Medium	12	--	193.33	55.88	62.30	37.38	348.89	34.89
Fast	14	--	202.86	71.58	85.09	25.17	384.70	38.47
Flock wallpaper, medium rooms (bedrooms, dining rooms)								
Walls								
Slow	10	--	180.00	43.20	42.41	42.50	308.11	30.81
Medium	13	--	178.46	51.58	57.51	34.50	322.05	32.20
Fast	16	--	177.50	62.65	74.45	22.02	336.62	33.66
Foil wallpaper, medium rooms (bedrooms, dining rooms)								
Walls								
Slow	9	--	200.00	48.00	47.12	47.22	342.34	34.23
Medium	12	--	193.33	55.88	62.30	37.38	348.89	34.89
Fast	15	--	189.33	66.81	79.41	23.49	359.04	35.91

	Labor rolls per day	Material by others	Labor cost per 10 rolls	Labor burden 10 rolls	Overhead per 10 rolls	Profit per 10 rolls	Total cost per 10 rolls	Total price per roll
Canvas wallpaper, medium rooms (bedrooms, dining rooms)								
Walls								
Slow	8	--	225.00	54.00	53.01	53.12	385.13	38.51
Medium	10	--	232.00	67.04	74.76	44.86	418.66	41.87
Fast	12	--	236.67	83.55	99.27	29.36	448.85	44.88
Scenic wallpaper, medium rooms (bedrooms, dining rooms)								
Walls								
Slow	12	--	150.00	36.01	35.34	35.41	256.76	25.68
Medium	14	--	165.71	47.88	53.40	32.04	299.03	29.90
Fast	16	--	177.50	62.65	74.45	22.02	336.62	33.66

The table above assumes that residential rolls are hand pasted. ADD for surface preparation time from page 425 or from the tables in Part II, Preparation costs beginning on page 295, as needed. For heights above 8 feet, use the High Time Difficulty Factors on page 139. "Slow" work is based on an hourly wage of $22.50, "Medium" work on an hourly wage of $29.00, and "Fast" work on an hourly wage of $35.50. Add for ready-mix paste.

Index

Practical References for Builders

Builder's Guide to Accounting Revised

Step-by-step, easy-to-follow guidelines for setting up and maintaining records for your building business. This practical guide to all accounting methods shows how to meet state and federal accounting requirements, explains the new depreciation rules, and describes how the Tax Reform Act can affect the way you keep records. Full of charts, diagrams, simple directions and examples to help you keep track of where your money is going. Recommended reading for many state contractor's exams. Each chapter ends with a set of test questions, and a CD-ROM included FREE has all the questions in interactive self-test software. Use the Study Mode to make studying for the exam much easier, and Exam Mode to practice your skills.
360 pages, 8½ x 11, $35.50.
Also available as an eBook (PDF), $17.75 at www.craftsman-book.com

Insurance Restoration Contracting: Startup to Success

Insurance restoration — the repair of buildings damaged by water, fire, smoke, storms, vandalism and other disasters — is an exciting field of construction that provides lucrative work that's immune to economic downturns. And, with insurance companies funding the repairs, your payment is virtually guaranteed. But this type of work requires special knowledge and equipment, and that's what you'll learn about in this book. It covers fire repairs and smoke damage, water losses and specialized drying methods, mold remediation, content restoration, even damage to mobile and manufactured homes. You'll also find information on equipment needs, training classes, estimating books and software, and how restoration leads to lucrative remodeling jobs. It covers all you need to know to start and succeed as the restoration contractor that both homeowners and insurance companies call on first for the best jobs. **640 pages, 8½ x 11, $69.00.**
Also available as an eBook (PDF), $34.50 at www.craftsman-book.com

Estimating Home Building Costs, Revised

Estimate every phase of residential construction from site costs to the profit margin you include in your bid. Shows how to keep track of manhours and make accurate labor cost estimates for site clearing and excavation, footings, foundations, framing and sheathing finishes, electrical, plumbing, and more. Provides and explains sample cost estimate worksheets with complete instructions for each job phase. This practical guide to estimating home construction costs has been updated with digital Excel estimating forms and worksheets that ensure accurate and complete estimates for your residential projects. Enter your project information on the worksheets and Excel automatically totals each material and labor cost from every stage of construction to a final cost estimate worksheet. Load the enclosed CD-ROM into your computer and create your own estimate as you follow along with the step-by-step techniques in this book. **336 pages, 8½ x 11, $38.00.**
Also available as an eBook (PDF), $19.00 at www.craftsman-book.com

National Renovation & Insurance Repair Estimator

Current prices in dollars and cents for hard-to-find items needed on most insurance, repair, remodeling, and renovation jobs. All price items include labor, material, and equipment breakouts, plus special charts that tell you exactly how these costs are calculated. Includes a free download of an electronic version of the book with *National Estimator*, a standalone Windows™ estimating program. Additional information and *National Estimator* ShowMe tutorial video is available on our website under the "Support" dropdown tab.
488 pages, 8½ x 11, $99.50. Revised annually.
Also available as an eBook (PDF), $49.75 at www.craftsman-book.com

How to Succeed With Your Own Construction Business

Everything you need to start your own construction business: setting up the paperwork, finding the jobs, advertising, using contracts, dealing with lenders, estimating, scheduling, finding and keeping good employees, keeping the books, and coping with success. If you're considering starting your own construction business, all the knowledge, tips, and blank forms you need are here. **336 pages, 8½ x 11, $28.50.**
eBook (PDF) also available, $14.25 at www.craftsman-book.com

Contractor's Plain-English Legal Guide

For today's contractors, legal problems are like snakes in the swamp — you might not see them, but you know they're there. This book tells you where the snakes are hiding and directs you to the safe path. With the directions in this easy-to-read handbook you're less likely to need a $200-an-hour lawyer. Includes simple directions for starting your business, writing contracts that cover just about any eventuality, collecting what's owed you, filing liens, protecting yourself from unethical subcontractors, and more. For about the price of 15 minutes in a lawyer's office, you'll have a guide that will make many of those visits unnecessary. Includes a CD-ROM with blank copies of all the forms and contracts in the book. **272 pages, 8½ x 11, $49.50.**

Contractor's Guide to QuickBooks by Online Accounting

This book is designed to help a contractor, bookkeeper and their accountant set up and use QuickBooks Desktop specifically for the construction industry. No use re-inventing the wheel, we have used this system with contractors for over 30 years. It works and is now the national standard. By following the steps we outlined in the book you, too, can set up a good system for job costing as well as financial reporting.
156 pages, 8½ x 11, $68.50.

Contractor's Math Short Cuts Quick-Cards

In this single, 4-page laminated card, you get all the math essentials you need in contracting. The formulas and rules of thumb for calculating dimensions, surface areas, volume, etc. **4 pages, 8½ x 11, $7.95.**

Craftsman's Construction Installation Encyclopedia

Step-by-step installation instructions for just about any residential construction, remodeling or repair task, arranged alphabetically, from *Acoustic tile* to *Wood flooring*. Includes hundreds of illustrations that show how to build, install, or remodel each part of the job, as well as manhour tables for each work item so you can estimate and bid with confidence. Also includes a CD-ROM with all the material in the book, handy look-up features, and the ability to capture and print out for your crew the instructions and diagrams for any job. **792 pages, 8½ x 11, $65.00.**
Also available as an eBook (PDF), $32.50 at www.craftsman-book.com

Finish Carpenter's Manual

Everything you need to know to be a finish carpenter: assessing a job before you begin, and tricks of the trade from a master finish carpenter. Easy-to-follow instructions for installing doors and windows, ceiling treatments (including fancy beams, corbels, cornices and moldings), wall treatments (including wainscoting and sheet paneling), and the finishing touches of chair, picture, and plate rails. Specialized interior work includes cabinetry and built-ins, stair finish work, and closets. Also covers exterior trims and porches. Includes manhour tables for finish work, and hundreds of illustrations and photos. **208 pages, 8½ x 11, $22.50.**

CD Estimator

If your computer has *Windows*™ and a CD-ROM drive, CD Estimator puts at your fingertips over 150,000 construction costs for new construction, remodeling, renovation & insurance repair, home improvement, framing & finish carpentry, electrical, concrete & masonry, painting, earthwork & heavy equipment and plumbing & HVAC. Quarterly cost updates are available at no charge on the Internet. You'll also have the *National Estimator* program — a stand-alone estimating program for *Windows*™ that *Remodeling* magazine called a "computer wiz," and *Job Cost Wizard*, a program that lets you export your estimates to *QuickBooks Pro* for actual job costing. A 60-minute interactive video teaches you how to use this CD-ROM to estimate construction costs. And to top it off, to help you create professional-looking estimates, the disk includes over 40 construction estimating and bidding forms in a format that's perfect for nearly any *Windows*™ word processing or spreadsheet program.
CD Estimator is $149.50.

Construction Forms for Contractors

This practical guide contains 78 practical forms, letters and checklists, guaranteed to help you streamline your office, organize your jobsites, gather and organize records and documents, keep a handle on your subs, reduce estimating errors, administer change orders and lien issues, monitor crew productivity, track your equipment use, and more. Includes accounting forms, change order forms, forms for customers, estimating forms, field work forms, HR forms, lien forms, office forms, bids and proposals, subcontracts, and more. All are also on the CD-ROM included, in *Excel* spreadsheets, as formatted Rich Text that you can fill out on your computer, and as PDFs. **360 pages, 8½ x 11, $48.50.**
Also available as an eBook (PDF), $24.25 at www.craftsman-book.com

Profits in Buying & Renovating Homes

Step-by-step instructions for selecting, repairing, improving, and selling highly profitable "fixer-uppers." Shows which price ranges offer the highest profit-to-investment ratios, which neighborhoods offer the best return, practical directions for repairs, and tips on dealing with buyers, sellers, and real estate agents. Shows you how to determine your profit before you buy, what "bargains" to avoid, and how to make simple, profitable, inexpensive upgrades. **304 pages, 8½ x 11, $24.75.**

Rough Framing Carpentry

If you'd like to make good money working outdoors as a framer, this is the book for you. Here you'll find shortcuts to laying out studs; speed cutting blocks, trimmers and plates by eye; quickly building and blocking rake walls; installing ceiling backing, ceiling joists, and truss joists; cutting and assembling hip trusses and California fills; arches and drop ceilings — all with production line procedures that save you time and help you make more money. Over 100 on-the-job photos of how to do it right and what can go wrong.
304 pages, 8½ x 11, $26.50.

National Home Improvement Estimator

Current labor and material prices for home improvement projects. Provides manhours for each job, recommended crew size, and the labor cost for removal and installation work. Material prices are current, with location adjustment factors and free monthly updates on the Web. Gives step-by-step instructions for the work, with helpful diagrams, and home improvement shortcuts and tips from experts. Includes a free download of an electronic version of the book, and *National Estimator*, a stand-alone *Windows*™ estimating program. Additional information and *National Estimator* ShowMe tutorial video is available on our website under the "Support" dropdown tab.
568 pages, 8½ x 11, $98.75. Revised annually.
Also available as an eBook (PDF), $49.38 at www.craftsman-book.com

National Appraisal Estimator

An Online Appraisal Estimating Service. Produce credible single-family residence appraisals – in as little as five minutes. A smart resource for appraisers using the cost approach. Reports consider all significant cost variables and both physical and functional depreciation. For more information, visit www.craftsman-book.com/national-appraisal-estimator-online-software

Markup & Profit: A Contractor's Guide, Revisited

In order to succeed in a construction business, you have to be able to price your jobs to cover all labor, material and overhead expenses, and make a decent profit. But calculating markup is only part of the picture. If you're going to beat the odds and stay in business — profitably, you also need to know how to write good contracts, manage your crews, work with subcontractors and collect on your work. This book covers the business basics of running a construction company, whether you're a general or specialty contractor working in remodeling, new construction or commercial work. The principles outlined here apply to all construction-related businesses. You'll find tried and tested formulas to guarantee profits, with step-by-step instructions and easy-to-follow examples to help you learn how to operate your business successfully. Includes a link to free downloads of blank forms and checklists used in this book. **336 pages, 8½ x 11, $52.50.**
Also available as an eBook (ePub, mobi for Kindle), $39.95 at www.craftsman-book.com

National Construction Estimator

Current building costs for residential, commercial, and industrial construction. Estimated prices for every common building material. Provides manhours, recommended crew, and gives the labor cost for installation. Includes a free download of an electronic version of the book with *National Estimator*, a stand-alone *Windows*™ estimating program. Additional information and *National Estimator* ShowMe tutorial video is available on our website under the "Support" dropdown tab.
672 pages, 8½ x 11, $97.50. Revised annually.
Also available as an eBook (PDF), $48.75 at www.craftsman-book.com

Estimating Excavation Revised eBook

How to calculate the amount of dirt you'll have to move and the cost of owning and operating the machines you'll do it with. Detailed, step-by-step instructions on how to assign bid prices to each part of the job, including labor and equipment costs. Also, the best ways to set up an organized and logical estimating system, take off from contour maps, estimate quantities in irregular areas, and figure your overhead. This revised edition includes a chapter on earthwork estimating software. As with any tool, you have to pick the right one. Written by an experienced dirt contractor and instructor of computer estimating software, this chapter covers the program types, explains how they work, gives the basics of how to use them, and discusses what will work best for the type of work you handle. This e-Book is the download version of the book in text searchable, PDF format. Craftsman eBooks are for use in the freely distributed Adobe Reader and are compatible with Reader 6.0 or above.
550 pages, eBook (PDF), $21.75, available at www.craftsman-book.com

Wood-Frame House Construction

Step-by-step construction details, from the layout of the outer walls, excavation and formwork, to finish carpentry and painting. Packed with clear illustrations and explanations updated for modern construction methods. Everything you need to know about framing, roofing, siding, interior finishings, floor covering and stairs — your complete book of wood-frame homebuilding. **320 pages, 8½ x 11, $25.50. Revised edition**

Estimating & Bidding for Builders & Remodelers

This 5th edition has all the information you need for estimating and bidding new construction and home improvement projects. It shows how to select jobs that will be profitable, do a labor and materials take-off from the plans, calculate overhead and figure your markup, and schedule the work. Includes a CD with an easy-to-use construction estimating program and a database of 50,000 current labor and material cost estimates for new construction and home improvement work, with area modifiers for every zip code. Price updates on the Web are free and automatic.
272 pages, 8½ x 11, $89.50.
Also available as an eBook (PDF), $44.75 at www.craftsman-book.com

Painter's Handbook

Loaded with "how-to" information you'll use every day to get professional results on any job: the best way to prepare a surface for painting or repainting; selecting and using the right materials and tools (including airless spray); tips for repainting kitchens, bathrooms, cabinets, eaves and porches; how to match and blend colors; why coatings fail and what to do about it. Lists 30 profitable specialties in the painting business.
320 pages, 8½ x 11, $33.00.

Building Code Compliance for Contractors & Inspectors

An answer book for both contractors and building inspectors, this manual explains what it takes to pass inspections under the 2009 *International Residential Code*. It includes a code checklist for every trade, covering some of the most common reasons why inspectors reject residential work — footings, foundations, slabs, framing, sheathing, plumbing, electrical, HVAC, energy conservation and final inspection. The requirement for each item on the checklist is explained, and the code section cited so you can look it up or show it to the inspector. Knowing in advance what the inspector wants to see gives you an (almost unfair) advantage. To pass inspection, do your own pre-inspection before the inspector arrives. If your work requires getting permits and passing inspections, put this manual to work on your next job. If you're considering a career in code enforcement, this can be your guidebook. **8½ x 11, 232 pages, $32.50.**
Also available as an eBook (PDF), $16.25 at www.craftsman-book.com

Contractor's Survival Manual Revised

The "real skinny" on the down-and-dirty survival skills that no one likes to talk about – unique, unconventional ways to get through a debt crisis: what to do when the bills can't be paid, finding money and buying time, conserving income, transferring debt, setting payment priorities, cash float techniques, dealing with judgments and liens, and laying the foundation for recovery. Here you'll find out how to survive a downturn and the key things you can do to pave the road to success. Have this book as your insurance policy; when hard times come to your business it will be your guide. **336 pages, 8½ x 11, $38.00.**
Also available as an eBook (PDF), $19.00 at www.craftsman-book.com

Home Building Mistakes & Fixes

This is an encyclopedia of practical fixes for real-world home building and repair problems. There's never an end to "surprises" when you're in the business of building and fixing homes, yet there's little published on how to deal with construction that went wrong - where out-of-square or non-standard or jerry-rigged turns what should be a simple job into a nightmare. This manual describes jaw-dropping building mistakes that actually occurred, from disastrous misunderstandings over property lines, through basement floors leveled with an out-of-level instrument, to a house collapse when a siding crew removed the old siding. You'll learn the pitfalls the painless way, and real-world working solutions for the problems every contractor finds in a home building or repair jobsite. Includes dozens of those "surprises" and the author's step-by-step, clearly illustrated tips, tricks and workarounds for dealing with them. **384 pages, 8½ x 11, $52.50.**
Also available as an eBook (PDF), $26.25 at www.craftsman-book.com

Roofing Construction & Estimating

Installation, repair and estimating for nearly every type of roof covering available today in residential and commercial structures: asphalt shingles, roll roofing, wood shingles and shakes, clay tile, slate, metal, built-up, and elastomeric. Covers sheathing and underlayment techniques, as well as secrets for installing leakproof valleys. Many estimating tips help you minimize waste, as well as insure a profit on every job. Troubleshooting techniques help you identify the true source of most leaks. Over 300 large, clear illustrations help you find the answer to just about all your roofing questions. **448 pages, 8½ x 11, $38.00.**
Also available as an eBook (PDF), $19.00 at www.craftsman-book.com

Fences & Retaining Walls Revised eBook

Everything you need to know to run a profitable business in fence and retaining wall contracting. Takes you through layout and design, construction techniques for wood, masonry, and chain link fences, gates and entries, including finishing and electrical details. How to build retaining and rock walls. How to get your business off to the right start, keep the books, and estimate accurately. The book even includes a chapter on contractor's math. **400 pages.**
Available only as an eBook (PDF, EPUB & MOBI/Kindle), $23.00 at www.craftsman-book.com

Estimating Electrical Construction Revised

Estimating the cost of electrical work can be a very detailed and exacting discipline. It takes specialized skills and knowledge to create reliable estimates for electrical work. See how an expert estimates materials and labor for residential and commercial electrical construction. Learn how to use labor units, the plan take-off, and the bid summary to make an accurate estimate, how to deal with suppliers, use pricing sheets, and modify labor units. This book provides extensive labor unit tables and blank forms on a CD for estimating your next electrical job. **272 pages, 8½ x 11, $59.00.**
Also available as an eBook (PDF), $29.50 at www.craftsman-book.com

Drywall Contracting

How to set up and operate a drywall contracting business. Here you'll find how to get set up as a drywall contractor, the tools you'll need, how to do the work, and how to estimate and schedule jobs so that you keep the work flowing and the money coming in. How to prevent drywall problems, spot hidden problems before you begin, and how to install and repair drywall in new construction or remodeling work. Lists the eight essential steps in making a drywall estimate and includes estimating forms and manhour tables for your use. This book is filled with practical tips, illustrations, pictures, tables and forms to help you build your career as a successful drywall contractor. **256 pages, 8½ x 11, $34.95.**

Construction Contract Writer

Relying on a "one-size-fits-all" boilerplate construction contract to fit your jobs can be dangerous — almost as dangerous as a handshake agreement. *Construction Contract Writer* lets you draft a contract in minutes that precisely fits your needs and the particular job, and meets both state and federal requirements. You just answer a series of questions — like an interview — to construct a legal contract for each project you take on. Anticipate where disputes could arise and settle them in the contract before they happen. Include the warranty protection you intend, the payment schedule, and create subcontracts from the prime contract by just clicking a box. Includes a feedback button to an attorney on the Craftsman staff to help should you get stumped — *No extra charge*. **$149.95. Download the *Construction Contract Writer* program at** http://www.constructioncontractwriter.com

Easy Scheduling

Easy Scheduling presents you with a complete set of "real world"scheduling tools that are specifically tailored to meet the needs of small- to medium-sized construction businesses. Step by step, it shows you how to use *Microsoft Project* to build a schedule that will synchronize everyone's efforts into an organized system that becomes the foundation of all planning and communication for all your jobs. You'll see how to establish realistic project goals, set checkpoints, activities, relationships and time estimates for each task, as well as establish priorities. You'll learn how to create a project flowchart to keep everyone focused and on track, and see how to use CSI (Construction Specification Institute) coding to organize and sort tasks, methods, and materials across multiple projects. If you want an easy way to schedule your jobs, *Microsoft Project* and *Easy Scheduling* is the answer for you. (Does not include Microsoft Project.) Published by BNI.
316 pages, 8½ x 11, $59.95.

Basic Plumbing with Illustrations, Revised

This completely-revised edition brings this comprehensive manual fully up-to-date with all the latest plumbing codes. It is the journeyman's and apprentice's guide to installing plumbing, piping, and fixtures in residential and light commercial buildings: how to select the right materials, lay out the job and do professional-quality plumbing work, use essential tools and materials, make repairs, maintain plumbing systems, install fixtures, and add to existing systems. Includes extensive study questions at the end of each chapter, and a section with all the correct answers.
384 pages, 8½ x 11, $44.75.
Also available as an eBook (PDF), $22.37 at www.craftsman-book.com